Cell Polarity 1

Klaus Ebnet
Editor

Cell Polarity 1

Biological Role and Basic Mechanisms

Springer

Editor
Klaus Ebnet
Institute-associated Research Group "Cell Adhesion and Cell Polarity"
Institute of Medical Biochemistry
Center for Molecular Biology of Inflammation (ZMBE)
Münster
Germany

ISBN 978-3-319-14462-7 ISBN 978-3-319-14463-4 (eBook)
DOI 10.1007/978-3-319-14463-4

Library of Congress Control Number: 2015934475

Springer Cham Heidelberg New York Dordrecht London

Printed on acid-free paper

Springer International Publishing AG Switzerland is part of Springer Science+Business Media (www.springer.com)

Preface

Cell polarity refers to an asymmetric distribution of proteins, lipids, or RNA in the cell. Most animal cells are polarized. In many cell types, polarity is morphologically visible. Neurons for example contain a single, long axon and multiple, short dendrites. Migrating fibroblasts protrude multiple lamellipodia selectively at the leading edge. In other cases, cell polarity manifests itself at the molecular and functional level. Stem cells divide asymmetrically by unequally partitioning cell fate determinants into the two daughter cells, thus generating daughter cells with different cell fates. Epithelial cells contain two different membrane compartments characterized by a different molecular composition and different functions. Cellular polarization is a highly dynamic process, and the ability of individual cells to polarize is required for the early development of the zygote and for the generation of functional organs. It is not surprising that a loss of the ability of cells to develop and/or maintain a polarized state results in developmental defects and promotes tumor formation. Given the many facets of cellular polarization in different cell types and tissues, it is a central question how this diversity is generated at the molecular level. Remarkably, the molecular diversity is much smaller than expected. It turned out that a small set of polarity proteins, identified in *C. elegans* and conserved in all metazoans, acts as a hub to regulate cell polarity in many different contexts. Specificity in polarization is achieved by dynamic interactions of this molecular hub with other signaling complexes and the intersection between polarity-regulating pathways with other signaling networks.

This book on cell polarity is designed to provide a state-of-the-art overview on the most relevant aspects of cell polarity. It covers the relevant model organisms for the analysis of cell polarity including *C. elegans*, Drosophila, lower vertebrates, and mammalia. In the first volume, it describes the molecular tools that are used to generate cell polarity (Vol. 1, Part I) and introduces various aspects of cell polarity and the mode of their regulation (Vol. 1, Part II). In addition, the first volume underscores the influence of cell–cell adhesion to the generation of cell polarity in different types of cellular interactions (Vol. 1, Part III), and it illustrates the role of polarized protein trafficking during the establishment of apicobasal polarity in epithelial cells (Vol. 1, Part IV). The second volume of the book has a major focus on physiological and pathophysiological aspects of cell polarity. It describes processes of polarization during early development in various organisms (Vol. 2, Part I), and it illustrates the impact of cell polarity on the asymmetric division of stem cells (Vol. 2, Part II). Furthermore, it describes the important role of cell polarity for tissue homeostasis (Vol. 2, Part III) and provides examples of how pathogens target cell polarity signaling pathways for their own benefits (Vol. 2, Part IV).

This is the first book that describes cell polarity in a variety of cell types, tissues, and organisms. In its entity it provides a comprehensive overview on the universal biological phenomenon of cell polarity, and it illustrates a principle of evolution, i.e., the invention of core mechanisms and their adaptation to new functions to generate diversity and higher complexity. The book is of interest for both basic and applied research, for researchers at all levels, for lecturers, as well as for clinicians.

Münster, Germany K. Ebnet

Contents

Part I
Cell Polarity Regulators: Conserved Sets of Cell Polarity Proteins as Molecular Toolkits

Chapter 1
The PAR3-aPKC-PAR6 Complex

Shigeo Ohno, Spyros Goulas, and Tomonori Hirose

Abstract The PAR3-aPKC-PAR6 complex (the aPKC-PAR complex) is among the most well-studied "polarity proteins" that play fundamental roles in cell polarity in a variety of biological contexts. It is one of the core signaling cassettes containing a protein kinase and scaffold proteins that is conserved in multicellular organisms. One of the most important features of "polarity proteins" is that many of them localize to specific sites on the cytoplasmic side of the cell periphery and position other polarity proteins through antagonistic interactions and positive feedback mechanisms. Asymmetric distribution of polarity proteins in the cell periphery thus provides cellular landmarks required for the generation and maintenance of cell polarity. Here, we describe what we know about the mechanisms of how the aPKC-PAR complex is specifically positioned and activated and regulates overall cell polarity of epithelial cells with special attention to its molecular nature.

Keywords aPKC • Cell-cell junctions • Epithelial polarity • Membrane domain • PAR3 • PAR6

Abbreviations

AJ	Adherens junction
aPKC	Atypical protein kinase C
Crb	Crumbs
CRIB	Cdc42/Rac interactive binding
Dlg	Discs large
ECM	Extracellular matrix
EMT	Epithelial to mesenchymal transition
GTP	Guanosine 5′-triphosphate
Lgl	Lethal (2) giant larvae
LKB1/STK11	Liver kinase B1/serine-threonine protein kinase 11
MDCK	Madin-Darby canine kidney

S. Ohno (✉) • S. Goulas • T. Hirose
Department of Molecular Biology, Yokohama City University School of Medicine, 3-9 Fuku-ura, Kanazawa-ku, Yokohama 236-0004, Japan
e-mail: ohnos@med.yokohama-cu.ac.jp

K. Ebnet (ed.), *Cell Polarity 1*, DOI 10.1007/978-3-319-14463-4_1

Pals	Protein associated with *Caenorhabditis elegans* Lin-7 protein
PAR	Partitioning defective
Patj	Pals1-associated TJ protein
PB1	Phox and Bem1
PDK1	Phosphoinositide-dependent kinase 1
PDZ	Postsynaptic density 95/discs large/zona occludens-1
Pins	Partner of inscuteable
TJ	Tight junction

1.1 Identification of Members of the aPKC-PAR Complex: A Historical Background

Cell polarity is one of the most fundamental features of a cell. In multicellular organisms, it is essential not only for cell functions but also for the development and tissue maintenance (Rodriguez-Boulan and Nelson 1989; Bryant and Mostov 2008). However, the variety of shapes and functions of polarized cells had greatly hindered the discovery of the universal mechanism and genes/proteins regulating cell polarity. Breakthroughs for the discovery were made in the 1990s when several independent research fields converged to identify conserved cell polarity proteins that localized in an asymmetric manner to restricted regions of the cell periphery and were involved in the establishment and/or maintenance of cell polarity (Ohno 2001; Goldstein and Macara 2007) (Table 1.1). One such field was the developmental genetics during the early embryogenesis in *Drosophila*. In the late 1970s to early 1980s, the *Drosophila bazooka* gene was identified through a screen of mutations that affected embryonic patterning (Wieschaus et al. 1984). Although *bazooka* mutant embryos exhibit a phenotype characterized by the loss of apicobasal polarity, the analysis of the molecular nature of Bazooka had not been reported until 1998, when Kuchinke et al. (1998) indicated a sequence similarity between Bazooka and one of the polarity proteins in *C. elegans*. The genetic screening in *C. elegans* embryogenesis also identified several genes involved in generating multiple cell types through their appropriate spatial distribution (Kemphues et al. 1988). Among them, six *Par* genes were required for reading the polarity cue provided by the sperm to establish the polarity axis required for the asymmetric division leading to embryonic anterior-posterior polarity (Kemphues 2000). Importantly, four out of the six *Par* genes encoded proteins that transiently showed asymmetric distribution to the periphery of the one-cell embryo. One of the most important research fields that contributed to the identification of polarity proteins is epithelial cell biology in vertebrates. Polarized epithelial cells play fundamental roles in the development and function of a variety of tissues and organs (Rodriguez-Boulan and Nelson 1989; Bryant and Mostov 2008). The morphogenesis of a sheet of polarized epithelial cells is the first sign of cellular differentiation in early development. In the adult, polarized epithelial cells line all body cavities and exist in tissues that carry out specialized vectorial transport

Table 1.1 Components of the PAR complex, aPKC, PAR3, and PAR6, and their isoforms in multicellular organisms

	Mammals	Gene symbol	*Drosophila*	*C. elegans*
PAR3	PAR3 (ASIP)	PARD3	Bazooka (Baz)	PAR3
	PAR3L (PAR3β)	PARD3B		
aPKC	aPKCι (aPKCλ)	PRKCi	DaPKC	PKC-3
	aPKCζ	PRKCz		
PAR6	PAR6C (PAR6α)	PARD6A	Par6	PAR6
	PAR6B (PAR6β)	PARD6B		
	PAR6G (PAR6γ)	PARD6G		
Lateral polarity proteins mentioned in this chapter (see other chapters for details)				
PAR1	PAR1c (MARK1)	MARK1	PAR1	PAR1
	PAR1b (MARK2, EMK1)	MARK2		
	PAR1a (MARK3, EMK2, STK10, p78, c-TAK1)	MARK3		
	PAR1d (MARK4, MARKL1)	MARK4		
Lgl	Lgl1 (L2gl1, Hugl1)	LLGL1	Lethal giant larvae (Lgl)	–
	Lgl2 (L2GL2, Hgl)	LLGL2		
Scrb	Scrb1	SCRB1	Scribble (Scrib)	LET-413
Apical polarity proteins mentioned in this chapter (see other chapters for details)				
Crb	CRB1	CRB1	Crumbs (Crb)	Crb1
	CRB2	CRB2		
	CRB3	CRB3		
Pals1		MPP5	Stardust (Std)	–
Patj		INADL	Patj	–
Uniformly distributed polarity proteins				
LKB1	LKB1 (STK11)	STK11	Lkb1	PAR4
14-3-3	Many isoforms	Many isoforms	14-3-3	PAR5

functions of absorption and secretion. Epithelial apicobasal cell polarity is established by reading the spatial cues usually provided externally, such as by cell-cell and cell-matrix interactions (Yeaman et al. 1999). This leads to the reorganization of proteins in the cytoplasm and on the plasma membrane, followed by the segregation and retention of specific proteins and lipids in distinct apical and basolateral plasma membrane domains (Rodriguez-Boulan and Nelson 1989).

In the late 1990s, a search for binding partners of the mammalian signaling protein kinase, atypical protein kinase C (aPKC), identified ASIP (aPKC-specific interacting protein), a mammalian homologue of *C. elegans* PAR3 (Izumi et al. 1998). This finding led to the discovery of the *C. elegans* aPKC (PKC-3) as a key polarity protein involved in asymmetric cell division of the one-cell embryo (Tabuse et al. 1998). Consequently, this raised the question whether mammalian aPKC and PAR3 are involved in cell polarity signaling as they do in the *C. elegans* one-cell embryo. Interestingly, mammalian PAR3 showed an intriguing distribution in mammalian tissues; PAR3 and aPKC localize to tight junctions (TJs) in a variety of epithelial tissues. Tight junctions are cell-cell junctions specific to epithelial tissues and were implicated to function as a fence that inhibits intermixing of apical and basolateral membrane proteins by lateral diffusion (Tsukita et al. 2001). Subsequent studies established that PAR3, aPKC, PAR6, and PAR1 are involved in the establishment of epithelial cell polarity in vertebrates, reading the spatial cue provided by the cell-cell contact and reorganizing cellular materials (Suzuki and Ohno 2006).

1.2 The aPKC-PAR Complex: Its Physical and Functional Interaction with Other Polarity Regulators

C. elegans and *Drosophila* developmental genetics, together with mammalian epithelial cell biology, have identified conserved polarity proteins that are involved in epithelial cell polarity. Importantly, they localize asymmetrically and interact functionally and physically to each other. These polarity proteins are divided into three classes, the aPKC-PAR complex, the apical Crumbs complex, and the basolateral polarity proteins that localize in a mutually exclusive manner (Figs. 1.1 and 1.2). Positive feedback mechanisms through self-recruitment of polarity proteins seem to enforce localized concentration of these polarity proteins (Thompson 2013).

The aPKC-PAR polarity complex defined as a ternary complex is composed of PAR3, aPKC, and PAR6 (Ohno 2001) (Fig. 1.3). These three proteins transiently colocalize to the anterior and apical periphery of the *C. elegans* one-cell embryo (Fig. 1.2a) and *Drosophila* neuroblasts (Fig. 1.2b), respectively, prior to asymmetric division. Although in mammals such transient localization has been shown during cell migration (Fig. 1.2d) (Etienne-Manneville and Hall 2001), it has yet to be clearly demonstrated during cell division. However, in mammalian epithelial cells, a distinct asymmetric apico-lateral localization of the ternary complex can be observed at the level of the tight junctions (TJs) (Fig. 1.2c). In contrast to invertebrates, in mammals, in addition to multiple splicing isoforms, there are also paralogues encoded by different genes (Table 1.1). However, as the experimental data are limited for some of the isoforms and many studies use antibodies that react to several isoforms, this makes it generally difficult to distinguish the differences between isoforms. Thus, here, we refer to isoforms only when it is necessary.

PAR3 is a scaffold-like protein with several conserved domains implicated in protein-protein interactions, CR1 (conserved region 1), CR2, and CR3. CR1 is required for self-association and localization to the cell-cell contact region and TJs (Fig. 1.4) (Mizuno et al. 2003). The structural analyses of the CR1 revealed that this self-association is mediated by its PB1 domain-like fold that forms electrostatic interface for the "front-to-back" interaction (Feng et al. 2007; Zhang et al. 2013).

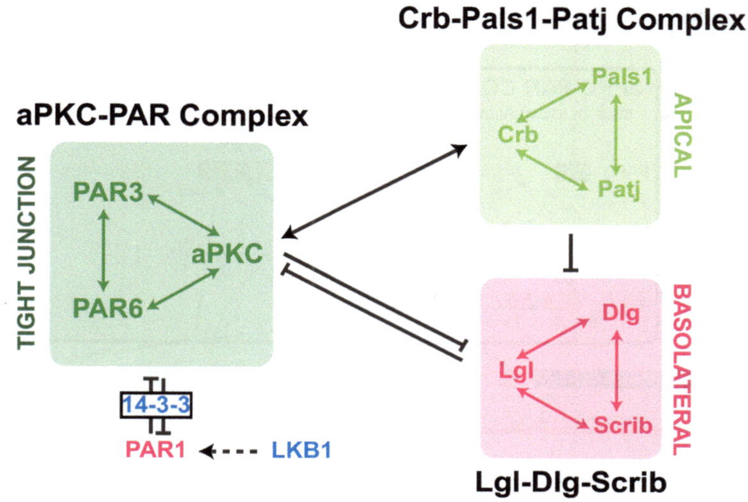

Fig. 1.1 The evolutionarily conserved polarity protein network in epithelial cells. A schematic diagram showing the functional interactions between the three core polarity complexes required for epithelial polarization: the aPKC-PAR complex, the Crb-Pals-Patj complex, and the Lgl-Dlg-Scrib group (Suzuki and Ohno 2006; Tepass 2012). The spatial and functional differences between these complexes are indicated by the different colored groups

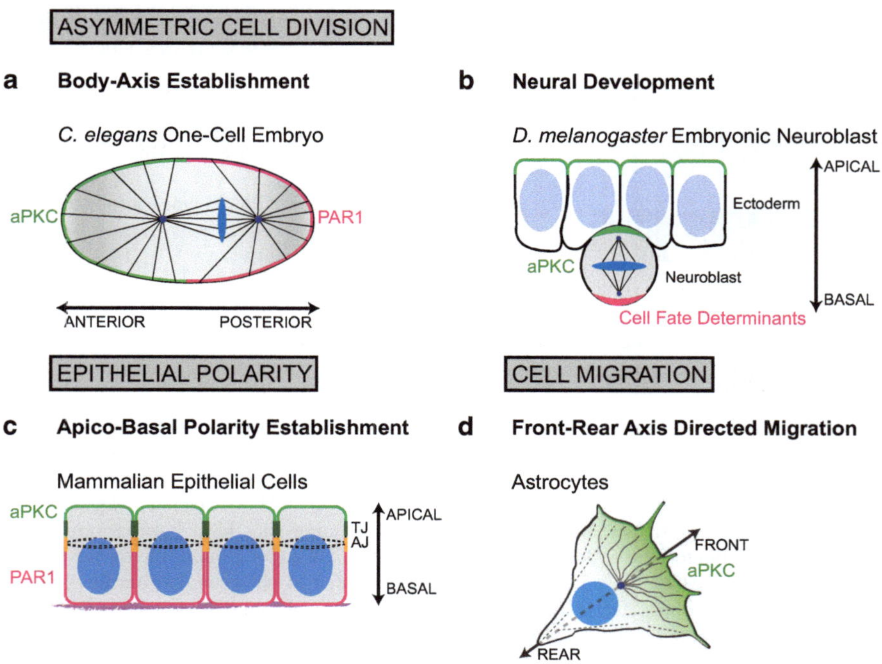

Fig. 1.2 Function of the aPKC-PAR complex in various cellular contexts. Localization of the aPKC-PAR complex during asymmetric cell division in *C. elegans* one-cell embryo (**a**) and *D. melanogaster* embryonic neuroblast (**b**). The localization of aPKC in polarized epithelial cells (**c**) and during astrocyte migration (**d**). *Arrows* indicate the axis of polarity. *Green*, aPKC or aPKC-PAR complex; *Magenta*, PAR1 and/or cell fate determinants

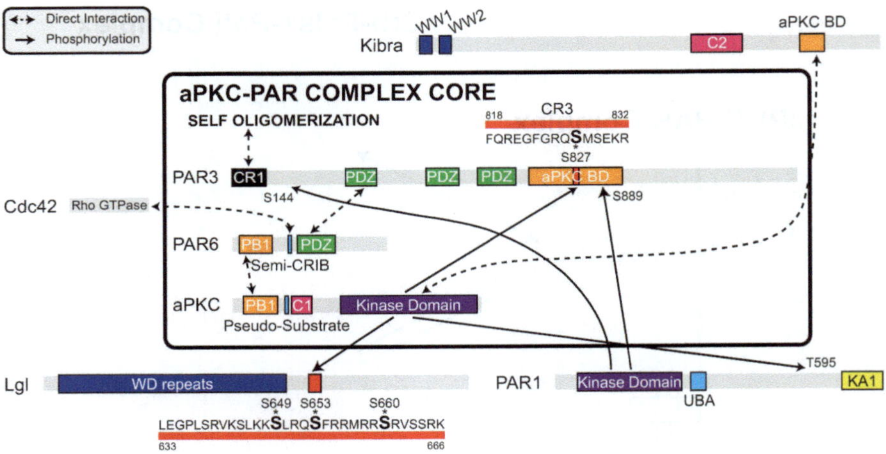

Fig. 1.3 Interaction map of core components of the aPKC-PAR complex and some of its key regulators/and or substrates. *aPKC BD* aPKC-binding domain, *C1* conserved domain 1, *C2* conserved domain 2, *CR1* conserved region 1, *CR3* conserved region 3, *KA1* kinase-associated domain 1, *PB1* Phox and Bem1p 1 domain, *PDZ* PSD95/Dlg/ZO-1 domain, *Semi-CRIB* Cdc42/Rac interactive binding motif, *WD repeats* WD40 domain repeats, *WW1* two conserved tryptophans domain 1, *WW2* two conserved tryptophans domain 2, *UBA* ubiquitin-associated domain

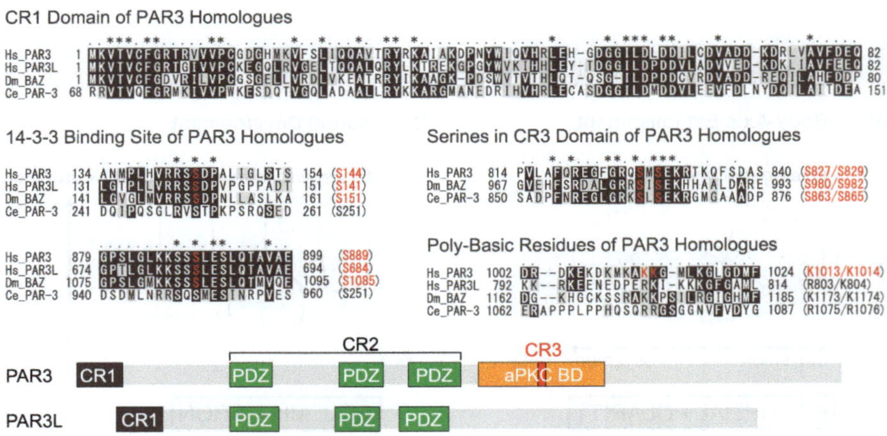

Fig. 1.4 Conserved sequence motifs in PAR3 homologues. Comparison of the 14-3-3-binding sites, conserved serines in CR3 domains, polybasic residues, and CR1 domains of PAR3 isoforms from different species. Amino acid sequences of human PAR3 (NP_062565.2) and PAR3L/PAR3β (NP_689739.4) are aligned with *D. melanogaster* Bazooka (NP_523383.2) and *C. elegans* PAR3 (CDH93174.1). The serines indicated in *red* in 14-3-3-binding sites have been reported to be phosphorylated (Benton and St Johnston 2003b; Hurd et al. 2003a) or responsible for the binding (Izaki et al. 2005). *Drosophila* PAR1 has been demonstrated to phosphorylate serines 151 and 1,085 in Bazooka (Ohno 2001). The serines indicated in *red* in the CR3 domains have been reported to be phosphorylated by aPKC (Nagai-Tamai et al. 2002; Morais-de-Sa et al. 2010). The CR3 domain is not conserved in PAR3L. The lysines indicated in *red* in the polybasic residues in human PAR3 have been identified as the phosphoinositide-binding site (Horikoshi et al. 2011). Self-association of PAR3 via its CR1 domains has been reported to control membrane localization of PAR3 and tight junction formation (Mizuno et al. 2003; Feng et al. 2007; Zhang et al. 2013). *Asterisks* and *dots* indicate identical and similar amino acids, respectively

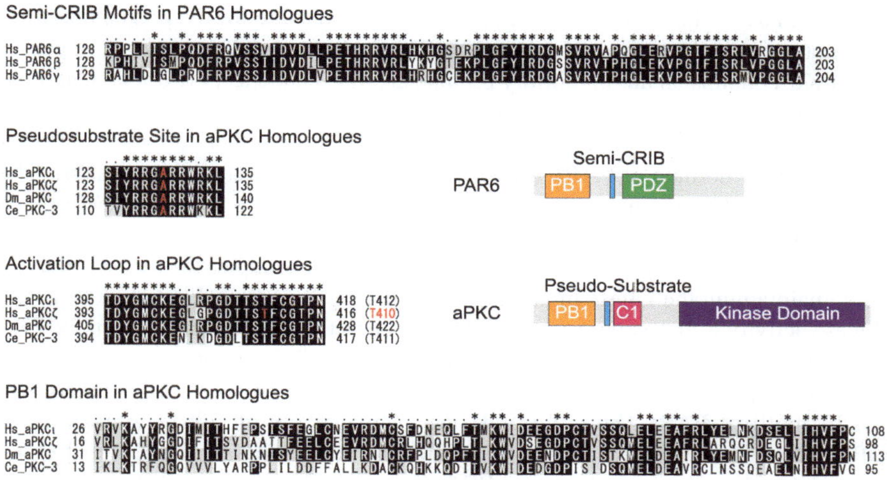

Fig. 1.5 Conserved semi-CRIB motif in PAR6 and sequence motifs in aPKC homologues. Comparison of the semi-CRIB motifs of human PAR6 isoforms. Amino acid sequences of human PAR6α (BAA96235.1), PAR6β (BAB40756.1), and PAR6γ (BAB40757.1) are aligned. *Asterisks* and *dots* indicate identical and similar amino acids, respectively. This semi-CRIB motif and continuing five amino acids within the PDZ domain form the interaction surface for Cdc42 (Garrard et al. 2003). Comparison of the pseudosubstrate sites, activation loops, and PB1 domains of aPKC isoforms from different species. Amino acid sequences of human aPKCι (P41743.2) and aPKCζ (Q05513.4) are aligned with *D. melanogaster* aPKC (NP_524892.2) and *C. elegans* PKC-3 (Q19266.1). The conserved alanines in the pseudosubstrate sites were indicated in *red*. It has been reported that PDK1 activates aPKCζ by phosphorylation on Thr410 (Chou et al. 1998; Le Good et al. 1998; Perander et al. 2001). *Asterisks* and *dots* indicate identical and similar amino acids, respectively

The CR2 contains three PDZ domains that can interact with a variety of proteins including PAR6 and several integral membrane proteins forming cell-cell junctions (Itoh et al. 2001; Takekuni et al. 2003; Iden et al. 2006; Ebnet 2008). PAR3 can bind not only proteins but also phosphoinositides, via the second PDZ domain in the CR2 and the conserved polybasic region (Wu et al. 2007; Horikoshi et al. 2011). The CR3 contains a conserved sequence motif involved in direct interaction with the catalytic domain of aPKC and can be phosphorylated at S827 (Figs. 1.3 and 1.4) (Izumi et al. 1998; Nagai-Tamai et al. 2002). Phosphorylation of PAR3 S827 by aPKC destabilizes aPKC-PAR3 interactions, resulting in the dissociation of the complex (Nagai-Tamai et al. 2002). Thus, the interaction between aPKC and PAR3 can be regulated. Although PAR3L (or PAR3b) shares many features with PAR3, it lacks the sequence and capacity for aPKC binding (Gao et al. 2002; Kohjima et al. 2002). In *Drosophila*, PAR1 can phosphorylate PAR3 on S151/S1085 and antagonize the interaction of aPKC with PAR3 via the binding of 14-3-3 to PAR3 (Figs. 1.3 and 1.4) (Benton and St Johnston 2003b). Although the binding of 14-3-3 is conserved in mammalian PAR3 and PAR3L, destabilization of the aPKC-PAR3 complex has not been observed, and the kinase responsible for the phosphorylation has not yet been identified (Hurd et al. 2003a; Izaki et al. 2005).

aPKC contains an autoinhibitory pseudosubstrate sequence, C1 domain, and the kinase domain conserved among the protein kinase C family (Fig. 1.5) (Suzuki et al. 2003). In addition, aPKC contains the PB1 domain that can stably interact with the PB1 domain of PAR6 (Figs. 1.3 and 1.5) (Hirano et al. 2004, 2005). The N-terminal half contains the structural features that regulate the kinase activity of aPKC. The pseudosubstrate sequence resembles the substrate sequence except that the phosphorylation site is not Ser or Thr but Ala and it occupies the kinase catalytic site, maintaining the kinase in an inactive state (Fig. 1.5) (Suzuki et al. 2003). The kinase activity is also suppressed by PAR6 through the interaction between the PB1 domains of aPKC and PAR6 (Yamanaka et al. 2001). This suppression seems to be alleviated upon the interaction of active, GTP-loaded Cdc42 (or Rac1) to the semi-CRIB motif of PAR6 (Figs. 1.3 and 1.5) (Yamanaka et al. 2001). Activation of aPKC is also mediated by phosphoinositide-dependent kinase 1 (PDK1) that phosphorylates the activation loop in aPKC (Fig. 1.5) (Chou et al. 1998; Le Good et al. 1998; Perander et al. 2001).

1.3 Converting the Spatial Cue to Apicobasal Polarity: The aPKC-PAR Complex

The principal function of epithelia is to form barriers that separate biological compartments. This is accomplished through the establishment of epithelial polarity along the apicobasal axis, defined by the apical "free" surface of the epithelium and the basal adhesive surface that is in direct contact with the extracellular matrix (ECM) (Yeaman et al. 1999; Bryant and Mostov 2008; Nelson 2009; Tepass 2012). Epithelia require spatial cues from the ECM and neighboring cells to orient their polarity (Bryant and Mostov 2008). The former acts as the first landmark to generate a premature basal domain, while the latter determine the border between apical and basal membrane domains to establish and maintain the orientation of apicobasal polarity. Although different types of epithelial cells can be observed within one organism or across species and how they arise may differ, the molecular machinery involved in generating epithelial polarity appears to be highly conserved. Simple columnar epithelia such as Madin-Darby canine kidney (MDCK) cells and *Drosophila* follicular epithelial cells have served as excellent model systems to clarify the molecular mechanism of epithelial polarization (Suzuki and Ohno 2006; Martin-Belmonte and Mostov 2008; Tepass 2012). MDCK cellular monolayers can be manipulated to induce cell-cell adhesion between epithelial cells through calcium depletion/repletion, the so-called calcium switch, to stimulate E-cadherin-mediated intracellular signaling and thereby allowing the step-by-step monitoring of epithelial polarization (Suzuki and Ohno 2006). The genetic amenability of *Drosophila*, on the other hand, has allowed for the identification of bona fide as well as novel polarity regulators through large-scale genetic screens (Tepass 2012). Together with the accumulating evidence from other systems, it has been demonstrated that an evolutionary conserved molecular mechanism exists that regulates apicobasal epithelial polarity.

The process of epithelial cell polarization has been precisely described in cultured mammalian epithelial cells by experimentally controlling cell-cell adhesion (e.g., Ca^{2+} switch, wound healing, and replating of the cells) (Ebnet et al. 2004; Suzuki and Ohno 2006; Martin-Belmonte and Mostov 2008) (Fig. 1.6). "Cell-cell contact naïve" cells first migrate and find neighboring cells with protruding lamellipodia and filopodia (Fig. 1.6a) (Yonemura et al. 1995). When cell-cell contact is established through these protrusions, E-cadherin is recruited to these sites and transduces this spatial cue to form primordial adherens junctions (AJs) or spot-like AJs (Fig. 1.6b) (Yonemura et al. 1995). Primordial AJs consist not only of the structural components for mature AJs and tight junctions (TJs) but also include the aPKC-PAR complex and regulatory molecules involved in vesicle trafficking and cytoskeletal organization (Fig. 1.6c) (Ebnet et al. 2004; Suzuki and Ohno 2006; Nelson 2009). Primordial AJs further coalesce into epithelium-specific belt-like AJs encircling each epithelial cell and eventually develop into mature AJs by segregating TJs (Fig. 1.6d) (Suzuki and Ohno 2006). Concurrently, TJs also mature, with the polymerization of TJ-specific integral membrane proteins into strands within the plasma membrane that act as a "fence" between the apical and basolateral membrane domains (Tsukita et al. 2001). Thus, epithelial apicobasal polarization is completed with the formation of distinct apical and basolateral membrane domains separated by mature TJs that are apical to the AJs (Fig. 1.6e) (Yeaman et al. 1999; Nelson 2009).

The aPKC-PAR complex has been established to have an essential role during epithelial polarization by converting the spatial cues received from cell-cell contacts into apicobasal polarity (Suzuki and Ohno 2006; Goldstein and Macara 2007). During the polarization process, members of the aPKC-PAR complex are sequentially recruited to the developing primordial AJs. PAR3 is initially recruited through its direct interaction with junctional adhesion molecules (JAMs) and nectin family proteins, providing a positional cue for the subsequent recruitment of aPKC-PAR6 (Ebnet et al. 2001; Itoh et al. 2001; Takekuni et al. 2003). PAR3 localization is also, at least in part, dependent on the p53-binding protein, ASPP2 in a mutually dependent manner (Cong et al. 2010), and its self-oligomerization through the N-terminal conserved domain (Benton and St Johnston 2003a; Mizuno et al. 2003). However, experimental induction of cell-cell adhesion through calcium depletion has demonstrated that aPKC-PAR6 is initially observed at primordial AJs together with Lgl (Fig. 1.3) (Yamanaka et al. 2003). This localization is only transient as the phosphorylation of Lgl, mediated by aPKC, leads to the loading of PAR3 into the aPKC-PAR6 complex (Yamanaka et al. 2006). Consistently, aPKC-PAR6 is found to interact with PAR3 and Lgl in a mutually exclusive manner (Betschinger et al. 2003; Yamanaka et al. 2003). The direct interaction between

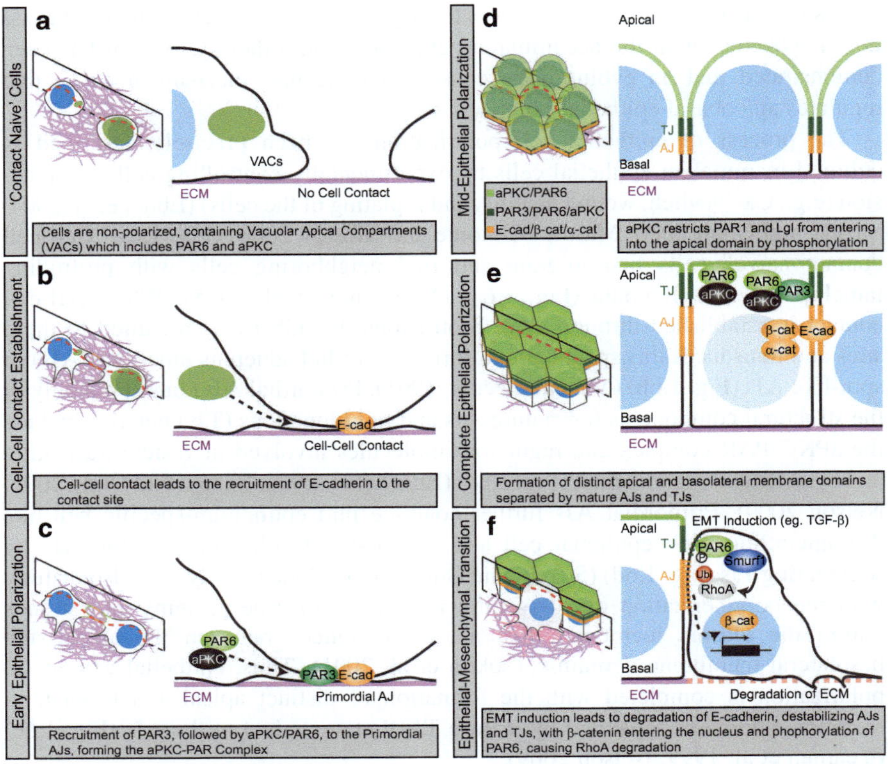

Fig. 1.6 The aPKC-PAR complex during epithelial polarization and epithelial to mesenchymal transition (EMT). Stages of epithelial polarization in vitro, initially with the so-called "contact naïve" cell stage (**a**), followed by the cell-cell contact establishment stage (**b**) and the formation of the primordial adherens junction (AJ) (**c**). The maturation of the primordial AJs leads to the segregation of distinct membrane compartments consisting of AJs and TJs (**d**) which develop until the epithelium becomes fully polarized (**e**). TGF-β signaling induces EMT through the phosphorylation of PAR6, leading to Smurf activation and subsequent ubiquitination and degradation of RhoA (**f**) (Ozdamar et al. 2005)

PAR3 and aPKC-PAR6 is required for the maturation of both AJs (Imai et al. 2006) and TJs (Suzuki et al. 2001, 2002). In addition, the PAR3-bound form of the aPKC-PAR6 complex has also been shown to have a role in the proper development of the apical domain, as the loss of PAR3 results in multi-lumen formation as opposed to the typical single lumen during epithelial morphogenesis (Horikoshi et al. 2009).

PAR3 distinctly localizes to TJs (equivalent to *Drosophila* AJs) as observed by electron microscopy of mammalian renal tubule epithelial cells as well as in the *Drosophila* embryonic ectoderm and follicle cells (Hirose et al. 2002; Harris and Peifer 2005; Morais-de-Sa et al. 2010). Apart from its localization to TJs, the aPKC-PAR6 complex can also localize to the apical domain independently of PAR3 (Harris and Peifer 2005; Martin-Belmonte et al. 2007; Morais-de-Sa

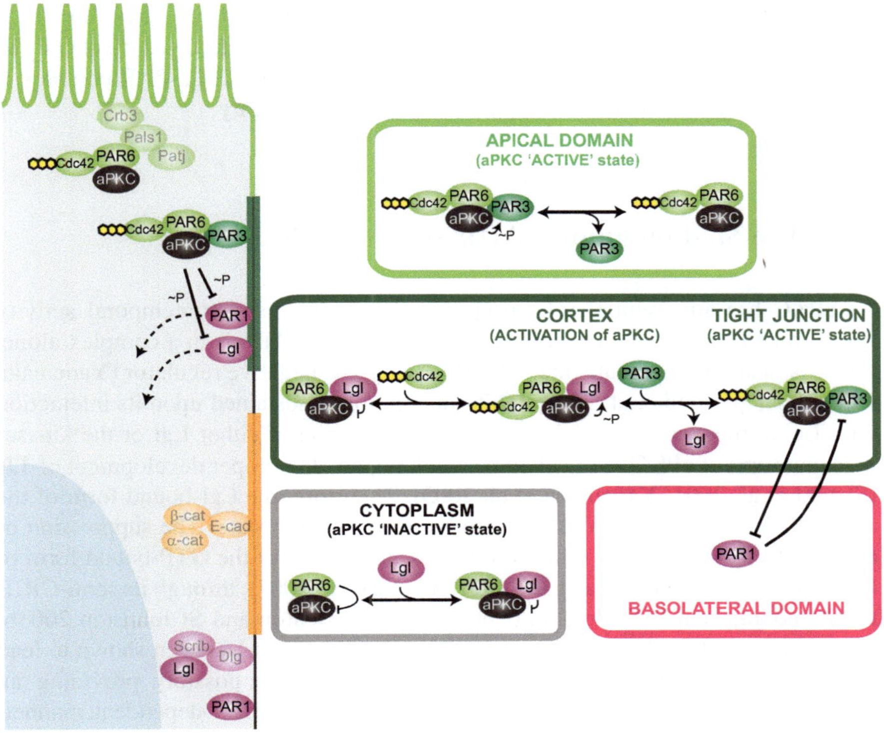

Fig. 1.7 The molecular mechanism of epithelial polarity establishment and aPKC activation. In the cytoplasm, aPKC is in an inactive state due to its interaction with PAR6 (Yamanaka et al. 2001). Interaction with Lgl is also inhibitory (Yamanaka et al. 2006). Binding of the GTP-bound form of Cdc42 to PAR6 results in the activation of aPKC and the phosphorylation of Lgl (Yamanaka et al. 2001, 2003). The release of Lgl allows PAR3 to enter aPKC-PAR6 complex to form the active aPKC-PAR complex at the tight junctions. The active aPKC-PAR complex can phosphorylate not only PAR1, leading to its dissociation from the apico-lateral domain, but also PAR3, leading to the disassembly of the aPKC-PAR complex (Nagai-Tamai et al. 2002; Hurov et al. 2004; Suzuki et al. 2004). The phosphorylation of PAR1 restricts its localization to the basolateral domain. Conversely, the basolateral PAR1 prevents the aPKC-PAR complex from entering the basolateral domain through the phosphorylation of PAR3 (Benton and St Johnston 2003b). The aPKC-PAR6 complex can also interact with the apical Crb-Pals-Patj complex (Hurd et al. 2003b). *Orange*, adherens junction; *Dark green*, tight junction; *Light green*, apical domain

et al. 2010). The apical localization of aPKC-PAR6 is mediated through the direct interaction of PAR6 to protein associated with Lin seven 1 (Pals1), a member of the Crumbs-Pals-Patj complex and the active GTP-bound form of small Rho GTPase Cdc42, both of which are known to localize to the apical domain (Fig. 1.7) (Hurd et al. 2003b; Martin-Belmonte et al. 2007). The recruitment of aPKC-PAR6 to the apical domain is likely to be further aided by the aPKC-dependent phosphorylation of PAR3 at the apical edge of primordial AJs, leading to the dissociation of the aPKC-PAR complex (Hirose et al. 2002; Nagai-Tamai et al. 2002). Therefore, these multiple transitions that aPKC and PAR6 undergo during the changes in subcellular localization and subunit composition demonstrate how the aPKC-PAR complex is able to convert spatial cues into apicobasal epithelial polarity.

1.4 The Spatiotemporal Activation of aPKC

As aPKC has an essential role in epithelial polarity, its spatiotemporal activity needs to be tightly regulated (Fig. 1.7). When aPKC-PAR6 is in a complex alone, aPKC is in an inactive state due to PAR6 acting as a negative regulator (Yamanaka et al. 2001; Atwood et al. 2007). This suppression is sustained upon its interaction with Lgl at the primordial AJs, as the overexpression of either Lgl or the kinase-deficient form of aPKC after calcium switch impairs the proper development of TJs (Suzuki et al. 2001; Yamanaka et al. 2003). Therefore, the Lgl-bound form of the aPKC-PAR6 complex is considered to be in an inactive state. The suppression of aPKC activity by PAR6 is relieved through the binding of the GTP-bound form of Cdc42 to PAR6 (Yamanaka et al. 2001; Atwood et al. 2007), through its semi-CRIB motif and adjacent PDZ domain (Lin et al. 2000; Benton and St Johnston 2003b; Garrard et al. 2003). E-cadherin-mediated cell-cell contact has been shown to lead to the activation of Cdc42 (Kim et al. 2000), thereby possibly providing an explanation for this temporal cascade. Although in a context-dependent manner, in *Drosophila* sensory organ precursor (SOP) cells and neuroblasts undergoing asymmetric cell division, the mitotic kinase Aurora A has also been shown to be involved in the activation of aPKC, through the phosphorylation of PAR6 (Wirtz-Peitz et al. 2008). Upon its activation, aPKC phosphorylates Lgl, releasing Lgl from the aPKC-PAR6 complex (Betschinger et al. 2003; Yamanaka et al. 2003). This leads to the loading of PAR3 to form the active aPKC-PAR complex (Yamanaka et al. 2006; Wirtz-Peitz et al. 2008). Consistently, the loss of Lgl results in the increased interaction of the GTP-bound Cdc42 and PAR3 with aPKC-PAR6 (Yamanaka et al. 2006). The PAR3-bound, active form of the aPKC-PAR complex is also considered to be transient, as the aPKC-dependent phosphorylation of PAR3 results in the dissociation of PAR3, from the aPKC-PAR6 complex (Nagai-Tamai et al. 2002). The dissociation from PAR3 allows the aPKC-PAR6 complex to also localize to the apical domain through the GTP-bound form of Cdc42 (Martin-Belmonte et al. 2007). This is also consistent with the binding of PAR6 to Pals1, which is enhanced in the presence of a dominant-active but not dominant-negative mutant of Cdc42 (Hurd et al. 2003b). Taken together, this demonstrates that the precise spatiotemporal activation of aPKC is required for the polarization of epithelia, which is mediated through the sequential subunit exchange of the aPKC-PAR complex.

1.5 The Effector of the aPKC-PAR Complex: aPKC Activity in Epithelial Polarity Establishment and Other Systems

Due to the spatial transitions of aPKC, its kinase activity has been implicated in various processes during epithelial polarization. aPKC kinase activity is required to restrict the localization of the basolateral determinant, phospholipid phosphatidylinositol (3,4,5)-trisphosphate, thereby regulating the size of the basolateral membrane domain during cell-cell contact-mediated polarization (Takahama et al. 2008). During epithelial morphogenesis, the proper specification of the apical domain requires the formation of the active aPKC-PAR complex (Horikoshi et al. 2009) and the phosphorylation of partner of inscuteable (Pins) by aPKC, to maintain the correct spindle orientation during cell division (Hao et al. 2010). Furthermore, the size of the apical domain has also been attributed to the kinase activity of aPKC (Ishiuchi and Takeichi 2011). Here, active aPKC localizing to the TJs prevents the localization of the Rho-associated kinase I (ROCKI) at the TJs through its phosphorylation, thereby regulating the apical actin/myosin constriction and the size of the apical domain. In the *Drosophila* ectoderm, aPKC kinase activity has also been shown to stabilize the apical localization of Crumbs (Crb) through its phosphorylation, to maintain the apical domain (Sotillos et al. 2004). Indeed, aPKC activity requires tight regulation at the apical domain, which, in part, is mediated by Kibra (Fig. 1.3) (Yoshihama et al. 2011), an upstream component of the Hippo tumor-suppressor pathway (Baumgartner et al. 2010; Genevet et al. 2010; Yu et al. 2010). Kibra suppresses the exocytosis of the apical domain compartment and through a highly conserved aPKC-binding domain, that is similar to the pseudosubstrate motif of aPKC, negatively regulates its activity (Yoshihama et al. 2011). Therefore, the loss of Kibra leads to increased apical exocytosis and the expansion of the apical domain through the hyperactivation of aPKC, although this is thought to be in a Crb-independent manner.

In other contexts, such as the *Drosophila* neuroblast, the aPKC-PAR complex and aPKC kinase activity have also been implicated in cell proliferation during asymmetric cell division (Fig. 1.2b). The *Drosophila* neuroblast delaminates from a polarized ectoderm and inherits this apicobasal polarity to undergo asymmetric cell division, generating another neuroblast and a ganglion mother cell (GMC) that further produces terminally differentiated neurons (Knoblich 2010). During mitosis, the apical polarization of the aPKC-PAR complex is coupled to the alignment of mitotic spindle machinery, through the interaction between Bazooka (the *Drosophila* homologue of PAR3) and the adapter protein Inscuteable (Kraut et al. 1996; Schober et al. 1999; Wodarz et al. 1999; Schaefer et al. 2000), leading to the asymmetric segregation of the so-called cell fate determinants, such as the Notch repressor Numb (Rhyu et al. 1994; Spana et al. 1995), to only the basal GMC. This exclusion from the apical domain is dependent on aPKC activity, as the overexpression of the membrane-tethered form of aPKC leads to the cytoplasmic mislocalization of fate determinants and the overproliferation of neuroblasts (Lee et al. 2006), while the loss of aPKC results in uniform cortical localization of fate determinants and the premature depletion of neuroblasts (Rolls et al. 2003).

Therefore, aPKC kinase activity is essential in various contexts and in epithelia and is required not only for the maturation of epithelial junctional structures but also for the proper establishment of apical and basolateral domains.

1.6 Molecular Antagonism: The Mechanism Behind Epithelial Polarization

During epithelial polarization, membrane compartments are generated consisting of distinct protein complexes (Bryant and Mostov 2008). These complexes can be divided by their subcellular localization into the Crb-Pals-Patj complex and the aPKC-PAR complex, localizing to the apical domain and the apico-lateral TJs, respectively, and the Lgl-Dlg-Scrib group which localizes to the basolateral domain (Fig. 1.1) (Suzuki and Ohno 2006; Yamanaka and Ohno 2008). While the Crb-Pals-Patj complex and the aPKC-PAR complex are required for establishing apical membrane domains, the Lgl-Dlg-Scrib group is responsible for establishing basolateral membrane domains. Genetic interactions in the *Drosophila* embryo have demonstrated that a mutual, negative regulatory feedback mechanism between these complexes is responsible for the establishment of proper membrane domains and epithelial polarization (Bilder et al. 2003; Tanentzapf and Tepass 2003). Consistent with this, altering these membrane domains, such as by the overexpression of Crb (Wodarz et al. 1995) or loss of Scrib (Bilder and Perrimon 2000), leads to epithelial disorganization due to the expansion of the apical domain at the expense of the basolateral domain. The MDCK cyst formation assay in collagen gel, which mimics epithelial tissue organization in vivo, also demonstrated the essential roles of this negative feedback for establishing three-dimensional epithelial structures (Martin-Belmonte and Mostov 2008). This mechanism of mutual antagonism is thought to be, at least in part, through the aPKC-PAR complex and the activity of aPKC.

The Lgl-Dlg-Scrib group, which localizes to the basolateral membrane domain of epithelia, restricts the localization of the apical membrane compartment (Fig. 1.7) (Bilder and Perrimon 2000). However, Lgl is also able to interact with the aPKC-PAR6 complex (Yamanaka et al. 2003). As Lgl competes with Cdc42 and PAR3 for binding to the aPKC-PAR6 complex, it antagonizes the formation of the active aPKC-PAR complex (Yamanaka et al. 2006). Upon the activation of aPKC, the phosphorylation of Lgl leads to its dissociation from aPKC-PAR6 and also prevents its localization to the apical membrane domain (Musch et al. 2002; Yamanaka et al. 2003; Wirtz-Peitz et al. 2008). Indeed, the overexpression of Lgl phospho-mutants leads to its mislocalization to the apical domain (Yamanaka et al. 2006; Wirtz-Peitz et al. 2008). Therefore, this negative feedback loop between Lgl and aPKC-PAR6 appears to be essential not only for the spatiotemporal regulation of aPKC activity but also for the generation of proper membrane compartments required during epithelial polarization.

Similar to Lgl, in contexts such as during body axis establishment in *C. elegans* and epithelial polarity establishment, the aPKC-PAR complex and PAR1 form mutually exclusive cortical domains (Fig. 1.2a, c) (Suzuki and Ohno 2006; Yamanaka and Ohno 2008). In epithelial cells, PAR1 localizes to the lateral domain and is essential for its proper development (Bohm et al. 1997; Suzuki et al. 2004; Masuda-Hirata et al. 2009). This localization is dependent on its phosphorylation by aPKC, which excludes PAR1 from the apical domain, as the overexpression of the phospho-mutant form of PAR1 results in its leakage into the apical domain (Fig. 1.7) (Hurov et al. 2004; Suzuki et al. 2004). The phosphorylation of PAR1 also leads to the increased binding of 14-3-3, allowing its dissociation from the lateral membrane (Suzuki et al. 2004). Consistent with its localization, PAR1 has also been described in a context-dependent manner, to have antagonistic functions to aPKC (Ossipova et al. 2007, 2009). PAR1, on the other hand, is able to phosphorylate PAR3 (Benton and St Johnston 2003b). The phosphorylation of PAR3 not only generates binding sites for 14-3-3 proteins on PAR3 but is also thought to prevent its membrane association and binding to aPKC and thereby inhibits the further assembly of the aPKC-PAR complex (Benton and St Johnston 2003b). Taken together, the mutual antagonism of PAR1 and aPKC leads to the proper establishment of apical and lateral membrane compartments and hence generates apicobasal polarity.

1.7 Concluding Remarks and Future Perspectives

The aPKC-PAR complex is an evolutionary conserved cassette of proteins that are required not only for the establishment and maintenance of epithelial polarity but also during the development and homeostasis of an organism. This is supported by the fact that members of the aPKC-PAR complex in vertebrates are largely redundant, and in invertebrates such as *Drosophila*, which only have one homologue of each, the loss of function of these genes leads to lethality (Wieschaus et al. 1984; Wodarz et al. 2000; Petronczki and Knoblich 2001; Rolls et al. 2003). Consistently, in vertebrates, although aPKCι/λ and aPKCζ are thought to be partially redundant, the mouse complete knockout of aPKCλ leads to lethality, demonstrating that it is essential for embryonic development (Soloff et al. 2004). Therefore, the loss of members of the aPKC-PAR complex has been also associated with the cause of diseases such as in the renal glomerular epithelial cells, in which the loss of aPKCλ leads to proteinuria and eventually causes glomerulosclerosis (Hirose et al. 2009; Huber et al. 2009).

As most tumors arise from epithelial-derived carcinomas (Lim and Thiery 2012), this would imply that maintenance of epithelial polarity has an essential role in preventing tumorigenesis. Indeed, the loss of polarity is thought to be one of the hallmark events of cancer (Aranda et al. 2008; Lee and Vasioukhin 2008).

Furthermore, increasing evidence suggests that there are links between members of the aPKC-PAR complex and cancer (Eder et al. 2005; Regala et al. 2005; Aranda et al. 2008; Kojima et al. 2008; Takagawa et al. 2010). In TGF-β-induced EMT, which mimics many aspects of cancer metastasis, PAR6 is phosphorylated, leading to the degradation of tight junctions and RhoA, through the activation of the ubiquitin ligase Smurf1, ultimately causing the loss of epithelial characteristics (Fig. 1.6f) (Ozdamar et al. 2005). Therefore, understanding how the aPKC-PAR complex is regulated, and identifying its downstream targets in various contexts will give further insights into not only how apicobasal polarity is established but also how diseases associated with polarity defects occur.

References

Aranda V, Nolan ME, Muthuswamy SK (2008) Par complex in cancer: a regulator of normal cell polarity joins the dark side. Oncogene 27(55):6878–6887. doi:10.1038/onc.2008.340

Atwood SX, Chabu C, Penkert RR, Doe CQ, Prehoda KE (2007) Cdc42 acts downstream of Bazooka to regulate neuroblast polarity through Par-6 aPKC. J Cell Sci 120(Pt 18):3200–3206. doi:10.1242/jcs.014902

Baumgartner R, Poernbacher I, Buser N, Hafen E, Stocker H (2010) The WW domain protein Kibra acts upstream of Hippo in Drosophila. Dev Cell 18(2):309–316. doi:10.1016/j.devcel.2009.12.013

Benton R, St Johnston D (2003a) A conserved oligomerization domain in drosophila Bazooka/PAR-3 is important for apical localization and epithelial polarity. Curr Biol 13(15):1330–1334. doi:10.1016/S0960-9822(03)00508-6

Benton R, St Johnston D (2003b) Drosophila PAR-1 and 14-3-3 inhibit Bazooka/PAR-3 to establish complementary cortical domains in polarized cells. Cell 115(6):691–704

Betschinger J, Mechtler K, Knoblich JA (2003) The Par complex directs asymmetric cell division by phosphorylating the cytoskeletal protein Lgl. Nature 422(6929):326–330. doi:10.1038/nature01486

Bilder D, Perrimon N (2000) Localization of apical epithelial determinants by the basolateral PDZ protein Scribble. Nature 403(6770):676–680

Bilder D, Schober M, Perrimon N (2003) Integrated activity of PDZ protein complexes regulates epithelial polarity. Nat Cell Biol 5(1):53–58. doi:10.1038/ncb897

Bohm H, Brinkmann V, Drab M, Henske A, Kurzchalia TV (1997) Mammalian homologues of C. elegans PAR-1 are asymmetrically localized in epithelial cells and may influence their polarity. Curr Biol 7(8):603–606. doi:10.1016/S0960-9822(06)00260-0

Bryant DM, Mostov KE (2008) From cells to organs: building polarized tissue. Nat Rev Mol Cell Biol 9(11):887–901. doi:10.1038/nrm2523

Chou MM, Hou W, Johnson J, Graham LK, Lee MH, Chen CS, Newton AC, Schaffhausen BS, Toker A (1998) Regulation of protein kinase C zeta by PI 3-kinase and PDK-1. Curr Biol 8(19):1069–1077. doi:10.1016/S0960-9822(98)70444-0

Cong W, Hirose T, Harita Y, Yamashita A, Mizuno K, Hirano H, Ohno S (2010) ASPP2 regulates epithelial cell polarity through the PAR complex. Curr Biol. doi:10.1016/j.cub.2010.06.024

Ebnet K (2008) Organization of multiprotein complexes at cell-cell junctions. Histochem Cell Biol 130(1):1–20. doi:10.1007/s00418-008-0418-7

Ebnet K, Suzuki A, Horikoshi Y, Hirose T, Meyer Zu Brickwedde MK, Ohno S, Vestweber D (2001) The cell polarity protein ASIP/PAR-3 directly associates with junctional adhesion molecule (JAM). EMBO J 20(14):3738–3748. doi:10.1093/emboj/20.14.3738

Ebnet K, Suzuki A, Ohno S, Vestweber D (2004) Junctional adhesion molecules (JAMs): more molecules with dual functions? J Cell Sci 117(Pt 1):19–29. doi:10.1242/jcs.00930

Eder AM, Sui X, Rosen DG, Nolden LK, Cheng KW, Lahad JP, Kango-Singh M, Lu KH, Warneke CL, Atkinson EN, Bedrosian I, Keyomarsi K, Kuo WL, Gray JW, Yin JC, Liu J, Halder G, Mills GB (2005) Atypical PKCiota contributes to poor prognosis through loss of apical-basal polarity and cyclin E overexpression in ovarian cancer. Proc Natl Acad Sci U S A 102(35): 12519–12524. doi:10.1073/pnas.0505641102

Etienne-Manneville S, Hall A (2001) Integrin-mediated activation of Cdc42 controls cell polarity in migrating astrocytes through PKCzeta. Cell 106(4):489–498

Feng W, Wu H, Chan LN, Zhang M (2007) The Par-3 NTD adopts a PB1-like structure required for Par-3 oligomerization and membrane localization. EMBO J 26(11):2786–2796. doi:10.1038/sj.emboj.7601702

Gao L, Macara IG, Joberty G (2002) Multiple splice variants of Par3 and of a novel related gene, Par3L, produce proteins with different binding properties. Gene 294(1–2):99–107

Garrard SM, Capaldo CT, Gao L, Rosen MK, Macara IG, Tomchick DR (2003) Structure of Cdc42 in a complex with the GTPase-binding domain of the cell polarity protein, Par6. EMBO J 22(5):1125–1133. doi:10.1093/emboj/cdg110

Genevet A, Wehr MC, Brain R, Thompson BJ, Tapon N (2010) Kibra is a regulator of the Salvador/Warts/Hippo signaling network. Dev Cell 18(2):300–308. doi:10.1016/j.devcel.2009.12.011

Goldstein B, Macara IG (2007) The PAR proteins: fundamental players in animal cell polarization. Dev Cell 13(5):609–622. doi:10.1016/j.devcel.2007.10.007

Hao Y, Du Q, Chen X, Zheng Z, Balsbaugh JL, Maitra S, Shabanowitz J, Hunt DF, Macara IG (2010) Par3 controls epithelial spindle orientation by aPKC-mediated phosphorylation of apical Pins. Curr Biol 20(20):1809–1818. doi:10.1016/j.cub.2010.09.032

Harris TJ, Peifer M (2005) The positioning and segregation of apical cues during epithelial polarity establishment in Drosophila. J Cell Biol 170(5):813–823. doi:10.1083/jcb.200505127

Hirano Y, Yoshinaga S, Ogura K, Yokochi M, Noda Y, Sumimoto H, Inagaki F (2004) Solution structure of atypical protein kinase C PB1 domain and its mode of interaction with ZIP/p62 and MEK5. J Biol Chem 279(30):31883–31890

Hirano Y, Yoshinaga S, Takeya R, Suzuki NN, Horiuchi M, Kohjima M, Sumimoto H, Inagaki F (2005) Structure of a cell polarity regulator, a complex between atypical PKC and Par6 PB1 domains. J Biol Chem 280(10):9653–9661

Hirose T, Izumi Y, Nagashima Y, Tamai-Nagai Y, Kurihara H, Sakai T, Suzuki Y, Yamanaka T, Suzuki A, Mizuno K, Ohno S (2002) Involvement of ASIP/PAR-3 in the promotion of epithelial tight junction formation. J Cell Sci 115(Pt 12):2485–2495

Hirose T, Satoh D, Kurihara H, Kusaka C, Hirose H, Akimoto K, Matsusaka T, Ichikawa I, Noda T, Ohno S (2009) An essential role of the universal polarity protein, aPKClambda, on the maintenance of podocyte slit diaphragms. PLoS One 4(1):e4194. doi:10.1371/journal.pone.0004194

Horikoshi Y, Hamada S, Ohno S, Suetsugu S (2011) Phosphoinositide binding by par-3 involved in par-3 localization. Cell Struct Funct 36(1):97–102. doi:10.1247/csf.11005

Horikoshi Y, Suzuki A, Yamanaka T, Sasaki K, Mizuno K, Sawada H, Yonemura S, Ohno S (2009) Interaction between PAR-3 and the aPKC-PAR-6 complex is indispensable for apical domain development of epithelial cells. J Cell Sci 122(Pt 10):1595–1606. doi:10.1242/jcs.043174

Huber TB, Hartleben B, Winkelmann K, Schneider L, Becker JU, Leitges M, Walz G, Haller H, Schiffer M (2009) Loss of podocyte aPKClambda/iota causes polarity defects and nephrotic syndrome. J Am Soc Nephrol 20(4):798–806, doi:ASN.2008080871 [pii] 10.1681/ASN.2008080871

Hurd TW, Fan S, Liu CJ, Kweon HK, Hakansson K, Margolis B (2003a) Phosphorylation-dependent binding of 14-3-3 to the polarity protein Par3 regulates cell polarity in mammalian epithelia. Curr Biol 13(23):2082–2090. doi:10.1016/j.cub.2003.11.020

Hurd TW, Gao L, Roh MH, Macara IG, Margolis B (2003b) Direct interaction of two polarity complexes implicated in epithelial tight junction assembly. Nat Cell Biol 5(2):137–142. doi:10.1038/ncb923

Hurov JB, Watkins JL, Piwnica-Worms H (2004) Atypical PKC phosphorylates PAR-1 kinases to regulate localization and activity. Curr Biol 14(8):736–741. doi:10.1016/j.cub.2004.04.007

Iden S, Rehder D, August B, Suzuki A, Wolburg-Buchholz K, Wolburg H, Ohno S, Behrens J, Vestweber D, Ebnet K (2006) A distinct PAR complex associates physically with VE-cadherin in vertebrate endothelial cells. EMBO Rep 7(12):1239–1246

Imai F, Hirai S, Akimoto K, Koyama H, Miyata T, Ogawa M, Noguchi S, Sasaoka T, Noda T, Ohno S (2006) Inactivation of aPKClambda results in the loss of adherens junctions in neuroepithelial cells without affecting neurogenesis in mouse neocortex. Development 133(9):1735–1744

Ishiuchi T, Takeichi M (2011) Willin and Par3 cooperatively regulate epithelial apical constriction through aPKC-mediated ROCK phosphorylation. Nat Cell Biol 13(7):860–866. doi:10.1038/ncb2274

Itoh M, Sasaki H, Furuse M, Ozaki H, Kita T, Tsukita S (2001) Junctional adhesion molecule (JAM) binds to PAR-3: a possible mechanism for the recruitment of PAR-3 to tight junctions. J Cell Biol 154(3):491–497. doi:10.1083/jcb.200103047

Izaki T, Kamakura S, Kohjima M, Sumimoto H (2005) Phosphorylation-dependent binding of 14-3-3 to Par3beta, a human Par3-related cell polarity protein. Biochem Biophys Res Commun 329(1):211–218

Izumi Y, Hirose T, Tamai Y, Hirai S, Nagashima Y, Fujimoto T, Tabuse Y, Kemphues KJ, Ohno S (1998) An atypical PKC directly associates and colocalizes at the epithelial tight junction with ASIP, a mammalian homologue of Caenorhabditis elegans polarity protein PAR-3. J Cell Biol 143(1):95–106

Kemphues K (2000) PARsing embryonic polarity. Cell 101(4):345–348

Kemphues KJ, Priess JR, Morton DG, Cheng NS (1988) Identification of genes required for cytoplasmic localization in early C. elegans embryos. Cell 52(3):311–320

Kim SH, Li Z, Sacks DB (2000) E-cadherin-mediated cell-cell attachment activates Cdc42. J Biol Chem 275(47):36999–37005. doi:10.1074/jbc.M003430200

Knoblich JA (2010) Asymmetric cell division: recent developments and their implications for tumour biology. Nat Rev Mol Cell Biol 11(12):849–860. doi:10.1038/nrm3010

Kohjima M, Noda Y, Takeya R, Saito N, Takeuchi K, Sumimoto H (2002) PAR3beta, a novel homologue of the cell polarity protein PAR3, localizes to tight junctions. Biochem Biophys Res Commun 299(4):641–646

Kojima Y, Akimoto K, Nagashima Y, Ishiguro H, Shirai S, Chishima T, Ichikawa Y, Ishikawa T, Sasaki T, Kubota Y, Inayama Y, Aoki I, Ohno S, Shimada H (2008) The overexpression and altered localization of the atypical protein kinase C lambda/iota in breast cancer correlates with the pathologic type of these tumors. Hum Pathol 39(6):824–831

Kraut R, Chia W, Jan LY, Jan YN, Knoblich JA (1996) Role of inscuteable in orienting asymmetric cell divisions in Drosophila. Nature 383(6595):50–55. doi:10.1038/383050a0

Kuchinke U, Grawe F, Knust E (1998) Control of spindle orientation in Drosophila by the Par-3-related PDZ-domain protein Bazooka. Curr Biol 8(25):1357–1365

Le Good JA, Ziegler WH, Parekh DB, Alessi DR, Cohen P, Parker PJ (1998) Protein kinase C isotypes controlled by phosphoinositide 3-kinase through the protein kinase PDK1. Science 281 (5385):2042–2045

Lee CY, Robinson KJ, Doe CQ (2006) Lgl, Pins and aPKC regulate neuroblast self-renewal versus differentiation. Nature 439(7076):594–598

Lee M, Vasioukhin V (2008) Cell polarity and cancer-cell and tissue polarity as a non-canonical tumor suppressor. J Cell Sci 121(Pt 8):1141–1150

Lim J, Thiery JP (2012) Epithelial-mesenchymal transitions: insights from development. Development 139(19):3471–3486. doi:10.1242/dev.071209

Lin D, Edwards AS, Fawcett JP, Mbamalu G, Scott JD, Pawson T (2000) A mammalian PAR-3-PAR-6 complex implicated in Cdc42/Rac1 and aPKC signalling and cell polarity. Nat Cell Biol 2(8):540–547

Martin-Belmonte F, Gassama A, Datta A, Yu W, Rescher U, Gerke V, Mostov K (2007) PTEN-mediated apical segregation of phosphoinositides controls epithelial morphogenesis through Cdc42. Cell 128(2):383–397

Martin-Belmonte F, Mostov K (2008) Regulation of cell polarity during epithelial morphogenesis. Curr Opin Cell Biol 20(2):227–234

Masuda-Hirata M, Suzuki A, Amano Y, Yamashita K, Ide M, Yamanaka T, Sakai M, Imamura M, Ohno S (2009) Intracellular polarity protein PAR-1 regulates extracellular laminin assembly by regulating the dystroglycan complex. Genes Cells. doi:10.1111/j.1365-2443.2009.01315.x

Mizuno K, Suzuki A, Hirose T, Kitamura K, Kutsuzawa K, Futaki M, Amano Y, Ohno S (2003) Self-association of PAR-3-mediated by the conserved N-terminal domain contributes to the development of epithelial tight junctions. J Biol Chem 278(33):31240–31250

Morais-de-Sa E, Mirouse V, St Johnston D (2010) aPKC phosphorylation of Bazooka defines the apical/lateral border in Drosophila epithelial cells. Cell 141(3):509–523. doi:10.1016/j.cell.2010.02.040

Musch A, Cohen D, Yeaman C, Nelson WJ, Rodriguez-Boulan E, Brennwald PJ (2002) Mammalian homolog of Drosophila tumor suppressor lethal (2) giant larvae interacts with basolateral exocytic machinery in Madin-Darby canine kidney cells. Mol Biol Cell 13(1):158–168. doi:10.1091/mbc.01-10-0496

Nagai-Tamai Y, Mizuno K, Hirose T, Suzuki A, Ohno S (2002) Regulated protein-protein interaction between aPKC and PAR-3 plays an essential role in the polarization of epithelial cells. Genes Cells 7(11):1161–1171

Nelson WJ (2009) Remodeling epithelial cell organization: transitions between front-rear and apical-basal polarity. Cold Spring Harb Perspect Biol 1(1):a000513. doi:10.1101/cshperspect.a000513

Ohno S (2001) Intercellular junctions and cellular polarity: the PAR-aPKC complex, a conserved core cassette playing fundamental roles in cell polarity. Curr Opin Cell Biol 13(5):641–648

Ossipova O, Ezan J, Sokol SY (2009) PAR-1 phosphorylates Mind bomb to promote vertebrate neurogenesis. Dev Cell 17(2):222–233. doi:10.1016/j.devcel.2009.06.010

Ossipova O, Tabler J, Green JB, Sokol SY (2007) PAR1 specifies ciliated cells in vertebrate ectoderm downstream of aPKC. Development 134(23):4297–4306. doi:10.1242/dev.009282

Ozdamar B, Bose R, Barrios-Rodiles M, Wang HR, Zhang Y, Wrana JL (2005) Regulation of the polarity protein Par6 by TGFbeta receptors controls epithelial cell plasticity. Science 307(5715):1603–1609

Perander M, Bjorkoy G, Johansen T (2001) Nuclear import and export signals enable rapid nucleocytoplasmic shuttling of the atypical protein kinase C lambda. J Biol Chem 276(16):13015–13024

Petronczki M, Knoblich JA (2001) DmPAR-6 directs epithelial polarity and asymmetric cell division of neuroblasts in Drosophila. Nat Cell Biol 3(1):43–49. doi:10.1038/35050550

Regala RP, Weems C, Jamieson L, Khoor A, Edell ES, Lohse CM, Fields AP (2005) Atypical protein kinase C iota is an oncogene in human non-small cell lung cancer. Cancer Res 65(19):8905–8911

Rhyu MS, Jan LY, Jan YN (1994) Asymmetric distribution of numb protein during division of the sensory organ precursor cell confers distinct fates to daughter cells. Cell 76(3):477–491

Rodriguez-Boulan E, Nelson WJ (1989) Morphogenesis of the polarized epithelial cell phenotype. Science 245(4919):718–725

Rolls MM, Albertson R, Shih HP, Lee CY, Doe CQ (2003) Drosophila aPKC regulates cell polarity and cell proliferation in neuroblasts and epithelia. J Cell Biol 163(5):1089–1098. doi:10.1083/jcb.200306079

Schaefer M, Shevchenko A, Knoblich JA (2000) A protein complex containing Inscuteable and the Galpha-binding protein Pins orients asymmetric cell divisions in Drosophila. Curr Biol 10(7): 353–362

Schober M, Schaefer M, Knoblich JA (1999) Bazooka recruits Inscuteable to orient asymmetric cell divisions in Drosophila neuroblasts. Nature 402(6761):548–551. doi:10.1038/990135

Soloff RS, Katayama C, Lin MY, Feramisco JR, Hedrick SM (2004) Targeted deletion of protein kinase C lambda reveals a distribution of functions between the two atypical protein kinase C isoforms. J Immunol 173(5):3250–3260

Sotillos S, Diaz-Meco MT, Caminero E, Moscat J, Campuzano S (2004) DaPKC-dependent phosphorylation of Crumbs is required for epithelial cell polarity in Drosophila. J Cell Biol 166(4):549–557

Spana EP, Kopczynski C, Goodman CS, Doe CQ (1995) Asymmetric localization of numb autonomously determines sibling neuron identity in the Drosophila CNS. Development 121(11):3489–3494

Suzuki A, Akimoto K, Ohno S (2003) Protein kinase C lambda/iota (PKClambda/iota): a PKC isotype essential for the development of multicellular organisms. J Biochem (Tokyo) 133(1): 9–16

Suzuki A, Hirata M, Kamimura K, Maniwa R, Yamanaka T, Mizuno K, Kishikawa M, Hirose H, Amano Y, Izumi N, Miwa Y, Ohno S (2004) aPKC acts upstream of PAR-1b in both the establishment and maintenance of mammalian epithelial polarity. Curr Biol 14(16):1425–1435

Suzuki A, Ishiyama C, Hashiba K, Shimizu M, Ebnet K, Ohno S (2002) aPKC kinase activity is required for the asymmetric differentiation of the premature junctional complex during epithelial cell polarization. J Cell Sci 115(Pt 18):3565–3573

Suzuki A, Ohno S (2006) The PAR-aPKC system: lessons in polarity. J Cell Sci 119(Pt 6):979–987

Suzuki A, Yamanaka T, Hirose T, Manabe N, Mizuno K, Shimizu M, Akimoto K, Izumi Y, Ohnishi T, Ohno S (2001) Atypical protein kinase C is involved in the evolutionarily conserved par protein complex and plays a critical role in establishing epithelia-specific junctional structures. J Cell Biol 152(6):1183–1196

Tabuse Y, Izumi Y, Piano F, Kemphues KJ, Miwa J, Ohno S (1998) Atypical protein kinase C cooperates with PAR-3 to establish embryonic polarity in Caenorhabditis elegans. Development 125(18):3607–3614

Takagawa R, Akimoto K, Ichikawa Y, Akiyama H, Kojima Y, Ishiguro H, Inayama Y, Aoki I, Kunisaki C, Endo I, Nagashima Y, Ohno S (2010) High expression of atypical protein kinase C lambda/iota in gastric cancer as a prognostic factor for recurrence. Ann Surg Oncol 17(1): 81–88. doi:10.1245/s10434-009-0708-x

Takahama S, Hirose T, Ohno S (2008) aPKC restricts the basolateral determinant PtdIns(3,4,5)P3 to the basal region. Biochem Biophys Res Commun 368(2):249–255

Takekuni K, Ikeda W, Fujito T, Morimoto K, Takeuchi M, Monden M, Takai Y (2003) Direct binding of cell polarity protein PAR-3 to cell-cell adhesion molecule nectin at neuroepithelial cells of developing mouse. J Biol Chem 278(8):5497–5500

Tanentzapf G, Tepass U (2003) Interactions between the crumbs, lethal giant larvae and bazooka pathways in epithelial polarization. Nat Cell Biol 5(1):46–52

Tepass U (2012) The apical polarity protein network in Drosophila epithelial cells: regulation of polarity, junctions, morphogenesis, cell growth, and survival. Annu Rev Cell Dev Biol 28: 655–685. doi:10.1146/annurev-cellbio-092910-154033

Thompson BJ (2013) Cell polarity: models and mechanisms from yeast, worms and flies. Development 140(1):13–21. doi:10.1242/dev.083634

Tsukita S, Furuse M, Itoh M (2001) Multifunctional strands in tight junctions. Nat Rev Mol Cell Biol 2(4):285–293. doi:10.1038/35067088

Wieschaus E, Nüsslein-Volhard C, Jürgens G (1984) Mutations affecting the pattern of the larval cuticle in Drosophila melanogaster. III. Zygotic loci on the X-chromosome and fourth chromosome. Roux Arch Dev Biol 193(5):296–307. doi:10.1007/BF00848158

Wirtz-Peitz F, Nishimura T, Knoblich JA (2008) Linking cell cycle to asymmetric division: Aurora-A phosphorylates the Par complex to regulate Numb localization. Cell 135(1):161–173

Wodarz A, Hinz U, Engelbert M, Knust E (1995) Expression of crumbs confers apical character on plasma membrane domains of ectodermal epithelia of Drosophila. Cell 82(1):67–76

Wodarz A, Ramrath A, Grimm A, Knust E (2000) Drosophila atypical protein kinase C associates with Bazooka and controls polarity of epithelia and neuroblasts. J Cell Biol 150(6):1361–1374

Wodarz A, Ramrath A, Kuchinke U, Knust E (1999) Bazooka provides an apical cue for Inscuteable localization in Drosophila neuroblasts. Nature 402(6761):544–547. doi:10.1038/990128

Wu H, Feng W, Chen J, Chan LN, Huang S, Zhang M (2007) PDZ domains of Par-3 as potential phosphoinositide signaling integrators. Mol Cell 28(5):886–898

Yamanaka T, Horikoshi Y, Izumi N, Suzuki A, Mizuno K, Ohno S (2006) Lgl mediates apical domain disassembly by suppressing the PAR-3-aPKC-PAR-6 complex to orient apical membrane polarity. J Cell Sci 119(Pt 10):2107–2118

Yamanaka T, Horikoshi Y, Sugiyama Y, Ishiyama C, Suzuki A, Hirose T, Iwamatsu A, Shinohara A, Ohno S (2003) Mammalian Lgl forms a protein complex with PAR-6 and aPKC independently of PAR-3 to regulate epithelial cell polarity. Curr Biol 13(9):734–743

Yamanaka T, Horikoshi Y, Suzuki A, Sugiyama Y, Kitamura K, Maniwa R, Nagai Y, Yamashita A, Hirose T, Ishikawa H, Ohno S (2001) PAR-6 regulates aPKC activity in a novel way and mediates cell-cell contact-induced formation of the epithelial junctional complex. Genes Cells 6(8):721–731

Yamanaka T, Ohno S (2008) Role of Lgl/Dlg/Scribble in the regulation of epithelial junction, polarity and growth. Front Biosci 13:6693–6707

Yeaman C, Grindstaff KK, Nelson WJ (1999) New perspectives on mechanisms involved in generating epithelial cell polarity. Physiol Rev 79(1):73–98

Yonemura S, Itoh M, Nagafuchi A, Tsukita S (1995) Cell-to-cell adherens junction formation and actin filament organization: similarities and differences between non-polarized fibroblasts and polarized epithelial cells. J Cell Sci 108(Pt 1):127–142

Yoshihama Y, Sasaki K, Horikoshi Y, Suzuki A, Ohtsuka T, Hakuno F, Takahashi S, Ohno S, Chida K (2011) KIBRA suppresses apical exocytosis through inhibition of aPKC kinase activity in epithelial cells. Curr Biol 21(8):705–711. doi:10.1016/j.cub.2011.03.029

Yu J, Zheng Y, Dong J, Klusza S, Deng WM, Pan D (2010) Kibra functions as a tumor suppressor protein that regulates Hippo signaling in conjunction with Merlin and expanded. Dev Cell 18(2):288–299. doi:10.1016/j.devcel.2009.12.012

Zhang Y, Wang W, Chen J, Zhang K, Gao F, Gao B, Zhang S, Dong M, Besenbacher F, Gong W, Zhang M, Sun F, Feng W (2013) Structural insights into the intrinsic self-assembly of Par-3 N-terminal domain. Structure 21(6):997–1006. doi:10.1016/j.str.2013.04.004

Chapter 2
PAR-1 Kinase and Cell Polarity

Atsushi Suzuki

Abstract PAR-1 kinase was identified as one of the protein products of the *par* (partition defective) genes essential for the establishment of the anterior-posterior polarity of the *Caenorhabditis elegans* one-cell embryo. Subsequent studies have revealed the ubiquitous importance of the kinase in regulating cell polarity observed in various biological contexts of many organisms. As a membrane-localized kinase, PAR-1 antagonizes the atypical protein kinase C (aPKC)/PAR-3/PAR-6 complex to establish the mutually exclusive membrane domains along the polarity axis. Based on this asymmetric membrane localization, its kinase activity is utilized to phosphorylate various other target proteins that regulate cellular functions. The major function of PAR-1 is the regulation of microtubule dynamics, of which one of the underlying mechanisms was identified as the microtubule affinity-regulating kinase (MARK) activity in mammalian neurons. In addition, accumulating data have revealed the presence of various kinds of other target proteins involved in the regulation of actin cytoskeleton and protein stability. Although the significant versatility of PAR-1 has made it increasingly difficult to obtain a simple view of its function, the evolutional insight on PAR-1 provides a powerful tool for integrating increasing data on this kinase in the light of cell polarity.

Keywords Anterior-posterior polarity • Apicobasal polarity • MARK • Microtubules • Neuronal polarity • PAR-1

2.1 Introduction

The serine/threonine protein kinase, PAR-1, has two faces. One is being a component of the cell polarity machinery, the PAR-atypical protein kinase C (aPKC) system, which plays essential roles in establishing cell polarity in various biological backgrounds (Suzuki and Ohno 2006). This face was first uncovered by genetic studies on the embryonic development of *Caenorhabditis elegans* and *Drosophila* and further elucidated by cell biology studies on cultured mammalian epithelial

A. Suzuki (✉)
Molecular Cellular Biology Laboratory, Graduate School of Medical Life Science,
Yokohama City University, 1-7-29 Suehiro-cho, Tsurumi, Yokohama 230-0045, Japan
e-mail: abell@tsurumi.yokohama-cu.ac.jp

K. Ebnet (ed.), *Cell Polarity 1*, DOI 10.1007/978-3-319-14463-4_2

cells or neurons. The other face of PAR-1 is the microtubule (MT) affinity-regulating kinase (MARK), which phosphorylates classical microtubule-associated proteins (MAPs), such as tau, MAP2, and MAP4, and induces MT destabilization by inhibiting their MT-binding abilities (Drewes et al. 1997). This face was identified in studies searching for the kinases responsible for the hyperphosphorylation of tau in Alzheimer patients. Because MTs are essential for regulating and maintaining cell polarity, these two faces of PAR-1 were, at first, thought to be easily integrated. However, subsequent studies have not succeeded in superimposing these faces completely, but rather uncovered more complicated faces of this kinase. To date, several reviews on PAR-1 have been published, mainly focusing on the MARK face (Marx et al. 2010; Matenia and Mandelkow 2009; Timm et al. 2008b; Hayashi et al. 2012). Here, I summarize the accumulating data on this kinase, mainly from the viewpoint of cell polarity.

2.2 Essential Features of PAR-1

The structural and biochemical nature of PAR-1 has been analyzed mainly by Mandelkow and coworkers, who identified mammalian PAR-1 as MARK (Drewes et al. 1997). Briefly, PAR-1s are relatively large kinases (nearly 90 kDa) (Fig. 2.1a), which reside at the end of the AMP-activated protein kinase (AMPK) branch of the calcium/calmodulin-regulated protein kinase (CAMK) group (Fig. 2.1b) (Marx et al. 2010; Manning et al. 2002). Structurally, the N-terminal kinase domain is followed by a ubiquitin-associated (UBA) domain, a spacer domain, and a C-terminal kinase-associated domain (KA1). The UBA domain is a small ($\approx$40 residues), mostly α-helical domain, which has been suggested to exert an autoregulatory function by analogy to many other kinases (Panneerselvam et al. 2006). The KA1 domain is a protein domain that was first identified in the C-terminus ($\approx$50 residues) of the PAR-1 kinases on the basis of strong sequence homology (Fig. 2.2b). Recent studies based on structural comparisons with other kinases have suggested that this domain is a C-terminal folded domain comprising of more extended sequence with $\approx$100 residues (Fig. 2.1a) (Tochio et al. 2006; Marx et al. 2010). *C. elegans* and *Drosophila* have a single PAR-1 member, whereas mammals contain four PAR-1 paralogs (MARK1/PAR-1c, MARK2/PAR-1b, MARK3/PAR-1a, and MARK4/PAR-1d) (Fig. 2.2a, b) (Drewes et al. 1998; Shulman et al. 2000; Guo and Kemphues 1995). Furthermore, the amino acid similarity in the kinase and the KA1 domains identified the distantly related kinase, KIN1, as an ortholog of PAR-1 in fission yeast (Figs. 2.1 and 2.2), which is a plasma membrane-associated kinase regulating polarized cell growth (Guo and Kemphues 1995; Tassan and Le Goff 2004; Levin and Bishop 1990; Drewes and Nurse 2003). Interestingly, structural prediction suggests that, albeit low sequence similarities, a similar domain organization can be identified in many kinases belonging to the AMPK subfamily (Timm et al. 2008b; Marx et al. 2010). Furthermore, recent analysis has revealed that the C-terminal anionic lipid-binding

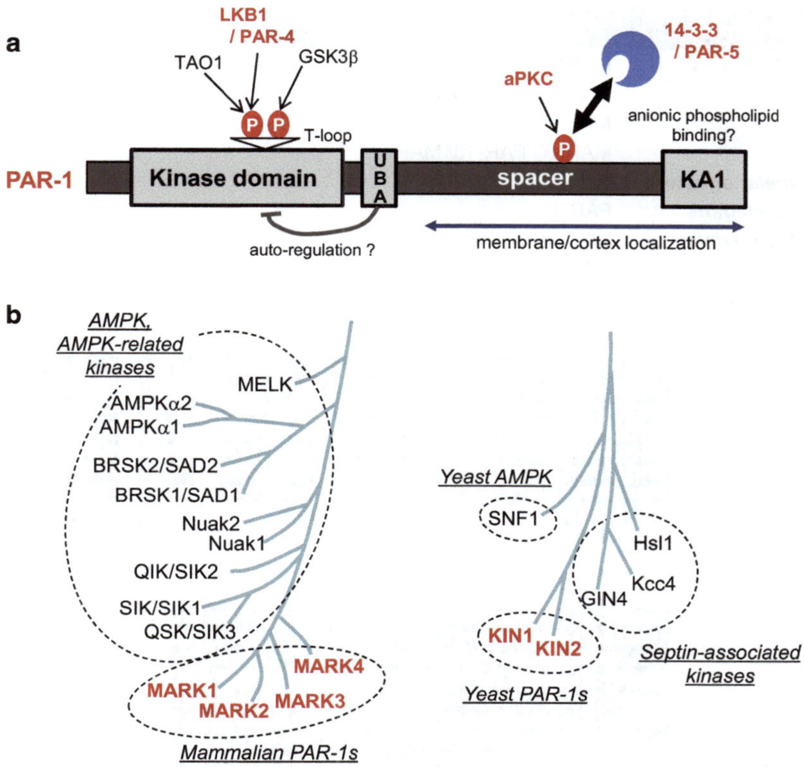

Fig. 2.1 Domain structure and phylogenetic tree of PAR-1. (**a**) All PAR-1 homologs have a similar domain structure consisting of the N-terminal kinase domain, followed by the ubiquitin-associated (UBA) domain, the spacer domain, and the kinase-associated (KA1) domain. The kinase domain and the KA1 domain show strong conservation between the homologs (see Fig. 2.2b). The C-terminal half of the protein is responsible for the membrane/cortex localization (*bidirectional arrow*), and the KA1 domain has recently been shown to have an anionic phospholipid-binding ability. aPKC phosphorylates the conserved threonine residue in the spacer region and excludes PAR-1 from its own territory with the aid of 14-3-3/PAR-5. LKB1/PAR-4 and TAO1 activate PAR-1 by phosphorylating the serine residue in the T-loop of the kinase domain, whereas GSK3β phosphorylates the downstream threonine residue in the T-loop and inhibits PAR-1. (**b**) Phylogenetic branches of the AMPK subfamily in the calcium-/calmodulin-regulated protein kinase (CAMK) group (*left*, human; *right*, *S. cerevisiae*). Trees are roughly depicted based on the data published in kinase.com (http://kinase.com/)

domains of the septin-associated kinases of *Saccharomyces cerevisiae*, Kcc4p, Gin4p, and Hsl1p (Fig. 2.1b), also adopt a KA1 fold, whereas the KA1 domains of MARK1 and MARK3 directly bind to anionic phospholipids in vitro and mediate their membrane localization in the cell (Moravcevic et al. 2010). The septin-associated kinases have a UBA domain following the N-terminal kinase domain and are involved in coordinating cell cycle progression with bud formation. Therefore, these results suggest that PAR-1 members have diverged and evolved from an

Organism	PAR-1 parallogues
Mammals	MARK1 /PAR-1c MARK2/ PAR-1b /EMK1 MARK3/ PAR-1a/ C-TAK/ Kp78 MARK4/ PAR-1d/ MARKL1
D. melanogaster	PAR-1
C. elegans	PAR-1
S. cerevisiae	KIN1 KIN2
S. pombe	KIN1

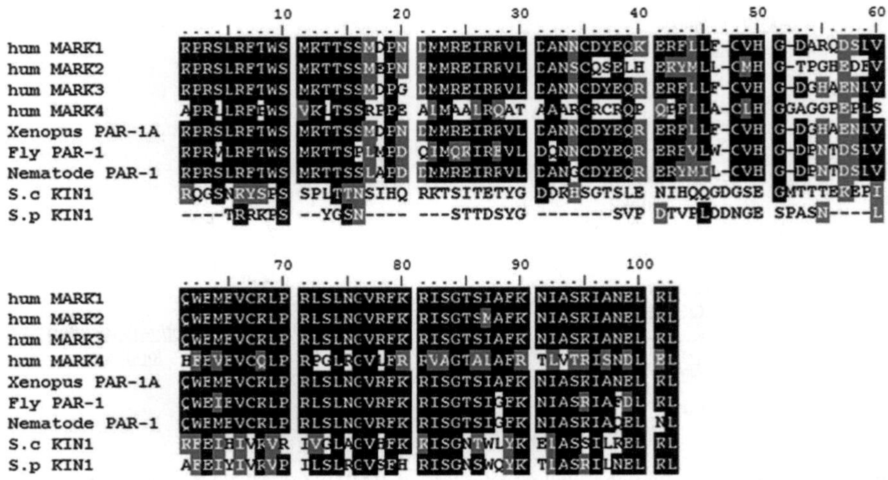

Fig. 2.2 PAR-1 homologs in different organisms. (**a**) PAR-1 homologs in various organisms. We omitted mammalian MELK/pEg3/MPK38, *C. elegans* MELK/PIG1/W03G1.6, and *Drosophila* Kp78, because of the extremely low conservation of the amino acid sequence of the KA1 domain. (**b**) Alignment of the amino acid sequence of the KA1 domain of PAR-1 homologs. Note that MARK4 shows significant divergence compared with the other mammalian members

ancestral membrane-bound kinase in unicellular organisms involved in regulating polarized cell growth, independently of the other PAR-aPKC components (except for PAR-5/14-3-3), which do not have clear counterparts in yeast (Suzuki and Ohno 2006). Importantly, in mammals, not only PAR-1 homologs (MARKs) but also most of the AMPK subfamily members have been shown to be activated by a common kinase, LKB1, which corresponds to another component of the PAR-aPKC system, PAR-4 (Lizcano et al. 2004). Furthermore, some members of the AMPK subfamily have been shown to function in cell polarity regulation after activation by LKB1/PAR-4 (Barnes et al. 2007; Chien et al. 2013; Lee et al. 2007). Therefore, these results provide important clues for analyzing the convergence process of this protein kinase subfamily and other PAR-aPKC components, although I confine the present descriptions to PAR-1 homologs hereafter and do not discuss this issue further. It has been shown that mammalian PAR-1s (MARKs) are activated not only by LKB1/PAR-4 but also by TAO1 and suppressed by glycogen synthase kinase-3 β (GSK3β) through phosphorylation of the T-loop within the kinase domain (Fig. 2.1a); however, the upstream events activating these signaling cascades have not been clarified (Timm et al. 2003, 2008a). The protein structure of the kinase domain with the UBA domain has been determined for MARK1, MARK2, and MARK3 (Panneerselvam et al. 2006; Marx et al. 2006; Murphy et al. 2007), and a candidate for a PAR-1-specific inhibitor has also been reported (Timm et al. 2011). It is also noteworthy that CagA, the toxin of *Helicobacter pylori*, has been shown to inhibit PAR-1 kinase activity by specifically binding to the substrate-binding pocket of the kinase domain (Nesic et al. 2010).

2.3 Role of PAR-1 in the PAR-aPKC System

2.3.1 Complimentary Localizations of the PAR-aPKC System Components

Historically, PAR-1 kinase was first identified in *C. elegans* as one of the protein products of the *par* (partition defective) genes that are essential for the development of the anterior-posterior (A-P) asymmetry of the one-cell embryo, which is triggered by sperm entry (Fig. 2.3a) (Kemphues et al. 1988). Genetic analysis of this model organism clarified the functional interactions between the PAR proteins including aPKC (the seventh PAR protein that was identified independently; see Chap. 1) (Tabuse et al. 1998) and established the concept that these proteins constitute the PAR-aPKC system, which translates the polarity cue (sperm entry, in the case of *C. elegans* embryo) to the establishment of the reciprocal membrane/cortex domains along the polarity axis by differentially localizing its components to these domains. For example, in the *C. elegans* embryo, PAR-3, PAR-6, and PKC-3 (*C. elegans* aPKC) asymmetrically localize to the anterior cortex of the embryo, whereas PAR-1 localizes to the posterior cortex (Fig. 2.3a) (Guo and Kemphues 1996a; Pellettieri and Seydoux 2002; Tabuse et al. 1998). Because the anteriorly localized PAR proteins, aPKC, PAR-6, and PAR-3, dynamically form a complex that functions as a single functional unit (see Chap. 1) (Suzuki and Ohno 2006), I hereafter describe this unit as the aPKC complex. Subsequent studies have demonstrated the generality of this complimentary localization of PAR-1 with the aPKC complex in other polarized cells. For example, the *Drosophila* PAR-1 has been shown to localize to the posterior cortex of the late phase of oocytes (stages 7–10) (Shulman et al. 2000), and the anterior localization of Bazooka (*Drosophila* PAR-3) was confirmed later (Benton and St Johnston 2003; Vaccari and Ephrussi 2002) (Fig. 2.3b). The basolateral localization of PAR-1 complementary to the apical localization of the aPKC complex has been shown in epithelial cells, first in cultured mammalian cells [Madin-Darby canine kidney (MDCK) cells] (Izumi et al. 1998; Bohm et al. 1997) (Fig. 2.3c) and then in follicle cells (Shulman

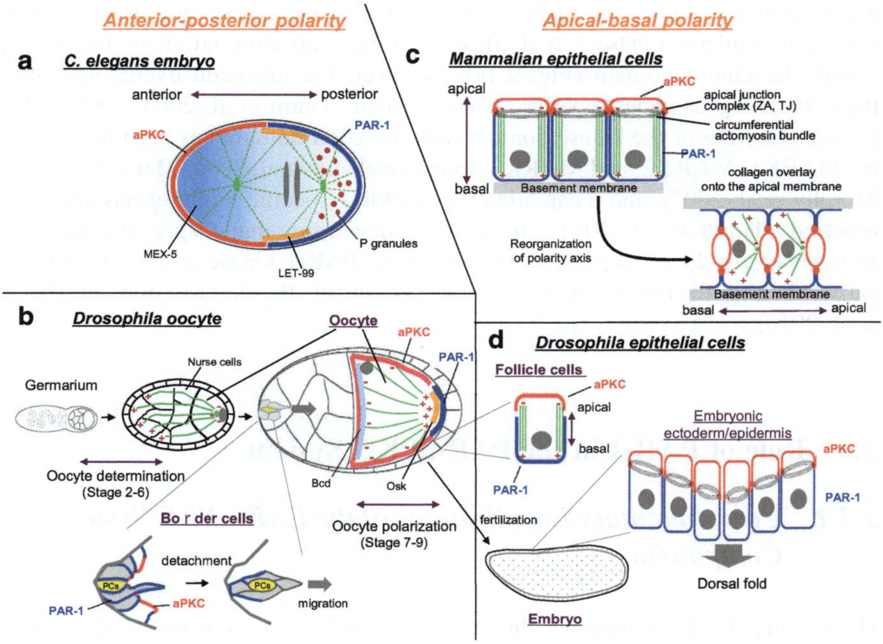

Fig. 2.3 Various types of cell polarity in which PAR-1 is involved. The subcellular localization of PAR-1 is indicated by *blue thick lines*, whereas that of the aPKC complex is indicated by *red thick lines*. *Gray ovals* represent nuclei. *Green lines* and *ovals* represent microtubules (MTs) and microtubule-organizing center (MTOC), respectively. (**a**) Anterior-posterior polarization of the *C. elegans* embryo just after fertilization. (**b**) Anterior-posterior polarity observed during *Drosophila* oogenesis. (**c**) Apical-basal polarity observed in cultured mammalian epithelial cells (*top*) and its reorganization after collagen overlay on the apical membrane (*bottom*). (**d**) Apical-basal polarity of *Drosophila* follicle cells in the egg chamber and the blastoderm epithelium in the area undergoing dorsal folding

et al. 2000; Cox et al. 2001) and the blastoderm (Bayraktar et al. 2006) of *Drosophila* (Fig. 2.3d).

2.3.2 Antagonistic Interaction Between the aPKC Complex and PAR-1

The complement localization of the aPKC complex and PAR-1 is maintained via mutually antagonistic interactions, involving kinase activity of aPKC and PAR-1. First, aPKC phosphorylates the conserved threonine of PAR-1 in the spacer region between the kinase domain and the C-terminal KA1 domain (Fig. 2.1a). This phosphorylation induces the dissociation of PAR-1 from the cell cortex/membrane (Suzuki et al. 2004; Hurov et al. 2004; Kusakabe and Nishida 2004) and thereby inhibits the entry of PAR-1 into the aPKC territory. Initially, this was shown in mammalian epithelial cells (Suzuki et al. 2004) and confirmed in *Drosophila* stage 7 oocyte (Doerflinger et al. 2006) and recently in *C. elegans* embryo (Motegi et al. 2011). On the contrary, PAR-1 phosphorylates the conserved serine residues of PAR-3/Bazooka, disrupts the aPKC/PAR-3/PAR-6 complex, and thereby inhibits the entry of the aPKC/PAR-3/PAR-6 complex into the PAR-1 territory. This was shown first in *Drosophila* follicle epithelial cells and stage 7 oocyte (Benton and St Johnston 2003), and recently suggested in *C. elegans* embryo (Motegi et al. 2011), but not been fully validated in mammalian cells. Importantly, the role of PAR-5, one of the other PAR components, is to support these antagonistic interactions between the aPKC complex and PAR-1(Fig. 2.1a) (Benton and St Johnston 2003; Suzuki et al. 2004). In general, PAR-5, which corresponds to the mammalian 14-3-3 proteins, recognizes serine/threonine-phosphorylated residues within the consensus RXXpS/TXP or RXXXpS/TXP (X represents any amino acid) sequence and affects functions and/or localization of target proteins (Tzivion et al. 2001). Within the PAR-aPKC system, 14-3-3/PAR-5 specifically interacts with PAR-3, which is phosphorylated by PAR-1, or with PAR-1, which is phosphorylated by aPKC, and thereby disrupts the PAR-3 interaction with aPKC or the membrane localization of PAR-1, respectively. In addition, PAR-1 regulates many target proteins other than PAR-3 by cooperating with 14-3-3/PAR-5 in a similar way (Table 2.1). Actually, several studies have succeeded in finding the PAR-1 substrates by utilizing their ability to bind to 14-3-3/PAR-5 after phosphorylation (Riechmann and Ephrussi 2004; Sung et al. 2008; Cohen et al. 2011; Muller et al. 2003). Considering that PAR-1 and 14-3-3/PAR-5 are unique members of the PAR-aPKC system, as they have a counterpart in yeast, the general tight coupling of these two proteins may give us some insight into the evolutional origin of the PAR-aPKC system in multicellular organisms.

2.3.3 Molecular Mechanisms Supporting the Membrane/Cortex Localization of PAR-1

Recently, the ring finger protein, PAR-2, was shown to recruit PAR-1 to the posterior cell cortex of the *C. elegans* embryo (Motegi et al. 2011). However, in other species lacking counterparts of PAR-2, the molecular mechanisms underlying the membrane/cortex recruitment of PAR-1 are still unclear. Several studies have demonstrated that not only the KA1 domain but also the preceding spacer domain is required for the membrane/cortex localization of PAR-1 (Bohm et al. 1997; Vaccari and Ephrussi 2002; Doerflinger et al. 2006). Considering that the phosphorylation by aPKC occurs in the spacer region and not in the KA1 domain (Fig. 2.1a), it is reasonable to assume that while the KA1 domain mediates the nonspecific phospholipid binding itself, another mechanism through the spacer region, probably involving protein-protein interactions, is required for the correct asymmetric localization

Table 2.1 List of reported substrates of PAR-1

	Substrate	14-3-3 Binding	Function	References
C. elegans	LET-99		G-protein regulator, spindle positioning	Wu and Rose (2007)
	MEX-5		Somatic fate determinant	Griffin et al. (2011)
	PAR-3	[a]	Component of the PAR-aPKC system	Motegi et al. (2011)
Drosophila	Dishevelled		Wnt signaling mediator	Sun et al. (2001)
	Oskar		Pole plasm organizer	Riechmann et al. (2002), Morais-de-Sa et al. (2013)
	PAR-3	+	Component of the PAR-aPKC system	Benton and St Johnston (2003)
	Exuperantia	[a]	Mediator of bicoid mRNA localization	Riechmann and Ephrussi (2004)
	Dlg		Scaffold protein for membrane protein	Zhang et al. (2007,b)
	Ensconsin	[a]	MT-binding protein	Sung et al. (2008)
	Myosin II phosphatase		Non-muscle myosin inactivation	Majumder et al. (2012)
Xenopus	Dishevelled		Wnt signaling mediator	Ossipova et al (2005)
	Mind bomb		E3 ligase involved in Notch signaling	Ossipova et al. (2009)
Mammals				
MARK2	tau/MAP2, 4		MT stabilization	Drewes et al. (1997)
	Doublecortin		MT-binding protein involved in neuron migration	Schaar et al. (2004)
	HDAC	+	Class IIa histone deacetylase	Dequiedt et al. (2006)
	Rab11-FIP2		Rab11 interacting protein, recycling regulation	Ducharme et al. (2006)
	Dishevelled		Wnt signaling mediator	Elbert et al. (2006)
	CRTC2/ TORC2	+	Transducer of regulated CREB activity	Fu and Screaton (2008)
	Utrophin		Component of laminin receptor complex	Yamashita et al. (2010)
	GAKIN	+	Kinesin motor	Yoshimura et al. (2010)
	GEF-H1	+	MT-associated Rho-specific GEF	Yoshimura and Miki (2011), Yamahashi et al. (2011)
	IRSp53	+	Rho-GTPase signaling platform	Cohen et al. (2011)
	PSD95		Scaffold protein for membrane protein	Wu et al. (2012)

(continued)

Table 2.1 (continued)

	Substrate	14-3-3 Binding	Function	References
MARK3	PTPH1	+	Tyrosine phosphatase specific for p38g MAPK	Zhang et al. (1997)
	Cdc25	+	Protein phosphatase that control entry into mitosis	Peng et al. (1998)
	KSR	+	MAPK scaffolding protein	Muller et al. (2003)
	PKP2	+	Desmosomal adaptor protein	Muller et al. (2003)
	HDAC	+		Dequiedt et al. (2006)

[a]14-3-3 binding is predicted, but not verified

of PAR-1 (Doerflinger et al. 2006). Several studies have suggested the involvement of actin filaments or MTs (Doerflinger et al. 2006; Sato et al. 2013).

2.3.4 *Functional Hierarchy of the aPKC Complex and PAR-1*

Although the aPKC complex and PAR-1 mutually regulate each other, PAR-1 generally works downstream of the aPKC complex within the PAR-aPKC system. This hierarchy was originally demonstrated in the initial genetic studies on the *C. elegans* embryos, in which mutations in one of the anterior PARs (PAR-3, PAR-6, and aPKC) disrupted asymmetric localization of PAR-1, whereas PAR-1 mutation did not severely affect the anterior localization of these proteins (Guo and Kemphues 1996a; Etemad-Moghadam et al. 1995). It is now clear that the anterior accumulation of the aPKC complex is primarily driven by the actomyosin-dependent cortical flow, which is triggered by the initial polarity cue (sperm entry) (Munro et al. 2004), and the exclusion by PAR-1 plays a secondary supportive role to finally stabilize the localization of the aPKC complex. The role of PAR-1 in antagonizing the aPKC complex localization was not detected until recently, when the role of PAR-1 was examined under suppressing actomyosin activity (Motegi et al. 2011). Similarly, the apical accumulation of Baz/PAR-3 in the *Drosophila* ectoderm is redundantly ensured during the cellularization process (initial stage of epithelialization) by the dynein-dependent basal-to-apical transport system (McKinley and Harris 2012). In various epithelial cells, the Dlg/Scribble/Lgl basolateral complex (see Chap. 4) and the Crb3/Pals1/PATJ apical complex (see Chap. 3) have also been shown to support the apical localization of the aPKC complex (Knust and Bossinger 2002). The similar role of Lgl in redundantly antagonizing the aPKC complex has also been demonstrated in *C. elegans* embryos and *Drosophila* oocytes (Tian and Deng 2008; Beatty et al. 2010).

2.4 Roles of PAR-1 in the *C. elegans* Embryo

The above results suggest that one of the essential roles of PAR-1 in cell polarity is to support the asymmetric localization of the aPKC complex, which is vital for establishing various cell polarities (Suzuki and Ohno 2006). However, lack of PAR-1 frequently causes severe defects in cell polarity without significant defects in aPKC localization, suggesting active roles of PAR-1 in regulating cell polarity. For example, mutations in the *par-1* gene strongly affect the asymmetric localization of cell fate determinants in *C. elegans* embryos but only weakly affect spindle positioning and orientation compared with the effects of *pkc-3*, *par-3*, or *par-6* mutations (Kemphues et al. 1988). These results suggest that the aPKC complex regulates asymmetric spindle positioning relatively independently of PAR-1 but regulates asymmetric distribution of cell fate determinants mainly via PAR-1. One of the direct targets of PAR-1 in this process is the RNA-binding protein, MEX-5 (somatic determinant), which upon fertilization concentrates in the anterior cytoplasm, where it induces degradation of germline proteins (Fig. 2.3a, Table 2.1). PAR-1 plays an essential role in establishing the anterior-posterior gradient of MEX-5 by directly phosphorylating MEX-5 and releasing it from a slow-diffusive, RNA-containing complex in the posterior region (Griffin et al. 2011). On the other hand, the DEP domain G-protein regulator, LET-99, is speculated to be another direct substrate of PAR-1 (Table 2.1) (Wu and Rose 2007). The regulation of spindle orientation and position in the *C. elegans* embryo depends on the polarized cortical dynein activation, which is regulated by the receptor-independent heteromeric G-protein pathway. LET-99 plays a central role in this process by locally inhibiting the G-protein pathway at the lateral cortical belt of the embryo (Fig. 2.3a). PAR-1 has been suggested to phosphorylate LET-99 and exclude it from the posterior cortex.

2.5 Roles of PAR-1 in the *Drosophila* Embryo

2.5.1 Phenotypes of the par-1 Mutants

PAR-1 plays essential roles in the *Drosophila* oogenesis in two distinct stages, and MTs have been suggested to be a critical target in both stages. First, the posterior localization of PAR-1 has been shown to be crucial for the establishment of the A-P polarity at the late stages of the oocyte (stages 7–9; Fig. 2.3b) (Shulman et al. 2000). At the beginning of this stage, a putative signal from the posterior follicle cells has been suggested to gradually induce the anterior localization of the aPKC complex and the posterior localization of PAR-1 (Riechmann and Ephrussi 2001; Vaccari and Ephrussi 2002). Then, the MT-organizing center (MTOC), located at the posterior pole of the oocyte (see stage 6 oocyte in Fig. 2.3b), disassembles, and instead, MTs are nucleated or anchored at the anterolateral cortex of the oocyte, with the plus-ends focused on the posterior pole (Fig. 2.3b). This polarized MT array directs the localization of *bicoid* (*bcd*) mRNA to the anterior of the oocyte, and *oskar* (*osk*) mRNA and Staufen protein to the posterior, and thus defines the subsequent development of the A-P body axis (Fig. 2.3b). In *par-1* hypomorphic mutant oocytes, in which fate determination of the oocyte proceeds normally (see below), MTs are nucleated all around the oocyte cortex, resulting in the ectopic localization of *bcd* mRNA to the lateral cortex and the accumulation of *osk* mRNA and Staufen protein with the MT plusends in the oocyte center (Shulman et al. 2000). The posterior localization of PAR-1 is indispensable for rescuing these defects (Doerflinger et al. 2006), and PAR-3/Baz is required for this PAR-1 localization as an upstream factor (Doerflinger et al. 2006, 2010). A recent live imaging analysis has demonstrated that all MTs in the oocyte are highly dynamic and nucleated randomly in the nucleus and cortex, but PAR-1 excludes the MT nucleation from the posterior end of the oocyte and thereby produces biased MT polarity toward the posterior (Parton et al. 2011).

In addition to the roles at the late phase of oogenesis, PAR-1 has also been shown to be involved in the oocyte determination process, which occurs in the very early stages of oogenesis in the germarium, a specific part of the ovary (Fig. 2.3a). In this process, a single oocyte is selected from 16 descendants of a germline stem cell, while the remaining cells become highly biosynthetically active nurse cells (Riechmann and Ephrussi 2001). MTs are organized in the selected oocyte and extend their plus-ends through cytoplasmic bridges (ring canals) into the nurse cells, and oocyte-specific factors are transported from the nurse cells to the oocyte by minusend-directed motors. In a PAR-1-deficient germline clone, MTOC failed to stably form at the posterior cortex of the prospective oocyte, and thus the accumulation of oocyte-specific factors was not maintained, albeit transiently induced (Huynh et al. 2001; Cox et al. 2001).

2.5.2 *Downstream Targets of PAR-1 in* Drosophila *Oogenesis*

Defects in MT organization in *par-1* mutant cells implied the involvement of the MARK activity of PAR-1 in both oogenesis processes. However, it is unlikely, because null mutants of *tau*, a single classic MAP in *Drosophila*, are fully fertile and have no effect on the arrangement of MTs in the oocyte (Doerflinger et al. 2003). Although a recent paper argued the involvement of tau in PAR-1 function in the oocyte (Tian and Deng 2009), the data were not strong enough to override the above results. Instead, the data on the *Drosophila* oocyte suggest the presence of unknown downstream targets that control MT nucleation, anchoring, and stability. It may be noteworthy that the MT-associated protein, Ensconsin, has been reported to localize to the anterior-lateral cortex of the oocyte in a PAR-1-dependent manner and play an essential role in the posterior determinants' localization (Table 2.1) (Sung et al. 2008). The authors suggested that, Ensconsin is regulated, at least partially, through direct phosphorylation by PAR-1.

In addition to regulating MTs, PAR-1 has been demonstrated to regulate the amount of fate determinants through direct phosphorylation. During oogenesis, *osk* mRNA is transported into the oocyte from nurse cells in a translationally repressed state and locally expressed when it correctly accumulates at the posterior pole of the matured oocyte. PAR-1 not only affects the posterior accumulation of *osk* mRNA, as described above, but also regulates the expression level of the Oskar protein by controlling its degradation (Riechmann et al. 2002; Morais-de-Sa et al. 2013). A recent study has clarified the phosphorylation site of PAR-1 on Oskar and demonstrated that this phosphorylation induces the recruitment of the SCF$^{\text{-slimb}}$ ubiquitin ligase, which limits the level of Oskar protein at the posterior and degrades mislocalized Oskar (Table 2.1) (Morais-de-Sa et al. 2013). On the other hand, Exuperantia, an essential mediator of *bcd* mRNA localization, has also been reported to be a direct target of PAR-1 in the oocyte (Table 2.1) (Riechmann and Ephrussi 2004). Although the underlying molecular mechanism is unclear, this phosphorylation has been suggested to occur in nurse cells and regulate the Exuperantia function.

2.6 Roles of PAR-1 in Epithelial Cell Polarity

2.6.1 Effects of PAR-1 Depletion in Epithelial Polarity

The important roles of PAR-1 in epithelial cell polarity were first revealed in mammalian MDCK cells by demonstrating that overexpression of the C-terminal region of MARK2, which localizes to the basolateral membrane, disrupts the epithelial monolayer (Bohm et al. 1997). Subsequently, PAR-1 has been genetically shown to be indispensable for epithelial polarity in *Drosophila* (Fig. 2.2d) (Doerflinger et al. 2003). When *par-1* mutations were somatically induced in follicle cells using the FLP/FRT technique, these epithelial cells showed uniform localization of the aPKC complex around the cell cortex or disruption of the epithelial monolayer. However, these severe phenotypes may reflect PAR-1 functions in the developmental process of follicle cells from their precursor cells, because small clones in which *par-1* mutations were induced at a later stage of development showed largely normal epithelial polarity but displayed several alterations in cytoskeletal organization. This is consistent with the results in *Drosophila* embryonic ectoderm, where PAR-1 depletion only induced slight apical-to-basal shifts of adherens junctions and Baz/PAR-3 localization and a relatively modest perturbation of epithelial architecture (Fig. 2.3d) (Bayraktar et al. 2006; McKinley and Harris 2012). Interestingly, a recent study has further demonstrate that the balance between PAR-1 and the aPKC complex is utilized to determine the position of the zonula adherens (ZA) along the lateral membrane, which critically affects the apical constriction required for dorsal folding of the gastrulating *Drosophila* embryo (Fig. 2.3d) (Wang et al. 2012). Essential establishment of epithelial polarity without PAR-1 has also been observed in MDCK cells, although the accumulation of the apical and basolateral membrane proteins, the ZA development associated with the circumferential F-actin bundles, and the asymmetric organization of MTs were all reduced, and the characteristic columnar shape was impaired (Cohen et al. 2004; Suzuki et al. 2004). These results indicate that PAR-1 is not required for the establishment of epithelial polarity itself, but is essential for the qualities of epithelial cells, which affect their behavior. This idea is consistent with the results showing that MARK2-depleted MDCK cells are insensitive to collagen overlaid on the apical membrane, which normally induces reorganization of the polarity axis (Fig. 2.3c) (Cohen et al. 2004; Masuda-Hirata et al. 2009).

2.6.2 Downstream Targets of PAR-1 in Regulating Epithelial Polarity

Regulation of MT organization and dynamics has also been implied in epithelial cells as one of the important targets of PAR-1. In *Drosophila* follicle cells, most of the drastic effects induced by *par-1* mutations have been observed in MTs (Doerflinger et al. 2003; Cox et al. 2001), although actin filament-related structures were also affected. For example, MTs disappeared in *par-1* mutant cells, if the cells were fixed by weak conditions (4 % formaldehyde). On the other hand, They were more densely observed than in wild-type cells, if the cells were fixed by stronger conditions (8 % formaldehyde) (Doerflinger et al. 2003). Together with the findings that MTs in *par-1* mutant cells become highly sensitive to treatment with the MT-depolymerizing drug, colcemid, or cold shock (Doerflinger et al. 2003), these results suggest that MTs in mutant cells are extremely unstable/dynamic. In addition, MT organization was also perturbed in *par-1* mutants. In epithelial cells, most MTs exhibit a uniform apicobasal polarity, with their minus-ends localized at the apical membrane and their plus-ends extending toward the basal membrane (Fig. 2.3c, d). In *par-1* mutant follicle cells, the minus-ends still accumulated at the apical membrane, but the plus-ends were not detected at the basal sides as they accumulated in the center in the stage 9 mutant oocyte (Doerflinger et al. 2003). A similar alteration of the epithelial-specific MT organization has also been observed in mammalian MDCK cells. When MARK2 was knocked down, the lateral MTs disappeared, leaving the MT network at the subapical cortex, which showed high sensitivity to the MT-disrupting drug, nocodazole (Cohen et al. 2004).

The fact that MTs are highly affected by PAR-1 depletion in various epithelial cells implied that PAR-1 primarily regulates MT dynamics and organization to promote the epithelial polarity development as suggested in *Drosophila* oogenesis. However, the underlying molecular mechanisms are still unknown in this case as well, as the MARK activity of PAR-1, which is expected to destabilize MTs, does not simply explain why PAR-1 depletion decreased MT stability in epithelial cells. Furthermore, tau has been shown to be dispensable for MT organization in *Drosophila* follicle cells, suggesting that MARK activity is not involved in the PAR-1 functions in epithelial polarity, at least in *Drosophila* (Doerflinger et al. 2003). It should be noted that most of the MT organization in epithelial cells is composed of specifically stabilized/bundled noncentrosomal MTs, the developmental process of which is still under investigation (Bartolini and Gundersen 2006). A recent study has demonstrated that PAR-1 interacts with a novel MT-regulating protein, named microtubule cross-linking factor 1 (MTCL1), which cross-links and stabilizes MTs, and cooperates with MARK2 for the development of lateral MTs in MDCK cells (Sato et al. 2013). Therefore, the MARK activity of PAR-1 may be involved in MT regulation in cooperation with MT-stabilizing/-bundling factors in a spatially and temporally dependent manner. The role of PAR-1 in the epithelial-specific MT dynamics and organization probably cannot be investigated only by simple depletion. We have to wait for progress in this field to completely understand the role of PAR-1 in epithelial cells.

2.7 Roles of PAR-1 in Neuron

2.7.1 Roles of PAR-1 in Neuronal Cell Polarity

Neurons are another type of cell, which develops an extremely polarized morphology, typically with one axon and multiple dendrites. Consistent with the fact that tau specifically localizes to the axon and stabilizes MTs, not only MTs but also tau is known to play critical roles in developing process of this neuronal polarity (Brandt 1996). The involvement of the MARK activity of PAR-1 has been also demonstrated in this process. Primary cultured hippocampal neurons of rodents have been used to investigate the establishing process of this neuronal polarity, in which one axon differentiated from multiple minor processes with dynamic growth cones (Fig. 2.4a). At first, the aPKC complex was shown to asymmetrically accumulate at the tip of axons and play essential roles in axon specification (Shi et al. 2003; Nishimura et al. 2004). Subsequently, the complex has been shown to exert this function by locally suppressing the inhibitory activity of PAR-1 on axon elongation (Chen et al. 2006). This conclusion was based on the findings that MARK2 knockdown induced multiple axon formations, whereas MARK2 overexpression inhibited the formation of an elongated axon. In addition, the aPKC complex suppressed the above effect of wild-type MARK2 overexpression, but not of T595A mutant overexpression, which is resistant to aPKC phosphorylation. Importantly, this inhibitory role of PAR-1 in axon elongation was consistently explained by its MARK activity, because tau phosphorylation on Ser262 consistently changed according to MARK2 levels.

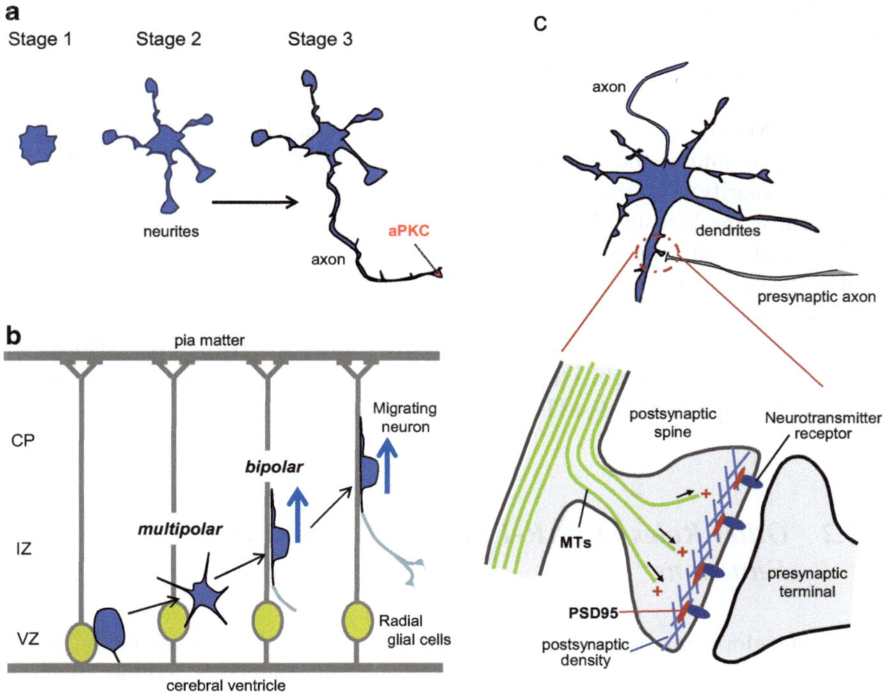

Fig. 2.4 Various neuronal activities in which PAR-1 is involved. (**a**) The axon specification process observed in primary cultures of hippocampal neurons proceeds in three stages. At the transition between stages 2 and 3, one of the several initially indistinguishable processes (neuritis) is selected to become the axon, at the tip of which the aPKC complex is asymmetrically localized. (**b**) Schematic illustration of radial migration of neurons observed in mammalian corticogenesis. Note that a neuron produced by asymmetric division of a ventral progenitor cell (radial glial cell) transiently takes a multipolar shape at the intermediate zone (IZ) and transforms to a bipolar shape for further migration to the pia surface. (**c**) Schematic illustration of the dendritic spine (*bottom*), of which the tip makes contact with the presynaptic terminus of another neuron (*top*), to develop the postsynaptic density structure. *Green lines* indicate MTs, of which the plusends (+) stochastically enter the spine to promote the maturation of the postsynaptic structure

It is noteworthy that despite the above consistent results, there are some contradictory results reporting that PAR-1 works positively in neurite elongation. For example, overexpression of MARK2 has been reported to increase the tau-positive axon length in rat hippocampal neurons (Uboha et al. 2007; Terabayashi et al. 2007) and promote neurite elongation from neuroblastoma n2a cells (Biernat et al. 2002). To decipher these paradoxical results, it is important to note that overexpression of tau also promoted neurite elongation, while its mutant lacking the PAR-1 phosphorylation site (Ser262) did not (Biernat et al. 2002). These results indicate that PAR-1-dependent phosphorylation is essential for tau to work in neurite extension. Actually, it has been shown that MTs must be dynamic to allow axon elongation (Tanaka et al. 1995): at the initial phase of neurite extension, dynamic instability of MTs is essential for making temporary excursions into the transiently formed lamellipodia and filopodia. On the other hand, MTs should be stabilized and bundled when the direction is finally determined and neurite elongation advances. As discussed above, this indicates that simple PAR-1 knockdown or overexpression can affect MTs differently depending on cell conditions. In fact, MARK2 has been suggested to inhibit the development of dendrites rather than axons after the neuronal polarity is established in hippocampal neurons (Terabayashi et al. 2007).

Recently, the kinesin-like motor protein, GAKIN/KIF13B, has been suggested to be another downstream target of PAR-1 for regulating neuronal polarity (Table 2.1) (Yoshimura et al. 2010). GAKIN/KIF13B accumulates at the tip of the nascent axon and positively regulates the axon specification by transporting phosphatidylinositol (3,4,5)-trisphosphate-containing vesicles (Horiguchi et al. 2006). MARK2 inhibits these GAKIN/KIF13B activities via phosphorylation, probably by suppressing its MT binding (Yoshimura et al. 2010).

2.7.2 Other Roles of PAR-1 in Regulating Neuronal Functions

In the development process of the layered structure of the mammalian brain, cortical neurons formed at the apical ventricular zone (VZ) exhibit dynamic polarity changes during their radial migration toward the deep basal layers (Fig. 2.4b). The neurons exit from VZ with directional polarity to the intermediate zone (IZ), where they transiently acquire a multipolar phenotype, and later transit to a bipolar morphology extending the leading and tailing processes, of which the latter corresponds to the future axon. The roles of MARK2 in this in vivo process have been analyzed by introducing knockdown or expression vectors into the developing mouse brain using an in utero electroporation technique (Sapir et al. 2008). The results indicated that elevated expression at the early stage induced round neurons losing polarity, whereas reduced expression caused cells to stall in the IZ boundary with multipolar morphology. These results appear to be consistent with the data obtained from cultured primary neurons described above. However, because the expression of a kinase dead mutant in knockdown cells only restored the morphological phenotypes but failed to rescue the migration defects, the role of PAR-1 in polarity transition from multipolar to bipolar has been attributed to a putative scaffold activity of MARK2 (Sapir et al. 2008). In addition, the effect of MARK2 on migration has been attributed to its role in centrosome dynamics through its MARK activity. The role of MARK2 in neuronal migration has already been suggested in 2004 by demonstrating that MARK2 inhibits the MT binding of doublecortin (Dcx), the mutation of which results in a neuronal migration disorder (X-linked lissencephaly: X-LIS). PAR-1 has been suggested to regulate neuronal migration by directly phosphorylating a serine residue, which is mutated in X-LIS cases (Schaar et al. 2004).

The important roles of PAR-1 in neurons are also demonstrated in the development of synapses, in which the highly polarized local structures exhibit an asymmetric distribution of proteins and mRNAs (Fig. 2.4c). This was demonstrated first in the *Drosophila* larval neuromuscular junction, to which PAR-1 is prominently localized (Zhang et al. 2007b). PAR-1 negatively regulates synapse formation by directly phosphorylating Dlg (Table 2.1), which recruits various synaptic proteins and assembles them into large protein complexes. Phosphorylation of the GUK domain of Dlg results in its reduced postsynaptic targeting. One of the mammalian homologs of Dlg, PSD-95, has also been shown to be phosphorylated by MARK2 at the corresponding serine residue (Table 2.1) (Wu et al. 2012). However, in this case, the phosphorylation did not suppress the postsynaptic targeting of PSD-95, but rather promoted the development of dendritic spines, at the tip of which the postsynaptic structures develop (Fig. 2.4c). Because spine maturation requires a dynamic state of MTs (Jaworski et al. 2009), the MARK2 activity of PAR-1 also contributes to the observed positive effect of MARK2, as demonstrated by an independent study (Hayashi et al. 2011a). In any case, it should be noted that PAR-3 and the aPKC/PAR-6 complex have also been demonstrated to play a crucial role in spine development (Hayashi et al. 2011b). This may indicate that dendritic spines, the very local asymmetric structures, provide another example where PAR-1 functions as a component of the PAR-aPKC system.

Finally, several studies on *Drosophila* and *Xenopus* have demonstrated that PAR-1 depletion promoted embryonic neurogenesis, because of reduced lateral inhibition mediated by Notch signaling (Bayraktar et al. 2006; Ossipova et al. 2009). PAR-1 is thought to function in signal-sending cells by affecting the membrane localization of the Notch ligand, Delta. Particularly, Mind bomb (Mib), the E3-ligase of Delta, has been shown to be a direct target of PAR-1 in this pathway, as PAR-1-dependent phosphorylation stimulates proteasome-dependent degradation of Mib and thereby attenuates Delta activity (Table 2.1) (Ossipova et al. 2009). This provides another example of PAR-1 function in regulating protein stability in addition to the effect on Oskar.

2.8 Roles of PAR-1 in Actin-Related Activities

PAR-1 has been shown to regulate cell migration and tissue morphogenesis (Hurd and Kemphues 2003). These effects of PAR-1 may reflect its activity not only on MTs but also on actin cytoskeleton. In fact, the interaction between PAR-1 and non-muscle myosin-II (NMY-II) has been reported in an early study in *C. elegans* (Guo and Kemphues 1996b). Although the physiological significance of this interaction remains to be clarified, a recent study on *Drosophila* has demonstrated another link between PAR-1 and myosin activity (Majumder et al. 2012). In the *Drosophila* egg chamber, the 6–10 border cells derived from the anterior follicle cells detach and show collective migration to the anterior border of the oocyte (Fig. 2.3b). In these cells, PAR-1 has been shown to enhance NMY-II activity by phosphorylating a known inactivation site of the myosin phosphatase subunit (Mbs/MYPT1) (Table 2.1). The basolateral localization of PAR-1 in follicle cells is inherited by migrating border cells as asymmetric localization to the rear end, where PAR-1 locally activates NMY-II and promotes detachment of border cells from follicle cells. Although a similar activity has not been observed in mammals, yet, it may provide the molecular basis for the phenomena that MARK2-lacking MDCK cells lose the epithelial-specific columnar shape (Cohen et al. 2004; Suzuki et al. 2004), which is maintained by the contraction of actomyosin bundles surrounding the apex of the epithelial cells.

The protein interaction cloning system in mammalian cells has identified several PAR-1-binding proteins involved in other actin-related activities. One of them is GEF-H1, an MT-associated guanine nucleotide exchange factor specific for Rho GTPase, which is activated when released from MTs. A recent study has demonstrated that MARK2 inhibits its GEF activity by inducing MT binding through direct phosphorylation of a C-terminal serine residue (Yamahashi et al. 2011). Because another study has demonstrated that MARK2 phosphorylates N-terminal serine residues and thus suppresses MT binding of GEF-H1 (which should result in GEF-H1 activation) (Yoshimura and Miki 2011), the effects of PAR-1 on GEF-H1 activity seem to be rather complicated and fine-tuned. On the other hand, MARK2 has also been shown to interact with IRSp53, a platform of the Rho family GTPase signaling pathway, and inhibit its interaction with the actin filament regulator, WAVE2, via direct phosphorylation (Table 2.1) (Cohen et al. 2011). Another MARK2-binding protein, PAK5, a neuronal member of the p21-acivated kinase family, has been reported to reside upstream, not downstream, of MARK2 and to inhibit the MARK2 kinase activity, not through its catalytic activity but through its binding activity (Matenia et al. 2005).

The involvement of PAR-1 in the regulation of the cell-extracellular matrix (ECM) interaction has also been demonstrated in MDCK epithelial cells (Masuda-Hirata et al. 2009). MARK2 interacts with the laminin receptor, the dystroglycan (DG) complex, and promotes basal lamina accumulation underneath the epithelial cells by enhancing the basolateral localization of the DG complex. The interaction is mediated by direct binding of PAR-1 to utrophin, an actin-binding protein that belongs to the spectrin superfamily, which anchors to the DG complex and mediates the linkage between the intracellular actin cytoskeleton and ECM (Yamashita et al. 2010). This role of PAR-1 in regulating basal lamina formation has been confirmed in organ cultures of mouse embryonic salivary glands and has been suggested to be regulated by Rho kinase (Daley et al. 2012). Taken together, these accumulating data suggest that PAR-1 regulates not only MTs but also the actin cytoskeleton, to control cell migration and tissue morphogenesis.

2.9 Roles of PAR-1 in Wnt Signaling and Dishevelled Functions

The findings that PAR-1 phosphorylates Dishevelled (Dvl) and positively regulates Wnt signaling stimulated the initial stage of the PAR-1 study (Sun et al. 2001), as they implied an unexpected link between two important developmental pathways regulating body axes. Although the physiological significance of the Dvl phosphorylation remains to be clarified (Matsubayashi et al. 2004; Bernatik et al. 2011; Elbert et al. 2006; Ossipova et al. 2005; Kusakabe and Nishida 2004), the studies on *Xenopus* embryos have demonstrated that PAR-1 is involved in the noncanonical Wnt signaling regulating the convergent extension and neural crest specification (Ossipova et al. 2005; Ossipova and Sokol 2011; Kusakabe and Nishida 2004). PAR-1BY, one of the *Xenopus* PAR-1 isoforms, but not other isoforms, has been shown to promote the membrane association of Dvl through phosphorylation and thereby regulate noncanonical Wnt signaling. On the other hand, noncanonical Wnt, Wnt5a and Wnt11, have been shown to trigger the redistribution of PAR-1 from the membrane to the cytosol in embryonic ectoderm (Ossipova and Sokol 2011), which is important for PAR-1 function during gastrulation (Kusakabe and Nishida 2004). Because mammalian Dvl1 has been shown to mediate Wnt5a-dependent activation of aPKC in primary neurons (Zhang et al. 2007a), this Wnt-induced membrane dissociation of PAR-1 may be caused by aPKC phosphorylation. It is also noteworthy that, albeit contradicting the above results, mammalian Dvl1 enhances the membrane recruitment of MARK2 by inducing its phosphorylation in the UBA domain by an unknown kinase (Terabayashi et al. 2008). In any case, these accumulating results suggest that PAR-1 and Dvl may mutually regulate the membrane localization of each other to mediate the noncanonical Wnt signaling.

2.10 Concluding Remarks

Cell polarity regulation is one of the fundamental functions of the cell, which is achieved by integrating various cellular activities. In reflection of this requirement, PAR-1 kinase, the main downstream effector of the topmost regulator of cell polarity, the aPKC complex, exerts significantly versatile and complicated activities. The impact of PAR-1 on cell activity is the sum of its effects on various downstream targets (Table 2.1), making it difficult to obtain a simple view of PAR-1 functions. This is particularly true in mammals, in which a single knockout of the four PAR-1 paralogs does not result in embryonic lethality with developmental defects, but leads to defects in spatial learning and memory, immune system dysfunction, and unregulated metabolism (Hurov et al. 2001, 2007; Lennerz et al. 2010). Nevertheless, it is safe to say that one of the main regulatory targets of PAR-1 is MT dynamics, and the MARK activity of PAR-1 represents one of the underlying mechanisms of this PAR-1 function, particularly in mammalian neurons. Accumulating evidence also suggests that the regulation of actin-related activity, involving myosin and Rho GTPase, represents another pathway by which PAR-1 regulates cell polarity. In this regard, it is very interesting that several other kinases belonging to the AMPK subfamily have been shown to regulate MTs or myosin activity (Lee et al. 2007; Barnes et al. 2007; Zagorska et al. 2010). These results may support the notion that PAR-1 represents a specific member of the AMPK subfamily, which is dominantly selected by the aPKC complex as the downstream target, because of its ability to bind phospholipids through the KA1 domain (Figs. 2.1 and 2.2). No matter whether this notion is correct or not, there is no doubt that the evolutional insight on the PAR-aPKC system will provide powerful tools to integrate increasing data on PAR-1.

References

Barnes AP, Lilley BN, Pan YA, Plummer LJ, Powell AW, Raines AN, Sanes JR, Polleux F (2007) LKB1 and SAD kinases define a pathway required for the polarization of cortical neurons. Cell 129(3):549–563. doi:10.1016/j.cell.2007.03.025

Bartolini F, Gundersen GG (2006) Generation of noncentrosomal microtubule arrays. J Cell Sci 119(Pt 20):4155–4163. doi:10.1242/jcs.03227

Bayraktar J, Zygmunt D, Carthew RW (2006) Par-1 kinase establishes cell polarity and functions in Notch signaling in the Drosophila embryo. J Cell Sci 119(Pt 4):711–721. doi:10.1242/jcs.02789

Beatty A, Morton D, Kemphues K (2010) The C. elegans homolog of Drosophila Lethal giant larvae functions redundantly with PAR-2 to maintain polarity in the early embryo. Development 137(23):3995–4004. doi:10.1242/dev.056028

Benton R, St Johnston D (2003) Drosophila PAR-1 and 14-3-3 inhibit Bazooka/PAR-3 to establish complementary cortical domains in polarized cells. Cell 115(6):691–704

Bernatik O, Ganji RS, Dijksterhuis JP, Konik P, Cervenka I, Polonio T, Krejci P, Schulte G, Bryja V (2011) Sequential activation and inactivation of dishevelled in the Wnt/beta-catenin pathway by casein kinases. J Biol Chem 286(12):10396–10410. doi:10.1074/jbc.M110.169870

Biernat J, Wu YZ, Timm T, Zheng-Fischhofer Q, Mandelkow E, Meijer L, Mandelkow EM (2002) Protein kinase MARK/PAR-1 is required for neurite outgrowth and establishment of neuronal polarity. Mol Biol Cell 13(11):4013–4028. doi:10.1091/mbc.02-03-0046

Bohm H, Brinkmann V, Drab M, Henske A, Kurzchalia TV (1997) Mammalian homologues of C. elegans PAR-1 are asymmetrically localized in epithelial cells and may influence their polarity. Curr Biol 7(8):603–606

Brandt R (1996) The tau proteins in neuronal growth and development. Front Biosci 1:d118–d130

Chen YM, Wang QJ, Hu HS, Yu PC, Zhu J, Drewes G, Piwnica-Worms H, Luo ZG (2006) Microtubule affinity-regulating kinase 2 functions downstream of the PAR-3/PAR-6/atypical PKC complex in regulating hippocampal neuronal polarity. Proc Natl Acad Sci U S A 103(22): 8534–8539. doi:10.1073/pnas.0509955103

Chien SC, Brinkmann EM, Teuliere J, Garriga G (2013) Caenorhabditis elegans PIG-1/MELK acts in a conserved PAR-4/LKB1 polarity pathway to promote asymmetric neuroblast divisions. Genetics 193(3):897–909. doi:10.1534/genetics.112.148106

Cohen D, Brennwald PJ, Rodriguez-Boulan E, Musch A (2004) Mammalian PAR-1 determines epithelial lumen polarity by organizing the microtubule cytoskeleton. J Cell Biol 164(5): 717–727. doi:10.1083/jcb.200308104

Cohen D, Fernandez D, Lazaro-Dieguez F, Musch A (2011) The serine/threonine kinase Par1b regulates epithelial lumen polarity via IRSp53-mediated cell-ECM signaling. J Cell Biol 192(3):525–540. doi:10.1083/jcb.201007002

Cox DN, Lu B, Sun TQ, Williams LT, Jan YN (2001) Drosophila par-1 is required for oocyte differentiation and microtubule organization. Curr Biol 11(2):75–87

Daley WP, Gervais EM, Centanni SW, Gulfo KM, Nelson DA, Larsen M (2012) ROCK1-directed basement membrane positioning coordinates epithelial tissue polarity. Development 139(2): 411–422. doi:10.1242/dev.075366

Dequiedt F, Martin M, Von Blume J, Vertommen D, Lecomte E, Mari N, Heinen MF, Bachmann M, Twizere JC, Huang MC, Rider MH, Piwnica-Worms H, Seufferlein T, Kettmann R (2006) New role for hPar-1 kinases EMK and C-TAK1 in regulating localization and activity of class IIa histone deacetylases. Mol Cell Biol 26(19):7086–7102. doi:10.1128/MCB. 00231-06

Doerflinger H, Benton R, Shulman JM, St Johnston D (2003) The role of PAR-1 in regulating the polarised microtubule cytoskeleton in the Drosophila follicular epithelium. Development 130(17):3965–3975

Doerflinger H, Benton R, Torres IL, Zwart MF, St Johnston D (2006) Drosophila anterior-posterior polarity requires actin-dependent PAR-1 recruitment to the oocyte posterior. Curr Biol 16(11): 1090–1095. doi:10.1016/j.cub.2006.04.001

Doerflinger H, Vogt N, Torres IL, Mirouse V, Koch I, Nusslein-Volhard C, St Johnston D (2010) Bazooka is required for polarisation of the Drosophila anterior-posterior axis. Development 137(10):1765–1773. doi:10.1242/dev.045807

Drewes G, Ebneth A, Mandelkow EM (1998) MAPs, MARKs and microtubule dynamics. Trends Biochem Sci 23(8):307–311

Drewes G, Ebneth A, Preuss U, Mandelkow EM, Mandelkow E (1997) MARK, a novel family of protein kinases that phosphorylate microtubule-associated proteins and trigger microtubule disruption. Cell 89(2):297–308

Drewes G, Nurse P (2003) The protein kinase kin1, the fission yeast orthologue of mammalian MARK/PAR-1, localises to new cell ends after mitosis and is important for bipolar growth. FEBS Lett 554(1–2):45–49

Ducharme NA, Hales CM, Lapierre LA, Ham AJ, Oztan A, Apodaca G, Goldenring JR (2006) MARK2/EMK1/Par-1Balpha phosphorylation of Rab11-family interacting protein 2 is necessary for the timely establishment of polarity in Madin-Darby canine kidney cells. Mol Biol Cell 17(8):3625–3637. doi:10.1091/mbc.E05-08-0736

Elbert M, Cohen D, Musch A (2006) PAR1b promotes cell-cell adhesion and inhibits dishevelled-mediated transformation of Madin-Darby canine kidney cells. Mol Biol Cell 17(8):3345–3355. doi:10.1091/mbc.E06-03-0193

Etemad-Moghadam B, Guo S, Kemphues KJ (1995) Asymmetrically distributed PAR-3 protein contributes to cell polarity and spindle alignment in early *C. elegans* embryos. Cell 83(5): 743–752

Fu A, Screaton RA (2008) Using kinomics to delineate signaling pathways: control of CRTC2/TORC2 by the AMPK family. Cell Cycle 7(24):3823–3828

Griffin EE, Odde DJ, Seydoux G (2011) Regulation of the MEX-5 gradient by a spatially segregated kinase/phosphatase cycle. Cell 146(6):955–968. doi:10.1016/j.cell.2011.08.012

Guo S, Kemphues KJ (1995) par-1, a gene required for establishing polarity in C. elegans embryos, encodes a putative Ser/Thr kinase that is asymmetrically distributed. Cell 81(4):611–620

Guo S, Kemphues KJ (1996a) Molecular genetics of asymmetric cleavage in the early *Caenorhabditis elegans* embryo. Curr Opin Genet Dev 6(4):408–415

Guo S, Kemphues KJ (1996b) A non-muscle myosin required for embryonic polarity in *Caenorhabditis elegans*. Nature 382(6590):455–458

Hayashi K, Suzuki A, Hirai S, Kurihara Y, Hoogenraad CC, Ohno S (2011a) Maintenance of dendritic spine morphology by partitioning-defective 1b through regulation of microtubule growth. J Neurosci 31(34):12094–12103. doi:10.1523/JNEUROSCI. 0751-11.2011

Hayashi K, Suzuki A, Ohno S (2011b) A novel function of the cell polarity-regulating kinase PAR-1/MARK in dendritic spines. Bioarchitecture 1(6):261–266. doi:10.4161/bioa.1.6.19199

Hayashi K, Suzuki A, Ohno S (2012) PAR-1/MARK: a kinase essential for maintaining the dynamic state of microtubules. Cell Struct Funct 37(1):21–25

Horiguchi K, Hanada T, Fukui Y, Chishti AH (2006) Transport of PIP3 by GAKIN, a kinesin-3 family protein, regulates neuronal cell polarity. J Cell Biol 174(3):425–436. doi:10.1083/jcb.200604031

Hurd DD, Kemphues KJ (2003) PAR-1 is required for morphogenesis of the Caenorhabditis elegans vulva. Dev Biol 253(1):54–65

Hurov JB, Huang M, White LS, Lennerz J, Choi CS, Cho YR, Kim HJ, Prior JL, Piwnica-Worms D, Cantley LC, Kim JK, Shulman GI, Piwnica-Worms H (2007) Loss of the Par-1b/MARK2 polarity kinase leads to increased metabolic rate, decreased adiposity, and insulin hypersensitivity in vivo. Proc Natl Acad Sci U S A 104(13):5680–5685. doi:10.1073/pnas.0701179104

Hurov JB, Stappenbeck TS, Zmasek CM, White LS, Ranganath SH, Russell JH, Chan AC, Murphy KM, Piwnica-Worms H (2001) Immune system dysfunction and autoimmune disease in mice lacking Emk (Par-1) protein kinase. Mol Cell Biol 21(9):3206–3219. doi:10.1128/MCB. 21.9.3206-3219.2001

Hurov JB, Watkins JL, Piwnica-Worms H (2004) Atypical PKC phosphorylates PAR-1 kinases to regulate localization and activity. Curr Biol 14(8):736–741. doi:10.1016/j.cub.2004.04.007

Huynh JR, Shulman JM, Benton R, St Johnston D (2001) PAR-1 is required for the maintenance of oocyte fate in Drosophila. Development 128(7):1201–1209

Izumi Y, Hirose T, Tamai Y, Hirai S, Nagashima Y, Fujimoto T, Tabuse Y, Kemphues KJ, Ohno S (1998) An atypical PKC directly associates and colocalizes at the epithelial tight junction with ASIP, a mammalian homologue of *Caenorhabditis elegans* polarity protein PAR-3. J Cell Biol 143(1):95–106

Jaworski J, Kapitein LC, Gouveia SM, Dortland BR, Wulf PS, Grigoriev I, Camera P, Spangler SA, Di Stefano P, Demmers J, Krugers H, Defilippi P, Akhmanova A, Hoogenraad CC (2009) Dynamic microtubules regulate dendritic spine morphology and synaptic plasticity. Neuron 61(1):85–100. doi:10.1016/j.neuron.2008.11.013

Kemphues KJ, Priess JR, Morton DG, Cheng NS (1988) Identification of genes required for cytoplasmic localization in early C. elegans embryos. Cell 52(3):311–320

Knust E, Bossinger O (2002) Composition and formation of intercellular junctions in epithelial cells. Science 298(5600):1955–1959. doi:10.1126/science.1072161

Kusakabe M, Nishida E (2004) The polarity-inducing kinase Par-1 controls Xenopus gastrulation in cooperation with 14-3-3 and aPKC. EMBO J 23(21):4190–4201. doi:10.1038/sj.emboj.7600381

Lee JH, Koh H, Kim M, Kim Y, Lee SY, Karess RE, Lee SH, Shong M, Kim JM, Kim J, Chung J (2007) Energy-dependent regulation of cell structure by AMP-activated protein kinase. Nature 447(7147):1017–1020. doi:10.1038/nature05828

Lennerz JK, Hurov JB, White LS, Lewandowski KT, Prior JL, Planer GJ, Gereau RW, Piwnica-Worms D, Schmidt RE, Piwnica-Worms H (2010) Loss of Par-1a/MARK3/C-TAK1 kinase leads to reduced adiposity, resistance to hepatic steatosis, and defective gluconeogenesis. Mol Cell Biol 30(21):5043–5056. doi:10.1128/MCB. 01472-09

Levin DE, Bishop JM (1990) A putative protein kinase gene (kin1+) is important for growth polarity in Schizosaccharomyces pombe. Proc Natl Acad Sci U S A 87(21):8272–8276

Lizcano JM, Goransson O, Toth R, Deak M, Morrice NA, Boudeau J, Hawley SA, Udd L, Makela TP, Hardie DG, Alessi DR (2004) LKB1 is a master kinase that activates 13 kinases of the AMPK subfamily, including MARK/PAR-1. EMBO J 23(4):833–843. doi:10.1038/sj.emboj.7600110

Majumder P, Aranjuez G, Amick J, McDonald JA (2012) Par-1 controls myosin-II activity through myosin phosphatase to regulate border cell migration. Curr Biol 22(5):363–372. doi:10.1016/j.cub.2012.01.037

Manning G, Whyte DB, Martinez R, Hunter T, Sudarsanam S (2002) The protein kinase complement of the human genome. Science 298(5600):1912–1934. doi:10.1126/science.1075762

Marx A, Nugoor C, Muller J, Panneerselvam S, Timm T, Bilang M, Mylonas E, Svergun DI, Mandelkow EM, Mandelkow E (2006) Structural variations in the catalytic and ubiquitin-associated domains of microtubule-associated protein/microtubule affinity regulating kinase (MARK) 1 and MARK2. J Biol Chem 281(37):27586–27599. doi:10.1074/jbc.M604865200

Marx A, Nugoor C, Panneerselvam S, Mandelkow E (2010) Structure and function of polarity-inducing kinase family MARK/Par-1 within the branch of AMPK/Snf1-related kinases. FASEB J 24(6):1637–1648. doi:10.1096/fj.09-148064

Masuda-Hirata M, Suzuki A, Amano Y, Yamashita K, Ide M, Yamanaka T, Sakai M, Imamura M, Ohno S (2009) Intracellular polarity protein PAR-1 regulates extracellular laminin assembly by regulating the dystroglycan complex. Genes Cells 14(7):835–850. doi:10.1111/j.1365-2443.2009.01315.x

Matenia D, Griesshaber B, Li XY, Thiessen A, Johne C, Jiao J, Mandelkow E, Mandelkow EM (2005) PAK5 kinase is an inhibitor of MARK/Par-1, which leads to stable microtubules and dynamic actin. Mol Biol Cell 16(9):4410–4422. doi:10.1091/mbc.E05-01-0081

Matenia D, Mandelkow EM (2009) The tau of MARK: a polarized view of the cytoskeleton. Trends Biochem Sci 34(7):332–342. doi:10.1016/j.tibs.2009.03.008

Matsubayashi H, Sese S, Lee JS, Shirakawa T, Iwatsubo T, Tomita T, Yanagawa S (2004) Biochemical characterization of the Drosophila wingless signaling pathway based on RNA interference. Mol Cell Biol 24(5):2012–2024

McKinley RF, Harris TJ (2012) Displacement of basolateral Bazooka/PAR-3 by regulated transport and dispersion during epithelial polarization in Drosophila. Mol Biol Cell 23(22): 4465–4471. doi:10.1091/mbc.E12-09-0655

Morais-de-Sa E, Vega-Rioja A, Trovisco V, St Johnston D (2013) Oskar is targeted for degradation by the sequential action of Par-1, GSK-3, and the SCF(-Slimb) ubiquitin ligase. Dev Cell 26(3):303–314. doi:10.1016/j.devcel.2013.06.011

Moravcevic K, Mendrola JM, Schmitz KR, Wang YH, Slochower D, Janmey PA, Lemmon MA (2010) Kinase associated-1 domains drive MARK/PAR1 kinases to membrane targets by binding acidic phospholipids. Cell 143(6):966–977. doi:10.1016/j.cell.2010.11.028

Motegi F, Zonies S, Hao Y, Cuenca AA, Griffin E, Seydoux G (2011) Microtubules induce self-organization of polarized PAR domains in Caenorhabditis elegans zygotes. Nat Cell Biol 13(11):1361–1367. doi:10.1038/ncb2354

Muller J, Ritt DA, Copeland TD, Morrison DK (2003) Functional analysis of C-TAK1 substrate binding and identification of PKP2 as a new C-TAK1 substrate. EMBO J 22(17):4431–4442. doi:10.1093/emboj/cdg426

Munro E, Nance J, Priess JR (2004) Cortical flows powered by asymmetrical contraction transport PAR proteins to establish and maintain anterior-posterior polarity in the early C. elegans embryo. Dev Cell 7(3):413–424. doi:10.1016/j.devcel.2004.08.001

Murphy JM, Korzhnev DM, Ceccarelli DF, Briant DJ, Zarrine-Afsar A, Sicheri F, Kay LE, Pawson T (2007) Conformational instability of the MARK3 UBA domain compromises ubiquitin recognition and promotes interaction with the adjacent kinase domain. Proc Natl Acad Sci U S A 104(36):14336–14341. doi:10.1073/pnas.0703012104

Nesic D, Miller MC, Quinkert ZT, Stein M, Chait BT, Stebbins CE (2010) Helicobacter pylori CagA inhibits PAR1-MARK family kinases by mimicking host substrates. Nat Struct Mol Biol 17(1):130–132. doi:10.1038/nsmb.1705

Nishimura T, Kato K, Yamaguchi T, Fukata Y, Ohno S, Kaibuchi K (2004) Role of the PAR-3-KIF3 complex in the establishment of neuronal polarity. Nat Cell Biol 6(4):328–334. doi:10.1038/ncb1118

Ossipova O, Dhawan S, Sokol S, Green JB (2005) Distinct PAR-1 proteins function in different branches of Wnt signaling during vertebrate development. Dev Cell 8(6):829–841. doi:10.1016/j.devcel.2005.04.011

Ossipova O, Ezan J, Sokol SY (2009) PAR-1 phosphorylates mind bomb to promote vertebrate neurogenesis. Dev Cell 17(2):222–233. doi:10.1016/j.devcel.2009.06.010

Ossipova O, Sokol SY (2011) Neural crest specification by noncanonical Wnt signaling and PAR-1. Development 138(24):5441–5450. doi:10.1242/dev.067280

Panneerselvam S, Marx A, Mandelkow EM, Mandelkow E (2006) Structure of the catalytic and ubiquitin-associated domains of the protein kinase MARK/Par-1. Structure 14(2):173–183. doi:10.1016/j.str.2005.09.022

Parton RM, Hamilton RS, Ball G, Yang L, Cullen CF, Lu W, Ohkura H, Davis I (2011) A PAR-1-dependent orientation gradient of dynamic microtubules directs posterior cargo transport in the Drosophila oocyte. J Cell Biol 194(1):121–135. doi:10.1083/jcb.201103160

Pellettieri J, Seydoux G (2002) Anterior-posterior polarity in C. elegans and Drosophila–PARallels and differences. Science 298(5600):1946–1950. doi:10.1126/science.1072162

Peng CY, Graves PR, Ogg S, Thoma RS, Byrnes MJ 3rd, Wu Z, Stephenson MT, Piwnica-Worms H (1998) C-TAK1 protein kinase phosphorylates human Cdc25C on serine 216 and promotes 14-3-3 protein binding. Cell Growth Differ 9(3):197–208

Riechmann V, Ephrussi A (2001) Axis formation during Drosophila oogenesis. Curr Opin Genet Dev 11(4):374–383

Riechmann V, Ephrussi A (2004) Par-1 regulates bicoid mRNA localisation by phosphorylating Exuperantia. Development 131(23):5897–5907. doi:10.1242/dev.01515

Riechmann V, Gutierrez GJ, Filardo P, Nebreda AR, Ephrussi A (2002) Par-1 regulates stability of the posterior determinant Oskar by phosphorylation. Nat Cell Biol 4(5):337–342. doi:10.1038/ncb782

Sapir T, Sapoznik S, Levy T, Finkelshtein D, Shmueli A, Timm T, Mandelkow EM, Reiner O (2008) Accurate balance of the polarity kinase MARK2/Par-1 is required for proper cortical neuronal migration. J Neurosci 28(22):5710–5720. doi:10.1523/jneurosci. 0911-08.2008

Sato Y, Akitsu M, Amano Y, Yamashita K, Ide M, Shimada K, Yamashita A, Hirano H, Arakawa N, Maki T, Hayashi I, Ohno S, Suzuki A (2013) The novel PAR-1-binding protein MTCL1 has crucial roles in organizing microtubules in polarizing epithelial cells. J Cell Sci 126(Pt 20):4671–4683. doi:10.1242/jcs.127845

Schaar BT, Kinoshita K, McConnell SK (2004) Doublecortin microtubule affinity is regulated by a balance of kinase and phosphatase activity at the leading edge of migrating neurons. Neuron 41(2):203–213

Shi SH, Jan LY, Jan YN (2003) Hippocampal neuronal polarity specified by spatially localized mPar3/mPar6 and PI 3-kinase activity. Cell 112(1):63–75

Shulman JM, Benton R, St Johnston D (2000) The Drosophila homolog of C. elegans PAR-1 organizes the oocyte cytoskeleton and directs oskar mRNA localization to the posterior pole. Cell 101(4):377–388

Sun TQ, Lu B, Feng JJ, Reinhard C, Jan YN, Fantl WJ, Williams LT (2001) PAR-1 is a dishevelled-associated kinase and a positive regulator of Wnt signalling. Nat Cell Biol 3(7):628–636. doi:10.1038/35083016

Sung HH, Telley IA, Papadaki P, Ephrussi A, Surrey T, Rorth P (2008) Drosophila ensconsin promotes productive recruitment of Kinesin-1 to microtubules. Dev Cell 15(6):866–876. doi:10.1016/j.devcel.2008.10.006

Suzuki A, Hirata M, Kamimura K, Maniwa R, Yamanaka T, Mizuno K, Kishikawa M, Hirose H, Amano Y, Izumi N, Miwa Y, Ohno S (2004) aPKC acts upstream of PAR-1b in both the establishment and maintenance of mammalian epithelial polarity. Curr Biol 14(16):1425–1435. doi:10.1016/j.cub.2004.08.021

Suzuki A, Ohno S (2006) The PAR-aPKC system: lessons in polarity. J Cell Sci 119(Pt 6):979–987. doi:10.1242/jcs.02898

Tabuse Y, Izumi Y, Piano F, Kemphues KJ, Miwa J, Ohno S (1998) Atypical protein kinase C cooperates with PAR-3 to establish embryonic polarity in *Caenorhabditis elegans*. Development 125(18):3607–3614

Tanaka E, Ho T, Kirschner MW (1995) The role of microtubule dynamics in growth cone motility and axonal growth. J Cell Biol 128(1–2):139–155

Tassan JP, Le Goff X (2004) An overview of the KIN1/PAR-1/MARK kinase family. Biol Cell 96(3):193–199. doi:10.1016/j.biolcel.2003.10.009

Terabayashi T, Funato Y, Miki H (2008) Dishevelled-induced phosphorylation regulates membrane localization of Par1b. Biochem Biophys Res Commun 375(4):660–665. doi:10.1016/j.bbrc.2008.08.098

Terabayashi T, Itoh TJ, Yamaguchi H, Yoshimura Y, Funato Y, Ohno S, Miki H (2007) Polarity-regulating kinase partitioning-defective 1/microtubule affinity-regulating kinase 2 negatively regulates development of dendrites on hippocampal neurons. J Neurosci 27(48):13098–13107. doi 10.1523/JNEUROSCI. 3986-07.2007

Tian AG, Deng WM (2008) Lgl and its phosphorylation by aPKC regulate oocyte polarity formation in Drosophila. Development 135(3):463–471. doi:10.1242/dev.016253

Tian AG, Deng WM (2009) Par-1 and Tau regulate the anterior-posterior gradient of microtubules in Drosophila oocytes. Dev Biol 327(2):458–464. doi:10.1016/j.ydbio.2008.12.031

Timm T, Balusamy K, Li X, Biernat J, Mandelkow E, Mandelkow EM (2008a) Glycogen synthase kinase (GSK) 3beta directly phosphorylates Serine 212 in the regulatory loop and inhibits microtubule affinity-regulating kinase (MARK) 2. J Biol Chem 283(27):18873–18882. doi:10.1074/jbc.M706596200

Timm T, Li XY, Biernat J, Jiao J, Mandelkow E, Vandekerckhove J, Mandelkow EM (2003) MARKK, a Ste20-like kinase, activates the polarity-inducing kinase MARK/PAR-1. EMBO J 22(19):5090–5101. doi:10.1093/emboj/cdg447

Timm T, Marx A, Panneerselvam S, Mandelkow E, Mandelkow EM (2008b) Structure and regulation of MARK, a kinase involved in abnormal phosphorylation of Tau protein. BMC Neurosci 9(Suppl 2):9. doi:10.1186/1471-2202-9-S2-S9

Timm T, von Kries JP, Li X, Zempel H, Mandelkow E, Mandelkow EM (2011) Microtubule affinity regulating kinase activity in living neurons was examined by a genetically encoded fluorescence resonance energy transfer/fluorescence lifetime imaging-based biosensor: inhibitors with therapeutic potential. J Biol Chem 286(48):41711–41722. doi:10.1074/jbc.M111.257865

Tochio N, Koshiba S, Kobayashi N, Inoue M, Yabuki T, Aoki M, Seki E, Matsuda T, Tomo Y, Motoda Y, Kobayashi A, Tanaka A, Hayashizaki Y, Terada T, Shirouzu M, Kigawa T, Yokoyama S (2006) Solution structure of the kinase-associated domain 1 of mouse microtubule-associated protein/microtubule affinity-regulating kinase 3. Protein Sci 15(11): 2534–2543. doi:10.1110/ps.062391106

Tzivion G, Shen YH, Zhu J (2001) 14-3-3 proteins; bringing new definitions to scaffolding. Oncogene 20(44):6331–6338. doi:10.1038/sj.onc.1204777

Uboha NV, Flajolet M, Nairn AC, Picciotto MR (2007) A calcium- and calmodulin-dependent kinase Ialpha/microtubule affinity regulating kinase 2 signaling cascade mediates calcium-dependent neurite outgrowth. J Neurosci 27(16):4413–4423. doi:10.1523/JNEUROSCI. 0725-07.2007

Vaccari T, Ephrussi A (2002) The fusome and microtubules enrich Par-1 in the oocyte, where it effects polarization in conjunction with Par-3, BicD, Egl, and dynein. Curr Biol 12(17): 1524–1528

Wang YC, Khan Z, Kaschube M, Wieschaus EF (2012) Differential positioning of adherens junctions is associated with initiation of epithelial folding. Nature 484(7394):390–393. doi:10.1038/nature10938

Wu JC, Rose LS (2007) PAR-3 and PAR-1 inhibit LET-99 localization to generate a cortical band important for spindle positioning in Caenorhabditis elegans embryos. Mol Biol Cell 18(11): 4470–4482. doi:10.1091/mbc.E07-02-0105

Wu Q, DiBona VL, Bernard LP, Zhang H (2012) The polarity protein partitioning-defective 1 (PAR-1) regulates dendritic spine morphogenesis through phosphorylating postsynaptic density protein 95 (PSD-95). J Biol Chem 287(36):30781–30788. doi:10.1074/jbc.M112.351452

Yamahashi Y, Saito Y, Murata-Kamiya N, Hatakeyama M (2011) Polarity-regulating kinase partitioning-defective 1b (PAR1b) phosphorylates guanine nucleotide exchange factor H1 (GEF-H1) to regulate RhoA-dependent actin cytoskeletal reorganization. J Biol Chem 286(52):44576–44584. doi:10.1074/jbc.M111.267021

Yamashita K, Suzuki A, Satoh Y, Ide M, Amano Y, Masuda-Hirata M, Hayashi YK, Hamada K, Ogata K, Ohno S (2010) The 8th and 9th tandem spectrin-like repeats of utrophin cooperatively form a functional unit to interact with polarity-regulating kinase PAR-1b. Biochem Biophys Res Commun 391(1):812–817. doi:10.1016/j.bbrc.2009.11.144

Yoshimura Y, Miki H (2011) Dynamic regulation of GEF-H1 localization at microtubules by Par1b/MARK2. Biochem Biophys Res Commun 408(2):322–328. doi:10.1016/j.bbrc.2011.04.032

Yoshimura Y, Terabayashi T, Miki H (2010) Par1b/MARK2 phosphorylates kinesin-like motor protein GAKIN/KIF13B to regulate axon formation. Mol Cell Biol 30(9):2206–2219. doi:10.1128/MCB. 01181-09

Zagorska A, Deak M, Campbell DG, Banerjee S, Hirano M, Aizawa S, Prescott AR, Alessi DR (2010) New roles for the LKB1-NUAK pathway in controlling myosin phosphatase complexes and cell adhesion. Sci Signal 3(115):ra25. doi:10.1126/scisignal.2000616

Zhang SH, Kobayashi R, Graves PR, Piwnica-Worms H, Tonks NK (1997) Serine phosphorylation-dependent association of the band 4.1-related protein-tyrosine phosphatase PTPH1 with 14-3-3beta protein. J Biol Chem 272(43):27281–27287

Zhang X, Zhu J, Yang GY, Wang QJ, Qian L, Chen YM, Chen F, Tao Y, Hu HS, Wang T, Luo ZG (2007a) Dishevelled promotes axon differentiation by regulating atypical protein kinase C. Nat Cell Biol 9(7):743–754. doi:10.1038/ncb1603

Zhang Y, Guo H, Kwan H, Wang JW, Kosek J, Lu B (2007b) PAR-1 kinase phosphorylates and regulates its postsynaptic targeting at the Drosophila neuromuscular junction. Neuron 53 (2):201–215. doi:10.1016/j.neuron.2006.12.016

Chapter 3
The Crumbs3 Complex

Barbara Vacca, Magali Barthélémy-Requin, Céline Burcklé, Dominique Massey-Harroche, and André Le Bivic

Abstract Epithelial layers have allowed the evolution of metazoans by promoting complex multicellularity and protection from the environment. This essential role for epithelia relies on their ability to build an apical surface facing the outside world and a basolateral surface connecting cells together to coordinate their movements and resistance to stress. These epithelial features are dependent on several protein complexes for their establishment and maintenance. Among these protein complexes, one complex contains an apical transmembrane protein, Crumbs, that plays an essential role for the proper organization of tight junctions, apical morphology, or cell and tissue growth. In this chapter we will review how Crumbs3, one of the Crumbs proteins, both recruits adapter proteins essential for the building of tight or adherens junctions in vertebrates and interacts with the subapical actin cytoskeleton. These functions are broadly conserved in animals from flies to human, and it remains a challenge to understand all the molecular mechanisms by which Crumbs3 and its partners participate in the building of a functional epithelial sheet to keep body homeostasis and to allow for morphogenetic events as essential as gastrulation.

Keywords Actin cytoskeleton • Adherens junctions • Crumbs3 • Epithelial polarity • Human (all on page 1)

Abbreviations

aPKC	Atypical protein kinase C
Crb	Crumbs
FERM	4.1-Ezrin-radixin-moesin
IMCD3	Inner medullary collecting duct
KO	Knockout

B. Vacca
UCL – Institute of Ophthalmology, 11-43 Bath Street, London EC1V 9EL, UK

M. Barthélémy-Requin • C. Burcklé • D. Massey-Harroche • A. Le Bivic (✉)
Aix-Marseille Université, CNRS, UMR 7288, Developmental Biology Institute of Marseille (IBDM), Case 907, Marseille Cedex 09 13288, France
e-mail: andre.le-bivic@univ-amu.fr

K. Ebnet (ed.), *Cell Polarity 1*, DOI 10.1007/978-3-319-14463-4_3

MAGUK	Membrane-associated guanylate kinase
MDCK	Madin-Darby canine kidney
MPP5	Membrane-associated palmitoylated proteins
OLM	Outer limiting membrane
PALS1	Protein associated with Lin-7
PAR6	Partition defective-6
PATJ	PALS1-associated tight junction protein
PDZ	PSD-95, discs large, zona occludens 1
RP12	Retinitis pigmentosa group 12
Sdt	Stardust
TJ	Tight junction

3.1 Introduction

Crumbs was first identified in *Drosophila melanogaster*, where mutations in this gene produced an absent or dispersed cuticle, hence giving the name "Crumbs" (DCrb in the case of *Drosophila* Crumbs) (Tepass et al. 1990). This phenotype indicated a role of DCrb in epithelium organization as the cuticle is secreted by the underlying epithelial epidermis. DCrb is a transmembrane protein with a small cytoplasmic domain consisting of 37 amino acids and a large extracellular domain consisting of 28 EGF-like repeats and three laminin G-like repeats (Fig. 3.1a). A few years later, Wodarz et al. (1995) established that the cytoplasmic domain of DCrb is crucial for its function in epithelial morphogenesis. This intracellular domain shows a FERM (4.1-ezrin-radixin-moesin)-binding motif, a C-terminal PDZ (PSD-95, discs large, ZO1)-binding motif (called ERLI for the last four amino acids), and a putative DaPKC (Drosophila atypical protein kinase C.) phosphorylation motif (for review see Bazellieres et al. (2009)).

The first evidence of a conservation of Crumbs between invertebrates and vertebrates was presented 10 years later by den Hollander et al. (den Hollander et al. 1999). These authors discovered that a human gene mutated in retinitis pigmentosa group 12 (RP12) disease was very similar to *DCrb*. This gene was called *CRB1* and the largest protein isoform expressed has the same domain architecture as both the extra and intracellular domains of DCrb (Fig. 3.1a). Two additional genes, *CRB2* and *CRB3*, were then discovered (Katoh and Katoh 2004; Makarova et al. 2003) and found to encode proteins with a highly conserved cytoplasmic domain (showing 51 and 56 % of identity with human CRB1) (Fig. 3.1b). CRB3 is the only homolog exhibiting a very short extracellular sequence with no recognizable protein domain (as shown in Fig. 3.1a). Several splice variants have been described for CRB1 and CRB2, which encode proteins lacking the transmembrane and intracellular domains, which are therefore presumably secreted (CRB1s and CRB2s) (Katoh and Katoh 2004; Watanabe et al. 2004). This chapter will focus on the CRB3 core complex made of PALS1 (for protein

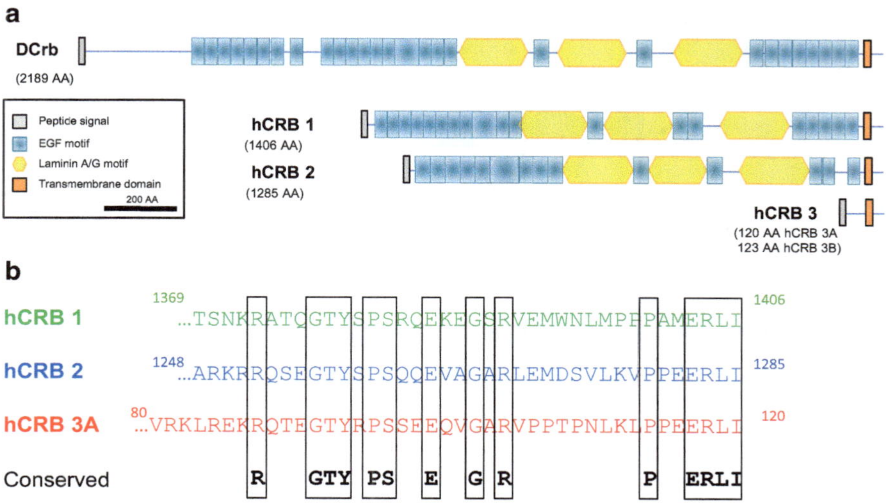

		Identities		
		hCRB 1 cyt.	hCRB 2 cyt.	hCRB 3A cyt.
Positives	hCRB 1cyt.		51%	56% (2 gaps)
	hCRB 2cyt.	77%		59%
	hCRB 3Acyt.	66%	75%	

Fig. 3.1 Schematic representation of DCrb and human CRB1, CRB2, and CRB3. (**a**) Extracellular regions of DCrb, CRB1, and CRB2 are composed of EGF and laminin A/G domains. Domains are shown according to the SMART program (simple modular architecture research tool, http://smart.embl-heidelberg.de). (**b**) Alignment of human CRB cytoplasmic domain sequences. The signal peptide is not considered for the amino acid numbers. In the table, the percentage of identical amino acids is indicated in *dark blue* color compartments (one gap for each CRB was necessary for alignment of CRB1 and CRB3A) and the conserved positive charges in *light blue* (http://blast.ncbi.nlm.nih.gov/Blast.cgi)

associated with Lin-7) and PATJ (for PALS1-associated tight junction protein) (Fig. 3.2).

3.1.1 Crumbs3 Diversity and Evolution

From the recent genomic studies, it is now becoming more and more obvious that the polarity complexes appeared at the onset of animal evolution, and in particular

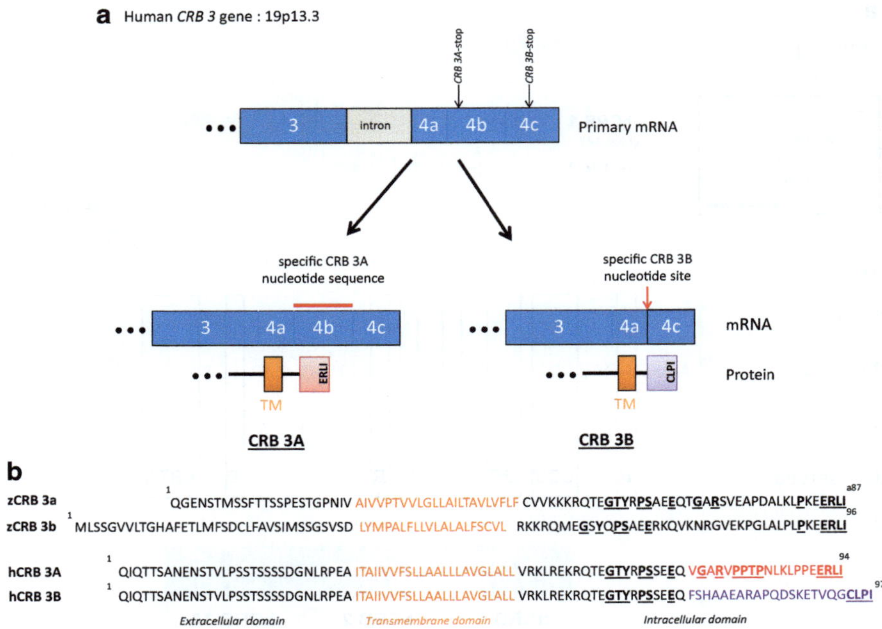

Fig. 3.2 hCRB3. (**a**) Schematic diagram of splicing CRB3 mRNA to produce CRB3A and CRB3B proteins. (**b**) Sequences of zebra fish (z) and human CRB3 proteins. Conserved amino acids are in *underlined bold*

Crumbs homologues have been identified in sponges such as *Amphimedon queenslandica* and are present in all metazoan clades while they are not found in more ancient phyla (for review see Le Bivic (2013); Fahey and Degnan (2010)). Among the Crumbs proteins, CRB3 isoforms however are so far found only in vertebrates and are likely derived from a partial duplication of a Crumbs gene. The loss of a typical Crumbs extracellular domain with its EGF- and laminin-like repeats had a likely consequence upon CRB3 function in extracellular signaling by preventing interactions which have yet to be clearly identified and demonstrated. Genomic analysis clearly shows that CRB3 appeared with the vertebrate lineage more than 500 million years ago and has been duplicated in species such as the *Danio rerio* (Omori and Malicki 2006). The fact that mammalians shared two new evolution traits not found in other vertebrates indicates that this duplication in fishes likely occurred after the emergence of the mammals. The first new feature is the presence of a PPxP motif in the cytoplasmic tail of CRB3A probably giving more flexibility to the C-terminal ERLI for yet unknown reasons (see Fig. 3.1b). The second one is the existence of a spliced isoform called CRB3B leading to a slightly longer cytoplasmic tail and a CLPI motif at the C-terminal end but lacking the proline-rich region (Fan et al. 2007) (Fig. 3.2a, b).

3.1.2 The Crumbs3 Complex

In mammalian epithelial cells, CRB3 makes a complex with several adaptors or scaffold proteins, namely, PALS1, PAR6 (for partition defective-6), aPKC, and PATJ or MUPP1 (multi-PDZ domain protein) (Fig. 3.3).

PALS1 was first identified as a Lin7-binding protein (Kamberov et al. 2000) that localizes to tight junctions (TJs) in Madin-Darby canine kidney (MDCK) cells (Roh et al. 2002b). At least two isoforms of PALS1 were identified in that study, but their functional relevance has not yet been identified. In mammals, PALS1 is also called MPP5 (for membrane-associated palmitoylated proteins), and in *Drosophila melanogaster*, it was identified as the product of the *Stardust* (*Sdt*) mutant which was found in a genetic screen that also uncovered *Crumbs* (for review see Muller (2000)) but was only identified molecularly in 2001 (Bachmann et al. 2001; Hong et al. 2001). PALS1 belongs to the MAGUK (for membrane-associated guanylate kinase) family of adaptor proteins which contain PDZ, Src homology 3, and guanylate kinase domains. PALS1 binds to the C-terminal end (ERLI) of Crumbs by its PDZ domain and has two L27 C and N domains that bind to Lin7 and PATJ, respectively (Roh et al. 2002b). Later, it was demonstrated that CRB3A is ubiquitously expressed in mammalian epithelial cells where it binds to PALS1 with its conserved ERLI motif (Makarova et al. 2003).

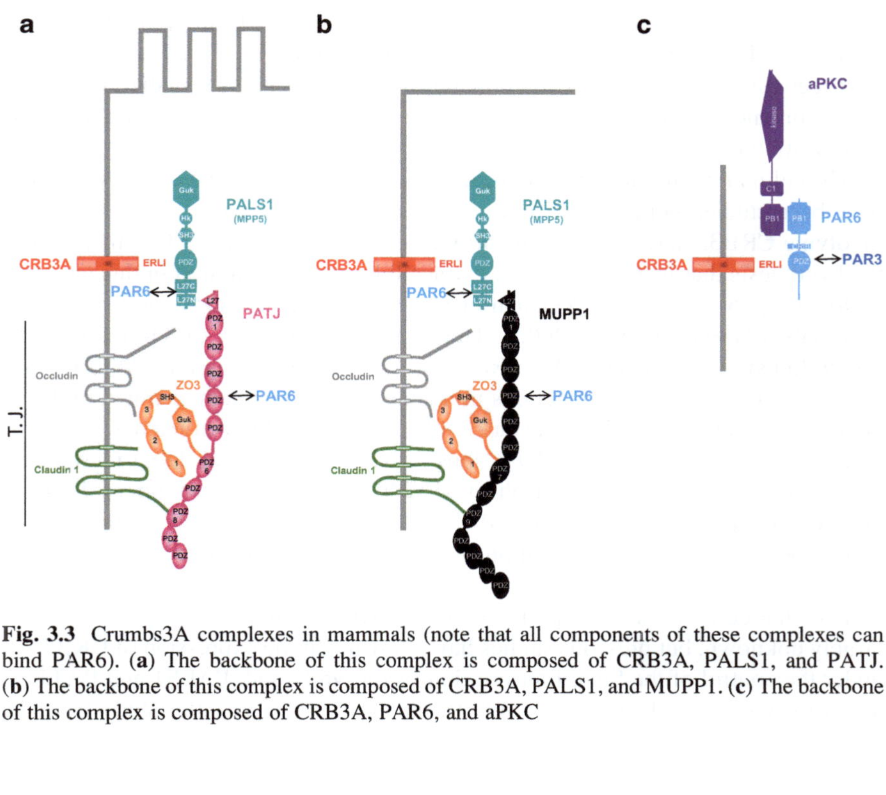

Fig. 3.3 Crumbs3A complexes in mammals (note that all components of these complexes can bind PAR6). (**a**) The backbone of this complex is composed of CRB3A, PALS1, and PATJ. (**b**) The backbone of this complex is composed of CRB3A, PALS1, and MUPP1. (**c**) The backbone of this complex is composed of CRB3A, PAR6, and aPKC

PALS1 with its L27N domain recruits two multi-PDZ proteins named PATJ and MUPP1 (Ullmer et al. 1998) via their own L27 domains. PALS1 binding affinity for MUPP1 is weaker than that for PATJ (Adachi et al. 2009; Roh et al. 2002b). PATJ and MUPP1 contain from 9 to 13 PDZ domains, respectively, and most of these domains share a high similarity both in sequence and in binding partners (Adachi et al. 2009; Assemat et al. 2013; Bazellieres et al. 2009). Since both PATJ and MUPP1 are scaffold proteins with multiple PDZ domains, the list of their partners is impressive and will not be further discussed here (for review see Assemat et al. (2008)). PATJ and MUPP1 form exclusive complexes with CRB3A and PALS1 in human epithelial cells, but how the balance between these two complexes is maintained remains unknown (Assemat et al. 2013). PATJ and MUPP1 also interact with PAR6 although the binding affinity between MUPP1 and PAR6 seemed to be much lower than between PATJ and PAR6 (Adachi et al. 2009). This later interaction involves the PDZ4 domain of PATJ and the amino acids 98–132 of PAR6, but its functional importance is not known so far (Adachi et al. 2009).

CRB3A by its ERLI motif binds directly to the PDZ domain of PAR6 (Lemmers et al. 2004), and this interaction is conserved in *D. melanogaster* (Kempkens et al. 2006). PAR6 also makes a conserved complex with aPKC in metazoans (Le Bivic 2013), and aPKC has been shown to phosphorylate Crumbs in *D. melanogaster* (Sotillos et al. 2004). It is thus considered that the CRB3A complex is minimally made of PALS1/PATJ and PAR6/aPKC. Indeed it was shown that the two polarity complexes are molecularly linked through multiple interactions both in mammalian and fly epithelial cells (Adachi et al. 2009; Hurd et al. 2003; Lemmers et al. 2004; Nam and Choi 2003) (Fig. 3.3). Thus, one of the next technological challenges in epithelial biology will be to visualize in situ the formation and composition of these apical polarity complexes to better understand their spatiotemporal evolution.

The cellular expression and localization of the members of the CRB3A complex have been studied both in vitro and in vivo, and while the apical targeting signals involving CRB3A have not been identified yet (Makarova et al. 2003), there is still the hypothesis that its *O*-glycan-rich extracellular stalk is required in this targeting as for the $p75^{NTR}$receptor (Yeaman et al. 1997) through a galectin-3 dependent mechanism (Delacour et al. 2006). The targeting of PALS1 to TJs requires the L27N domain that binds to PATJ (Roh et al. 2002b) indicating that PATJ recruits PALS1 to the apico-lateral junctions in mammals as it was demonstrated by PATJ depletion in Caco2 cells (Michel et al. 2005). Finally, PATJ is targeted to TJs by its PDZ6 domain which binds to ZO-3, thus providing a potential direct recruitment mechanism for PATJ at TJs (Roh et al. 2002a). This localization of PATJ is however independent of CRB3A or PALS1 localization since their overexpression had no effect on PATJ localization at TJs while it delocalized aPKC (Michel et al. 2005).

How the co-expression of proteins of the CRB3A complex is regulated is still largely unknown, but previous studies have shown that the expression of CRB3A is under the control of the Snail transcription factor while both PALS1 and PATJ are less sensitive to Snail expression (Whiteman et al. 2008). A latter study from the same group showed that CRB3A in addition to being a transcriptional target of Snail is also regulated by this factor at the posttranscriptional level (Harder et al. 2012). CRB3A expression is not directly regulated by PALS1 levels since its knockdown (KD) does not modify CRB3 expression in MDCK cells (Wang et al. 2007). CRB3A expression levels are also not affected by PATJ depletion both in MDCK and Caco-2 cells (Michel et al. 2005; Shin et al. 2005), indicating that these two cytosolic proteins are not essential for CRB3A synthesis and stabilization. In vivo CRB3A expression is however essential for normal levels of PALS1 and PATJ since both proteins are strongly reduced in the lungs of *CRB3-/-* mice (Whiteman et al. 2014), suggesting a non-reciprocal dependence.

PALS1 and PATJ on the other hand show a complex co-regulation that seems to depend on the cell type or organization. In MDCK and Caco2 cells (two highly polarized epithelial cell lines), PATJ KD has no effect on PALS1 expression (Michel et al. 2005; Shin et al. 2005), while in MCF10A cells (weakly polarized epithelial cell line), PATJ KD resulted in a reduction of PALS1 levels (Assemat et al. 2013). PALS1 KD on the other hand strongly decreased PATJ levels both in MDCK (Wang et al. 2007) and MCF10A cells (Assemat et al. 2013). In this latter cell line it was shown that PALS1 depletion induced a shorter half-life for PATJ suggesting that if PATJ is required for PALS1 localization, PALS1 is required for PATJ stabilization. In addition, in the same MCF10A cells, it was shown that there is a balance between the levels of PATJ and MUPP1 and that in the absence of MUPP1, PATJ is stabilized by forming more complexes with PALS1 (Assemat et al. 2013). How general this opposite regulation is remains to be determined.

3.1.3 Role of the Crumbs3 Complex in Cell Polarity and Tight Junction Formation

In *D. melanogaster*, *Crumbs* was first described as a gene essential for the organization of the first epithelial layer at the onset of gastrulation (Tepass et al. 1990). Since the general organization of the primary epithelium and the secretion of the apical cuticle material were severely compromised, Dcrumbs was later associated with other genes producing similar phenotypes such as the PAR complexes and labeled as a polarity protein together with Stardust (Bachmann et al. 2001; Hong et al. 2001; Hurd et al. 2003; Johnson and Wodarz 2003). It is however puzzling that in *C. elegans*, depletion of both *crb-1* and *crb*-like genes does not show any obvious phenotype on epithelial organization or polarity (Bossinger et al. 2001; Shibata et al. 2000) while the other polarity complexes such as the PAR complexes have a conserved function at least in cell polarity besides asymmetric cell divisions (Pellettieri and Seydoux 2002). In fact even in *crb* mutant flies, while the global organization of the primary epithelium is severally affected, there are little evidences that DCrumbs or Stardust are bona fide polarity proteins since basolateral proteins are not found on the apical side of epithelial cells in *sdt* embryos, for example (Hong et al. 2001). Further reexamination of *crb* and *sdt* phenotypes in several species will be necessary to clarify this important issue.

Crumbs and Stardust however are essential for the formation and maintenance of adherens junction (AJ) in flies (for review see Tepass (2012)). This function in junction formation seems to be conserved in vertebrates for CRB3A since it has been shown that exogenous expression of CRB3A in MCF10A human epithelial cells induced the formation of TJs and that both the PDZ- and FERM-binding domains are necessary for this process (Fogg et al. 2005). In addition, overexpression of CRB3A in either MDCK or Caco2 epithelial cells inhibited or delayed TJ formation, thus reinforcing the hypothesis that CRB3A plays a role in TJ assembly or maintenance (Lemmers et al. 2004; Roh et al. 2003). TJ formation and maintenance rely on the integrity of AJ, and one more evidence that the CRB3A/PALS1 complex is involved in AJ regulation was brought by the knockdown (KD) of PALS1 in MDCK cells showing severe TJ and AJ defects and a failure to deliver E-cadherin to the cell surface (Wang et al. 2007). Finally, PATJ KD in MDCK, Caco2, or EpH4 (from mouse mammary gland) consistently leads to defects in TJ formation and sealing in epithelial layers (Adachi et al. 2009; Michel et al. 2005; Shin et al. 2005). Altogether these data strongly indicate a role for the CRB3A/PALS1/PATJ complex in the building and/or stabilization of AJs and TJs in epithelial cells in culture. In addition PALS1, PATJ, and MUPP1 PDZ domains are targets of viruses that induce their degradation or their mislocalization and then defects in TJ (Javier and Rice 2011). It is thus intriguing that in a recent study, a mouse *CRB3-/-* shows no signs of TJ alterations both by immunostaining and electron microscopy in organs such as lungs, kidney, or intestine (Whiteman et al. 2014). The contribution of other Crumbs (namely, 1 and 2) was however not tested, and further work using double-knockout mouse (KO) with the already published *CRB1-/-* (van de Pavert et al. 2004) or *CRB2-/-* (Alves et al. 2013; Xiao et al. 2011) will be necessary to clarify this crucial point. On the other hand, depletion of PALS1 in vivo induced a loss of AJs both in the neuroepithelium of the neural tube in a KO mouse (Kim et al. 2010) and in the outer limiting membrane (OLM) of the retina in a KD mouse (Park et al. 2011). Finally, nagie oko, the homolog of PALS1 in the zebra fish, is also essential for proper epithelial integrity (Bit-Avragim et al. 2008; Zou et al. 2008). These in vivo data confirm that PALS1 is indeed essential for the correct assembly and/or maintenance of AJs.

3.1.4 Role of the Crumbs3 Complex in Cell Migration and Ciliogenesis

Proteins important for epithelial organization such as E-cadherin are known to be essential to prevent carcinogenesis (for review see Canel et al. (2013)). Accordingly, it has been shown that CRB3A expression is lost during in vivo acquisition of tumorigenicity and that restoration of its expression suppressed cell migration and metastasis while rescuing epithelial organization (Karp et al. 2008). This role of CRB3A in epithelial cell migration was however not confirmed using MDCK cells. The same study showed in addition that both PALS1 and PATJ are necessary for correct polarized cell migration (Shin et al. 2007). A recent study using MCF10A cells confirmed that PATJ is essential for polarized cell migration while neither PALS1 nor MUPP1 were needed (Assemat et al. 2013). Thus, more studies using different models in vitro and in vivo will be necessary to clarify the exact role of the CRB3A complex in cell migration.

One very intriguing role of the CRB3 complex concerns its possible implication in primary ciliogenesis. Using antibodies against the C-terminal end (including the ERLI motif), it was shown that CRB3A is detected in the cilia on adult rat kidney sections and in the primary cilium of MDCK and IMCD3 (inner medullary collecting duct) cells that are both of renal origin (Fan et al. 2004). Depletion of CRBA/B from MDCK cells using an RNA interference strategy targeting both isoforms prevented the formation of the primary cilium without disrupting PATJ staining at the TJ level (Fan et al. 2004). In a later study, the same group showed that CRB3B is also detected in the MDCK primary cilium and that its KD with specific shRNA also blocked primary ciliogenesis albeit the re-expression of a tagged version of CRB3B could not reverse this phenotype (Fan et al. 2007). The role of CRB3A in ciliogenesis however was not investigated in that context and remains to be clarified in mammalian cells in culture. It must be noted however that cilia are still present in the *CRB3-/-* mouse kidney or lungs (Whiteman et al. 2014), indicating that CRB3A and B are either not essential or compensated by other Crumbs (1 or 2). Further studies with the *CRB1-/-* (van de Pavert et al. 2004) and *CRB2-/-* (Alves et al. 2013; Xiao et al. 2011) mice will help to investigate the contribution of each Crumbs in ciliogenesis.

In other species such as the zebra fish, Crumbs3a localizes to the basal bodies of auditory hair cells, and its depletion resulted in a shortening of the kinocilium (Omori and Malicki 2006), indicating that indeed Crumbs3 proteins might have a role in ciliogenesis. It has been postulated that this role in ciliogenesis was mediated through the aPKC/PAR6/PAR3 complex since these proteins interact with CRB3A (Hurd et al. 2003; Lemmers et al. 2004) and that PAR3 and PAR6 co-localized with CRB3A in the primary cilium of MDCK cells (Fan et al. 2004). PALS1 and PATJ on the other side have not been reported to accumulate in the cilium, but in a PALS1 mouse KO, there is a strong reduction of the number of basal bodies and cilia in the neural epithelium (Kim et al. 2010). This last finding should lead to revisiting the role of the CRB3A/PALS1 in ciliogenesis both in vitro and in vivo.

3.2 Conclusions and Perspectives

Since the discovery of CRB1, the first vertebrate homolog of DCrumbs in 1999, a considerable amount of work has been devoted to understand the cellular and tissue roles of Crumbs and its partners in epithelial morphogenesis and physiology.

In particular the functions of the CRB3A complex have been uncovered in epithelial cells in culture which, with the development of RNAi, offered suitable systems to investigate cell polarity, junction formation, primary ciliogenesis, and cell division. While it seems that some of these cellular roles might be confirmed in vivo, the recent development of mouse models for CRB1, CRB2, and CRB3 will allow to dissect in vivo the role of CRB3. These mouse models, in conjonction with the use of appropriate cell lines in vitro, will help to further test the link between CRB3 and the subapical cytoskeleton as it has been already shown in flies (Medina et al. 2002). It is thus exciting to see that in the *CRB3-/-* mouse intestinal cells show similar defects as those observed in the Ezrin KO and that CRB3A and Ezrin can form a complex (Whiteman et al. 2014). Further studies will dissect this interaction and its functional and morphological significance. Another promising line of research will be to investigate in vivo the role of CRB3A in cell proliferation and carcinogenesis through the Hippo and the mTOR pathways (Massey-Harroche et al. 2007; Varelas et al. 2010). In addition, the development of a specific mouse model depleted in either CRB3A or B will also allow understanding the respective roles of these two isoforms. The use of the zebra fish model will offer, in addition to the mouse model, the possibility to investigate the role of the CRB3A complex in cell migration and neuronal development. This is essential since it has been described in the mouse that PALS1 plays an important role in the morphogenesis of the myelin sheet in peripheral nerves (Ozcelik et al. 2010). The use of a developmental animal model allowing live imaging such as the zebra fish will be of first importance to better understand the pleiotropic roles of the CRB3A complex that is likely conserved in vertebrates.

Acknowledgments We thank the Le Bivic Lab members for the critical reading of the manuscript. ALB laboratory is supported by the CNRS (UMR7288) and Aix-Marseille University and is an "Equipe labellisée 2008 de la Ligue Nationale contre le Cancer." This work has been carried out in the framework of the Labex INFORM (ANR-11-LABX-0054). B. Vacca is supported by Biotechnology and Biological Sciences Research Council (BBSRC).

References

Adachi M, Hamazaki Y, Kobayashi Y, Itoh M, Tsukita S, Furuse M, Tsukita S (2009) Similar and distinct properties of MUPP1 and PATJ, two homologous PDZ domain-containing tight-junction proteins. Mol Cell Biol 29:2372–2389

Alves CH, Sanz AS, Park B, Pellissier LP, Tanimoto N, Beck SC, Huber G, Murtaza M, Richard F, Sridevi Gurubaran I et al (2013) Loss of CRB2 in the mouse retina mimics human retinitis pigmentosa due to mutations in the CRB1 gene. Hum Mol Genet 22:35–50

Assemat E, Bazellieres E, Pallesi-Pocachard E, Le Bivic A, Massey-Harroche D (2008) Polarity complex proteins. Biochim Biophys Acta 1778:614–630

Assemat E, Crost E, Ponserre M, Wijnholds J, Le Bivic A, Massey-Harroche D (2013) The multi-PDZ domain protein-1 (MUPP-1) expression regulates cellular levels of the PALS-1/PATJ polarity complex. Exp Cell Res 319:2514–2525

Bachmann A, Schneider M, Theilenberg E, Grawe F, Knust E (2001) Drosophila Stardust is a partner of Crumbs in the control of epithelial cell polarity. Nature 414:638–643

Bazellieres E, Assemat E, Arsanto JP, Le Bivic A, Massey-Harroche D (2009) Crumbs proteins in epithelial morphogenesis. Front Biosci 14:2149–2169

Bit-Avragim N, Hellwig N, Rudolph F, Munson C, Stainier DY, Abdelilah-Seyfried S (2008) Divergent polarization mechanisms during vertebrate epithelial development mediated by the Crumbs complex protein Nagie oko. J Cell Sci 121:2503–2510

Bossinger O, Klebes A, Segbert C, Theres C, Knust E (2001) Zonula adherens formation in Caenorhabditis elegans requires dlg-1, the homologue of the Drosophila gene discs large. Dev Biol 230:29–42

Canel M, Serrels A, Frame MC, Brunton VG (2013) E-cadherin-integrin crosstalk in cancer invasion and metastasis. J Cell Sci 126:393–401

Delacour D, Cramm-Behrens CI, Drobecq H, Le Bivic A, Naim HY, Jacob R (2006) Requirement for galectin-3 in apical protein sorting. Curr Biol 16:408–414

den Hollander AI, ten Brink JB, de Kok YJ, van Soest S, van den Born LI, van Driel MA, van de Pol DJ, Payne AM, Bhattacharya SS, Kellner U et al (1999) Mutations in a human homologue of Drosophila crumbs cause retinitis pigmentosa (RP12). Nat Genet 23:217–221

Fahey B, Degnan BM (2010) Origin of animal epithelia: insights from the sponge genome. Evol Dev 12:601–617

Fan S, Fogg V, Wang Q, Chen XW, Liu CJ, Margolis B (2007) A novel Crumbs3 isoform regulates cell division and ciliogenesis via importin beta interactions. J Cell Biol 178:387–398

Fan S, Hurd TW, Liu CJ, Straight SW, Weimbs T, Hurd EA, Domino SE, Margolis B (2004) Polarity proteins control ciliogenesis via kinesin motor interactions. Curr Biol 14:1451–1461

Fogg VC, Liu CJ, Margolis B (2005) Multiple regions of Crumbs3 are required for tight junction formation in MCF10A cells. J Cell Sci 118:2859–2869

Harder JL, Whiteman EL, Pieczynski JN, Liu CJ, Margolis B (2012) Snail destabilizes cell surface Crumbs3a. Traffic 13:1170–1185

Hong Y, Stronach B, Perrimon N, Jan LY, Jan YN (2001) Drosophila Stardust interacts with Crumbs to control polarity of epithelia but not neuroblasts. Nature 414:634–638

Hurd TW, Gao L, Roh MH, Macara IG, Margolis B (2003) Direct interaction of two polarity complexes implicated in epithelial tight junction assembly. Nat Cell Biol 5:137–142

Javier RT, Rice AP (2011) Emerging theme: cellular PDZ proteins as common targets of pathogenic viruses. J Virol 85:11544–11556

Johnson K, Wodarz A (2003) A genetic hierarchy controlling cell polarity. Nat Cell Biol 5:12–14

Kamberov E, Makarova O, Roh M, Liu A, Karnak D, Straight S, Margolis B (2000) Molecular cloning and characterization of Pals, proteins associated with mLin-7. J Biol Chem 275:11425–11431

Karp CM, Tan TT, Mathew R, Nelson D, Mukherjee C, Degenhardt K, Karantza-Wadsworth V, White E (2008) Role of the polarity determinant crumbs in suppressing mammalian epithelial tumor progression. Cancer Res 68:4105–4115

Katoh M, Katoh M (2004) Identification and characterization of Crumbs homolog 2 gene at human chromosome 9q33.3. Int J Oncol 24:743–749

Kempkens O, Medina E, Fernandez-Ballester G, Ozuyaman S, Le Bivic A, Serrano L, Knust E (2006) Computer modelling in combination with in vitro studies reveals similar binding affinities of Drosophila Crumbs for the PDZ domains of Stardust and DmPar-6. Eur J Cell Biol 85:753–767

Kim S, Lehtinen MK, Sessa A, Zappaterra MW, Cho SH, Gonzalez D, Boggan B, Austin CA, Wijnholds J, Gambello MJ et al (2010) The apical complex couples cell fate and cell survival to cerebral cortical development. Neuron 66:69–84

Le Bivic A (2013) Evolution and cell physiology. 4. Why invent yet another protein complex to build junctions in epithelial cells? Am J Physiol Cell Physiol 305:C1193–C1201

Lemmers C, Michel D, Lane-Guermonprez L, Delgrossi MH, Medina E, Arsanto JP, Le Bivic A (2004) CRB3 binds directly to Par6 and regulates the morphogenesis of the tight junctions in mammalian epithelial cells. Mol Biol Cell 15:1324–1333

Makarova O, Roh MH, Liu CJ, Laurinec S, Margolis B (2003) Mammalian Crumbs3 is a small transmembrane protein linked to protein associated with Lin-7 (Pals1). Gene 302:21–29

Massey-Harroche D, Delgrossi MH, Lane-Guermonprez L, Arsanto JP, Borg JP, Billaud M, Le Bivic A (2007) Evidence for a molecular link between the tuberous sclerosis complex and the Crumbs complex. Hum Mol Genet 16:529–536

Medina E, Williams J, Klipfell E, Zarnescu D, Thomas G, Le Bivic A (2002) Crumbs interacts with moesin and beta(Heavy)-spectrin in the apical membrane skeleton of Drosophila. J Cell Biol 158:941–951

Michel D, Arsanto JP, Massey-Harroche D, Beclin C, Wijnholds J, Le Bivic A (2005) PATJ connects and stabilizes apical and lateral components of tight junctions in human intestinal cells. J Cell Sci 118:4049–4057

Muller HA (2000) Genetic control of epithelial cell polarity: lessons from Drosophila. Dev Dyn 218:52–67

Nam SC, Choi KW (2003) Interaction of Par-6 and Crumbs complexes is essential for photoreceptor morphogenesis in Drosophila. Development 130:4363–4372

Omori Y, Malicki J (2006) oko meduzy and related crumbs genes are determinants of apical cell features in the vertebrate embryo. Curr Biol 16:945–957

Ozcelik M, Cotter L, Jacob C, Pereira JA, Relvas JB, Suter U, Tricaud N (2010) Pals1 is a major regulator of the epithelial-like polarization and the extension of the myelin sheath in peripheral nerves. J Neurosci 30:4120–4131

Park B, Alves CH, Lundvig DM, Tanimoto N, Beck SC, Huber G, Richard F, Klooster J, Andlauer TF, Swindell EC et al (2011) PALS1 is essential for retinal pigment epithelium structure and neural retina stratification. J Neurosci 31:17230–17241

Pellettieri J, Seydoux G (2002) Anterior-posterior polarity in C. elegans and Drosophila – PARallels and differences. Science 298:1946–1950

Roh MH, Fan S, Liu CJ, Margolis B (2003) The Crumbs3-Pals1 complex participates in the establishment of polarity in mammalian epithelial cells. J Cell Sci 116:2895–2906

Roh MH, Liu CJ, Laurinec S, Margolis B (2002a) The carboxyl terminus of zona occludens-3 binds and recruits a mammalian homologue of discs lost to tight junctions. J Biol Chem 277:27501–27509

Roh MH, Makarova O, Liu CJ, Shin K, Lee S, Laurinec S, Goyal M, Wiggins R, Margolis B (2002b) The Maguk protein, Pals1, functions as an adapter, linking mammalian homologues of Crumbs and Discs Lost. J Cell Biol 157:161–172

Shibata Y, Fujii T, Dent JA, Fujisawa H, Takagi S (2000) EAT-20, a novel transmembrane protein with EGF motifs, is required for efficient feeding in Caenorhabditis elegans. Genetics 154:635–646

Shin K, Straight S, Margolis B (2005) PATJ regulates tight junction formation and polarity in mammalian epithelial cells. J Cell Biol 168:705–711

Shin K, Wang Q, Margolis B (2007) PATJ regulates directional migration of mammalian epithelial cells. EMBO Rep 8:158–164

Sotillos S, Diaz-Meco MT, Caminero E, Moscat J, Campuzano S (2004) DaPKC-dependent phosphorylation of Crumbs is required for epithelial cell polarity in Drosophila. J Cell Biol 166:549–557

Tepass U (2012) The apical polarity protein network in Drosophila epithelial cells: regulation of polarity, junctions, morphogenesis, cell growth, and survival. Annu Rev Cell Dev Biol 28:655–685

Tepass U, Theres C, Knust E (1990) crumbs encodes an EGF-like protein expressed on apical membranes of Drosophila epithelial cells and required for organization of epithelia. Cell 61:787–799

Ullmer C, Schmuck K, Figge A, Lubbert H (1998) Cloning and characterization of MUPP1, a novel PDZ domain protein. FEBS Lett 424:63–68

van de Pavert SA, Kantardzhieva A, Malysheva A, Meuleman J, Versteeg I, Levelt C, Klooster J, Geiger S, Seeliger MW, Rashbass P et al (2004) Crumbs homologue 1 is required for maintenance of photoreceptor cell polarization and adhesion during light exposure. J Cell Sci 117:4169–4177

Varelas X, Samavarchi-Tehrani P, Narimatsu M, Weiss A, Cockburn K, Larsen BG, Rossant J, Wrana JL (2010) The Crumbs complex couples cell density sensing to Hippo-dependent control of the TGF-beta-SMAD pathway. Dev Cell 19:831–844

Wang Q, Chen XW, Margolis B (2007) PALS1 regulates E-cadherin trafficking in mammalian epithelial cells. Mol Biol Cell 18:874–885

Watanabe T, Miyatani S, Katoh I, Kobayashi S, Ikawa Y (2004) Expression of a novel secretory form (Crb1s) of mouse Crumbs homologue Crb1 in skin development. Biochem Biophys Res Commun 313:263–270

Whiteman EL, Fan S, Harder JL, Walton KD, Liu CJ, Soofi A, Fogg VC, Hershenson MB, Dressler GR, Deutsch GH et al (2014) Crumbs3 is essential for proper epithelial development and viability. Mol Cell Biol 34:43–56

Whiteman EL, Liu CJ, Fearon ER, Margolis B (2008) The transcription factor snail represses Crumbs3 expression and disrupts apico-basal polarity complexes. Oncogene 27:3875–3879

Wodarz A, Hinz U, Engelbert M, Knust E (1995) Expression of crumbs confers apical character on plasma membrane domains of ectodermal epithelia of Drosophila. Cell 82:67–76

Xiao Z, Patrakka J, Nukui M, Chi L, Niu D, Betsholtz C, Pikkarainen T, Vainio S, Tryggvason K (2011) Deficiency in Crumbs homolog 2 (Crb2) affects gastrulation and results in embryonic lethality in mice. Dev Dyn 240:2646–2656

Yeaman C, Le Gall AH, Baldwin AN, Monlauzeur L, Le Bivic A, Rodriguez-Boulan E (1997) The O-glycosylated stalk domain is required for apical sorting of neurotrophin receptors in polarized MDCK cells. J Cell Biol 139:929–940

Zou J, Lathrop KL, Sun M, Wei X (2008) Intact retinal pigment epithelium maintained by Nok is essential for retinal epithelial polarity and cellular patterning in zebrafish. J Neurosci 28:13684–13695

Chapter 4
The Scribble–Dlg–Lgl Module in Cell Polarity Regulation

Patrick O. Humbert, Sarah M. Russell, Lorey Smith, and Helena E. Richardson

P.O. Humbert (✉)
Cell Cycle and Cancer Genetics Laboratory, Research Division, Peter MacCallum Cancer Centre, Locked Bag 1, A'Beckett Street, Melbourne, VIC 8006, Australia

Sir Peter MacCallum Department of Oncology, The University of Melbourne, Parkville, Australia

Department of Pathology, The University of Melbourne, Parkville, Australia

Department of Biochemistry and Molecular Biology, The University of Melbourne, Parkville, Australia
e-mail: patrick.humbert@petermac.org

S.M. Russell
Immune Signalling Laboratory, Peter MacCallum Cancer Centre, Locked Bag 1, A'Beckett Street, Melbourne, VIC 8006, Australia

Sir Peter MacCallum Department of Oncology, The University of Melbourne, Parkville, Australia

Department of Pathology, The University of Melbourne, Parkville, Australia

Centre for Micro-Photonics, Faculty of Science, Engineering and Technology, Swinburne University of Technology, Hawthorn, Australia

L. Smith
Cell Cycle and Cancer Genetics Laboratory, Research Division, Peter MacCallum Cancer Centre, Locked Bag 1, A'Beckett Street, Melbourne, VIC 8006, Australia

Department of Pathology, The University of Melbourne, Parkville, Australia

H.E. Richardson
Cell Cycle and Development Laboratory, Peter MacCallum Cancer Centre, Locked Bag 1, A'Beckett Street, Melbourne, VIC 8006, Australia

Sir Peter MacCallum Department of Oncology, The University of Melbourne, Parkville, Australia

Department of Biochemistry and Molecular Biology, The University of Melbourne, Parkville, Australia

Department of Anatomy and Cell Biology, The University of Melbourne, Parkville, Australia

K. Ebnet (ed.), *Cell Polarity 1*, DOI 10.1007/978-3-319-14463-4_4

Abstract Although the Scribble polarity module has long been known as a key regulator of apicobasal polarity, it is only recently that its broader role in the control of near all polarity states and transitions is being appreciated. Here we review the Scribble module in the regulation of cell polarity and other cellular functions at the molecular and cellular level. The more recent detailed analysis of multiple vertebrate models for each of its component homologues, Scribble, Dlg and Lgl, has revealed specific but also common roles for individual homologues in a variety of developmental contexts. In addition, emerging data has also implicated the Scribble polarity module in human developmental syndromes and the etiology of human cancer, highlighting a need for a better understanding of this polarity module for therapeutic purposes. Unlocking the temporal and spatial coordination of the myriad interactions that these signaling scaffolds regulate is a major challenge for the field and will be key to resolve the function of Scribble, Dlg, and Lgl in the control of cell polarity and tissue architecture.

Keywords Asymmetric cell division • Cancer • Cell competition • Lymphocytes • Migration • Planar cell polarity • Protein localization

4.1 Introduction

Cell polarity can be defined as an asymmetric distribution of cellular protein, lipids, and organelles and is a fundamental property of all eukaryotic cells (Nelson 2003; Macara and Mili 2008; St Johnston and Ahringer 2010). Cell polarity appears in various different guises in multicellular organisms that include apical–basal polarity (ABP), a common form of polarity involved in epithelial tissue that allows spatial segregation of cellular components between lumen and the basement membrane; planar cell polarity (PCP), also known as tissue polarity, which is the coordinated asymmetries seen across multiple cells within a tissue sheet or organ; front–rear polarity (FRP), observed in migrating cells such as immune cells or in wound healing; asymmetric cell division (ACD), where spindle orientation and asymmetric cargo distribution during cell division is utilized to confer cell fate; as well as a number of cell specific polarities including immune synapse formation and the polarity seen in neuronal dendrite formation (Nelson 2003; Dow and Humbert 2007; St Johnston and Ahringer 2010; Rodriguez-Boulan and Macara 2014). A number of key core protein modules have been implicated in the regulation of this asymmetry at a cellular level and include the Par protein complex (see Ohno et al., Vol. 1, Part 1, Chap. 1), Crumbs complex (see Le Bivic et al., Vol. 1, Part 1, Chap. 3), the Pins complex (see Wodarz et al., Vol. 2, Part 2, Chap.1, and van den Heuvel et al., Vol. 2, Part 2, Chap. 2), and the subject of this chapter, the Scribble polarity module. Here we will focus on the role of Scribble, Dlg, and Lgl in cell polarity with an emphasis on epithelial systems. We direct the reader to excellent supporting reviews for further reading (Roberts et al. 2012; Grifoni et al. 2013; Su et al. 2012; Elsum et al. 2012; Yamanaka and Ohno 2008; Humbert et al. 2008).

4.2 The Scribble Polarity Module: Structure, Partners, and Localization

The Scribble polarity module consists of three distinct proteins, namely, Scribble (Scrib), discs large (Dlg), and lethal giant larvae (Lgl) that were first described as apicobasal polarity regulators and neoplastic tumor suppressors in *Drosophila melanogaster* (reviewed in Bilder 2004; Humbert et al. 2003). We will refer to these proteins as a module rather than complex, since although well-defined physical interactions occur between proteins of the Par and Crumbs polarity complexes, it is less clear for the Scribble polarity module. Of note Scribble, Dlg, and Lgl proteins play highly conserved roles in the regulation of cell polarity in multicellular organisms and have been found in the majority of metazoans examined to date from *Trichoplax adhaerens* (an organism only consisting of two epitheloid layers and lacking any organs or internal structures) to *Homo sapiens* either as single copy genes or duplicated gene sets (see OrthoDb; Waterhouse et al. 2013). Components of the Scribble polarity module are thought to operate as site-specific protein scaffolds that can regulate temporal and localized cell signaling to enable the establishment and maintenance of cell polarity in various epithelia and cell types. The major physical features that define and influence their function therefore include their individual conformational structure, their intermolecular interactions within the module, and how the physical link with other protein complexes and signaling pathways is regulated. Importantly, in view of their critical role in establishing and maintaining asymmetry within cells through local positioning of signaling pathway components, the regulation of their cellular localization is central to their function.

4.2.1 Structural Determinants of Scribble, Dlg, and Lgl

The three core members of the Scribble polarity module are large proteins that are rich in protein–protein interaction domains (see Fig. 4.1 for a schematic representation). Scribble is a LAP family protein (*L*RR *a*nd *P*DZ) that contains an LRR domain (leucine-rich repeat), a protein recognition motif that consists of repeated 20–30 amino acid stretches that give rise to a "horseshoe"-shaped domain (Kobe and Kajava 2001), and four PDZ (*P*SD-95/*D*lg/*Z*O-1 homology) domains, regions of 80–90 amino acids that are common protein motifs and mediate protein–protein interactions generally through binding to peptide motifs at the C-terminal end of proteins (Santoni et al. 2002; Nourry et al. 2003). Discs large (Dlg) is a member of the MAGUK family of proteins and contains 3 PDZ domains as well as Src homology 3 (SH3) and guanylate kinase-like (GUK) domains (Funke et al. 2005) (Fig. 4.1). The GUK domain of MAGUK proteins is catalytically inactive due to the absence of an ATP-binding site and is thought to regulate protein interactions (Funke et al. 2005), while the SH3 domain is a protein–protein interaction domain

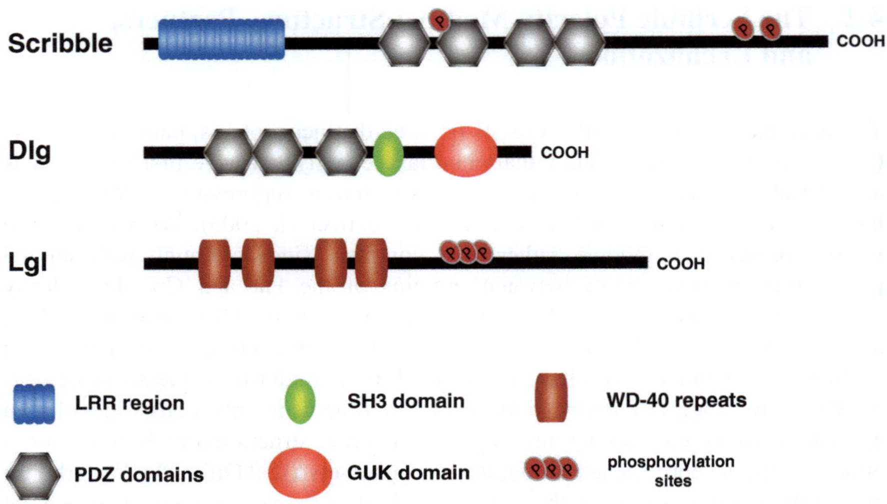

Fig. 4.1 *Structural representation of Scribble, Dlg, and Lgl*. Scribble is a member of the LAP protein consisting of 16 LRRs and four PDZ domains. Dlg is a MAGUK protein and contains three PDZ domains, an SH3 domain, and a GUK domain. Lgl contains a number of WD-40 repeat domains as well as conserved phosphorylation sites. Each of the polarity proteins are structurally highly conserved in all metazoa

that is selective for proline-rich ligands and can enhance the assembly of large multiprotein complexes (Mayer 2001). Lgl contains four WD40 repeats that are structural motifs composed of approximately 40–60 amino acids that usually end with a tryptophan–aspartic acid (WD) dipeptide and assemble in four-stranded antiparallel sheet conformation (Stirnimann et al. 2010) (Fig. 4.1). Of note, WD40-containing proteins often regulate the assembly of macromolecular complexes and invariably appear to lack enzymatic activities (Stirnimann et al. 2010).

4.2.2 Interaction Partners

As proteins with multiple protein–protein interaction domains, it is not surprising that members of the Scribble polarity module display a wide variety of protein interactions that allow control of the concentration, localization, and composition of macromolecular complexes. Although it is beyond the scope of this review to provide a detailed list of all the interactions that have so far been described for each of Scribble, Dlg, and Lgl family members, these interactions are wide and varied and range from transmembrane receptors, enzymes, and transcription factors to cytoskeletal proteins with many of these interactions hijacked by viral proteins.

At the core of the Scribble polarity module, intermolecular interactions have been described in mammals between Scribble and Lgl (Kallay et al. 2006) and

Dlg and Lgl (Zhu et al. 2014). Importantly, the Dlg interaction with Lgl is regulated and requires phosphorylation of one of the three conserved Serine residues of the central linker region of Lgl (Zhu et al. 2014). Furthermore, in *Drosophila*, the protein GUK-holder (NHS in vertebrates) serves to scaffold Scribble and Dlg (Mathew et al. 2002). This interaction is likely to be important in vertebrates as Scribble can interact both genetically and physically with NHS to regulate neuronal migration in zebrafish, while zebrafish Dlg also binds to NHS (Walsh et al. 2011).

The Scribble polarity module can further interact directly with other core cell polarity complexes in a wide variety of cellular and developmental contexts. This includes planar cell polarity (PCP) components with Scribble binding physically to PCP protein Vangl2, Dlg with Frizzled receptors, and Lgl with Dishevelled, as well as components of the asymmetric cell division (ACD) complex, with Dlg and Scribble being required for Pins/Gpsm2 protein localization in *Drosophila* during ACD developmental decisions (reviewed in Humbert et al, 2008; Bergstralh et al. 2013). Another critical inter-complex regulation is that of Lgl and aPKC which provides an important mutual antagonism between the Scribble module and the PAR complex. Thus, the physical interaction of Lgl with aPKC can regulate aPKC activity, and in turn phosphorylation of Lgl by aPKC can regulate Lgl localization and function in a wide variety of cell types and organisms (see Humbert et al. 2006). The Scribble polarity module can also physically interact with regulators of cytoskeleton and junctional complexes such as APC and β-catenin (Scribble, Dlg), E-cadherin (Scribble), and Myosin II (Lgl) (reviewed in Humbert et al. 2008; Roberts et al. 2012; Grifoni et al. 2013).

The major set of Scribble polarity module interacting proteins are the regulators of intracellular signaling. The broad range of interactions described so far attests to this, with both Scribble and Dlg being able to bind and localize a wide variety of signaling components from the PTEN/PI3K, MAPK, Wnt, and Hippo pathways as well as GPCR, neurotransmitter, and RhoGTPase signaling (Humbert et al. 2008). The functional implication of these interactions will be discussed in more detail below. Finally, both mammalian Dlg and Scribble are also targeted by viral oncoproteins such as the E6 protein of the herpes simplex virus (HPV) and human T cell lymphotropic virus (HTLV)-1 Tat protein that can lead to their abnormal stability, localization, and/or interactions with normal endogenous partners (Banks et al. 2012; Javier and Rice 2011). Because of the large interest in the specific regulation of protein–protein interaction networks by common protein motifs such as PDZ, LRR, and WD40, novel interactions are also continually being revealed for Scribble, Dlg, and Lgl through the use of genome-wide interaction studies such as yeast-two-hybrid, proteomic mass spectrometry, and phage display screening (see Ivarsson et al. 2014; Belotti et al. 2013; Tonikian et al. 2008; Stiffler et al. 2007). The challenge will be to determine which of these myriad of interactions are critical in any given context and how the Scribble polarity module can discriminate and coordinate these interactions at any given time or location within the cell to establish and maintain various polarity states and to regulate signaling pathways.

4.2.3 Localization

Although some information is now becoming available on the transcriptional regulation of individual Scribble polarity module members (e.g., Scribble: Wells et al. 2010; Vaira et al. 2012; Warzecha et al. 2010), by far the most important mode of regulation of the Scribble polarity module is through the regulation of the localization of its individual protein components. Although Scribble, Dlg, and Lgl colocalize to the basolateral membrane of epithelial cells, Lgl can also show substantial cytoplasmic staining, and the various isoforms of human Dlg can also show differential cytoplasmic/membrane localization in epithelial cells (Humbert et al. 2008; Roberts et al. 2012; Grifoni et al. 2013; Van Campenhout et al. 2011). Both LRR and PDZ domains appear to be required for proper localization of mammalian Scribble in epithelial cells with the LRR domain being key for localization in flies and worms (Navarro et al. 2005; Zeitler et al. 2004; Legouis et al. 2003). Importantly, mislocalization alleles of Scribble in vivo lead to loss of function-like phenotypes in *C. elegans* (Legouis et al. 2003), *Drosophila* (Zeitler et al. 2004), mice (Zarbalis et al. 2004), and likely human germline mutant patients (Robinson et al. 2012), highlighting the importance of proper localization of Scribble for its activity. Furthermore, Scribble mislocalization is prevalent in human epithelial cancers and is associated with poorer outcome in prostate and breast cancer (Pearson et al. 2011; Feigin et al. 2014).

Localization may be further regulated through phosphorylation or altered interactions of Scribble, Dlg, or Lgl with their binding partners. Indeed, phosphorylation of Scribble following CD74 amplification in epithelial cancer leads to its phosphorylation and mislocalization to the cytosol (Metodieva et al. 2013), Dlg phosphorylation by JNK can regulate its localization in mammalian cells (Massimi et al. 2006), and Lgl is hyperphosphorylated and mislocalized in glioblastoma in a PTEN-dependent manner (Gont et al. 2013). The regulation of Lgl function by phosphorylation is well documented, whereby phosphorylation by aPKC causes conformational changes in Lgl that leads to its dissociation from the cell cortex (Grifoni et al. 2013). Studies in mammalian cells have also indicated phosphorylation-dependent localization of Dlg in a cell cycle-dependent manner (Narayan et al. 2009) with functional studies in *Drosophila* identifying that phosphorylation by Par1 kinase and CaMKII regulates Dlg localization at the synapse of neuromuscular junctions (Zhang et al. 2007; Koh et al. 1999). The localization of the Scribble polarity module can also be further regulated by osmotic stress and through alterations in vesicle transport (Massimi et al. 2008). Finally both Scribble and Dlg interactions with binding partners such as E-cadherin or by viral oncoproteins may alter their localization at the basolateral surface (Navarro et al. 2005; Reuver and Garner 1998; Nakagawa and Huibregtse 2000; Elsum et al. 2012).

4.3 Animal Models of Scribble, Dlg, and Lgl Function

The study of the Scribble polarity module has been facilitated through the careful functional examination of these proteins in a number of diverse animal models. Although distinct phenotypes have been observed in different organisms as outlined below, the evidence for a conserved role for Scribble, Dlg, and Lgl paralogues across various polarity states and transitions is overwhelming and demonstrates a central function for the Scribble polarity module in the control of tissue morphogenesis, organ development, and tumor suppression in multicellular organism. The wide variety of settings utilized to study the Scribble module also bodes well for rapid progress in the field and has allowed approaches as disparate as rapid genetic screens to detailed in vivo imaging to bring to light their molecular and cellular functions.

4.3.1 Drosophila melanogaster

The first model to study and describe the Scribble polarity module, and by far the best understood to date, is *Drosophila melanogaster*, the common fruit fly. Loss of Scribble, Dlg, or Lgl in the whole *Drosophila* organism gives rise to similar phenotypes, including overproliferation and loss of cellular junctions and results in the disorganization of rapidly growing epithelial larval tissues known as imaginal discs (hence, "Discs large" phenotype) (Woods and Bryant 1989; Gateff and Schneiderman 1974; Mechler et al. 1985; Bilder and Perrimon 2000; Bilder et al. 2000). This leads to the inability to pupate, continuous overgrowth, and the eventual death at the larval stage ("Lethal giant larvae" phenotype). In addition, concomitant loss of maternal stores of Scribble, Dlg, or Lgl leads to embryonic lethality due to loss of organization of the embryonic epithelium and the associated abnormal cuticle deposition (the cuticle looking like a scribble under the microscope, hence "Scribble" phenotype) (Bilder and Perrimon 2000; Bilder et al. 2000). The Scribble polarity module has further been implicated in a wide variety of developmental contexts and systems as diverse as dorsal closure, neural stem cell differentiation, neuronal synapse formation, and ovarian follicle cell migration (see Humbert et al. 2008; Elsum et al. 2012). The similarity of the mutant phenotypes of Scribble, Dlg, or Lgl, together with the genetic interaction between loss-of-function alleles, was the first indication that these three proteins form a core polarity module (Bilder et al. 2000). Subsequent experiments using clonal analysis, whereby clones of Scribble, Dlg, and Lgl mutant tissue can be generated within a normal epithelium, also demonstrated that loss of function of these genes leads to disorganization of the tissue and JNK-dependent apoptosis and clearance of the tissue in vivo (Brumby and Richardson 2003; Pagliarini and Xu 2003). Concomitant activation of the Ras or Notch signaling pathway could rescue this apoptosis and results in expansion of clonal tumors within the tissue (Brumby and Richardson 2003).

This 2-step model has become a gold standard for the study of oncogene-driven tumorigenesis in *Drosophila* and has recently been harnessed for the rapid screening of anticancer compounds (Gladstone and Su 2011; Gonzalez 2013; Willoughby et al. 2012). Because of the ease of genetic manipulation and the decreased redundancy compared to more complex genomes such as those of mammals, *Drosophila* has been a critical animal model to examine the molecular mechanisms by which the Scribble module regulates cell polarity.

4.3.2 Caenorhabditis elegans

C. elegans has a single orthologue for each of the members of the Scribble polarity module known as Let-413 (Scribble), DLG-1, and LGL-1. Mutation in *let-413* or *dlg-1* results in disruption of junction formation, with mutant embryos arresting during the elongation stage of development (Legouis et al. 2000; Koppen et al. 2001; McMahon et al. 2001; reviewed in Armenti and Nance 2012). Despite these similarities, Let-413 and DGL-1 appear to have divergent molecular functions in *C. elegans*. DLG-1 regulates junction formation through AJM- 1, a protein only found in *C. elegans*, with only minor effects on apicobasal polarity, whereas *let-413* mutants have gross expansion of apical proteins basolaterally and a more severe failure of junction maturation (Legouis et al. 2000; Koppen et al. 2001; McMahon et al. 2001; Firestein and Rongo 2001). *lgl-1* mutants have no defects in junction formation or epithelial polarity; however LGL-1 functions redundantly with Par-2 in the maintenance of anterior–posterior polarity in the early embryo (Hoege et al. 2010; Beatty et al. 2010; Fichelson et al. 2010). Interestingly, no members of the Scribble polarity module have so far been implicated in proliferation control in *C. elegans*.

4.3.3 Danio rerio

Analysis of zebrafish models has revealed roles for the Scribble polarity module in the development of epithelial organs and includes functions in the control of migration, planar cell polarity (PCP), apicobasal polarity, as well as asymmetric cell division. Initial studies using the *landlocked* (Scribble) mutant and gene depletion using morpholinos ("morphants") demonstrated a dual role in the migration of facial branchiomotor neurons and in the convergence and extension movements that occur during gastrulation, suggesting a role in PCP signaling (Wada et al. 2005). As with other PCP mediators, Scribble was also required for pronephros development with morphants displaying kidney cysts (Skouloudaki et al. 2009). In different settings, Scribble was shown to play a role in developmental angiogenesis (Michaelis et al. 2013), as well as being required in the neural keel, with Scribble morphants exhibiting neural tube closure defects due to abnormal oriented cell divisions (Zigman et al. 2011). In contrast to Scribble, the tumor suppressive function of Lgl appears to be conserved in zebrafish, as loss of *llgl2* was sufficient to induce the formation of epidermal tumors in a cell-autonomous manner (Reischauer et al. 2009). Llgl2 mutant zebrafish (*Penner* mutant) show defects in tissue architecture of the epidermis that can lead to epithelial-to-mesenchymal transition (EMT) of basal epidermal cells (Reischauer et al. 2009; Sonawane et al. 2005, 2009). Of note, zebrafish Lgl was shown to be required for the proper architecture of the zebrafish neuromast, a mechanoreceptive organ consisting of a central hair cell bundle that allows the sensing of mechanical changes in water (Hava et al. 2009). Llgl2 is also required for the development of other ciliated epithelial organs in zebrafish including Kupffer's vesicle, otic vesicles, and the pronephric ducts of the kidney, with Llgl2 depletion resulting in disruption of lumen formation, apicobasal polarity, and cilia formation (Tay et al. 2013). In contrast to Llgl2, Llgl1 loss leads to defective neurogenesis, associated with defects in cell cycle exit and apical membrane domain expansion of retinal neuroepithelial progenitor cells (Clark et al. 2012). Although Dlg1-4 orthologues are present in zebrafish, only expression data have been reported to date (Meyer et al. 2005).

4.3.4 Mus musculus

Mice homozygous for Scribble mutant alleles, including *circletail (Crc)*, and the ENU mutants *rumpelstiltzchen (rumz)*, *crn2*, and *5120-6B*, or an engineered null mouse, all display multiple failures of developmental migration processes resulting in severe neural tube closure defects (*craniorachischisis*), abdominal wall defects, and an "eyes-open-at-birth" phenotype, which are all phenotypes characteristic of defects in PCP mutants (Murdoch et al. 2003; Zarbalis et al. 2004; Wansleeben et al. 2011; Stottmann et al. 2011; Pearson et al. 2011). Interestingly, mutations in Scribble are associated in humans with the most severe type of neural tube closure defects, craniorachischisis, as well as spina bifida (Robinson et al. 2012; Lei et al. 2013). Consistent with the phenotypes found in PCP mutant mice, Scribble loss also resulted in disorganization of the stereociliary bundles in the cochlea (Montcouquiol et al. 2003). Scribble has been shown to be required for the development of a number of other organ systems in vivo, including the lung and gonads, as well as the cardiovascular system, with *Crc* mutant mice displaying heart malformations and impaired angiogenesis (Yates et al. 2013; Murdoch et al. 2003; Zarbalis et al. 2004; Pearson et al. 2011; Phillips et al. 2007; Michaelis et al. 2013). To bypass the embryonic lethality of loss of Scribble, conditional deletion strategies have recently been utilized and revealed additional requirements for Scribble in a number of epithelial compartments. For example, specific loss of Scribble in the corneal epithelium and lens is associated with epithelial-to-mesenchymal transition and results in cataracts and lens malformation (Yamben et al. 2013). Similarly loss of Scribble in the prostate or mammary gland leads to abnormal ductal structures including loss of tissue architecture, disruption of AB polarity, multilayering, and hyperproliferation (Pearson et al. 2011; Godde et al. 2014).

With 4 paralogues of Dlg in the mouse (Dlg1-4), comparative analysis of mouse models to examine individual gene function has become possible. Because of their initial characterization as major controllers of receptor density and signal strength at neuronal synapses, much work has focused on the role of Dlg paralogues in brain function (see Nithianantharajah et al. 2013). Dlg2 (PSD93), Dlg3 (SAP102), and Dlg4 (PSD-95) knockout mice are viable, display a variety of behavioral defects and abnormal cognitive functions (McGee et al. 2001; Cuthbert et al. 2007; Migaud et al. 1998), and may provide good models to study human mental disorders, as mutations in *DLG2* and *DLG3* in humans have been associated with schizophrenia and mental retardation (Nithianantharajah et al. 2013; Walsh et al. 2008; Tarpey et al. 2004). More recent work has revealed direct roles for mouse Dlg homologues in cell polarity and developmental processes. In contrast to other family members, Dlg1 (SAP97) mutant mice are neonatal lethal, displaying craniofacial defects and kidney, urogenital, and lens development defects (Caruana and Bernstein 2001; Naim et al. 2005; Mahoney et al. 2006; Iizuka-Kogo et al. 2007; Rivera et al. 2009). More recently, Dlg1 null mice were shown to also display defects in skeletogenesis of the trunk and limb structures and, importantly, similarly to loss of Scribble, defects in neural tube, eyelid closure, and in the disorganization of the stereociliary bundles in the cochlea (Rivera et al. 2013). These studies indicate a novel role for Dlg1 in PCP signaling. Dlg1 loss in the mouse also impacts on non-epithelial tissues such as T lymphocytes (Stephenson et al. 2007; Humphries et al. 2012; Gmyrek et al. 2013). Interestingly, closer scrutiny of Dlg3 mice showed low penetrance embryonic lethality, associated with open brain phenotype and cochlear inner ear cell defects, also suggesting a role in PCP signaling (Van Campenhout et al. 2011).

As in zebrafish, mice have two paralogues for Lgl: Llgl1 and Llgl2. Llgl1 null mice develop severe brain dysplasia and die at birth from extensive hydrocephalus (Klezovitch et al. 2004). In addition, Lgl1 null mice show defects in the development of the retina (Clark et al. 2012), and loss of Llgl1 enhances hematopoietic cell numbers and activity (Heidel et al. 2013). In contrast to zebrafish, Llgl2 null mice are viable and show no gross phenotypes, albeit Llgl2 appears to be necessary for branching morphogenesis during placental development (Sripathy et al. 2011). In view of the ubiquitous expression of Llgl1 in the mouse, these findings have been interpreted as support for redundancy between Lgl1 and Lgl2 in vivo. There is no current evidence of increased tumorigenesis in mice carrying one or more null alleles for *Llgl1* or *Llgl2* (Sripathy et al. 2011) (Hawkins et al. 2014).

4.4 Regulation of Cell Polarity States by Scribble, Dlg and Lgl

A general function of the Scribble polarity module is to orchestrate cell behaviors such as cell migration, tissue organization, and fate decisions of the cells and their progeny, often by ensuring that cellular behavior is responsive to the environment. The Scribble module frequently exerts this control by compartmentalizing signaling complexes within the cell. Such intracellular compartmentalization can often, but not always, involve the establishment of and recruitment to membrane domains that are required for cell–cell or cell–substrate interactions. The adaptor function of the Scribble module proteins is particularly important for these activities, allowing large complexes of surface receptors and signaling molecules to be linked to each other and to regulate signaling responses to the extracellular environment. A particular attribute of the Scribble module, that of mutual antagonism between it and the PAR complex to facilitate their assembly in different regions of the cell, means that both direct and indirect interactions of Scribble can control cellular behavior.

4.4.1 Apicobasal Polarity

The *raison d'être* of epithelial cells, that is, lining a tissue to ensure protection of the tissue and to control exit and entry of molecules and cells, almost invariably dictates that the cells must adopt apicobasal polarity, such that the side of the cell contacting the tissue is molecularly distinct from the opposite (luminal) side and must form adherens junctions with which to connect with each other along their lateral sides. The Scribble module is essential for the adoption of apicobasal polarity and the formation and maintenance of adherens junctions. Scribble module proteins are recruited to and define the basolateral region of the epithelial cell, through antagonistic interactions with the PAR complex (and Crumbs complex in some instances) (Rodriguez-Boulan and Macara 2014) (Fig. 4.2a).

4.4.2 Planar Cell Polarity

Mouse studies initially revealed an important role for Scribble in noncanonical Wnt/PCP signaling (Fig. 4.2b). Mice mutant for Scribble display the characteristic PCP mutant phenotype, namely, neural tube closure defects, cochlear hair cell orientation defects, and "eyes-open-at-birth" phenotypes (Montcouquiol et al. 2003; Murdoch et al. 2003; Zarbalis et al. 2004; Pearson et al. 2011; Wansleeben et al. 2011; Stottmann et al. 2011). In addition, Scribble interacts both physically and genetically with the core PCP protein Vangl2 during neural

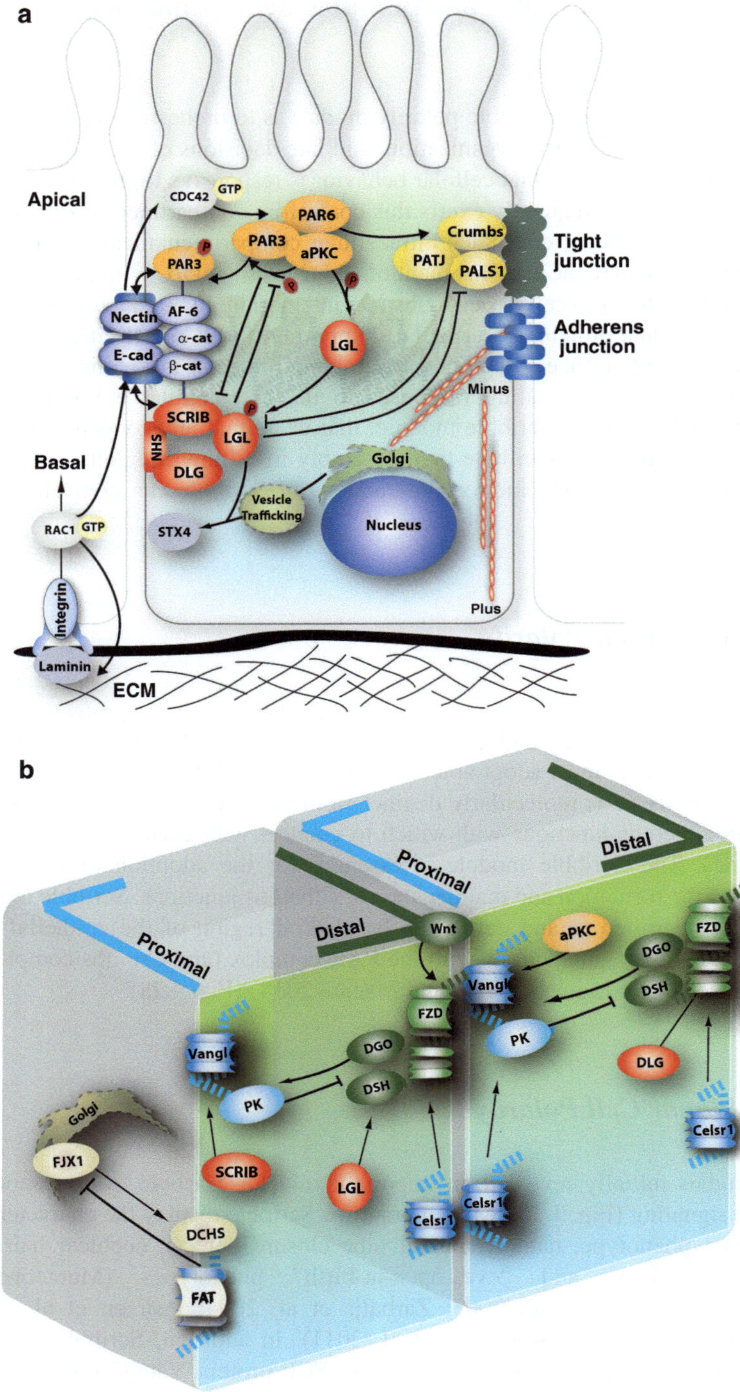

Fig. 4.2 (Continued)

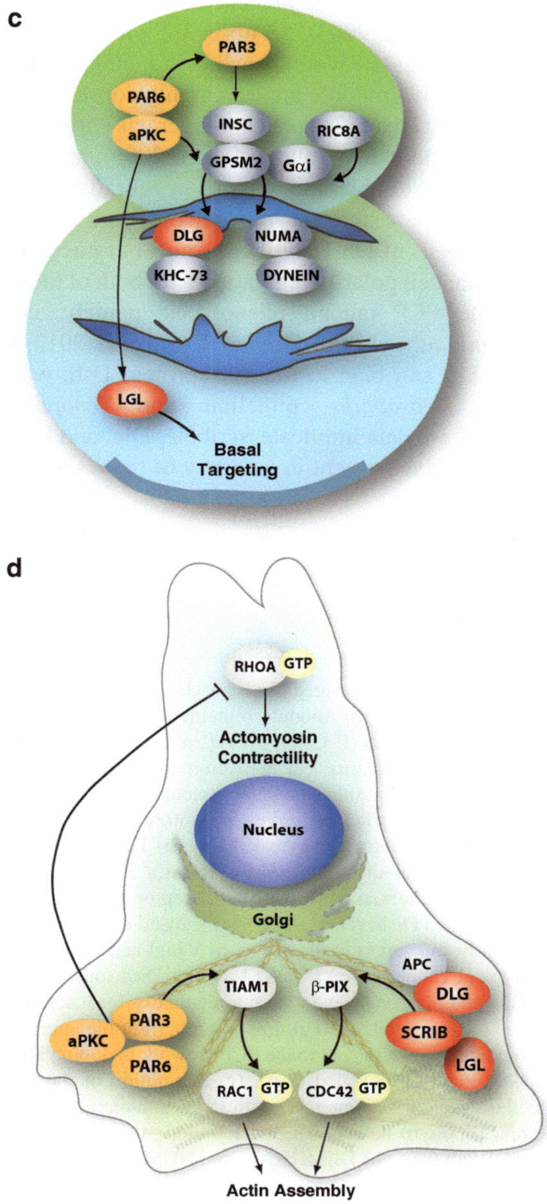

Fig. 4.2 *Regulation of cell polarity states by the Scribble polarity module.* (**a**) Epithelial cells are polarized along their apicobasal axis through the action of three core polarity complexes. The PAR complex (PAR3, PAR6, aPKC) (shown in *orange*) is localized to the apical domain where it promotes the apical recruitment and activity of the Crumbs complex (CRB, PALS1, PATJ) (shown in *yellow*). Both complexes act through mutual antagonistic interactions to maintain basolateral localization of the Scribble complex (Scribble, Dlg, and Lgl) (shown in *red*, with adaptor NHS;

tube closure and in the development of branched structures such as the lung (Yates et al. 2013; Montcouquiol et al. 2003). As observed with other PCP genes such as Vangl2, Scribble can also regulate wound healing in mammals (Caddy et al. 2010). The PCP role for the Scribble module appears highly conserved across species. In *Drosophila*, Scribble physically and genetically interacts with Vangl2 to direct PCP processes in the eye and the wing (Courbard et al. 2009), and similar interactions have been found between Scribble and Vangl2 in zebrafish in the control of the PCP-mediated convergence extension movements during gastrulation (Wada et al. 2005). Recent analysis of Dlg1 and Dlg3 knockout mice has also implicated mammalian Dlg in planar cell polarity regulation with both of these mice displaying at low penetrance the PCP mutant phenotypes, namely, cochlear hair cell misorientation and neural tube closure defects (Rivera et al. 2013; Van Campenhout et al. 2011). Finally, Lgl can bind to the PCP protein Dsh, which regulates Lgl localization, in both *Drosophila* epithelium and *Xenopus* tissues (Dollar et al. 2005). Together, these data implicate Scribble, Dlg, and Lgl in the regulation of PCP signaling during development in a variety of tissues and organisms.

Fig. 4.2 (continued) see text). aPKC-mediated phosphorylation of Lgl excludes it from the apical cortex and ensures the Scribble polarity module remains basally localized. The nectin–afadin (nectin, AF-6) and E-cadherin–catenin (E-CAD, β-CAT, α-CAT) adhesion complexes serve as a spatial cue to initiate apicobasal polarity via activation of CDC42, while the integrin–ECM adhesion complex directs cell orientation along the apicobasal axis in collaboration with the GTPase RAC1. Polarized vesicle trafficking serves to reinforce these spatial landmarks in association with the microtubule and actomyosin cytoskeleton. (**b**) The core PCP complexes distribute asymmetrically along the proximal–distal axis. Vangl forms a complex with Prickle (PK) at the proximal end of the cell and interacts with the Wnt receptor Frizzled (FZD), which forms a complex with Diego (DGO) and Dishevelled (DSH) at the distal end of the cell. CELSR1 is found both distally and proximally and serves to localize both Vangl and FZD. A second PCP complex comprised of large atypical cadherins FAT (FAT1-4) and Dachsous (DCHS1-2) also regulate planar polarization. This complex engages in an autoregulatory feedback loop with four-jointed (FJX1), a protein kinase localized to the Golgi. Importantly, multiple interactions directly couple the Scribble polarity module (shown in *red*) to the PCP network including interactions with Vangl (Scribble), Frizzled (Dlg), and Dishevelled (Lgl). (**c**) Integration of the core polarity modules during asymmetric cell division. PAR3 interacts with INSC and recruits the INSC-LGN/GPSM2-Gαi complex to the apical pole to induce cell division along the apicobasal axis. LGN/GPSM2 functions via two parallel pathways, the Dlg-KHC-73 and NUMA-Dynein pathway, to both anchor astral microtubules and generate pulling forces required for cytokinesis, respectively. Lgl functions downstream of aPKC and is required for appropriate localization of basal cell fate determinants. Scribble polarity module proteins are shown in *red*. (**d**) Core polarity modules in front–rear polarization and cell migration. RHO family GTPases have different functions inducing actin polymerization at the leading edge (RAC1, CDC42) and actomyosin contraction at the rear (RHOA), and these functions are coordinated by the core PAR and Scribble polarity modules, which localize to the leading edge during cell migration. Scribble polarity module proteins are shown in *red*

4.4.3 Asymmetric Cell Division

A role for the Scribble polarity module in ACD was first identified in studies of *D. melanogaster* neuroblasts (Albertson and Doe 2003), where it was shown that Scribble, Dlg, and Lgl are transiently polarized during mitosis and are required for the asymmetric distribution of cell fate determinants. Par3, whose name derives from the first studies showing that it was involved in asymmetric partitioning in the dividing *C. elegans* zygote (Kemphues et al. 1988), plays a central role in asymmetric cell division in many tissues, raising the possibility that the Scribble module plays a general role in ACD by controlling the location of the Par3 complex. The limited evidence to date, that supports such a role, is restricted to a few tissues. In addition to neuroblasts, Lgl is polarized in the *Drosophila* oocyte and plays a role in restricting the Par3 complex to the opposite side (Fichelson et al. 2010; St Johnston and Ahringer 2010).

In vertebrates, Scribble has been implicated in the orientation of cell division during zebrafish neural tube formation (Zigman et al. 2011), as well as the ACD of mammalian T lymphocytes (Chang et al. 2007; Oliaro et al. 2010; Chang et al. 2011; King et al. 2012; Pham et al. 2014). A number of studies have implicated Dlg in the regulation of spindle orientation during ACD, particularly through interactions with the Pins complex (Morin and Bellaiche 2011) (Fig. 4.2c), to anchor astral microtubules at the cortex and influence centrosome positioning (Manneville et al. 2010), and it was recently shown that both Dlg and Scribble are key regulators of spindle positioning in *Drosophila* epithelia (Nakajima et al. 2013).

4.4.4 Directed Migration

A fundamental requirement of directed cell migration is the definition of a front and back end of the cell, and this is determined by antagonistic actions of the Scribble and Par polarity modules in many cell types (Humbert et al. 2006). Both Scribble and Par polarity modules coordinate the cytoskeleton by orchestrating the differential activity of Rho-GTPases in microdomains in different regions of the cell, and this is particularly well characterized in cell migration (see Fig. 4.2d). All three members of the Scribble polarity module have been implicated in cell migration in a number of cell types including epithelial cells, endothelial cells, lymphocytes, astrocytes, and fibroblasts, both as single cells and in collective migration (Michaelis et al. 2013; Vaira et al. 2012; Li et al. 2011a; Dow et al. 2007; Zhao et al. 2008; Osmani et al. 2006; Ludford-Menting et al. 2005; Eastburn et al. 2012; Bahri et al. 2010), where they coordinate differential Rho-GTPases activity at the leading and trailing edge of the cell, polarization of the Golgi complex, capture of microtubule plus ends at the cell cortex, and E-cadherin activity (Osmani et al. 2006; Etienne-Manneville et al. 2005; Dow et al. 2007; Qin et al. 2005; Manneville et al. 2010; Lasserre et al. 2010; Dahan et al. 2012; Phua et al. 2009).

4.4.5 Neuronal Polarity

The Scribble polarity module plays a key role in several aspects of neuronal polarity, including asymmetric cell division (see below), axon specification, and the assembly and function of the neuronal synapse. Mammalian Dlgs (also known as PSD proteins, due to their prominent presence in the postsynaptic density of the neuronal synapse) have long been known as a key regulator of neuronal synapse assembly, and this function is conserved as far back in evolution as the protosynapses of choanoflagellates (Ryan and Grant 2009; Cho et al. 1992; Sampedro et al. 1981). All four mammalian isoforms of Dlg are recruited to the postsynaptic density and exert both distinct and redundant functions, which make it difficult to discern their individual roles. However, it is clear that Dlg family members are essential for appropriate composition of the neuronal synapse and for its maintenance and plasticity. Following the seminal observation that Dlg4 (PSD95)-deficient mice have memory defects, numerous studies have implicated Dlg family members in synaptic plasticity, drug addiction, sensitivity to pain, mental retardation, neuronal signaling in response to ischemic challenge such as stroke, and autism spectrum disorders (Huganir and Nicoll 2013; Gardoni et al. 2009; Migaud et al. 1998).

Lgl is also recruited to the neuronal synapse and the neuromuscular junction, where it regulates polarized vesicle delivery and exocytosis (Staples and Broadie 2013). Intriguingly, despite the prominent role of Dlg in the neuronal synapse, a role for Scribble in the neuronal synapse is not well understood. However, Scribble can be found at neuronal synapses, and loss of Scribble has been associated with enhanced learning and memory abilities and impaired social behavior, suggesting a role that overlaps with the Dlg family members (Moreau et al. 2010; Sun and Bamji 2011; Sun et al. 2009; Mathew et al. 2002; Roche et al. 2002).

4.4.6 The Immune Synapse

Both major cell types of the adaptive immune system, T and B cells, form a signaling domain called an immunological synapse that controls their activation and recruitment into an antigen-specific immune response by antigen-presenting cells. The first indication that the Scribble polarity module might be important in T cell signaling came from findings that Dlg1 was recruited to sites of T cell receptor activation and impacted upon NFAT signaling (Xavier et al. 2004). It was subsequently found that Scribble, Dlg1, and Dlg4 are transiently recruited to the immunological synapse of T cells and Dlg1 to the immunological synapse of B cells

(Ludford-Menting et al. 2005; Liu et al. 2012). Many studies have shown, using knockdown approaches, that Scribble and Dlg play key roles in the response of T and B cells to antigen presentation (Liu et al. 2012; Round et al. 2005, 2007; Humphries et al. 2012; Ludford-Menting et al. 2005; Xavier et al. 2004). However, gene knockout studies have shown minimal and/or transient effects of deletion of Scribble, Lgl1, and Dlg1 on hematopoiesis and immune cell function (Gmyrek et al. 2013; Hawkins et al. 2013, 2014; Humphries et al. 2012; Heidel et al. 2013). The discrepancy between knockdown and knockout studies is thought to reflect compensatory mechanisms by which lymphocytes might adjust to deletion of the Scribble module genes over time (Humphries et al. 2012; Pham et al. 2014), and further studies are required to resolve this discrepancy.

4.4.7 The Actin Cytoskeleton

The Scribble polarity module also has several links to the regulation of the actin cytoskeleton, either directly (direct binding) or indirectly (through effects on accumulation/activity). One important direct link between the Scribble module and an actin cytoskeletal regulator is a tripartite complex between Scribble, Dlg, and GUK-holder/NHS1 (Nance–Horan syndrome-like 1), which is a regulator of the WAVE/SCAR-ARP2/3-branched F-actin pathway (Katoh and Katoh 2004; Mathew et al. 2002; Walsh et al. 2011; Brooks et al. 2010). In mammalian cells, this interaction has roles in PCP and neuronal cell migration (Walsh et al. 2011; Brooks et al. 2010), while in *Drosophila* this interaction is important for Scribble localization at synaptic junctions. Furthermore, through its interaction with Git1 (G-protein-coupled receptor kinase-interacting Arf (ADP-ribosylation factor)-GAP (GTPase-activating protein) 1), and the GEF (guanine nucleotide exchange factor) β-Pix (Pak-interactive exchange factor), Scribble regulates the Rac-GTPase resulting in modulation of F-actin dynamics in cell migration (Lahuna et al. 2005; Nola et al. 2008). In mammalian cells, Scribble forms a multiprotein complex with Myosin II and MCC (mutated in colorectal cancer) where it is thought to regulate lamellipodia formation (Pangon et al. 2012). Lgl also binds to Myosin II and negatively regulates Myosin II function in *Drosophila* and mammalian cells (Strand et al. 1994, 1995; Betschinger et al. 2005). In mammalian cells, a recent study has revealed that Lgl forms alternative complexes with aPKC (PKCζ) or Myosin II and that phosphorylation of Lgl by aPKC affects its ability to inhibit Myosin II filament formation in polarized migrating cells (Dahan et al. 2014). In *Drosophila* the interaction of Lgl with Myosin II has a role in PCP in *Drosophila* embryo ventral epidermis (Kaplan and Tolwinski 2010); however the function for this regulation in other types of polarity is unclear.

Indirect effects on F-actin regulators occur through Scribble's involvement in PCP, where Scribble indirectly regulates the noncanonical Wnt signaling pathway, which signals through the Rho-GTPase pathway to regulate F-actin assembly and contractility (Courbard et al. 2009; Yan et al. 2009). Furthermore, in *Drosophila* epithelial tissue, *scrib* mutant Ras^{V12}-expressing cells show accumulation of the branched F-actin regulator, Filamin (Cher), Myosin II heavy chain, and phosphorylated (and activated) Myosin regulatory light chain, pMRLC (Kulshammer and Uhlirova 2013). This study showed that Cher, which binds to Myosin II, is required for pMRLC accumulation and Myosin II is required for *scrib* mutant Ras^{V12} tumorigenesis. *Drosophila* Lgl is also linked, via aPKC, to the regulation of Rho1–Rok–Myosin II-Jun kinase (JNK) signaling in compensatory proliferation following epithelial damage (Warner et al. 2010). In contrast, in MDCK cells, increased Lgl (or decreased Par activity) leads to activation of the Rock-Myosin II pathway, which remarkably results in cell–cell internalization (entosis) during cell–cell junction formation (Wan et al. 2012). It is possible that this unusual phenomenon is related to Lgl's role in endocytosis (see Sect. 4.4.8). Furthermore in mammalian cells, through its inhibition of aPKC, Lgl is linked to the regulation of Tiam1 (a Rac1-GEF) and activation of the Rac1-GTPase, which has roles in tight junction formation, dendritic spine morphogenesis and cell migration (Chen and Macara 2005; Mertens et al. 2005; Zhang and Macara 2006; Narayanan et al. 2013; Wang et al. 2012), and also the JNK or p38 stress response pathways (Lambert et al. 2002; Buchsbaum et al. 2002).

4.4.8 Vesicle Trafficking

The Scribble polarity module has roles in several aspects of vesicle trafficking, including endocytosis (internalization of lipids/proteins), recycling (shuffling of lipids/proteins in endosomes to return to the cell surface), retrograde trafficking (in which proteins/lipids are shuttled between endosomes and biosynthetic/secretory compartments, such as the Golgi apparatus), or exocytosis (secretion of proteins/lipids).

Endocytosis, Recycling, and Retrograde Trafficking to the Golgi: In *Drosophila*, various studies have revealed roles for Lgl and Scribble in endocytosis. In *Drosophila* asymmetric division of sensory organ precursor cells, Lgl is required for endocytosis of Sanpodo (a four-pass transmembrane protein and regulator of Notch signaling; see Sect. 4.5) (Roegiers et al. 2005). Furthermore, we have recently found that Lgl colocalizes with endosomes and *lgl* mutants result in the accumulation of early endosomal, recycling endosomal and multivesicular body markers in *Drosophila* epithelial tissue, suggesting that Lgl has a role in endosomal maturation (Parsons et al. 2014a). In *Drosophila* epithelial tissues, a role for Scribble in blocking endocytosis has been revealed; *scrib* mutant cells show increased endocytosis, and the internalization of TNF (tumor necrosis factor, Eiger in *Drosophila*) and apoptosis induction is dependent on the function of early endosomal regulator, Rab5 (Igaki et al. 2009). Consistent with this finding, in mammalian PC12 cells, Scribble binds to the TSHR (thyroid-stimulating hormone receptor), inhibits basal receptor endocytosis, and promotes recycling (Lahuna et al. 2005). A recent study has also revealed a role for Scribble in negatively regulating retromer-mediated diversion of E-cadherin to the Golgi (Lohia et al. 2012). Depletion of Scribble enhances endocytosis of E-cadherin by weakening the E-cadherin–p120catenin interaction and enhances E-cadherin binding to the retromer, which is a complex required for protein cargo trafficking from endocytic vesicles to the *trans*-Golgi network for recycling. Similarly in *Drosophila*, the Scribble polarity module is important in the trafficking of retromer cargo, such as Crumbs (de Vreede et al. 2014). However, in *Drosophila* there appears to be differences in the mechanism, as there was no evidence of Golgi trapping of E-cadherin in *scrib* mutants, as there was in mammalian cells. In *Drosophila*, the regulation of the retromer by the Scribble polarity module was shown to have both aPKC-dependent and aPKC-independent mechanisms (de Vreede et al. 2014).

Exocytosis: In addition to roles in endocytosis and regulation of the retromer, Scribble also has roles in exocytosis. Scribble, through its association with the Arf-GAP, GIT1, and the GEF, β-Pix, which are regulators of the Rac-GTPase, has an important role in regulating exocytosis in neuroendocrine cells; mutants in human Scribble suppress high KCl-induced exocytosis in these cells (Audebert et al. 2004). Furthermore, Scribble functions downstream of β-catenin to cluster synaptic vesicles at developing synapses (Sun et al. 2009).

Roles for Lgl in exocytosis were first revealed in yeast, where the Lgl orthologues (Sro7 and Sro77) bind to Sec9 (a t-SNARE, target SNARE, involved in membrane fusion) via their C terminus, to Sec4 (Rab-GTPase involved in exocytosis), and to Exo84 (component of the exocyst complex) to regulate polarized exocytosis (Gangar et al. 2005; Zhang et al. 2005; Grosshans et al. 2006; Hattendorf et al. 2007). Consistent with this finding, in MDCK mammalian epithelial cells, Llgl2 forms a complex with the t-SNARE, syntaxin 4, a component of the basolateral exocyst machinery (Musch et al. 2002), and is thought to regulate basolateral exocytosis. Furthermore, in mammalian cells, a complex between Llgl1 and syntaxin 4 forms in response to osmotic stress (Massimi et al. 2008). Of note, Scribble and Dlg were also found in this complex with Lgl and syntaxin 4, and this interaction was required for their membrane localization. Interestingly, Llgl1 has also been shown to interact with and regulate the Rab10-GTPase in directional membrane insertion during rat axonal development (Wang et al. 2011). Here, Llgl1 promotes the release of the GDP dissociation inhibitor (GDI) from Rab10, thereby promoting Rab10 activity and regulating membrane insertion in growing neurites. Whether the Scribble polarity module is involved in exocytosis in *Drosophila* is unknown. However, this is certainly possible in light of the requirement for Lgl in secretion of Dpp (decapentaplegic, TGFβ homologue) in embryonic ectodermal cells (Arquier et al. 2001) and genetic interactions observed between *Drosophila*, *scrib*, *dlg*, or *lgl* mutants and mutants in a core component of the exocytic machinery, *exo84*, which is required for the apical delivery of proteins, such as Crumbs (Blankenship et al. 2007).

4.4.9 Cell Competition

Cell competition is a surveillance mechanism selecting for the fittest cells in a tissue, where the slower growing cells (losers) are eliminated by apoptosis that is induced by the faster growing/proliferating fitter cells (winners) (de Beco et al. 2012; Amoyel and Bach 2014), leading to their engulfment in *Drosophila* by the macrophage-like hemocytes (Lolo et al. 2012). This mechanism was initially discovered in *Drosophila* but also functions in mammalian mosaic cell cultures and in vivo. Interestingly, in *Drosophila*, this mechanism does not necessarily require cell division, as in postmitotic tissues, competitive cell hypertrophy occurs, whereby winner cells increase cell mass without cell division to occupy a larger tissue space (Tamori and Deng 2013). With regard to *lgl, dlg,* or *scrib,* although individual depletion in *Drosophila* within a whole epithelial tissue will result in overgrowth, when patches (clones) of mutant tissue are generated in a wild-type background, the process of cell competition and JNK-mediated cell death of the mutant tissue ensues, despite ectopic proliferation (Brumby and Richardson 2003; Froldi et al. 2010; Grzeschik et al. 2010b; Uhlirova et al. 2005; Igaki et al. 2006). Oncogenic Ras (Ras^{V12}) expression in *scrib* (*dlg* or *lgl*) mutant tissue can overcome this cell death mechanism in *Drosophila* eye epithelium (Brumby and Richardson 2003; Doggett et al. 2011; Leong et al. 2009; Pagliarini and Xu 2003). This cooperation appears to depend on the size of the mutant tissue patch, since when smaller clones of *lgl* mutant Ras^{V12}-expressing wing epithelial tissue are induced, despite higher proliferation rates, Ras activation is not sufficient to prevent cell competition mechanisms from operating (Menendez et al. 2010; Chen et al. 2012). Interestingly, cell competition is context dependent, since *lgl* mutant *Drosophila* ovarian follicle cells, which have high levels of Myc, are not outcompeted (Froldi et al. 2010), and *dlg* or *lgl* (but not *scrib*) mutant stem cells in the ovary niche have a greater competitive advantage (Kronen et al. 2014).

The mechanism of cell competition of *scrib, dlg,* or *lgl* mutant loser cells occurs via the downregulation of the cell growth/proliferation/survival proteins, Myc and Yorkie (Yki, the co-transcriptional activator, negatively regulated by the Hippo negative tissue growth control pathway; see Sect. 4.5), and the upregulation of JNK in the loser cells (Menendez et al. 2010; Chen et al. 2012; Froldi et al. 2010). In the case of *scrib* mutant clones, JNK activation is induced by the production of TNF (tumor necrosis factor, Eiger (Egr)) by macrophage-like cells (hemocytes) or surrounding wild-type cells that results in the elimination and engulfment of $scrib^{-}$ cells (Cordero et al. 2010; Ohsawa et al. 2011; Lolo et al. 2012). However, the induction of JNK in *lgl* mutant cells appears to be independent of TNF (Egr) signaling (see Sect. 4.5) (Froldi et al. 2010), and one way this may occur is through the regulation of Mahjong/VprBP (a Cul4 ubiquitin ligase complex interacting protein) that binds to Lgl in *Drosophila* and mammalian cells (Tamori et al. 2010). Tamori and colleagues propose that Mahjong is a negative regulator of the JNK pathway, since when overexpressed Mahjong can prevent the JNK-mediated cell death of lgl^{-} clones in the developing wing disc. Furthermore, the transcription factor, Stat (downstream component of the JAK signaling pathway), is required in the surrounding normal tissue for *scrib* mutant tissue to be eliminated, and this mechanism is independent of the proliferation-inducing role for Stat (Schroeder et al. 2013).

The phenomenon of cell competition is an emerging area in mammalian cell biology; however there are clear examples of its role in vitro and in vivo for physiological function and in cancer (Hogan 2012; Amoyel and Bach 2014; de Beco et al. 2012). In regard to the Scribble polarity module, in MDCK mammalian cells, Scribble-depleted cells surrounded by normal cells undergo apoptosis and are extruded apically from the epithelium (Norman et al. 2012). Here this mechanism is dependent upon upregulation of the p38 (MAPK) stress response pathway; however in Mahjong-depleted MDCK cells, it is likely that this is JNK dependent, since apoptosis is prevented by treatment with a JNK-specific inhibitor (Tamori et al. 2010). Clearly, further analysis in mammalian cells is required to reveal whether Myc (Claveria et al. 2013), Yki, and Stat orthologues are also involved in this process.

4.5 Regulation of Signaling by the Scribble Polarity Module

From studies, primarily in *Drosophila*, zebrafish, mouse models, and mammalian cells, the Scribble polarity module is emerging as a modulator of many signaling pathways. The regulation of these signaling pathways links the Scribble polarity proteins to the control of tissue growth or differentiation, which highlights their central role in controlling tissue architecture and as tumor suppressors. Here we summarize recent findings on the regulation of the Notch, Wnt, Rtk–Ras–MAPK/JNK/P38, PI3K, Hippo, and other signaling pathways by Scribble, Dlg, or Lgl (illustrated in Fig. 4.3).

4.5.1 Notch

The Notch signaling pathway is an important pathway driving cell fate decisions during development and in T cell differentiation (Fortini 2009). Canonical Notch signaling involves interaction between cells expressing the Notch receptor and cells expressing ligands, such as Delta, which leads to the cleavage of the Notch receptor by Adam protease and γ-secretase to release the intracellular domain (Notch-ICD). Recently, it has become clear that the endocytosis of the Notch receptor is important in its activation (Baron 2012) and that the Lgl-aPKC polarity axis regulates Notch signaling via affecting the localization of Notch regulators, Numb and Sanpodo, or the intracellular trafficking of Notch (Fig. 4.3).

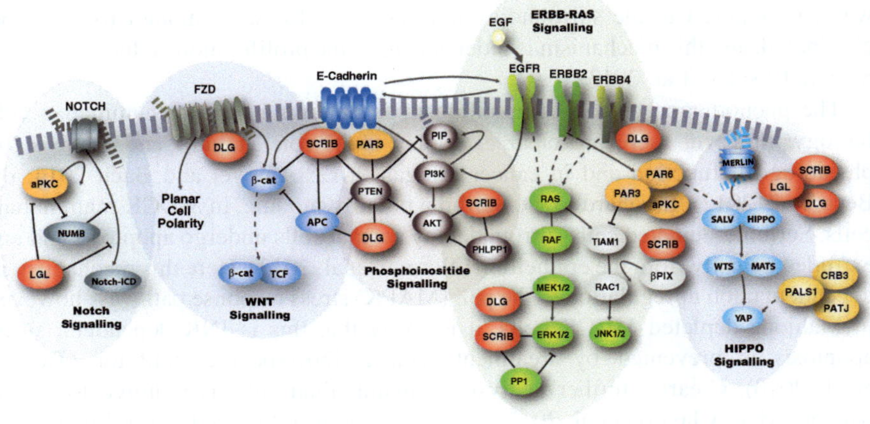

Fig. 4.3 *Molecular interactions of the Scribble polarity module with cellular signaling pathways.* The members of the Scribble polarity module function as a specific regulator of many diverse signal transduction pathways that regulate proliferation, differentiation, and survival. Scribble polarity module proteins are shown in *red*

In the *Drosophila* central and peripheral neural stem cells (neuroblasts), Lgl is required for the basal localization of the Numb protein (a membrane-associated regulator of endocytosis and proteosome-mediated degradation), which upon cell division is concentrated in the daughter cell, where it inhibits Notch signaling, enabling neural cell differentiation (Ohshiro et al. 2000). More recent analysis has shown that Lgl acts through blocking aPKC activity to control asymmetric localization of Numb and that when free from Lgl, aPKC binds Par3 (Baz) and can phosphorylate Numb, thereby excluding it from the cell cortex on one side of the cell (Wirtz-Peitz et al. 2008; Haenfler et al. 2012). Lgl also excludes the plasma membrane localization of Sanpodo (a transmembrane protein that binds Notch and promotes its endocytosis), which is required for Notch signaling. Since aPKC has been shown to directly phosphorylate another cell fate determinant, Miranda, in neuroblast ACD (Atwood and Prehoda 2009), it posits the question of whether the localization of Sanpodo is also regulated by aPKC-dependent phosphorylation. A new player in this process is Clueless, which regulates aPKC levels and therefore influences Numb localization and Notch signaling in the developing neuroblasts (Goh et al. 2013). A role for Lgl in Notch signaling via regulation of Numb is also conserved in mammals; in mouse Llgl1 knockout mice, the brain neuroprogenitor cells fail to exit the cell cycle and differentiate, which is associated with the failure to segregate Numb and inhibit Notch signaling in the progeny (Klezovitch et al. 2004).

Whether this mechanism of Notch regulation by Lgl through Numb localization occurs in other systems requires further study. However, in the cell fate divisions of another type of neural stem cell (sensory organ precursors (SOPs)), the regulation of Notch signaling by Lgl appears to be different; Lgl acts in parallel to Numb to block Notch function (Justice et al. 2003), which is thought to occur via its effect on the plasma membrane localization of Sanpodo (Langevin et al. 2005). Furthermore, in *Drosophila* eye neural epithelial tissue, our studies have recently revealed that Lgl regulates ligand-activated Notch signaling by affecting endosomal trafficking (see Sect. 4.4.8); *lgl* mutants show increased accumulation of acidified vesicles, Notch-ICD, and upregulation of Notch targets, which contribute to the proliferative and differentiation defects of the *lgl* mutant phenotype (Parsons et al. 2014a). Interestingly, this role for Lgl in Notch signaling in *Drosophila* is aPKC independent. This contrasts studies in vertebrates, where a role for aPKC in regulating Notch receptor intracellular trafficking has been revealed in the chick central nervous system in vivo and myogenic precursors in vitro (Sjoqvist et al. 2014). In this study aPKC (PKCζ) was shown to phosphorylate the Notch1 receptor on Serine-1791, which promoted trafficking of ligand-activated Notch to the nucleus thereby promoting signaling. However, when Notch1 was not activated, aPKC instead promoted its internalization from the cell surface and the secretory Golgi–ER pathway to intracellular vesicles. Whether this mechanism also occurs in other organisms is not clear, since the Notch protein in *Drosophila* and *C. elegans* lacks the conserved Serine-1791 residue, as does vertebrate Notch3 and Notch4.

The regulation of Notch signaling by Lgl also occurs in the zebrafish retinal neuroepithelial cells, where Llgl1 depletion results in increased Notch signaling and reduces neurogenesis, and blocking Notch activity rescues these defects (Clark et al. 2012). Here the mechanism is thought to be due to the expansion of the apical domain that occurs in this scenario, and the accumulation of the Notch receptor, although whether a more direct mechanism occurs requires further investigation.

4.5.2 Wnt

The Wnt signaling pathway is a critical pathway in development and tissue homeostasis, which is linked to PCP–noncanonical signaling (see Sect. 4.4.2) and to gene transcription, via β-catenin and TCF–canonical signaling (Fig. 4.3) (Clevers and Nusse 2012). The Wnt signaling pathway signals through the Frizzled (Fzd) family of receptors and regulates the stability of β-catenin through inactivation of the APC–axin complex, which induces ubiquitin-mediated degradation of β-catenin. Several physical interactions occur between the Scribble polarity module and Wnt signaling components; Lgl and Dsh, in controlling PCP (Dollar et al. 2005; Kaplan and Tolwinski 2010; see Sect. 4.4.2); Scrib and β-catenin, implicated in synaptic vesicle trafficking and recycling in neural cells (Sun et al. 2009); Dlg or Scrib and APC, important in cell cycle exit and cell migration (Takizawa et al. 2006; Sotelo et al. 2012; Matsumine et al. 1996; Ishidate et al. 2000; Etienne-Manneville et al. 2005); and Dlg and Fzd, thought to be important in establishing a scaffold in canonical or noncanonical Wnt signaling (Hering and Sheng 2002; Luyten et al. 2008). With the exception of the interaction of Lgl with Dsh, all of these studies have been undertaken in mammalian systems, and therefore further studies in invertebrate models are required to explore the universality of these interactions.

4.5.3 RTK–Ras–MAPK/JNK/P38

The Ras small GTPase–MAPK (ERK) signaling pathway acts downstream of RTKs (receptor tyrosine kinases), such as EGFR (epidermal growth factor receptor) to regulate cell growth proliferation during development and homeostasis, and is a key oncogenic pathway in cancer (Fernandez-Medarde and Santos 2011) (Fig. 4.3). The localization of the EGFR to basal lateral membranes in polarized epithelial cells is important for restricting pathway activation (Vermeer et al. 2003; Kuwada et al. 1998). Indeed, Dlg has been shown to bind to the EGFR family member ErbB4 in neuronal cells (Garcia et al. 2000). On the other hand, Scribble interacts indirectly with EGFR, by binding to the Arf-GAP, GIT1, and β-Pix (Pak-interactive exchange factor) that act as a MEK-ERK scaffold (Audebert et al. 2004; Yin et al. 2004). Since β-Pix together with the small GTPase, Cdc42, affects EGFR plasma membrane levels by blocking ubiquitination of the EGFR by the Cbl E3-ubiquitin ligase (Wu et al. 2003), Scribble, through its binding to β-Pix, might control EGFR downregulation. Scribble also impacts on downstream components of the Ras pathway; it interacts with RSK2 (ribosomal S6 kinase 2), a negative regulator of the EGFR pathway (Zeniou-Meyer et al. 2008), and also with ERK (Nagasaka et al. 2010a, b). Likewise, Dlg1 binds to MEK2, which phosphorylates and activates ERK (Gaudet et al. 2011; Maiga et al. 2011). While the functional consequences of all these interactions are not completely clear, studies in MCF10A human mammary epithelial cells have shown that Scribble negatively regulates the Ras–MAPK pathway in vitro and in vivo (Elsum et al. 2013a; Pearson et al. 2011; Dow et al. 2008; Elsum et al. 2013b; Godde et al. 2014; Nagasaka et al. 2010b), and overexpression of Scribble can also antagonize ectopic Ras signaling in *Drosophila* tissues (Dow et al. 2008). Human Scribble inhibits Ras signaling by direct binding to ERK at the kinase interaction motif (KIM) docking sites and preventing ERK phosphorylation and by anchoring ERK to membrane sites (Nagasaka et al. 2010a, b).

Scribble and Dlg also interact with phosphatases that impact on EGFR-Ras signaling. Dlg2 and Dlg3 bind the catalytic subunit of the PP1 phosphatase (Hendrickx et al. 2009), and human Scribble binds to protein phosphatase 1γ (PP1γ), through the PP1γ interaction motif on Scribble (Nagasaka et al. 2013). The interaction of Scribble with PP1γ is required for Scribble's ability to downregulate ERK phosphorylation and for Scribble to prevent oncogenic transformation of primary rodent cells. Scribble has also been shown to form a complex with SHOC2/SUR-8 (PP1 phosphatase regulator) and MRAS (RRas subgroup of Ras proteins), and Scribble blocks SHOC2-mediated dephosphorylation of RAF (a protein kinase downstream of Ras) at the conserved inhibitory site (S259), thereby inhibiting RAF activation (Young et al. 2013). This mechanism appears to be conserved in *Drosophila*, where the deregulation of PP1 activity contributes to the *scrib* mutant phenotype, but in this case it appears to function via regulation of the JNK stress response signaling (Jiang et al. 2011). As described above (see Sect. 4.4.9), *scrib* loss of function leads to the activation of the JNK pathway (Brumby and Richardson 2003; Uhlirova et al. 2005), which together with Ras activation (Ras^{V12}) leads to invasive tumor formation (Leong et al. 2009; Igaki et al. 2006; Uhlirova and Bohmann 2006). Overexpression of a PP1 regulator, Sds22, blocks *scrib-Ras*^{V12} tumorigenesis, by inhibiting Myosin II and JNK activity and preventing cell invasion/metastasis (Jiang et al. 2011).

There is also evidence for a role for Lgl in the regulation of Ras signaling. In zebrafish, Lgl2 has also been linked to regulation of the Ras signaling pathway (Reischauer et al. 2009). Furthermore, upregulation of Par6 in mammalian cells (Nolan et al. 2008) or aPKC in *Drosophila* epithelial tissues results in increased Erk activation (pErk) (Parsons et al. 2014b), indicating that Lgl via its effect on blocking PAR complex activity limits RTK–Ras signaling.

Despite the strong evidence that Scribble negatively regulates RTK–Ras–ERK signaling, a recent report has revealed that in response to nerve growth factor (NGF)–TrkA signaling, depletion of Scribble reduces sustained ERK activation in PC12 neural cells (Wigerius et al. 2013). Knockdown of Scribble decreases neurite numbers but increases neurite length. In this study it was also shown that Scribble binds to Ha-Ras and pERK1/2 in response to NGF. Thus in contrast to other studies, this work suggests that Scribble promotes Ras-ERK signaling, suggesting that the effect of Scribble on this signaling pathway is context dependent.

As described above (see Sect. 4.4.9), the JNK signaling pathway is upregulated in *Drosophila scrib* mutant tissue (Brumby and Richardson 2003; Uhlirova et al. 2005); however this is now thought to be primarily due to an extrinsic response due to the recruitment of the *Drosophila* macrophage-like cells (hemocytes) to the mutant tissue and their production of TNF (Egr) (Cordero et al. 2010), although surrounding epithelial cells may also contribute to Egr production (Igaki et al. 2009; Ohsawa et al. 2011). In mammalian tissue culture, Scribble knockdown also leads to JNK activation in MCF10A cells that contributes to the invasive phenotype with Ha-Ras^{V12} (Dow et al. 2008) and to p38 activation in mosaic MDCK cell cultures that contribute to apoptosis of the Scribble mutant cells (Norman et al. 2012). However, Lgl depletion appears to activate JNK signaling independently of Egr (Froldi et al. 2010) and may function via an aPKC-Rho1–Rok–MyoII pathway to lead to JNK activation (Warner et al. 2010) (see Sects. 4.4.7 and 4.4.9).

Given the above evidence, it is clear that the Scribble polarity network plays an important role in regulation of RTK–Ras–ERK, JNK, or P38 signaling pathways; however the precise regulatory impact on these signaling pathways appears to be organismal or context specific. These data suggest a model whereby the Scribble polarity module proteins act as scaffolds that modulate these pathways, which may also be influenced by environmental cues or cellular states.

4.5.4 PI3K

The PI3K (phosphoinositol-3-kinase)–AKT–mTOR (mechanistic target of rapamycin) pathway is a critical pathway in polarity regulation and in cell growth and proliferation control, which is often deregulated in cancer (Sabatini 2006; Bunney and Katan 2010; Shewan et al. 2011) (Fig. 4.3). In polarized epithelial cells, PI3K physically associates with E-cadherin at adherens junctions and PI3K is important in adherens junction function (Rivard 2009; Pece et al. 1999). Dlg and Scribble have been shown to modulate PI3K pathway signaling in several ways in different systems. Mammalian Dlg has been implicated in activation of the PI3K pathway; Dlg1 is required for the Adenovirus 9 E4-ORF1 (*E4* region-encoded open reading frame 1) oncoprotein to promote the constitutive activation of PI3K (Frese et al. 2006). This is consistent with studies in *Drosophila* where Dlg also appears to positively regulate PI3K signaling; PI3K signaling is downregulated in Dlg-depleted epithelial cells, and further knockdown of PI3K components results in synthetic lethality of Dlg-depleted tissue, even in the presence of oncogenic Ras (Willecke et al. 2011). Conversely, in human cells, Dlg1 can bind to PTEN (phosphatase and tensin homologue deleted on chromosome 10), a negative regulator of the PI3K pathway, leading to greater inhibition of PI3K signaling (Adey et al. 2000; Sotelo et al. 2012). PTEN is also bound by Par3 in mammalian MDCK cells (and by the Par3 homologue, Baz in *Drosophila*) where it plays an important role in polarity establishment (Feng et al. 2008; Pinal et al. 2006). This interaction between PTEN and Baz in *Drosophila* epithelial is also thought to have a role in the organization of the actin cytoskeleton, as discerned by the analysis of PTEN mutants (von Stein et al. 2005). However, how the interaction of PTEN with Baz and Dlg is coordinated during development or homeostasis, and how this might affect PI3K signaling, remains to be determined.

Scribble negatively regulates AKT activity via binding to PHLPP (PH (pleckstrin homology) domain and LRR protein phosphatase), a protein phosphatase that negatively regulates AKT, and localizes it to the plasma membrane (Li et al. 2011b). Scribble forms a tripartite complex with PHLPP and AKT, thereby inhibiting Akt activity, but when Scribble is downregulated, PHLPP is released, AKT activity is increased, and cell growth, proliferation, and survival are enhanced. Furthermore, high levels of mislocalized Scribble function as a neomorph to promote mammary tumorigenesis by affecting subcellular localization of PTEN and activating the AKT/mTOR/S6 kinase signaling pathway (Feigin et al. 2014). Since mislocalized Scribble is often observed in human cancer, this mechanism by which Scribble affects the PI3K pathway may be of particular importance in cancer.

There is also recent evidence for regulatory interactions between Lgl and PTEN. In glioblastoma tumor-initiating cells, PTEN loss leads to activation of aPKC, leading to the phosphorylation and inactivation of Lgl (Gont et al. 2013). Moreover, re-expression of PTEN, non-phosphorylable Lgl, or knockdown of aPKC results in neural differentiation of these glioblastoma cells, suggesting that loss of PTEN might promote an undifferentiated state by blocking Lgl function.

Considered together, these findings also highlight the context-dependent role of the Scribble polarity module in the regulation of the PI3K signaling pathway, similar to their effect on the RTK–Ras–ERK pathway. Indeed, the emerging theme is that Dlg and Scribble can form multiprotein complexes with several key signaling proteins, for example, Dlg with PTEN and APC (Sotelo et al. 2012) and Scrib with PHLPP and AKT (Li et al. 2011b). However, there is only limited evidence on the connection between Lgl and the PI3K pathway; therefore further investigation of this connection in different systems is clearly warranted.

4.5.5 Hippo

The Hippo signaling pathway negatively controls tissue growth (Grusche et al. 2011) (Fig. 4.3). The central mediators of this pathway are the Hippo and Warts protein kinases, which are regulated by several upstream regulators including the expanded and Merlin (FERM domain proteins) and Fat (atypical cadherin)–Dachs (nonconventional myosin) regulators. When activated, Wts induces the phosphorylation of the transcriptional co-activator, Yki (YAP and TAZ in mammalian cells), and results in its cytoplasmic retention. However, when the Hippo pathway is inactive, Yki enters the nucleus where it binds to the TEAD family of transcriptional regulators and leads to the upregulation of cell proliferation and survival genes. Recent studies have shown that the Scribble polarity module regulates the Hippo pathway in *Drosophila* and mammalian cells. Depletion of *Drosophila* Lgl (or upregulation of aPKC) leads to downregulation of the Hippo pathway and increased proliferation and cell survival (Grzeschik et al. 2010a). Lgl/aPKC affect the Hippo pathway by regulating the localization of Hippo and its negative regulators, RASSF and STRIPAK phosphatase complex; however RASSF and STRIPAK are individually not important in the control of Hippo activity by Lgl/aPKC (Parsons et al. 2014b; Grzeschik et al. 2010a). Thus, the most likely model is that Lgl/aPKC regulates the localization of Hippo and thereby its activity, which is consistent with data showing Lgl is involved in vesicle trafficking (see Sect. 4.4.8).

Despite their common role with Lgl in cell polarity regulation and also in the regulation of the retromer (de Vreede et al. 2014), Scribble and Dlg do not work in a common pathway with Lgl to regulate the Hippo pathway in *Drosophila*. Indeed in *Drosophila,* Scribble and Dlg only deregulate Hippo signaling when polarity is lost (Grzeschik et al. 2010a; Doggett et al. 2011). Epistasis experiments suggest that Hippo pathway inhibition in *scrib* mutants occurs downstream or in parallel to both the expanded and Fat-Dachs-Wts arms of Hippo pathway regulation (Doggett et al. 2011); however more complex interactions are suggested by another study (Verghese et al. 2012). In vertebrates, various direct interactions between Scribble with Hippo pathway regulation have been revealed. In mammalian cells, Scribble binds to TAZ and sequesters it to the cell cortex in breast cancer stem cells (Cordenonsi et al. 2011). Here, loss of Scribble or EMT induction frees TAZ from the Hippo (Mst)/Wts (Lats) inhibitory interaction. TAZ/YAP membrane localization in mammalian cells is also regulated by binding to the Crb complex components, Pals and Patj, thereby restricting their nuclear access (Varelas et al. 2010). However, in zebrafish, Scribble binds to Fat1, a regulator of the Hippo pathway, and works together with Fat1 to control Yap activity in pronephric cyst development (Skouloudaki et al. 2009).

While these interactions provide strong links between the Scribble polarity module and the Hippo pathway, clearly, further analysis is needed to provide a more comprehensive view of the regulatory mechanisms in different model organisms.

4.5.6 *Other Signaling Pathways*

There have also been limited reports for the involvement of the Scribble polarity module in several other signaling pathways. Scribble regulates the activity of G-protein-coupled receptor signaling, such as the thyroid-stimulating hormone receptor (TSHR), where direct binding of Scribble to TSHR recruits the β-PIX-GIT-ARF6 complex and affects vesicle trafficking of TSHR (Lahuna et al. 2005). Furthermore, recent analysis of Scribble conditional knockout mice in the lens epithelial also suggests other signaling pathways are regulated by Scribble, such as the TGFβ pathway transcription factors, Smad3 and Smad4, which is thought to contribute to the EMT in this tissue (Yamben et al. 2013). Notably, a role for Scribble in the TNFα–NF-κB signaling has also been suggested by the identification of human Scribble in a proteomics screen with Traf7 (a ring finger ubiquitin ligase) in human embryonic kidney Hek293 cells (Bouwmeester et al. 2004). Interestingly, Mcc1, which functions with Scribble in regulating cell migration (Arnaud et al. 2009) (see Sect. 4.4.7), was also identified as a regulator of TNFα–NF-κB signaling in this study.

In summary, as detailed here, the regulation of multiple signaling pathways by the Scribble polarity module suggests that it plays an overarching role in cellular signaling control in response to environmental inputs by acting as signaling scaffolds. In these signaling functions, the Scribble polarity module proteins' connection to the regulation of vesicle trafficking and the actin cytoskeleton has also been implicated in many cases. Furthermore, while the Scribble polarity module has a common function in cell polarity regulation, there is accumulating evidence that in many situations, Scribble, Dlg, and Lgl individually have specific roles in the regulation of various signaling pathways. The prevailing view emerging therefore is that although Scribble, Dlg, and Lgl can function together in some processes, they also have independent functions in other cellular processes.

4.6 The Scribble Polarity Module and Cancer

Since their identification as tumor suppressors in *Drosophila,* much work has been undertaken to probe the significance of the Scribble polarity module in human cancer progression (reviewed in Humbert et al. 2008; Facciuto et al. 2012; Banks et al. 2012; Halaoui and McCaffrey 2014). The first hint that the Scribble polarity module may be important in human cancer came from the discovery that both Dlg1 (Lee et al. 1997; Kiyono et al. 1997) and human Scribble (Nakagawa and Huibregtse 2000) were identified as targets of the oncogenic herpes simplex viruses (HPV), key drivers of cervical cancer progression (reviewed in Pim et al. 2012). Indeed, the HPV E6 proteins can bind to both Scribble and Dlg1 and lead to their degradation (Gardiol et al. 1999; Nakagawa and Huibregtse 2000), with altered expression of Scribble and Dlg1 in cervical cancer patient samples associated with more aggressive disease (reviewed in Banks et al. 2012). Significantly, other oncoviral proteins such as HTLV-1 (human T cell leukemia virus type 1) Tax, a causative agent for adult T cell leukemia, can also target Scribble and Dlg (Lee et al. 1997; Okajima et al. 2008), indicating that the Scribble polarity module may form a potent molecular focal point for these cancer-causing viruses. For example, targeting of the Scribble polarity module by these viruses could provide life cycle advantages through expansion of the number of cells capable of replicating the viral genome (Banks et al. 2012).

Initial surveys of the expression of components of the Scribble polarity module in human cancer patients indicated that loss or decreased expression of Scribble, Dlg, and Lgl correlated with more aggressive cancers (reviewed in Humbert et al, 2008). Indeed, analysis of genetically engineered mouse models (GEMMs) of human cancer have functionally supported these initial observations with loss of Scribble shown to enhance tumorigenesis in a number of epithelial tissues including the prostate, lung, and breast (Pearson et al. 2011; Elsum et al. 2013b; Godde et al. 2014). In addition, loss of Dlg1 can promote the development of B-cell acute lymphoblastic leukemia (B-ALL) in the mouse (Sandoval et al. 2013), and inactivation of LLgl1 may be important for maintenance of glioblastoma malignancy (Gont et al. 2013). Of note, the tissue organization defects following loss of Scribble are very similar to early lesions seen in breast and prostate cancers, namely, ductal carcinoma in situ (DCIS) and prostate intraepithelial neoplasia (PIN), respectively, which include hyperproliferation and multilayering of glandular epithelial tissue (Pearson et al. 2011; Godde et al. 2014). This suggests that loss of activity of the Scribble polarity module could be an early event in these epithelial tumors and drive the early steps of carcinogenesis. Altogether these provide evidence that the Scribble polarity module can act as tumor suppressor in mammals.

The tumor-suppressive properties of the Scribble polarity module however are likely to be context dependent. Indeed, recent evidence suggests that upregulation of Scribble is common in many cancers including hepatocellular carcinomas, with high levels of Scribble correlated with reduced patient survival (Vaira et al. 2011; Savi et al. 2014). Importantly, studies examining large cohort of prostate or breast cancer patients have now indicated that it is the mislocalization of Scribble that is associated with poor outcome (Pearson et al. 2011; Feigin et al. 2014; Savi et al. 2014). In particular, enforced expression of a mislocalization mutant of Scribble in transgenic mice can promote breast cancer progression, providing evidence that alterations in localization of Scribble polarity module components could be contributing to cancer progression (Feigin et al. 2014). Interestingly, the high expression of Scribble in some tumors may be regulated by miR-296, which is progressively lost during tumor progression and correlates with metastatic disease in multiple epithelial cancers (Vaira et al. 2012; Savi et al. 2014). Consistent with the notion that Scribble module components could act as oncogenes, Dlg1 has also been shown to have oncogenic properties in some situations, where, for example, it is required to localize the adenovirus E4-ORF1 oncoprotein to activate the PI3K pathway (Frese et al. 2006; Kong et al. 2014). The context-dependent properties of the Scribble polarity module on the regulation of cancer progression thus should signal some caution for the design of therapeutic strategies. A better understanding of the potential lesions in the Scribble polarity module and the specific interaction between the Scribble module members and oncogenic signaling pathways in any particular tumor will be critical for the development of effective drugs that block downstream consequences of altered Scribble polarity module function.

4.7 Conclusions and Future Challenges

Although initially identified as a core component of the machinery that regulates apicobasal polarity, it has become increasingly clear that the Scribble polarity module regulates a vast array of additional cell polarity states, from that involved in asymmetric cell divisions to the immunological synapse. In view of this conserved and more pervasive role, it is now imperative to understand how, when, and where the Scribble polarity module interacts with other key polarity complexes and signaling components to regulate these various polarity states. In addition, although Scribble, Dlg, and Lgl have common roles in cell polarity control, emerging data indicates they also have independent functions in other cellular processes related to their signaling modulatory properties. Careful biochemical, spatiotemporal, and structural information on the Scribble, Dlg, and Lgl interactions (including the Scribble–Dlg adaptor NHS/Gukh) and how these are altered in the presence of key signaling molecules will be required to fully understand the scaffolding function of the Scribble polarity module. The appreciation that the Scribble polarity module may be disrupted in human disease through developmental genetic defects such as spina bifida or during cancer progression also provides impetus to better understand how one might be able to reverse the effects of misregulation of the Scribble polarity module. Of note, the observation that blocking deregulated signaling, such as MAPK, following loss of Scribble in epithelial tissue in vivo can rescue tissue disorganization and proliferation defects (Pearson et al. 2011; Godde et al. 2014) suggests that a better understanding of the effects of deregulation of the Scribble polarity module may be utilized to therapeutically target diseases of tissue disorganization, such as cancer.

Acknowledgments We thank Marta Portela for critical reading of the manuscript. P.O.H., S.R., and H.E.R. were supported from funding from the ARC and NHMRC of Australia.

Glossary

Scrib Scribble
Dlg Discs large
Lgl Lethal giant larvae
AB Apical–basal
FR Front–rear
ACD Asymmetric cell division
PCP Planar cell polarity

References

Adey NB, Huang L, Ormonde PA, Baumgard ML, Pero R, Byreddy DV, Tavtigian SV, Bartel PL (2000) Threonine phosphorylation of the MMAC1/PTEN PDZ binding domain both inhibits and stimulates PDZ binding. Cancer Res 60(1):35–37

Albertson R, Doe CQ (2003) Dlg, Scrib and Lgl regulate neuroblast cell size and mitotic spindle asymmetry. Nat Cell Biol 5(2):166–170. doi:10.1038/ncb922

Amoyel M, Bach EA (2014) Cell competition: how to eliminate your neighbours. Development 141(5):988–1000. doi:10.1242/dev.079129, 141/5/988 [pii]

Armenti ST, Nance J (2012) Adherens junctions in C. elegans embryonic morphogenesis. Subcell Biochem 60:279–299. doi:10.1007/978-94-007-4186-7_12

Arnaud C, Sebbagh M, Nola S, Audebert S, Bidaut G, Hermant A, Gayet O, Dusetti NJ, Ollendorff V, Santoni MJ, Borg JP, Lecine P (2009) MCC, a new interacting protein for Scrib, is required for cell migration in epithelial cells. FEBS Lett 583(14):2326–2332

Arquier N, Perrin L, Manfruelli P, Semeriva M (2001) The Drosophila tumor suppressor gene lethal(2)giant larvae is required for the emission of the Decapentaplegic signal. Development 128(12):2209–2220

Atwood SX, Prehoda KE (2009) aPKC phosphorylates Miranda to polarize fate determinants during neuroblast asymmetric cell division. Curr Biol 19(9):723–729. doi:10.1016/j.cub.2009.03.056, S0960-9822(09)00910-5 [pii]

Audebert S, Navarro C, Nourry C, Chasserot-Golaz S, Lecine P, Bellaiche Y, Dupont JL, Premont RT, Sempere C, Strub JM, Van Dorsselaer A, Vitale N, Borg JP (2004) Mammalian Scribble forms a tight complex with the betaPIX exchange factor. Curr Biol 14(11):987–995

Bahri S, Wang S, Conder R, Choy J, Vlachos S, Dong K, Merino C, Sigrist S, Molnar C, Yang X, Manser E, Harden N (2010) The leading edge during dorsal closure as a model for epithelial plasticity: Pak is required for recruitment of the Scribble complex and septate junction formation. Development 137(12):2023–2032. doi:10.1242/dev.045088

Banks L, Pim D, Thomas M (2012) Human tumour viruses and the deregulation of cell polarity in cancer. Nat Rev Cancer 12(12):877–886. doi:10.1038/nrc3400

Baron M (2012) Endocytic routes to Notch activation. Semin Cell Dev Biol 23(4):437–442. doi:10.1016/j.semcdb.2012.01.008, S1084-9521(12)00012-2 [pii]

Beatty A, Morton D, Kemphues K (2010) The C. elegans homolog of Drosophila Lethal giant larvae functions redundantly with PAR-2 to maintain polarity in the early embryo. Development 137(23):3995–4004. doi:10.1242/dev.056028, dev.056028 [pii]

Belotti E, Polanowska J, Daulat AM, Audebert S, Thome V, Lissitzky JC, Lembo F, Biblek K, Omi S, Lenfant N, Gangar A, Montcouquiol M, Santoni MJ, Sebbagh M, Aurrand-Lions M, Angers S, Kodjabachian L, Reboul J, Borg JP (2013) The human PDZome: a gateway to PDZ mediated functions. Mol Cell Proteomics. doi:10.1074/mcp.O112.021022

Bergstralh DT, Lovegrove HE, St Johnston D (2013) Curr Biol 23:1707–1712

Betschinger J, Eisenhaber F, Knoblich JA (2005) Phosphorylation-induced autoinhibition regulates the cytoskeletal protein Lethal (2) giant larvae. Curr Biol 15(3):276–282

Bilder D (2004) Epithelial polarity and proliferation control: links from the Drosophila neoplastic tumor suppressors. Genes Dev 18(16):1909–1925

Bilder D, Li M, Perrimon N (2000) Cooperative regulation of cell polarity and growth by Drosophila tumor suppressors. Science 289(5476):113–116

Bilder D, Perrimon N (2000) Localization of apical epithelial determinants by the basolateral PDZ protein Scribble. Nature 403(6770):676–680

Blankenship JT, Fuller MT, Zallen JA (2007) The Drosophila homolog of the Exo84 exocyst subunit promotes apical epithelial identity. J Cell Sci 120(Pt 17):3099–3110

Bouwmeester T, Bauch A, Ruffner H, Angrand PO, Bergamini G, Croughton K, Cruciat C, Eberhard D, Gagneur J, Ghidelli S, Hopf C, Huhse B, Mangano R, Michon AM, Schirle M, Schlegl J, Schwab M, Stein MA, Bauer A, Casari G, Drewes G, Gavin AC, Jackson DB, Joberty G, Neubauer G, Rick J, Kuster B, Superti-Furga G (2004) A physical and functional map of the human TNF-alpha/NF-kappa B signal transduction pathway. Nat Cell Biol 6 (2):97–105. doi:10.1038/ncb1086, ncb1086 [pii]

Brooks SP, Coccia M, Tang HR, Kanuga N, Machesky LM, Bailly M, Cheetham ME, Hardcastle AJ (2010) The Nance-Horan syndrome protein encodes a functional WAVE homology domain (WHD) and is important for co-ordinating actin remodelling and maintaining cell morphology. Hum Mol Genet 19(12):2421–2432. doi:10.1093/hmg/ddq125, dq125 [pii]

Brumby AM, Richardson HE (2003) Scribble mutants cooperate with oncogenic Ras or Notch to cause neoplastic overgrowth in Drosophila. EMBO J 22(21):5769–5779

Buchsbaum RJ, Connolly BA, Feig LA (2002) Interaction of Rac exchange factors Tiam1 and Ras-GRF1 with a scaffold for the p38 mitogen-activated protein kinase cascade. Mol Cell Biol 22(12):4073–4085

Bunney TD, Katan M (2010) Phosphoinositide signalling in cancer: beyond PI3K and PTEN. Nat Rev Cancer 10(5):342–352. doi:10.1038/nrc2842, nrc2842 [pii]

Caddy J, Wilanowski T, Darido C, Dworkin S, Ting SB, Zhao Q, Rank G, Auden A, Srivastava S, Papenfuss TA, Murdoch JN, Humbert PO, Parekh V, Boulos N, Weber T, Zuo J, Cunningham JM, Jane SM (2010) Epidermal wound repair is regulated by the planar cell polarity signaling pathway. Dev Cell 19(1):138–147. doi:10.1016/j.devcel.2010.06.008, S1534-5807(10)00298-4 [pii]

Caruana G, Bernstein A (2001) Craniofacial dysmorphogenesis including cleft palate in mice with an insertional mutation in the discs large gene. Mol Cell Biol 21(5):1475–1483

Chang JT, Ciocca ML, Kinjyo I, Palanivel VR, McClurkin CE, Dejong CS, Mooney EC, Kim JS, Steinel NC, Oliaro J, Yin CC, Florea BI, Overkleeft HS, Berg LJ, Russell SM, Koretzky GA, Jordan MS, Reiner SL (2011) Asymmetric proteasome segregation as a mechanism for unequal partitioning of the transcription factor T-bet during T lymphocyte division. Immunity 34 (4):492–504. doi:10.1016/j.immuni.2011.03.017, S1074-7613(11)00125-7 [pii]

Chang JT, Palanivel VR, Kinjyo I, Schambach F, Intlekofer AM, Banerjee A, Longworth SA, Vinup KE, Mrass P, Oliaro J, Killeen N, Orange JS, Russell SM, Weninger W, Reiner SL (2007) Asymmetric T lymphocyte division in the initiation of adaptive immune responses. Science 315(5819):1687–1691. doi:10.1126/science.1139393

Chen CL, Schroeder MC, Kango-Singh M, Tao C, Halder G (2012) Tumor suppression by cell competition through regulation of the Hippo pathway. Proc Natl Acad Sci U S A 109 (2):484–489. doi:10.1073/pnas.1113882109, 1113882109 [pii]

Chen X, Macara IG (2005) Par-3 controls tight junction assembly through the Rac exchange factor Tiam1. Nat Cell Biol 7(3):262–269

Cho KO, Hunt CA, Kennedy MB (1992) The rat brain postsynaptic density fraction contains a homolog of the Drosophila discs-large tumor suppressor protein. Neuron 9(5):929–942

Clark BS, Cui S, Miesfeld JB, Klezovitch O, Vasioukhin V, Link BA (2012) Loss of Llgl1 in retinal neuroepithelia reveals links between apical domain size, Notch activity and neurogenesis. Development 139(9):1599–1610. doi:10.1242/dev.078097

Claveria C, Giovinazzo G, Sierra R, Torres M (2013) Myc-driven endogenous cell competition in the early mammalian embryo. Nature 500:39–44

Clevers H, Nusse R (2012) Wnt/beta-catenin signaling and disease. Cell 149(6):1192–1205. doi:10.1016/j.cell.2012.05.012, S0092-8674(12)00586-7 [pii]

Cordenonsi M, Zanconato F, Azzolin L, Forcato M, Rosato A, Frasson C, Inui M, Montagner M, Parenti AR, Poletti A, Daidone MG, Dupont S, Basso G, Bicciato S, Piccolo S (2011) The Hippo transducer TAZ confers cancer stem cell-related traits on breast cancer cells. Cell 147 (4):759–772. doi:10.1016/j.cell.2011.09.048, S0092-8674(11)01218-9 [pii]

Cordero JB, Macagno JP, Stefanatos RK, Strathdee KE, Cagan RL, Vidal M (2010) Oncogenic Ras diverts a host TNF tumor suppressor activity into tumor promoter. Dev Cell 18 (6):999–1011

Courbard JR, Djiane A, Wu J, Mlodzik M (2009) The apical/basal-polarity determinant Scribble cooperates with the PCP core factor Stbm/Vang and functions as one of its effectors. Dev Biol 333(1):67–77. doi:10.1016/j.ydbio.2009.06.024, S0012-1606(09)00993-2 [pii]

Cuthbert PC, Stanford LE, Coba MP, Ainge JA, Fink AE, Opazo P, Delgado JY, Komiyama NH, O'Dell TJ, Grant SG (2007) Synapse-associated protein 102/dlgh3 couples the NMDA receptor to specific plasticity pathways and learning strategies. J Neurosci 27(10):2673–2682. doi:10.1523/JNEUROSCI.4457-06.2007

Dahan I, Petrov D, Cohen-Kfir E, Ravid S (2014) The tumor suppressor Lgl1 forms discrete complexes with NMII-A and Par6alpha-aPKCzeta that are affected by Lgl1 phosphorylation. J Cell Sci 127(Pt 2):295–304. doi:10.1242/jcs.127357, jcs.127357 [pii]

Dahan I, Yearim A, Touboul Y, Ravid S (2012) The tumor suppressor Lgl1 regulates NMII-A cellular distribution and focal adhesion morphology to optimize cell migration. Mol Biol Cell. doi:10.1091/mbc.E11-01-0015, mbc.E11-01-0015 [pii]

de Beco S, Ziosi M, Johnston LA (2012) New frontiers in cell competition. Dev Dyn 241 (5):831–841. doi:10.1002/dvdy.23783

de Vreede G, Schoenfeld JD, Windler SL, Morrison H, Lu H, Bilder D (2014) The Scribble module regulates retromer-dependent endocytic trafficking during epithelial polarization. Development 141(14):2796–2802. doi:10.1242/dev.105403, 141/14/2796 [pii]

Doggett K, Grusche FA, Richardson HE, Brumby AM (2011) Loss of the Drosophila cell polarity regulator Scribbled promotes epithelial tissue overgrowth and cooperation with oncogenic Ras-Raf through impaired Hippo pathway signaling. BMC Dev Biol 11:57. doi:10.1186/1471-213X-11-57, 1471-213X-11-57 [pii]

Dollar GL, Weber U, Mlodzik M, Sokol SY (2005) Regulation of Lethal giant larvae by Dishevelled. Nature 437(7063):1376–1380

Dow LE, Elsum IA, King CL, Kinross KM, Richardson HE, Humbert PO (2008) Loss of human Scribble cooperates with H-Ras to promote cell invasion through deregulation of MAPK signalling. Oncogene 27(46):5988–6001. doi:10.1038/onc.2008.219, onc2008219 [pii]

Dow LE, Humbert PO (2007) Polarity regulators and the control of epithelial architecture, cell migration, and tumorigenesis. Int Rev Cytol 262:253–302

Dow LE, Kauffman JS, Caddy J, Zarbalis K, Peterson AS, Jane SM, Russell SM, Humbert PO (2007) The tumour-suppressor Scribble dictates cell polarity during directed epithelial migration: regulation of Rho GTPase recruitment to the leading edge. Oncogene 26 (16):2272–2282. doi:10.1038/sj.onc.1210016

Eastburn DJ, Zegers MM, Mostov KE (2012) Scrib regulates HGF-mediated epithelial morphogenesis and is stabilized by Sgt1-HSP90. J Cell Sci 125(Pt 17):4147–4157. doi:10.1242/jcs.108670

Elsum I, Yates L, Humbert PO, Richardson HE (2012) The Scribble-Dlg-Lgl polarity module in development and cancer: from flies to man. Essays Biochem 53:141–168. doi:10.1042/bse0530141, bse0530141 [pii]

Elsum IA, Martin C, Humbert PO (2013a) Scribble regulates an EMT-polarity pathway through modulation of MAPK-ERK signaling to mediate junction formation. J Cell Sci. doi:10.1242/jcs.129387, jcs.129387 [pii]

Elsum IA, Yates LL, Pearson HB, Phesse TJ, Long F, O'Donoghue R, Ernst M, Cullinane C, Humbert PO (2013b) Scrib heterozygosity predisposes to lung cancer and cooperates with KRas hyperactivation to accelerate lung cancer progression in vivo. Oncogene. doi:10.1038/onc.2013.498, onc2013498 [pii]

Etienne-Manneville S, Manneville JB, Nicholls S, Ferenczi MA, Hall A (2005) Cdc42 and Par6-PKCzeta regulate the spatially localized association of Dlg1 and APC to control cell polarization. J Cell Biol 170(6):895–901

Facciuto F, Cavatorta AL, Valdano MB, Marziali F, Gardiol D (2012) Differential expression of PDZ domain-containing proteins in human diseases - challenging topics and novel issues. FEBS J 279(19):3538–3548. doi:10.1111/j.1742-4658.2012.08699.x

Feigin ME, Akshinthala SD, Araki K, Rosenberg AZ, Muthuswamy LB, Martin B, Lehmann BD, Berman HK, Pietenpol JA, Cardiff RD, Muthuswamy SK (2014) Mislocalization of the cell polarity protein scribble promotes mammary tumorigenesis and is associated with basal breast cancer. Cancer Res 74(11):3180–3194. doi:10.1158/0008-5472.CAN-13-3415, 0008-5472.CAN-13-3415 [pii]

Feng W, Wu H, Chan LN, Zhang M (2008) Par-3-mediated junctional localization of the lipid phosphatase PTEN is required for cell polarity establishment. J Biol Chem 283 (34):23440–23449. doi:10.1074/jbc.M802482200, M802482200 [pii]

Fernandez-Medarde A, Santos E (2011) Ras in cancer and developmental diseases. Genes Cancer 2(3):344–358. doi:10.1177/1947601911411084, 10.1177_1947601911411084 [pii]

Fichelson P, Jagut M, Lepanse S, Lepesant JA, Huynh JR (2010) Lethal giant larvae is required with the par genes for the early polarization of the Drosophila oocyte. Development 137 (5):815–824. doi:10.1242/dev.045013

Firestein BL, Rongo C (2001) DLG-1 is a MAGUK similar to SAP97 and is required for adherens junction formation. Mol Biol Cell 12(11):3465–3475

Fortini ME (2009) Notch signaling: the core pathway and its posttranslational regulation. Dev Cell 16(5):633–647. doi:10.1016/j.devcel.2009.03.010, S1534-5807(09)00131-2 [pii]

Frese KK, Latorre IJ, Chung SH, Caruana G, Bernstein A, Jones SN, Donehower LA, Justice MJ, Garner CC, Javier RT (2006) Oncogenic function for the Dlg1 mammalian homolog of the Drosophila discs-large tumor suppressor. EMBO J 25(6):1406–1417

Froldi F, Ziosi M, Garoia F, Pession A, Grzeschik NA, Bellosta P, Strand D, Richardson HE, Grifoni D (2010) The lethal giant larvae tumour suppressor mutation requires dMyc oncoprotein to promote clonal malignancy. BMC Biol 8:33

Funke L, Dakoji S, Bredt DS (2005) Membrane-associated guanylate kinases regulate adhesion and plasticity at cell junctions. Annu Rev Biochem 74:219–245

Gangar A, Rossi G, Andreeva A, Hales R, Brennwald P (2005) Structurally conserved interaction of Lgl family with SNAREs is critical to their cellular function. Curr Biol 15(12):1136–1142

Garcia RA, Vasudevan K, Buonanno A (2000) The neuregulin receptor ErbB-4 interacts with PDZ-containing proteins at neuronal synapses. Proc Natl Acad Sci U S A 97(7):3596–3601. doi:10.1073/pnas.070042497, 070042497 [pii]

Gardiol D, Kuhne C, Glaunsinger B, Lee SS, Javier R, Banks L (1999) Oncogenic human papillomavirus E6 proteins target the discs large tumour suppressor for proteasome-mediated degradation. Oncogene 18(40):5487–5496

Gardoni F, Marcello E, Di Luca M (2009) Postsynaptic density-membrane associated guanylate kinase proteins (PSD-MAGUKs) and their role in CNS disorders. Neuroscience 158 (1):324–333. doi:10.1016/j.neuroscience.2008.07.068, S0306-4522(08)01130-5 [pii]

Gateff E, Schneiderman HA (1974) Developmental capacities of benign and malignant neoplasms of Drosophila. Roux's Arch Dev Biol 176:23–65

Gaudet S, Langlois MJ, Lue RA, Rivard N, Viel A (2011) The MEK2-binding tumor suppressor hDlg is recruited by E-cadherin to the midbody ring. BMC Cell Biol 12:55. doi:10.1186/1471-2121-12-55, 1471-2121-12-55 [pii]

Gladstone M, Su TT (2011) Chemical genetics and drug screening in Drosophila cancer models. J Genet Genomics 38(10):497–504. doi:10.1016/j.jgg.2011.09.003

Gmyrek GB, Graham DB, Sandoval GJ, Blaufuss GS, Akilesh HM, Fujikawa K, Xavier RJ, Swat W (2013) Polarity gene discs large homolog 1 regulates the generation of memory T cells. Eur J Immunol 43(5):1185–1194. doi:10.1002/eji.201142362

Godde NJ, Sheridan JM, Smith LK, Pearson HB, Britt KL, Galea RC, Yates LL, Visvader JE, Humbert PO (2014) Scribble modulates the MAPK/Fra1 pathway to disrupt luminal and ductal integrity and suppress tumour formation in the mammary gland. PLoS Genet 10(5):e1004323. doi:10.1371/journal.pgen.1004323, PGENETICS-D-13-03014 [pii]

Goh LH, Zhou X, Lee MC, Lin S, Wang H, Luo Y, Yang X (2013) Clueless regulates aPKC activity and promotes self-renewal cell fate in Drosophila lgl mutant larval brains. Dev Biol 381(2):353–364. doi:10.1016/j.ydbio.2013.06.031, S0012-1606(13)00364-3 [pii]

Gont A, Hanson JE, Lavictoire SJ, Parolin DA, Daneshmand M, Restall IJ, Soucie M, Nicholas G, Woulfe J, Kassam A, Da Silva VF, Lorimer IA (2013) PTEN loss represses glioblastoma tumor initiating cell differentiation via inactivation of Lgl1. Oncotarget 4(8):1266–1279, doi:1164 [pii]

Gonzalez C (2013) Drosophila melanogaster: a model and a tool to investigate malignancy and identify new therapeutics. Nat Rev Cancer 13(3):172–183. doi:10.1038/nrc3461

Grifoni D, Froldi F, Pession A (2013) Connecting epithelial polarity, proliferation and cancer in Drosophila: the many faces of lgl loss of function. Int J Dev Biol 57(9–10):677–687. doi:10.1387/ijdb.130285dg

Grosshans BL, Andreeva A, Gangar A, Niessen S, Yates JR 3rd, Brennwald P, Novick P (2006) The yeast lgl family member Sro7p is an effector of the secretory Rab GTPase Sec4p. J Cell Biol 172(1):55–66

Grusche FA, Degoutin JL, Richardson HE, Harvey KF (2011) The Salvador/Warts/Hippo pathway controls regenerative tissue growth in Drosophila melanogaster. Dev Biol 350(2):255–266

Grzeschik NA, Parsons LM, Allott ML, Harvey KF, Richardson HE (2010a) Lgl, aPKC, and Crumbs regulate the Salvador/Warts/Hippo pathway through two distinct mechanisms. Curr Biol 20(7):573–581. doi:10.1016/j.cub.2010.01.055, S0960-9822(10)00151-X [pii]

Grzeschik NA, Parsons LM, Richardson HE (2010b) Lgl, the SWH pathway and tumorigenesis: it's a matter of context & competition! Cell Cycle 9(16):3202–3212. doi:10.4161/cc.9.16.12633, 12633 [pii]

Haenfler JM, Kuang C, Lee CY (2012) Cortical aPKC kinase activity distinguishes neural stem cells from progenitor cells by ensuring asymmetric segregation of Numb. Dev Biol 365 (1):219–228. doi:10.1016/j.ydbio.2012.02.027, S0012-1606(12)00104-2 [pii]

Halaoui R, McCaffrey L (2014) Rewiring cell polarity signaling in cancer. Oncogene. doi:10.1038/onc.2014.59

Hattendorf DA, Andreeva A, Gangar A, Brennwald PJ, Weis WI (2007) Structure of the yeast polarity protein Sro7 reveals a SNARE regulatory mechanism. Nature 446(7135):567–571. doi:10.1038/nature05635, nature05635 [pii]

Hava D, Forster U, Matsuda M, Cui S, Link BA, Eichhorst J, Wiesner B, Chitnis A, Abdelilah-Seyfried S (2009) Apical membrane maturation and cellular rosette formation during morphogenesis of the zebrafish lateral line. J Cell Sci 122(Pt 5):687–695. doi:10.1242/jcs.032102

Hawkins ED, Oliaro J, Kallies A, Belz GT, Filby A, Hogan T, Haynes N, Ramsbottom KM, Van Ham V, Kinwell T, Seddon B, Davies D, Tarlinton D, Lew AM, Humbert PO, Russell SM (2013) Regulation of asymmetric cell division and polarity by Scribble is not required for humoral immunity. Nat Commun 4:1801. doi:10.1038/ncomms2796

Hawkins ED, Oliaro J, Ramsbottom KM, Ting SB, Sacirbegovic F, Harvey M, Kinwell T, Ghysdael J, Johnstone RW, Humbert PO, Russell SM (2014) Lethal giant larvae 1 tumour suppressor activity is not conserved in models of Mammalian T and B cell leukaemia. PLoS One 9(1):e87376. doi:10.1371/journal.pone.0087376

Heidel FH, Bullinger L, Arreba-Tutusaus P, Wang Z, Gaebel J, Hirt C, Niederwieser D, Lane SW, Dohner K, Vasioukhin V, Fischer T, Armstrong SA (2013) The cell fate determinant Llgl1 influences HSC fitness and prognosis in AML. J Exp Med 210(1):15–22. doi:10.1084/jem.20120596, jem.20120596 [pii]

Hendrickx A, Beullens M, Ceulemans H, Den Abt T, Van Eynde A, Nicolaescu E, Lesage B, Bollen M (2009) Docking motif-guided mapping of the interactome of protein phosphatase-1. Chem Biol 16(4):365–371. doi:10.1016/j.chembiol.2009.02.012, S1074-5521(09)00079-9 [pii]

Hering H, Sheng M (2002) Direct interaction of Frizzled-1, -2, -4, and -7 with PDZ domains of PSD-95. FEBS Lett 521(1–3):185–189

Hoege C, Constantinescu AT, Schwager A, Goehring NW, Kumar P, Hyman AA (2010) LGL can partition the cortex of one-cell Caenorhabditis elegans embryos into two domains. Curr Biol 20 (14):1296–1303. doi:10.1016/j.cub.2010.05.061, S0960-9822(10)00710-4 [pii]

Hogan C (2012) Impact of interactions between normal and transformed epithelial cells and the relevance to cancer. Cell Mol Life Sci 69(2):203–213. doi:10.1007/s00018-011-0806-3

Huganir RL, Nicoll RA (2013) AMPARs and synaptic plasticity: the last 25 years. Neuron 80 (3):704–717. doi:10.1016/j.neuron.2013.10.025

Humbert P, Russell S, Richardson H (2003) Dlg, Scribble and Lgl in cell polarity, cell proliferation and cancer. Bioessays 25(6):542–553

Humbert PO, Dow LE, Russell SM (2006) The Scribble and Par complexes in polarity and migration: friends or foes? Trends Cell Biol 16(12):622–630. doi:10.1016/j.tcb.2006.10.005

Humbert PO, Grzeschik NA, Brumby AM, Galea R, Elsum I, Richardson HE (2008) Control of tumourigenesis by the Scribble/Dlg/Lgl polarity module. Oncogene 27(55):6888–6907. doi:10.1038/onc.2008.341, onc2008341 [pii]

Humphries LA, Shaffer MH, Sacirbegovic F, Tomassian T, McMahon KA, Humbert PO, Silva O, Round JL, Takamiya K, Huganir RL, Burkhardt JK, Russell SM, Miceli MC (2012) Characterization of in vivo Dlg1 deletion on T cell development and function. PLoS One 7(9):e45276. doi:10.1371/journal.pone.0045276

Igaki T, Pagliarini RA, Xu T (2006) Loss of cell polarity drives tumor growth and invasion through JNK activation in Drosophila. Curr Biol 16(11):1139–1146. doi:10.1016/j.cub.2006.04.042, S0960-9822(06)01539-9 [pii]

Igaki T, Pastor-Pareja JC, Aonuma H, Miura M, Xu T (2009) Intrinsic tumor suppression and epithelial maintenance by endocytic activation of Eiger/TNF signaling in Drosophila. Dev Cell 16(3):458–465. doi:10.1016/j.devcel.2009.01.002, S1534-5807(09)00030-6 [pii]

Iizuka-Kogo A, Ishidao T, Akiyama T, Senda T (2007) Abnormal development of urogenital organs in Dlgh1-deficient mice. Development 134(9):1799–1807

Ishidate T, Matsumine A, Toyoshima K, Akiyama T (2000) The APC-hDLG complex negatively regulates cell cycle progression from the G0/G1 to S phase. Oncogene 19(3):365–372

Ivarsson Y, Arnold R, McLaughlin M, Nim S, Joshi R, Ray D, Liu B, Teyra J, Pawson T, Moffat J, Li SS, Sidhu SS, Kim PM (2014) Large-scale interaction profiling of PDZ domains through proteomic peptide-phage display using human and viral phage peptidomes. Proc Natl Acad Sci U S A 111(7):2542–2547. doi:10.1073/pnas.1312296111

Javier RT, Rice AP (2011) Emerging theme: cellular PDZ proteins as common targets of pathogenic viruses. J Virol 85(22):11544–11556. doi:10.1128/JVI.05410-11

Jiang Y, Scott KL, Kwak SJ, Chen R, Mardon G (2011) Sds22/PP1 links epithelial integrity and tumor suppression via regulation of myosin II and JNK signaling. Oncogene 30 (29):3248–3260. doi:10.1038/onc.2011.46, onc201146 [pii]

Justice N, Roegiers F, Jan LY, Jan YN (2003) Lethal giant larvae acts together with numb in notch inhibition and cell fate specification in the Drosophila adult sensory organ precursor lineage. Curr Biol 13(9):778–783, doi:S0960982203002884 [pii]

Kallay LM, McNickle A, Brennwald PJ, Hubbard AL, Braiterman LT (2006) Scribble associates with two polarity proteins, Lgl2 and Vangl2, via distinct molecular domains. J Cell Biochem 99(2):647–664

Kaplan NA, Tolwinski NS (2010) Spatially defined Dsh-Lgl interaction contributes to directional tissue morphogenesis. J Cell Sci 123(Pt 18):3157–3165

Katoh M, Katoh M (2004) Identification and characterization of human GUKH2 gene in silico. Int J Oncol 24(4):1033–1038

Kemphues KJ, Priess JR, Morton DG, Cheng NS (1988) Identification of genes required for cytoplasmic localization in early C. elegans embryos. Cell 52(3):311–320, doi:S0092-8674 (88)80024-2 [pii]

King CG, Koehli S, Hausmann B, Schmaler M, Zehn D, Palmer E (2012) T cell affinity regulates asymmetric division, effector cell differentiation, and tissue pathology. Immunity 37 (4):709–720. doi:10.1016/j.immuni.2012.06.021

Kiyono T, Hiraiwa A, Fujita M, Hayashi Y, Akiyama T, Ishibashi M (1997) Binding of high-risk human papillomavirus E6 oncoproteins to the human homologue of the Drosophila discs large tumor suppressor protein. Proc Natl Acad Sci U S A 94(21):11612–11616

Klezovitch O, Fernandez TE, Tapscott SJ, Vasioukhin V (2004) Loss of cell polarity causes severe brain dysplasia in Lgl1 knockout mice. Genes Dev 18(5):559–571

Kobe B, Kajava AV (2001) The leucine-rich repeat as a protein recognition motif. Curr Opin Struct Biol 11(6):725–732

Koh YH, Popova E, Thomas U, Griffith LC, Budnik V (1999) Regulation of DLG localization at synapses by CaMKII-dependent phosphorylation. Cell 98(3):353–363

Kong K, Kumar M, Taruishi M, Javier RT (2014) The human adenovirus E4-ORF1 protein subverts discs large 1 to mediate membrane recruitment and dysregulation of phosphatidylinositol 3-kinase. PLoS Pathog 10(5):e1004102. doi:10.1371/journal.ppat.1004102

Koppen M, Simske JS, Sims PA, Firestein BL, Hall DH, Radice AD, Rongo C, Hardin JD (2001) Cooperative regulation of AJM-1 controls junctional integrity in Caenorhabditis elegans epithelia. Nat Cell Biol 3(11):983–991

Kronen MR, Schoenfelder KP, Klein AM, Nystul TG (2014) Basolateral junction proteins regulate competition for the follicle stem cell niche in the Drosophila ovary. PLoS One 9(7):e101085. doi:10.1371/journal.pone.0101085, PONE-D-14-21899 [pii]

Kulshammer E, Uhlirova M (2013) The actin cross-linker Filamin/Cheerio mediates tumor malignancy downstream of JNK signaling. J Cell Sci 126(Pt 4):927–938. doi:10.1242/jcs.114462, jcs.114462 [pii]

Kuwada SK, Lund KA, Li XF, Cliften P, Amsler K, Opresko LK, Wiley HS (1998) Differential signaling and regulation of apical vs. basolateral EGFR in polarized epithelial cells. Am J Physiol 275(6 Pt 1):C1419–C1428

Lahuna O, Quellari M, Achard C, Nola S, Meduri G, Navarro C, Vitale N, Borg JP, Misrahi M (2005) Thyrotropin receptor trafficking relies on the hScrib-betaPIX-GIT1-ARF6 pathway. EMBO J 24(7):1364–1374

Lambert JM, Lambert QT, Reuther GW, Malliri A, Siderovski DP, Sondek J, Collard JG, Der CJ (2002) Tiam1 mediates Ras activation of Rac by a PI(3)K-independent mechanism. Nat Cell Biol 4(8):621–625. doi:10.1038/ncb833, ncb833 [pii]

Langevin J, Le Borgne R, Rosenfeld F, Gho M, Schweisguth F, Bellaiche Y (2005) Lethal giant larvae controls the localization of notch-signaling regulators numb, neuralized, and Sanpodo in

Drosophila sensory-organ precursor cells. Curr Biol 15(10):955–962. doi:10.1016/j.cub.2005.04.054, S0960-9822(05)00455-0 [pii]

Lasserre R, Charrin S, Cuche C, Danckaert A, Thoulouze MI, de Chaumont F, Duong T, Perrault N, Varin-Blank N, Olivo-Marin JC, Etienne-Manneville S, Arpin M, Di Bartolo V, Alcover A (2010) Ezrin tunes T-cell activation by controlling Dlg1 and microtubule positioning at the immunological synapse. EMBO J 29(14):2301–2314. doi:10.1038/emboj.2010.127, emboj2010127 [pii]

Lee SS, Weiss RS, Javier RT (1997) Binding of human virus oncoproteins to hDlg/SAP97, a mammalian homolog of the Drosophila discs large tumor suppressor protein. Proc Natl Acad Sci U S A 94(13):6670–6675

Legouis R, Gansmuller A, Sookhareea S, Bosher JM, Baillie DL, Labouesse M (2000) LET-413 is a basolateral protein required for the assembly of adherens junctions in Caenorhabditis elegans. Nat Cell Biol 2(7):415–422

Legouis R, Jaulin-Bastard F, Schott S, Navarro C, Borg JP, Labouesse M (2003) Basolateral targeting by leucine-rich repeat domains in epithelial cells. EMBO Rep 4(11):1096–1102

Lei Y, Zhu H, Duhon C, Yang W, Ross ME, Shaw GM, Finnell RH (2013) Mutations in planar cell polarity gene SCRIB are associated with spina bifida. PLoS One 8(7):e69262. doi:10.1371/journal.pone.0069262

Leong GR, Goulding KR, Amin N, Richardson HE, Brumby AM (2009) Scribble mutants promote aPKC and JNK-dependent epithelial neoplasia independently of Crumbs. BMC Biol 7:62

Li Q, Feng S, Yu L, Zhao G, Li M (2011a) Requirements of Lgl in cell differentiation and motility during Drosophila ovarian follicular epithelium morphogenesis. Fly (Austin) 5(2):81–87

Li X, Yang H, Liu J, Schmidt MD, Gao T (2011b) Scribble-mediated membrane targeting of PHLPP1 is required for its negative regulation of Akt. EMBO Rep. doi:10.1038/embor.2011.106, embor2011106 [pii]

Liu W, Chen E, Zhao XW, Wan ZP, Gao YR, Davey A, Huang E, Zhang L, Crocetti J, Sandoval G, Joyce MG, Miceli C, Lukszo J, Aravind L, Swat W, Brzostowski J, Pierce SK (2012) The scaffolding protein synapse-associated protein 97 is required for enhanced signaling through isotype-switched IgG memory B cell receptors. Sci Signal 5(235):ra54. doi:10.1126/scisignal.2002820

Lohia M, Qin Y, Macara IG (2012) The Scribble polarity protein stabilizes E-cadherin/p120-catenin binding and blocks retrieval of E-cadherin to the Golgi. PLoS One 7(11):e51130. doi:10.1371/journal.pone.0051130, PONE-D-12-27172 [pii]

Lolo FN, Casas-Tinto S, Moreno E (2012) Cell competition time line: winners kill losers, which are extruded and engulfed by hemocytes. Cell Rep 2(3):526–539. doi:10.1016/j.celrep.2012.08.012, S2211-1247(12)00237-9 [pii]

Ludford-Menting MJ, Oliaro J, Sacirbegovic F, Cheah ET, Pedersen N, Thomas SJ, Pasam A, Iazzolino R, Dow LE, Waterhouse NJ, Murphy A, Ellis S, Smyth MJ, Kershaw MH, Darcy PK, Humbert PO, Russell SM (2005) A network of PDZ-containing proteins regulates T cell polarity and morphology during migration and immunological synapse formation. Immunity 22(6):737–748. doi:10.1016/j.immuni.2005.04.009

Luyten A, Mortier E, Van Campenhout C, Taelman V, Degeest G, Wuytens G, Lambaerts K, David G, Bellefroid EJ, Zimmermann P (2008) The postsynaptic density 95/disc-large/zona occludens protein syntenin directly interacts with frizzled 7 and supports noncanonical Wnt signaling. Mol Biol Cell 19(4):1594–1604. doi:10.1091/mbc.E07-08-0832, E07-08-0832 [pii]

Macara IG, Mili S (2008) Polarity and differential inheritance–universal attributes of life? Cell 135(5):801–812. doi:10.1016/j.cell.2008.11.006, S0092-8674(08)01392-5 [pii]

Mahoney ZX, Sammut B, Xavier RJ, Cunningham J, Go G, Brim KL, Stappenbeck TS, Miner JH, Swat W (2006) Discs-large homolog 1 regulates smooth muscle orientation in the mouse ureter. Proc Natl Acad Sci U S A 103(52):19872–19877

Maiga O, Philippe M, Kotelevets L, Chastre E, Benadda S, Pidard D, Vranckx R, Walch L (2011) Identification of mitogen-activated protein/extracellular signal-responsive kinase kinase 2 as a

novel partner of the scaffolding protein human homolog of disc-large. FEBS J 278 (15):2655–2665. doi:10.1111/j.1742-4658.2011.08192.x
Manneville JB, Jehanno M, Etienne-Manneville S (2010) Dlg1 binds GKAP to control dynein association with microtubules, centrosome positioning, and cell polarity. J Cell Biol 191 (3):585–598. doi:10.1083/jcb.201002151, jcb.201002151 [pii]
Massimi P, Narayan N, Cuenda A, Banks L (2006) Phosphorylation of the discs large tumour suppressor protein controls its membrane localisation and enhances its susceptibility to HPV E6-induced degradation. Oncogene 25(31):4276–4285
Massimi P, Narayan N, Thomas M, Gammoh N, Strand S, Strand D, Banks L (2008) Regulation of the hDlg/hScrib/Hugl-1 tumour suppressor complex. Exp Cell Res 314(18):3306–3317
Mathew D, Gramates LS, Packard M, Thomas U, Bilder D, Perrimon N, Gorczyca M, Budnik V (2002) Recruitment of scribble to the synaptic scaffolding complex requires GUK-holder, a novel DLG binding protein. Curr Biol 12(7):531–539
Matsumine A, Ogai A, Senda T, Okumura N, Satoh K, Baeg GH, Kawahara T, Kobayashi S, Okada M, Toyoshima K, Akiyama T (1996) Binding of APC to the human homolog of the Drosophila discs large tumor suppressor protein. Science 272(5264):1020–1023
Mayer BJ (2001) SH3 domains: complexity in moderation. J Cell Sci 114(Pt 7):1253–1263
McGee AW, Topinka JR, Hashimoto K, Petralia RS, Kakizawa S, Kauer F, Aguilera-Moreno A, Wenthold RJ, Kano M, Bredt DS (2001) PSD-93 knock-out mice reveal that neuronal MAGUKs are not required for development or function of parallel fiber synapses in cerebellum. J Neurosci 21(9):3085–3091
McMahon L, Legouis R, Vonesch JL, Labouesse M (2001) Assembly of C. elegans apical junctions involves positioning and compaction by LET-413 and protein aggregation by the MAGUK protein DLG-1. J Cell Sci 114(Pt 12):2265–2277
Mechler BM, McGinnis W, Gehring WJ (1985) Molecular cloning of lethal(2)giant larvae, a recessive oncogene of Drosophila melanogaster. EMBO J 4(6):1551–1557
Menendez J, Perez-Garijo A, Calleja M, Morata G (2010) A tumor-suppressing mechanism in Drosophila involving cell competition and the Hippo pathway. Proc Natl Acad Sci U S A 107 (33):14651–14656
Mertens AE, Rygiel TP, Olivo C, van der Kammen R, Collard JG (2005) The Rac activator Tiam1 controls tight junction biogenesis in keratinocytes through binding to and activation of the Par polarity complex. J Cell Biol 170(7):1029–1037. doi:10.1083/jcb.200502129, jcb.200502129 [pii]
Metodieva G, Nogueira-de-Souza NC, Greenwood C, Al-Janabi K, Leng L, Bucala R, Metodiev MV (2013) CD74-dependent deregulation of the tumor suppressor scribble in human epithelial and breast cancer cells. Neoplasia 15(6):660–668
Meyer MP, Trimmer JS, Gilthorpe JD, Smith SJ (2005) Characterization of zebrafish PSD-95 gene family members. J Neurobiol 63(2):91–105. doi:10.1002/neu.20118
Michaelis UR, Chavakis E, Kruse C, Jungblut B, Kaluza D, Wandzioch K, Manavski Y, Heide H, Santoni MJ, Potente M, Eble JA, Borg JP, Brandes RP (2013) The polarity protein Scrib is essential for directed endothelial cell migration. Circ Res 112(6):924–934. doi:10.1161/CIRCRESAHA.112.300592
Migaud M, Charlesworth P, Dempster M, Webster LC, Watabe AM, Makhinson M, He Y, Ramsay MF, Morris RG, Morrison JH, O'Dell TJ, Grant SG (1998) Enhanced long-term potentiation and impaired learning in mice with mutant postsynaptic density-95 protein. Nature 396 (6710):433–439
Montcouquiol M, Rachel RA, Lanford PJ, Copeland NG, Jenkins NA, Kelley MW (2003) Identification of Vangl2 and Scrb1 as planar polarity genes in mammals. Nature 423 (6936):173–177. doi:10.1038/nature01618, nature01618 [pii]
Moreau MM, Piguel N, Papouin T, Koehl M, Durand CM, Rubio ME, Loll F, Richard EM, Mazzocco C, Racca C, Oliet SH, Abrous DN, Montcouquiol M, Sans N (2010) The planar polarity protein Scribble1 is essential for neuronal plasticity and brain function. J Neurosci 30 (29):9738–9752. doi:10.1523/JNEUROSCI.6007-09.2010

Morin X, Bellaiche Y (2011) Mitotic spindle orientation in asymmetric and symmetric cell divisions during animal development. Dev Cell 21(1):102–119. doi:10.1016/j.devcel.2011.06.012, S1534-5807(11)00246-2 [pii]

Murdoch JN, Henderson DJ, Doudney K, Gaston-Massuet C, Phillips HM, Paternotte C, Arkell R, Stanier P, Copp AJ (2003) Disruption of scribble (Scrb1) causes severe neural tube defects in the circletail mouse. Hum Mol Genet 12(2):87–98

Musch A, Cohen D, Yeaman C, Nelson WJ, Rodriguez-Boulan E, Brennwald PJ (2002) Mammalian homolog of Drosophila tumor suppressor lethal (2) giant larvae interacts with basolateral exocytic machinery in Madin-Darby canine kidney cells. Mol Biol Cell 13(1):158–168

Nagasaka K, Massimi P, Pim D, Subbaiah VK, Kranjec C, Nakagawa S, Yano T, Taketani Y, Banks L (2010a) The mechanism and implications of hScrib regulation of ERK. Small GTPases 1(2):108–112. doi:10.4161/sgtp.1.2.13649, 2154-1248-1-2-6 [pii]

Nagasaka K, Pim D, Massimi P, Thomas M, Tomaic V, Subbaiah VK, Kranjec C, Nakagawa S, Yano T, Taketani Y, Myers M, Banks L (2010b) The cell polarity regulator hScrib controls ERK activation through a KIM site-dependent interaction. Oncogene 29(38):5311–5321. doi:10.1038/onc.2010.265, onc2010265 [pii]

Nagasaka K, Seiki T, Yamashita A, Massimi P, Subbaiah VK, Thomas M, Kranjec C, Kawana K, Nakagawa S, Yano T, Taketani Y, Fujii T, Kozuma S, Banks L (2013) A novel interaction between hScrib and PP1gamma downregulates ERK signaling and suppresses oncogene-induced cell transformation. PLoS One 8(1):e53752. doi:10.1371/journal.pone.0053752, PONE-D-12-02664 [pii]

Naim E, Bernstein A, Bertram JF, Caruana G (2005) Mutagenesis of the epithelial polarity gene, discs large 1, perturbs nephrogenesis in the developing mouse kidney. Kidney Int 68 (3):955–965. doi:10.1111/j.1523-1755.2005.00489.x, KID489 [pii]

Nakagawa S, Huibregtse JM (2000) Human scribble (Vartul) is targeted for ubiquitin-mediated degradation by the high-risk papillomavirus E6 proteins and the E6AP ubiquitin-protein ligase. Mol Cell Biol 20(21):8244–8253

Nakajima YI, Meyer EJ, Kroesen A, McKinney SA, Gibson MC (2013) Epithelial junctions maintain tissue architecture by directing planar spindle orientation. Nature. doi:10.1038/nature12335

Narayan N, Subbaiah VK, Banks L (2009) The high-risk HPV E6 oncoprotein preferentially targets phosphorylated nuclear forms of hDlg. Virology 387(1):1–4. doi:10.1016/j.virol.2009.02.030

Narayanan AS, Reyes SB, Um K, McCarty JH, Tolias KF (2013) The Rac-GAP Bcr is a novel regulator of the PAR complex that controls cell polarity. Mol Biol Cell 24(24):3857–3868. doi:10.1091/mbc.E13-06-0333, mbc.E13-06-0333 [pii]

Navarro C, Nola S, Audebert S, Santoni MJ, Arsanto JP, Ginestier C, Marchetto S, Jacquemier J, Isnardon D, Le Bivic A, Birnbaum D, Borg JP (2005) Junctional recruitment of mammalian Scribble relies on E-cadherin engagement. Oncogene 24(27):4330–4339

Nelson WJ (2003) Adaptation of core mechanisms to generate cell polarity. Nature 422 (6933):766–774. doi:10.1038/nature01602, nature01602 [pii]

Nithianantharajah J, Komiyama NH, McKechanie A, Johnstone M, Blackwood DH, St Clair D, Emes RD, van de Lagemaat LN, Saksida LM, Bussey TJ, Grant SG (2013) Synaptic scaffold evolution generated components of vertebrate cognitive complexity. Nat Neurosci 16 (1):16–24. doi:10.1038/nn.3276

Nola S, Sebbagh M, Marchetto S, Osmani N, Nourry C, Audebert S, Navarro C, Rachel R, Montcouquiol M, Sans N, Etienne-Manneville S, Borg JP, Santoni MJ (2008) Scrib regulates PAK activity during the cell migration process. Hum Mol Genet 17(22):3552–3565. doi:10.1093/hmg/ddn248, ddn248 [pii]

Nolan ME, Aranda V, Lee S, Lakshmi B, Basu S, Allred DC, Muthuswamy SK (2008) The polarity protein Par6 induces cell proliferation and is overexpressed in breast cancer. Cancer Res 68(20):8201–8209. doi:10.1158/0008-5472.CAN-07-6567, 68/20/8201 [pii]

Norman M, Wisniewska KA, Lawrenson K, Garcia-Miranda P, Tada M, Kajita M, Mano H, Ishikawa S, Ikegawa M, Shimada T, Fujita Y (2012) Loss of Scribble causes cell competition in mammalian cells. J Cell Sci 125(Pt 1):59–66. doi:10.1242/jcs.085803, jcs.085803 [pii]

Nourry C, Grant SG, Borg JP (2003) PDZ domain proteins: plug and play! Sci STKE 2003(179): RE7

Ohsawa S, Sugimura K, Takino K, Xu T, Miyawaki A, Igaki T (2011) Elimination of oncogenic neighbors by JNK-mediated engulfment in Drosophila. Dev Cell 20(3):315–328. doi:10.1016/j.devcel.2011.02.007, S1534-5807(11)00076-1 [pii]

Ohshiro T, Yagami T, Zhang C, Matsuzaki F (2000) Role of cortical tumour-suppressor proteins in asymmetric division of Drosophila neuroblast. Nature 408(6812):593–596

Okajima M, Takahashi M, Higuchi M, Ohsawa T, Yoshida S, Yoshida Y, Oie M, Tanaka Y, Gejyo F, Fujii M (2008) Human T-cell leukemia virus type 1 Tax induces an aberrant clustering of the tumor suppressor Scribble through the PDZ domain-binding motif dependent and independent interaction. Virus Genes 37(2):231–240. doi:10.1007/s11262-008-0259-4

Oliaro J, Van Ham V, Sacirbegovic F, Pasam A, Bomzon Z, Pham K, Ludford-Menting MJ, Waterhouse NJ, Bots M, Hawkins ED, Watt SV, Cluse LA, Clarke CJ, Izon DJ, Chang JT, Thompson N, Gu M, Johnstone RW, Smyth MJ, Humbert PO, Reiner SL, Russell SM (2010) Asymmetric cell division of T cells upon antigen presentation uses multiple conserved mechanisms. J Immunol 185(1):367–375. doi:10.4049/jimmunol.0903627

Osmani N, Vitale N, Borg JP, Etienne-Manneville S (2006) Scrib controls Cdc42 localization and activity to promote cell polarization during astrocyte migration. Curr Biol 16(24):2395–2405. doi:10.1016/j.cub.2006.10.026

Pagliarini RA, Xu T (2003) A genetic screen in Drosophila for metastatic behavior. Science 302 (5648):1227–1231

Pangon L, Van Kralingen C, Abas M, Daly RJ, Musgrove EA, Kohonen-Corish MR (2012) The PDZ-binding motif of MCC is phosphorylated at position -1 and controls lamellipodia formation in colon epithelial cells. Biochim Biophys Acta 1823(6):1058–1067. doi:10.1016/j.bbamcr.2012.03.011, S0167-4889(12)00077-8 [pii]

Parsons L, Portela M, Grzeschik N, Richardson HE (2014a) Lgl regulates Notch signaling via endocytosis, independently of aPKC, in the Drosophila developing eye epithelia. Curr Biol 24 (18):2073–2084

Parsons LM, Grzeschik NA, Richardson HE (2014b) lgl Regulates the Hippo Pathway Independently of Fat/Dachs, Kibra/Expanded/Merlin and dRASSF/dSTRIPAK. Cancers (Basel) 6 (2):879–896. doi:10.3390/cancers6020879, cancers6020879 [pii]

Pearson HB, Perez-Mancera PA, Dow LE, Ryan A, Tennstedt P, Bogani D, Elsum I, Greenfield A, Tuveson DA, Simon R, Humbert PO (2011) SCRIB expression is deregulated in human prostate cancer, and its deficiency in mice promotes prostate neoplasia. J Clin Invest 121 (11):4257–4267. doi:10.1172/JCI58509, 58509 [pii]

Pece S, Chiariello M, Murga C, Gutkind JS (1999) Activation of the protein kinase Akt/PKB by the formation of E-cadherin-mediated cell-cell junctions. Evidence for the association of phosphatidylinositol 3-kinase with the E-cadherin adhesion complex. J Biol Chem 274 (27):19347–19351

Pham K, Sacirbegovic F, Russell SM (2014) Polarized cells, polarized views: asymmetric cell division in hematopoietic cells. Front Immunol 5:26. doi:10.3389/fimmu.2014.00026

Phillips HM, Rhee HJ, Murdoch JN, Hildreth V, Peat JD, Anderson RH, Copp AJ, Chaudhry B, Henderson DJ (2007) Disruption of planar cell polarity signaling results in congenital heart defects and cardiomyopathy attributable to early cardiomyocyte disorganization. Circ Res 101 (2):137–145. doi:10.1161/CIRCRESAHA.106.142406, CIRCRESAHA.106.142406 [pii]

Phua DC, Humbert PO, Hunziker W (2009) Vimentin regulates scribble activity by protecting it from proteasomal degradation. Mol Biol Cell 20(12):2841–2855. doi:10.1091/mbc.E08-02-0199

Pim D, Bergant M, Boon SS, Ganti K, Kranjec C, Massimi P, Subbaiah VK, Thomas M, Tomaic V, Banks L (2012) Human papillomaviruses and the specificity of PDZ domain targeting. FEBS J 279(19):3530–3537. doi:10.1111/j.1742-4658.2012.08709.x

Pinal N, Goberdhan DC, Collinson L, Fujita Y, Cox IM, Wilson C, Pichaud F (2006) Regulated and polarized PtdIns(3,4,5)P3 accumulation is essential for apical membrane morphogenesis in photoreceptor epithelial cells. Curr Biol 16(2):140–149. doi:10.1016/j.cub.2005.11.068, S0960-9822(05)01518-6 [pii]

Qin Y, Capaldo C, Gumbiner BM, Macara IG (2005) The mammalian Scribble polarity protein regulates epithelial cell adhesion and migration through E-cadherin. J Cell Biol 171 (6):1061–1071. doi:10.1083/jcb.200506094

Reischauer S, Levesque MP, Nusslein-Volhard C, Sonawane M (2009) Lgl2 executes its function as a tumor suppressor by regulating ErbB signaling in the zebrafish epidermis. PLoS Genet 5 (11):e1000720

Reuver SM, Garner CC (1998) E-cadherin mediated cell adhesion recruits SAP97 into the cortical cytoskeleton. J Cell Sci 111(Pt 8):1071–1080

Rivard N (2009) Phosphatidylinositol 3-kinase: a key regulator in adherens junction formation and function. Front Biosci 14:510–522, doi:3259 [pii]

Rivera C, Simonson SJ, Yamben IF, Shatadal S, Nguyen MM, Beurg M, Lambert PF, Griep AE (2013) Requirement for Dlgh-1 in planar cell polarity and skeletogenesis during vertebrate development. PLoS One 8(1):e54410. doi:10.1371/journal.pone.0054410, PONE-D-12-18879 [pii]

Rivera C, Yamben IF, Shatadal S, Waldof M, Robinson ML, Griep AE (2009) Cell-autonomous requirements for Dlg-1 for lens epithelial cell structure and fiber cell morphogenesis. Dev Dyn 238(9):2292–2308. doi:10.1002/dvdy.22036

Roberts S, Delury C, Marsh E (2012) The PDZ protein discs-large (DLG): the 'Jekyll and Hyde' of the epithelial polarity proteins. FEBS J 279(19):3549–3558. doi:10.1111/j.1742-4658.2012.08729.x

Robinson A, Escuin S, Doudney K, Vekemans M, Stevenson RE, Greene ND, Copp AJ, Stanier P (2012) Mutations in the planar cell polarity genes CELSR1 and SCRIB are associated with the severe neural tube defect craniorachischisis. Hum Mutat 33(2):440–447. doi:10.1002/humu.21662

Roche JP, Packard MC, Moeckel-Cole S, Budnik V (2002) Regulation of synaptic plasticity and synaptic vesicle dynamics by the PDZ protein Scribble. J Neurosci 22(15):6471–6479, doi:20026646

Rodriguez-Boulan E, Macara IG (2014) Organization and execution of the epithelial polarity programme. Nat Rev Mol Cell Biol 15(4):225–242. doi:10.1038/nrm3775

Roegiers F, Jan LY, Jan YN (2005) Regulation of membrane localization of Sanpodo by lethal giant larvae and neuralized in asymmetrically dividing cells of Drosophila sensory organs. Mol Biol Cell 16(8):3480–3487

Round JL, Humphries LA, Tomassian T, Mittelstadt P, Zhang M, Miceli MC (2007) Scaffold protein Dlgh1 coordinates alternative p38 kinase activation, directing T cell receptor signals toward NFAT but not NF-kappaB transcription factors. Nat Immunol 8(2):154–161

Round JL, Tomassian T, Zhang M, Patel V, Schoenberger SP, Miceli MC (2005) Dlgh1 coordinates actin polymerization, synaptic T cell receptor and lipid raft aggregation, and effector function in T cells. J Exp Med 201(3):419–430

Ryan TJ, Grant SG (2009) The origin and evolution of synapses. Nat Rev Neurosci 10 (10):701–712. doi:10.1038/nrn2717, nrn2717 [pii]

Sabatini DM (2006) mTOR and cancer: insights into a complex relationship. Nat Rev Cancer 6 (9):729–734. doi:10.1038/nrc1974, nrc1974 [pii]

Sampedro MN, Bussineau CM, Cotman CW (1981) Postsynaptic density antigens: preparation and characterization of an antiserum against postsynaptic densities. J Cell Biol 90(3):675–686

Sandoval GJ, Graham DB, Gmyrek GB, Akilesh HM, Fujikawa K, Sammut B, Bhattacharya D, Srivatsan S, Kim A, Shaw AS, Yang-Iott K, Bassing CH, Duncavage E, Xavier RJ, Swat W

(2013) Novel mechanism of tumor suppression by polarity gene discs large 1 (DLG1) revealed in a murine model of pediatric B-ALL. Cancer Immunol Res 1(6):426–437. doi:10.1158/2326-6066.CIR-13-0065

Santoni M, Pontarotti P, Birnbaum D, Borg J (2002) The LAP family: a phylogenetic point of view. Trends Genet 18(10):494

Savi F, Forno I, Faversani A, Luciani A, Caldiera S, Gatti S, Foa P, Ricca D, Bulfamante G, Vaira V, Bosari S (2014) MiR-296/Scribble axe is deregulated in human breast cancer and miR-296 restoration reduces tumor growth in vivo. Clin Sci. doi:10.1042/CS20130580

Schroeder MC, Chen CL, Gajewski K, Halder G (2013) A non-cell-autonomous tumor suppressor role for Stat in eliminating oncogenic scribble cells. Oncogene 32(38):4471–4479. doi:10.1038/onc.2012.476, onc2012476 [pii]

Shewan A, Eastburn DJ, Mostov K (2011) Phosphoinositides in cell architecture. Cold Spring Harb Perspect Biol 3(8):a004796. doi:10.1101/cshperspect.a004796, cshperspect.a004796 [pii]

Sjoqvist M, Antfolk D, Ferraris S, Rraklli V, Haga C, Antila C, Mutvei A, Imanishi SY, Holmberg J, Jin S, Eriksson JE, Lendahl U, Sahlgren C (2014) PKCzeta regulates Notch receptor routing and activity in a Notch signaling-dependent manner. Cell Res 24(4):433–450. doi:10.1038/cr.2014.34, cr201434 [pii]

Skouloudaki K, Puetz M, Simons M, Courbard JR, Boehlke C, Hartleben B, Engel C, Moeller MJ, Englert C, Bollig F, Schafer T, Ramachandran H, Mlodzik M, Huber TB, Kuehn EW, Kim E, Kramer-Zucker A, Walz G (2009) Scribble participates in Hippo signaling and is required for normal zebrafish pronephros development. Proc Natl Acad Sci U S A 106(21):8579–8584

Sonawane M, Carpio Y, Geisler R, Schwarz H, Maischein HM, Nuesslein-Volhard C (2005) Zebrafish penner/lethal giant larvae 2 functions in hemidesmosome formation, maintenance of cellular morphology and growth regulation in the developing basal epidermis. Development 132(14):3255–3265

Sonawane M, Martin-Maischein H, Schwarz H, Nusslein-Volhard C (2009) Lgl2 and E-cadherin act antagonistically to regulate hemidesmosome formation during epidermal development in zebrafish. Development 136(8):1231–1240. doi:10.1242/dev.032508, dev.032508 [pii]

Sotelo NS, Valiente M, Gil A, Pulido R (2012) A functional network of the tumor suppressors APC, hDlg, and PTEN, that relies on recognition of specific PDZ-domains. J Cell Biochem 113 (8):2661–2670. doi:10.1002/jcb.24141

Sripathy S, Lee M, Vasioukhin V (2011) Mammalian Llgl2 is necessary for proper branching morphogenesis during placental development. Mol Cell Biol 31(14):2920–2933. doi:10.1128/MCB.05431-11, MCB.05431-11 [pii]

St Johnston D, Ahringer J (2010) Cell polarity in eggs and epithelia: parallels and diversity. Cell 141(5):757–774. doi:10.1016/j.cell.2010.05.011, S0092-8674(10)00547-7 [pii]

Staples J, Broadie K (2013) The cell polarity scaffold Lethal Giant Larvae regulates synapse morphology and function. J Cell Sci 126(Pt 9):1992–2003. doi:10.1242/jcs.120139

Stephenson LM, Sammut B, Graham DB, Chan-Wang J, Brim KL, Huett AS, Miletic AV, Kloeppel T, Landry A, Xavier R, Swat W (2007) DLGH1 is a negative regulator of T-lymphocyte proliferation. Mol Cell Biol 27(21):7574–7581

Stiffler MA, Chen JR, Grantcharova VP, Lei Y, Fuchs D, Allen JE, Zaslavskaia LA, MacBeath G (2007) PDZ domain binding selectivity is optimized across the mouse proteome. Science 317 (5836):364–369. doi:10.1126/science.1144592

Stirnimann CU, Petsalaki E, Russell RB, Muller CW (2010) WD40 proteins propel cellular networks. Trends Biochem Sci 35(10):565–574. doi:10.1016/j.tibs.2010.04.003

Stottmann RW, Moran JL, Turbe-Doan A, Driver E, Kelley M, Beier DR (2011) Focusing forward genetics: a tripartite ENU screen for neurodevelopmental mutations in the mouse. Genetics 188 (3):615–624. doi:10.1534/genetics.111.126862, genetics.111.126862 [pii]

Strand D, Jakobs R, Merdes G, Neumann B, Kalmes A, Heid HW, Husmann I, Mechler BM (1994) The Drosophila lethal(2)giant larvae tumor suppressor protein forms homo-oligomers and is associated with nonmuscle myosin II heavy chain. J Cell Biol 127(5):1361–1373

Strand D, Unger S, Corvi R, Hartenstein K, Schenkel H, Kalmes A, Merdes G, Neumann B, Krieg-Schneider F, Coy JF et al (1995) A human homologue of the Drosophila tumour suppressor gene l(2)gl maps to 17p11.2-12 and codes for a cytoskeletal protein that associates with nonmuscle myosin II heavy chain. Oncogene 11(2):291–301

Su WH, Mruk DD, Wong EW, Lui WY, Cheng CY (2012) Polarity protein complex Scribble/Lgl/Dlg and epithelial cell barriers. Adv Exp Med Biol 763:149–170

Sun Y, Aiga M, Yoshida E, Humbert PO, Bamji SX (2009) Scribble interacts with beta-catenin to localize synaptic vesicles to synapses. Mol Biol Cell 20(14):3390–3400. doi:10.1091/mbc.E08-12-1172

Sun Y, Bamji SX (2011) Beta-Pix modulates actin-mediated recruitment of synaptic vesicles to synapses. J Neurosci 31(47):17123–17133. doi:10.1523/JNEUROSCI.2359-11.2011

Takizawa S, Nagasaka K, Nakagawa S, Yano T, Nakagawa K, Yasugi T, Takeuchi T, Kanda T, Huibregtse JM, Akiyama T, Taketani Y (2006) Human scribble, a novel tumor suppressor identified as a target of high-risk HPV E6 for ubiquitin-mediated degradation, interacts with adenomatous polyposis coli. Genes Cells 11(4):453–464

Tamori Y, Bialucha CU, Tian AG, Kajita M, Huang YC, Norman M, Harrison N, Poulton J, Ivanovitch K, Disch L, Liu T, Deng WM, Fujita Y (2010) Involvement of Lgl and Mahjong/VprBP in cell competition. PLoS Biol 8(7):e1000422

Tamori Y, Deng WM (2013) Tissue repair through cell competition and compensatory cellular hypertrophy in postmitotic epithelia. Dev Cell 25(4):350–363. doi:10.1016/j.devcel.2013.04.013, S1534-5807(13)00242-6 [pii]

Tarpey P, Parnau J, Blow M, Woffendin H, Bignell G, Cox C, Cox J, Davies H, Edkins S, Holden S, Korny A, Mallya U, Moon J, O'Meara S, Parker A, Stephens P, Stevens C, Teague J, Donnelly A, Mangelsdorf M, Mulley J, Partington M, Turner G, Stevenson R, Schwartz C, Young I, Easton D, Bobrow M, Futreal PA, Stratton MR, Gecz J, Wooster R, Raymond FL (2004) Mutations in the DLG3 gene cause nonsyndromic X-linked mental retardation. Am J Hum Genet 75(2):318–324

Tay HG, Schulze SK, Compagnon J, Foley FC, Heisenberg CP, Yost HJ, Abdelilah-Seyfried S, Amack JD (2013) Lethal giant larvae 2 regulates development of the ciliated organ Kupffer's vesicle. Development 140(7):1550–1559. doi:10.1242/dev.087130

Tonikian R, Zhang Y, Sazinsky SL, Currell B, Yeh JH, Reva B, Held HA, Appleton BA, Evangelista M, Wu Y, Xin X, Chan AC, Seshagiri S, Lasky LA, Sander C, Boone C, Bader GD, Sidhu SS (2008) A specificity map for the PDZ domain family. PLoS Biol 6(9):e239. doi:10.1371/journal.pbio.0060239, 07-PLBI-RA-4212 [pii]

Uhlirova M, Bohmann D (2006) JNK- and Fos-regulated Mmp1 expression cooperates with Ras to induce invasive tumors in Drosophila. EMBO J 25(22):5294–5304

Uhlirova M, Jasper H, Bohmann D (2005) Non-cell-autonomous induction of tissue overgrowth by JNK/Ras cooperation in a Drosophila tumor model. Proc Natl Acad Sci U S A 102(37):13123–13128

Vaira V, Faversani A, Dohi T, Maggioni M, Nosotti M, Tosi D, Altieri DC, Bosari S (2011) Aberrant overexpression of the cell polarity module scribble in human cancer. Am J Pathol 178:2478–2483

Vaira V, Faversani A, Dohi T, Montorsi M, Augello C, Gatti S, Coggi G, Altieri DC, Bosari S (2012) miR-296 regulation of a cell polarity-cell plasticity module controls tumor progression. Oncogene 31(1):27–38. doi:10.1038/onc.2011.209, onc2011209 [pii]

Van Campenhout CA, Eitelhuber A, Gloeckner CJ, Giallonardo P, Gegg M, Oller H, Grant SG, Krappmann D, Ueffing M, Lickert H (2011) Dlg3 trafficking and apical tight junction formation is regulated by nedd4 and nedd4-2 e3 ubiquitin ligases. Dev Cell 21(3):479–491. doi:10.1016/j.devcel.2011.08.003, S1534-5807(11)00315-7 [pii]

Varelas X, Samavarchi-Tehrani P, Narimatsu M, Weiss A, Cockburn K, Larsen BG, Rossant J, Wrana JL (2010) The Crumbs complex couples cell density sensing to Hippo-dependent control of the TGF-beta-SMAD pathway. Dev Cell 19(6):831–844. doi:10.1016/j.devcel.2010.11.012, S1534-5807(10)00539-3 [pii]

Verghese S, Waghmare I, Kwon H, Hanes K, Kango-Singh M (2012) Scribble acts in the Drosophila fat-hippo pathway to regulate warts activity. PLoS One 7(11):e47173. doi:10.1371/journal.pone.0047173, PONE-D-12-25889 [pii]

Vermeer PD, Einwalter LA, Moninger TO, Rokhlina T, Kern JA, Zabner J, Welsh MJ (2003) Segregation of receptor and ligand regulates activation of epithelial growth factor receptor. Nature 422(6929):322–326

von Stein W, Ramrath A, Grimm A, Muller-Borg M, Wodarz A (2005) Direct association of Bazooka/PAR-3 with the lipid phosphatase PTEN reveals a link between the PAR/aPKC complex and phosphoinositide signaling. Development 132(7):1675–1686

Wada H, Iwasaki M, Sato T, Masai I, Nishiwaki Y, Tanaka H, Sato A, Nojima Y, Okamoto H (2005) Dual roles of zygotic and maternal Scribble1 in neural migration and convergent extension movements in zebrafish embryos. Development 132(10):2273–2285

Walsh GS, Grant PK, Morgan JA, Moens CB (2011) Planar polarity pathway and Nance-Horan syndrome-like 1b have essential cell-autonomous functions in neuronal migration. Development 138(14):3033–3042. doi:10.1242/dev.063842, 138/14/3033 [pii]

Walsh T, McClellan JM, McCarthy SE, Addington AM, Pierce SB, Cooper GM, Nord AS, Kusenda M, Malhotra D, Bhandari A, Stray SM, Rippey CF, Roccanova P, Makarov V, Lakshmi B, Findling RL, Sikich L, Stromberg T, Merriman B, Gogtay N, Butler P, Eckstrand K, Noory L, Gochman P, Long R, Chen Z, Davis S, Baker C, Eichler EE, Meltzer PS, Nelson SF, Singleton AB, Lee MK, Rapoport JL, King MC, Sebat J (2008) Rare structural variants disrupt multiple genes in neurodevelopmental pathways in schizophrenia. Science 320 (5875):539–543. doi:10.1126/science.1155174

Wan Q, Liu J, Zheng Z, Zhu H, Chu X, Dong Z, Huang S, Du Q (2012) Regulation of myosin activation during cell-cell contact formation by Par3-Lgl antagonism: entosis without matrix detachment. Mol Biol Cell 23(11):2076–2091. doi:10.1091/mbc.E11-11-0940, mbc.E11-11-0940 [pii]

Wang S, Watanabe T, Matsuzawa K, Katsumi A, Kakeno M, Matsui T, Ye F, Sato K, Murase K, Sugiyama I, Kimura K, Mizoguchi A, Ginsberg MH, Collard JG, Kaibuchi K (2012) Tiam1 interaction with the PAR complex promotes talin-mediated Rac1 activation during polarized cell migration. J Cell Biol 199(2):331–345. doi:10.1083/jcb.201202041, jcb.201202041 [pii]

Wang T, Liu Y, Xu XH, Deng CY, Wu KY, Zhu J, Fu XQ, He M, Luo ZG (2011) Lgl1 activation of rab10 promotes axonal membrane trafficking underlying neuronal polarization. Dev Cell 21 (3):431–444. doi:10.1016/j.devcel.2011.07.007, S1534-5807(11)00299-1 [pii]

Wansleeben C, van Gurp L, Feitsma H, Kroon C, Rieter E, Verberne M, Guryev V, Cuppen E, Meijlink F (2011) An ENU-mutagenesis screen in the mouse: identification of novel developmental gene functions. PLoS One 6(4):e19357. doi:10.1371/journal.pone.0019357, 10-PONE-RA-20961 [pii]

Warner SJ, Yashiro H, Longmore GD (2010) The Cdc42/Par6/aPKC Polarity Complex Regulates Apoptosis-Induced Compensatory Proliferation in Epithelia. Curr Biol 20:677–686

Warzecha CC, Jiang P, Amirikian K, Dittmar KA, Lu H, Shen S, Guo W, Xing Y, Carstens RP (2010) An ESRP-regulated splicing programme is abrogated during the epithelial-mesenchymal transition. EMBO J 29(19):3286–3300. doi:10.1038/emboj.2010.195

Waterhouse RM, Tegenfeldt F, Li J, Zdobnov EM, Kriventseva EV (2013) OrthoDB: a hierarchical catalog of animal, fungal and bacterial orthologs. Nucleic Acids Res 41(Database issue): D358–D365. doi:10.1093/nar/gks1116

Wells J, Rivera MN, Kim WJ, Starbuck K, Haber DA (2010) The predominant WT1 isoform (+KTS) encodes a DNA-binding protein targeting the planar cell polarity gene Scribble in renal podocytes. Mol Cancer Res 8(7):975–985. doi:10.1158/1541-7786.MCR-10-0033, 1541-7786. MCR-10-0033 [pii]

Wigerius M, Asghar N, Melik W, Johansson M (2013) Scribble controls NGF-mediated neurite outgrowth in PC12 cells. Eur J Cell Biol 92(6–7):213–221. doi:10.1016/j.ejcb.2013.07.002, S0171-9335(13)00048-4 [pii]

Willecke M, Toggweiler J, Basler K (2011) Loss of PI3K blocks cell-cycle progression in a Drosophila tumor model. Oncogene 30(39):4067–4074. doi:10.1038/onc.2011.125, onc2011125 [pii]

Willoughby LF, Schlosser T, Manning SA, Parisot JP, Street IP, Richardson HE, Humbert PO, Brumby AM (2012) An in vivo large-scale chemical screening platform using Drosophila for anti-cancer drug discovery. Dis Model Mech. doi:10.1242/dmm.009985

Wirtz-Peitz F, Nishimura T, Knoblich JA (2008) Linking cell cycle to asymmetric division: Aurora-A phosphorylates the PAR complex to regulate Numb localization. Cell 135 (1):161–173. doi:10.1016/j.cell.2008.07.049, S0092-8674(08)01024-6 [pii]

Woods DF, Bryant PJ (1989) Molecular cloning of the lethal(1)discs large-1 oncogene of Drosophila. Dev Biol 134(1):222–235

Wu WJ, Tu S, Cerione RA (2003) Activated Cdc42 sequesters c-Cbl and prevents EGF receptor degradation. Cell 114(6):715–725, doi:S0092867403006883 [pii]

Xavier R, Rabizadeh S, Ishiguro K, Andre N, Ortiz JB, Wachtel H, Morris DG, Lopez-Ilasaca M, Shaw AC, Swat W, Seed B (2004) Discs large (Dlg1) complexes in lymphocyte activation. J Cell Biol 166(2):173–178

Yamanaka T, Ohno S (2008) Role of Lgl/Dlg/Scribble in the regulation of epithelial junction, polarity and growth. Front Biosci 13:6693–6707

Yamben IF, Rachel RA, Shatadal S, Copeland NG, Jenkins NA, Warming S, Griep AE (2013) Scrib is required for epithelial cell identity and prevents epithelial to mesenchymal transition in the mouse. Dev Biol 384(1):41–52. doi:10.1016/j.ydbio.2013.09.027, S0012-1606(13)00513-7 [pii]

Yan J, Lu Q, Fang X, Adler PN (2009) Rho1 has multiple functions in Drosophila wing planar polarity. Dev Biol 333(1):186–199

Yates LL, Schnatwinkel C, Hazelwood L, Chessum L, Paudyal A, Hilton H, Romero MR, Wilde J, Bogani D, Sanderson J, Formstone C, Murdoch JN, Niswander LA, Greenfield A, Dean CH (2013) Scribble is required for normal epithelial cell-cell contacts and lumen morphogenesis in the mammalian lung. Dev Biol 373(2):267–280. doi:10.1016/j.ydbio.2012.11.012, S0012-1606(12)00631-8 [pii]

Yin G, Haendeler J, Yan C, Berk BC (2004) GIT1 functions as a scaffold for MEK1-extracellular signal-regulated kinase 1 and 2 activation by angiotensin II and epidermal growth factor. Mol Cell Biol 24(2):875–885

Young LC, Hartig N, Munoz-Alegre M, Oses-Prieto JA, Durdu S, Bender S, Vijayakumar V, Vietri Rudan M, Gewinner C, Henderson S, Jathoul AP, Ghatrora R, Lythgoe MF, Burlingame AL, Rodriguez-Viciana P (2013) An MRAS, SHOC2, and SCRIB complex coordinates ERK pathway activation with polarity and tumorigenic growth. Mol Cell 52(5):679–692. doi:10.1016/j.molcel.2013.10.004, S1097-2765(13)00745-4 [pii]

Zarbalis K, May SR, Shen Y, Ekker M, Rubenstein JL, Peterson AS (2004) A focused and efficient genetic screening strategy in the mouse: identification of mutations that disrupt cortical development. PLoS Biol 2(8):E219

Zeitler J, Hsu CP, Dionne H, Bilder D (2004) Domains controlling cell polarity and proliferation in the Drosophila tumor suppressor Scribble. J Cell Biol 167(6):1137–1146

Zeniou-Meyer M, Liu Y, Begle A, Olanich ME, Hanauer A, Becherer U, Rettig J, Bader MF, Vitale N (2008) The Coffin-Lowry syndrome-associated protein RSK2 is implicated in calcium-regulated exocytosis through the regulation of PLD1. Proc Natl Acad Sci U S A 105 (24):8434–8439. doi:10.1073/pnas.0710676105, 0710676105 [pii]

Zhang H, Macara IG (2006) The polarity protein PAR-3 and TIAM1 cooperate in dendritic spine morphogenesis. Nat Cell Biol 8(3):227–237

Zhang X, Wang P, Gangar A, Zhang J, Brennwald P, TerBush D, Guo W (2005) Lethal giant larvae proteins interact with the exocyst complex and are involved in polarized exocytosis. J Cell Biol 170(2):273–283

Zhang Y, Guo H, Kwan H, Wang JW, Kosek J, Lu B (2007) PAR-1 kinase phosphorylates Dlg and regulates its postsynaptic targeting at the Drosophila neuromuscular junction. Neuron 53 (2):201–215. doi:10.1016/j.neuron.2006.12.016

Zhao M, Szafranski P, Hall CA, Goode S (2008) Basolateral junctions utilize warts signaling to control epithelial-mesenchymal transition and proliferation crucial for migration and invasion of Drosophila ovarian epithelial cells. Genetics 178(4):1947–1971. doi:10.1534/genetics.108.086983

Zhu J, Shang Y, Wan Q, Xia Y, Chen J, Du Q, Zhang M (2014) Phosphorylation-dependent interaction between tumor suppressors Dlg and Lgl. Cell Res 24(4):451–463. doi:10.1038/cr.2014.16

Zigman M, le Trinh A, Fraser SE, Moens CB (2011) Zebrafish neural tube morphogenesis requires Scribble-dependent oriented cell divisions. Curr Biol 21(1):79–86. doi:10.1016/j.cub.2010.12.005

Part II
Cell Polarity: Different Aspects and Basic Principles of Regulation

Chapter 5
Front-to-Rear Polarity in Migrating Cells

Flora Llense and Sandrine Etienne-Manneville

Abstract Cell migration is a polarised cellular function involved both during development and in the adult where it participates to immune reaction, wound healing, tissue renewal, as well as cancer spreading. Migrating cells display a characteristic protruding front, at the opposite of a retracting trailing edge. This front-to-rear functional polarity, paralleled by the polarised morphology of the cell, reflects the polarisation of the intracellular organisation and signalling cascades. Random migration only requires the establishment of front-to-rear polarity; in contrast, the directed and persistent migration, as observed in vivo, necessitates the orientation of the front-to-rear polarity axis in a direction governed by multiple polarity cues found in the cell environment and the maintenance of the front-to-rear axis over time. This chapter summarises the characteristics of polarised migrating cells and presents the molecular mechanisms at the heart of the initiation, the orientation and the maintenance of front-to-rear polarity.

Keywords Cell adhesion • Chemotaxis • Collective migration • Cytoskeleton • Phosphoinositides • Rho GTPases

5.1 Introduction

Migration is an active process that allows a cell to move from one place to another. Protozoa use migration to find food or escape predators, and in multicellular organisms, migration of single or groups of cells occurs in a wide range of circumstances. For instance, during development, gastrulation and dorsal and ventral closure involve the migration of sheets of cells. In the adult, the collective migration of epithelial cells contributes to tissue renewal in the intestine and the skin, while leukocyte migration plays a key role in immunological responses. Migration is also at the heart of important pathologies, including chronic inflammatory diseases, vascular diseases, mental retardation or cancer metastasis. With the exception of cilia and flagella-driven migration which allows single cells to

F. Llense • S. Etienne-Manneville (✉)
Cell Polarity, Migration and Cancer Unit, Institut Pasteur - CNRS UMR 3691, 25 rue du Dr Roux, 75724 Paris Cedex 15, France
e-mail: sandrine.etienne-manneville@pasteur.fr

K. Ebnet (ed.), *Cell Polarity 1*, DOI 10.1007/978-3-319-14463-4_5

move in a liquid environment, cell migration generally occurs on or within a relatively dense environment composed of extracellular matrix and possibly other cells.

Migration is a fundamentally polarised process that requires the organisation of the cell machinery along a front-to-rear axis. At the cell front, extension of a lamellipodium, rapidly followed by the formation of new adhesive contacts, promotes cell protrusion. Actual cell displacement only occurs when the cell body contracts and the trailing edge eventually detaches. Each of these steps requires major rearrangements of the cytoskeleton as well as of the vesicular traffic controlling the directed delivery of membrane components. These rearrangements must be coordinated in space and time to generate an effective forward movement. The initial front-to-rear polarisation determines the position of the front and the localisation of the rear at the opposite side of the cell. This step is absolutely needed for the initiation of migration. The establishment and maintenance of a front-to-rear polarity axis is necessary for all kinds of migration, whether the direction of migration is random as frequently observed in vitro or dictated by environmental cues (Ridley et al. 2003). In vivo, cells are generally guided by a complex environment that presents a combination of polarity cues. Immune cells follow gradients of inflammatory factors or bacterial products. Angiogenic endothelial cells migrate towards sources of growth factors such as VEGF (vascular endothelial growth factor), while fibroblasts, for instance, follow PDGF (platelet-derived growth factor) gradients. Gradients of attractants or repellents can initiate front-to-rear polarisation, but their main effect is to impact on the orientation of the polarity axis and to control the direction of migration. These gradients are particularly important to direct amoeboid movement, during which cells form little or no adhesions. In contrast, adherent cells, such as fibroblasts, astrocytes and endothelial cells, adopt a mesenchymal type of migration, which differs from amoeboid migration by the role of cell adhesion to the extracellular matrix as well as by the requirement of matrix remodelling in a 3D environment. In these situations, cell interactions with the extracellular matrix and also with the surrounding cells influence the orientation of the front-to-rear axis and the direction of migration, reinforcing or dampening the effect of soluble gradients.

5.2 The Structural and Molecular Characteristics of Front-to-Rear Polarity

The front and the rear of a moving cell are functionally different, and these differences rely on the segregation of molecular processes specific of the front and of the rear (Fig. 5.1).

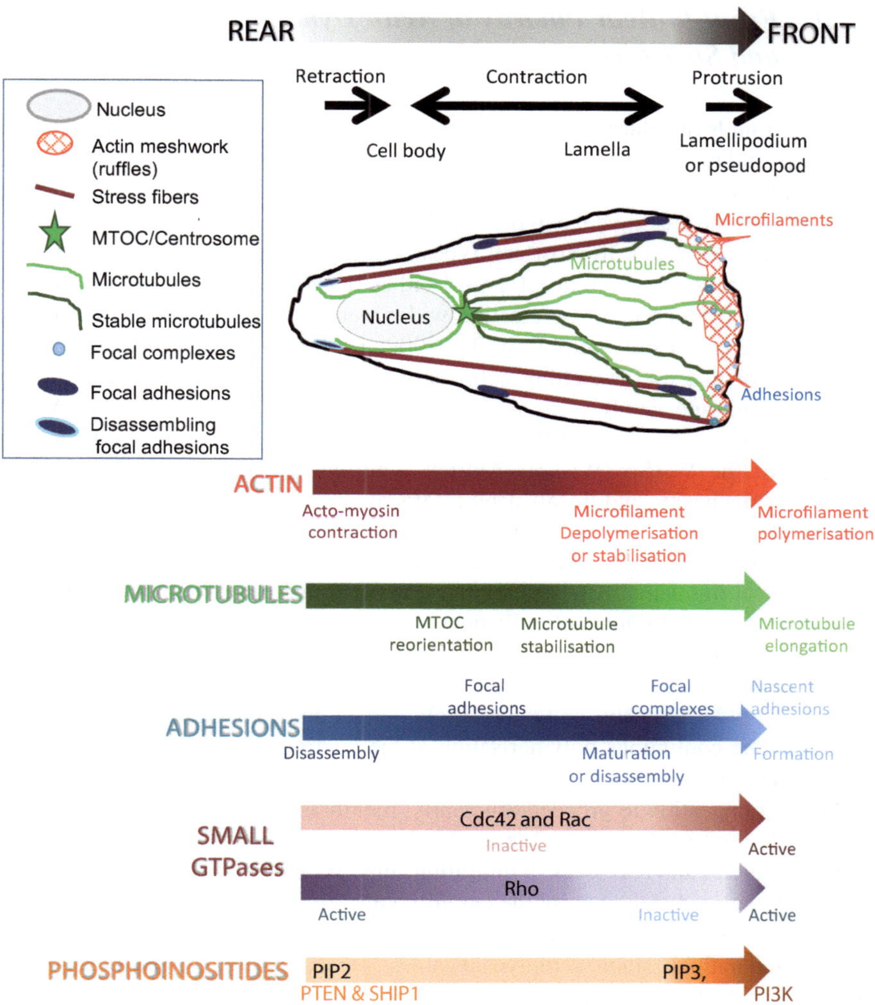

Fig. 5.1 Functional, structural and molecular differences between the front and the rear of migrating cells. The front-to-rear polarity axis extends along the major axis of the cell body from a protrusive leading edge to a retracting trailing edge. Actin microfilaments (*red*), microtubules (*green*) and cell adhesions (*blue*) on the extracellular matrix present different characteristics in the lamellipodia, the cell body and the cell rear. These structural differences parallel gradient of signalling molecules, such as Rho GTPases (*purple* and *pink*) and phosphoinositides (*orange*), whose activity and/or concentration varies from the front to the rear of the cells

5.2.1 *Front-to-Rear Polarity of Actin Dynamics and Structures*

The cell front is the place of active membrane protrusion promoted by cytoskeleton dynamics and membrane traffic. Depending on their shape and on the cell type, this highly dynamic region of migrating cells is called pseudopod or lamellipodia. It extends from the leading edge to the lamella and shows a characteristic cytoskeletal organisation (Fig. 5.1). In most cell types, actin polymerisation is the main force pushing the plasma membrane forward either by generating thin fingerlike protrusions called filopodia or by creating wider membrane protrusions called ruffles (for review, see Insall and Machesky 2009; Carlier and Pantaloni 2007). The formation of a dense branching actin network which effectively pushes on the plasma membrane involves the nucleation of new actin filaments by the Arp2/3 complex (Machesky et al. 1994). This molecular complex is strongly activated by proteins of the WASp (Wiskott–Aldrich syndrome protein) and SCAR/WAVE families (Machesky and Insall 1998; Stradal et al. 2001; Hahne et al. 2001; Kunda et al. 2003). Other actin filaments associate with Ena/VASP proteins which promote continuous elongation of actin polymers by protecting the barbed ends of actin filaments by cross-linking. The actin nucleator formins promote the assembly of parallel linear actin filaments from the plasma membrane (Evangelista et al. 2003). The polymerisation of actin beneath the plasma membrane is fuelled by actin monomers which are generated at the rear of the lamellipodia, following the severing of older filaments and their depolymerisation by the ADF/cofilin family of proteins. The continuous polymerisation at the leading edge leads to a retrograde flow of actin structures.

For cell migration to proceed efficiently, membrane protrusion must be limited to the cell front (Affolter and Weijer 2005). The protrusions are thus suppressed along the lateral and rear sides of migrating cells. Whereas a pool of actin filaments are depolymerised at the rear of the lamellipodia, others are stabilised as they move towards the rear of the cell (Fig. 5.1). The meshwork of actin filaments is reorganised in thick, stable actomyosin cables characteristic of the rear of migrating cells (Yoo et al. 2010). In cancer cells migrating in an amoeboid manner, the movement is largely dependent on cell contraction, and cells typically accumulate actomyosin fibres at the rear in a uropod-like structure (Lammermann et al. 2008; Hawkins et al. 2011; Poincloux et al. 2011). In this case, membrane protrusion is caused by actin polymerisation as in a regular lamellipodia or by membrane blebbing (Sahai 2005; Fackler and Grosse 2008). Actin polymerisation is not directly responsible for force generation and Arp2/3 is not involved in the formation of blebs (Eisenmann et al. 2007; Charras et al. 2006; Langridge and Kay 2006). In contrast, the intense contraction of actomyosin cables promotes the destabilisation of the actin cortex and the local disruption of cortical actin interactions with the plasma membrane (Paluch et al. 2006). Moreover, actomyosin contraction increases the hydrostatic pressure and leads to the rapid protrusion of the plasma membrane (Charras et al. 2005). The contractile peripheral network, which contributes to the translocation of the cell body, also suppresses the formation of lateral pseudopods.

5.2.2 *Front-to-Rear Organisation of the Microtubule Network*

Microtubules participate in the protrusion and the retraction of migrating cells (Etienne-Manneville 2013). Microtubule polymerisation can, in a few cases, for instance, in neurons or astrocytes, provide a pushing force required for cell protrusion (Etienne-Manneville 2004a). The formation of large microtubule bundles is also likely to sustain cell elongation along the axis of migration. However, microtubule functions in cell migration essentially involve microtubule association with actin and intermediate filaments and their function in vesicular trafficking. They act on transmembrane protein polarised recycling and recruitment, on membrane delivery and thus on membrane tension. Finally, microtubules interact with, deliver, and/or modulate the activity of signalling molecules, regulating the polarised organisation of cytoskeletal structures required for migration. Depolymerisation of microtubules can totally prevent the front-to-rear polarisation of some cell types such as astrocytes or neurons. However, in most cell types, disruption of the microtubule network does not prevent cell migration but dramatically affect the control of migration direction (Etienne-Manneville 2004a).

Like for actin, the fundamental differences of microtubule functions between the cell front and the cell rear imply that the microtubule network shows a front-to-rear polarity (Fig. 5.1). The microtubule network is composed of an array of polar microtubules emerging from microtubule-organising centre(s) (MTOC), which often corresponds to the centrosome. In migrating cells, the microtubule network is strikingly oriented in the direction of migration (Etienne-Manneville 2010; Manneville and Etienne-Manneville 2006). The microtubules align in the direction of migration and the MOTC frequently localises in front of the nucleus facing the leading edge. The position of the nucleus at the cell rear pushes most microtubules to grow towards the front of the cell (Fig. 5.1). In some cells, such as epithelial cells or hepatocytes, this can be reinforced by the presence of microtubules emerging from the Golgi apparatus and growing towards the leading edge (Efimov et al. 2007; Rivero et al. 2009). Microtubule dynamics is different at the front, the sides and the rear of the cell. This polarised regulation results from the recruitment of different plus-end tracking proteins (+TIP) which control microtubule growth, shrinkage, catastrophes and rescues, as well as microtubule interactions with intracellular organelles and the cell cortex (Etienne-Manneville 2010). A selection of +TIPs, including APC (adenomatous polyposis coli) and Clasps, associates specifically with leading edge microtubules and may promote their elongation. The maintenance of a microtubule network reaching the front edge of the cell despite the continuous retrograde flow of actin requires microtubule anchoring at the cell cortex. Interactions between microtubules and the cell cortex are mediated by various +TIP proteins found on leading edge microtubules, such as APC, Clips or Claps (Etienne-Manneville and Hall 2003a; Machesky et al. 1994; Akhmanova et al. 2001; Akhmanova and Steinmetz 2008; Fukata et al. 2002). +TIP interaction with microtubules is regulated by phosphorylation, and the inhibition of GSK3 has been shown to promote the recruitment of specific +TIPs, like APC and Clasps on leading edge microtubules (Akhmanova et al. 2001; Etienne-Manneville and Hall 2003a; Watanabe et al. 2009). GSK3 activity is controlled by phosphorylation in polarised manner both downstream of Rho GTPases and PI3K (phosphatidylinositol 3 kinase) (Zhou et al. 2004; Etienne-Manneville and Hall 2003a). This likely contributes to establish the specificity of microtubule dynamic instability at the leading edge (Drabek et al. 2006). APC-enriched microtubule tips bind to Dlg which accumulates at the cell front in response to aPKC activation (Etienne-Manneville 2009; Etienne-Manneville et al. 2005). Dlg is also required for the recruitment of dynein, leading to microtubule anchoring (Manneville et al. 2010). Clasps at the microtubule plus ends interact LL5β, which preferentially associates with the PI(3,4,5)P3-enriched plasma membrane of the cell front (Lansbergen et al. 2006; Mimori-Kiyosue et al. 2005). Microtubule plus ends are also captured by +TIPs interacting with actin partners, such as IQGAP1 (Fukata et al. 2002; Brandt and Grosse 2007). Forces exerted on microtubules captured at the cell cortex contribute to centrosome positioning (Manneville and Etienne-Manneville 2006). In particular, the microtubule-associated motor dynein, which preferentially associates with leading edge microtubules that are in close proximity to the cell cortex, may generate pulling forces to bring the centrosome in front of the nucleus (Manneville et al. 2010; Dujardin et al. 2003).

The global orientation of the microtubule network is paralleled by an increased stability of microtubules directed towards the leading edge. Several kinases, including GSK3, participate in the stabilisation of microtubules by controlling the binding of specific MAPs, like MAP1B or MAP2, along microtubules (Sanchez et al. 1996; Sanchez et al. 2000). These stable microtubules accumulate tubulin posttranslational modifications, such acetylation and detyrosination. These modified microtubules and the fact that the Golgi apparatus localises close to the centrosome facing the cell front may favour membrane traffic towards the cell front, bringing new or recycled membrane components where needed. Microtubule stabilisation involves both the association of microtubule-associated proteins (MAPs), like MAP1B or MAP2, and the control of microtubule dynamics by +TIPs (Sanchez et al. 1996; Sanchez et al. 2000; Palazzo et al. 2001a; Etienne-Manneville 2010).

5.2.3 Front-to-Rear Regulation of Cell Adhesions

The cytoskeletal networks not only provide pushing forces but also contribute to the formation and maturation of adhesive structures. During mesenchymal migration, adhesion to the extracellular matrix is crucial for cell polarisation and movement.

α-β integrin heterodimers are the major receptors for extracellular matrix fibres (Hynes 2002). Integrin engagement primarily occurs at the very leading edge during membrane protrusion (Huttenlocher and Horwitz 2011). Upon binding to extracellular ligands, integrins change their conformation and their affinity for the substrate (Huttenlocher et al. 1996). This change of conformation can also be induced by inside-out signalling (Luo et al. 2007; Luo and Springer 2006). Integrins first form small clusters (nascent adhesions) which progressively develop into focal complexes in the lamellipodia (Fig. 5.1). Integrins have short cytoplasmic tails that connect to the actin cytoskeleton (Hynes 2002). Their engagement with the extracellular matrix promotes the recruitment of adaptor proteins such as talin, Focal Adhesion Kinase (FAK), paxillin and vinculin. Integrin beta cytoplasmic domains (tails) mediate the activation of the Src family kinases and FAK and in parallel induce talin-dependent signals to promote cell adhesion and polarisation (Arias-Salgado et al. 2005). FAK and vinculin colocalise and bind directly to the Arp2/3 complex, promoting the actin polymerisation required for the formation of nascent adhesions and focal complexes. Talin can bind the Rac-exchange factor TIAM and further connect integrin complexes to the actin cytoskeleton (Wang et al. 2012). As the cell moves forward, focal complexes either dissociate at the rear of the lamellipodia or persist and mature into larger, more stable focal adhesions. Maturation into large focal adhesions is dependent on myosin II, the activity of the formins mDia, and the substrate stiffness (Huttenlocher and Horwitz 2011). Focal adhesions form larger protein complexes, progressively recruiting additional cytoplasmic proteins such as α-actinin, RhoGTPases and Src family kinases (SFKs) (Webb et al. 2002). Ultimately, this complex mediates a strong interaction between integrins and actomyosin cables, characterising mature focal adhesions (Ridley et al. 2003). In single cell, these focal adhesions serve as anchoring points for actomyosin cables at the front and the rear of the cell (Fig. 5.1). As a consequence focal adhesions are viewed as molecular clutches that hold the actin filaments which actively push on the leading edge plasma membrane (Giannone et al. 2009) and also support the contractile forces required for cell translocation (Beningo et al. 2001).

At the rear, focal adhesions eventually disassemble, contributing to the retraction of the cell rear (Fig. 5.1). The rearmost adhesions are strongly attached to the matrix, and increasing tension can sever the link between the cytoskeleton and the integrins, which are then left behind attached to the substrate while the cell moves on (Lauffenburger and Horwitz 1996). The tension can also open stretch-activated calcium channels (Lee et al. 1999). Increased intracellular calcium activates the protease calpain which cleaves major components of focal adhesion, including the integrins, talin and vinculin, leading to the disassembly of focal adhesions at the cell rear (Franco et al. 2004; Bhatt et al. 2002). Alternatively, the disassembly of focal adhesions can be induced by microtubule targeting and endocytosis (Stehbens and Wittmann 2012). The regulation of microtubule dynamics at focal adhesion involves the Focal Adhesion Kinase (FAK) and its substrate paxillin (Efimov et al. 2008) and leads to the recruitment of clathrin adaptors and clathrin around focal adhesions (Chao and Kunz 2009; Ezratty et al. 2009).

The front-to-rear polarised organisation of focal adhesions has been thoroughly described in the case of cell migrating on 2D surfaces (Lammermann et al. 2009). In vivo, however, the dynamics and the requirement of integrin clustering and activation are less clear. For example, leukocyte interstitial 3D amoeboid migration can be independent of integrin-mediated adhesion in vivo, and murine dendritic cells depleted of all integrin receptors can still migrate in interstitial tissues in three dimensions (Lammermann et al. 2008; Lammermann and Sixt 2009). In zebrafish, depletion of β1 integrins in germ cells does not impede cell polarisation and migration. Interestingly, in this context, the role of cell–cell adhesion can become predominant (Kardash et al. 2010). Finally, integrin-independent cell migration is frequently observed when cancer cells invade through a 3D environment and more generally upon alteration of cell adhesion to the extracellular matrix (Sahai 2005; Friedl 2004; Fackler and Grosse 2008).

5.2.4 Front and Rear Signalling Pathways

The central common features of all signalling pathways controlling front-to-rear polarity is the polarisation of the Rho family of small GTPases (Iden and Collard 2008). These small G proteins cycle between a GDP-bound inactive form and a GTP-bound active form. They act as molecular switches as they can, in their active form, bind and activate a large number of effectors. Among their vast panel of functions, they cooperate to control cytoskeleton dynamics at the front and the rear of migrating cells (Etienne-Manneville and Hall 2002; Ridley et al. 2003; Jaffe and Hall 2005).

Rac and Cdc42 are activated at the cell front (Nalbant et al. 2004; Kraynov et al. 2000). Both proteins are major regulators of the microtubule and actin cytoskeletons and crucial modulators of the speed and direction of cell migration (Szczur et al. 2009; Watanabe et al. 2005). Although Cdc42 does not appear essential for cell movement, it controls the direction and the persistence of migration. The polarised localisation and activation of Cdc42 is a central event in the control of cell polarity in a wide range of cellular contexts. In cell migration, Cdc42 mainly controls the orientation of the front-to-rear polarity of migrating cells (Etienne-Manneville 2004b). Cdc42 is recruited and activated at the leading edge of migrating cells (Etienne-Manneville and Hall 2001; Osmani et al. 2006) (Fig. 5.1). Acting at the cell front through several downstream effectors, Cdc42 promotes nucleus localisation at the rear of migrating cells by promoting the retrograde flow of actin cable and controls the position of the centrosome by participating in the anchoring of leading edge microtubules (Gomes et al. 2005; Palazzo et al. 2001b; Etienne-Manneville et al. 2005; Manneville et al. 2010). The localised activation of Cdc42 mediates a directional response via a pathway involving the evolutionary conserved polarity complex Par6-aPKC to correctly orient the microtubule network (Etienne-Manneville and Hall 2003b; Watanabe et al. 2005; Etienne-Manneville and Hall 2001). Activation of atypical PKCs at the leading edge leads to the local inhibition of GSK3, a serine kinase involved in the Wnt signalling (Etienne-Manneville and Hall 2003a). GSK3 can also be phosphorylated and inhibited by AKT downstream of PI3K signalling at the cell front (Zhou et al. 2004). Through the regulation of GSK3 activity, Cdc42 activation at the leading edge increases microtubules capture to the cell cortex, contributes to the stabilisation of microtubules reaching towards the leading edge, and promotes the reorientation of the centrosome and the Golgi apparatus towards the leading edge (see above). Cdc42 also participates in the regulation of actin dynamics at the leading edge. It promotes the formation of filopodia via the formin mDia1 and mDia2 (Goh and Ahmed 2012; Goh et al. 2012). It also contributes to Rac recruitment and/or activation leading to the formation of lamellipodia (Nobes and Hall 1995). Rac activity has been shown to gradually increase towards the leading edge (Chamberlain et al. 2000; Kraynov et al. 2000) (Fig. 5.1). At the cell leading edge, active Rac and Cdc42 induce the activation of WASP and of the Scar/WAVE complex which induces the formation of the actin meshwork that fills the lamellipodia. Rac also activates the LIM kinase, which phosphorylates ADF/cofilin leading to its inactivation at the leading edge, whereas at the rear of the lamellipodia the phosphatase slingshot activates ADF/cofilin to promote F-actin disassembly (Arber et al. 1998; Niwa et al. 2002). Rac together with Cdc42 induce the formation of focal complexes and lead to lamellipodium attachment to the extracellular matrix and its stabilisation. Like Cdc42, Rac can activate PAK which regulates actin and microtubule dynamics as well as focal adhesion turnover (Manabe et al. 2002; Obermeier et al. 1998; Daub et al. 2001; Kuntziger et al. 2001). This localised PAK activation is essential for the recruitment of βPIX and Rac activation at the front of the cell, restricting lamellipodium formation to the leading edge of migrating cells (Nola et al. 2008; Osmani et al. 2006).

Rac activity at the cell front is crucial for protrusion formation. However, whether Rac is required for cell migration per se is disputable as migration in absence of Rac in melanoblasts, fibroblasts and macrophages has been reported despite major defects in cell protrusion (Li et al. 2011; Vidali et al. 2006; Wheeler et al. 2006), suggesting that alternative pathways may also trigger cell migration. Although Rac and Cdc42 clearly appear as key players in the front edge protrusion, RhoA activity has also been observed at the leading edge (Pertz et al. 2006). RhoA activity peaks temporally before and spatially in front of Rac and Cdc42 activity (Machacek et al. 2009). The spatiotemporal distribution of RhoA and Rac activity reflects that Rac1 and RhoA are mutually inhibitory (Nimnual et al. 2003). This may involve the Par6-aPKC complex downstream of Cdc42 or Rac which inhibits RhoA via p190RhoGAP (Goldstein and Macara 2007).

The localisation of active RhoA at the front suggests that it participates in the initiation of the protrusion (Narumiya et al. 1997; Yamana et al. 2006). One important target of RhoA is the formins mDia, which facilitate actin assembly (Pruyne et al. 2002; Sagot et al. 2002). Rho and actomyosin contraction are also essential for the maturation of focal adhesions and the increased adhesion which sustains cell contraction. In contrast to the front, cytoskeleton dynamics in the cell body and at the cell rear is primarily controlled by Rho (Ridley et al. 2003). Rho promotes contraction and retraction forces in the cell body and the cell rear (Fig. 5.1). These events mainly occur through the Rho effector ROCK, a kinase which promotes actomyosin contraction by increasing the phosphorylation of the myosin light chain (Kawano et al. 1999). Rho and ROCK are required for retraction of the trailing edge in leukocytes (Alblas et al. 2001). Generally, Rho promotes the stabilisation of actin structures (Maekawa et al. 1999). Its effector mDia contributes to stress fibre assembly (Uehata et al. 1997). Moreover, Rho and mDia are also involved in the stabilisation of the microtubule network (Palazzo et al. 2001a; Palazzo et al. 2004). mDia promotes microtubule capture at focal adhesions and can facilitate focal adhesion disassembly (Wickstrom et al. 2010). Inhibition of Rho prevents the rear detachment in migrating leukocytes (Alblas et al. 2001). The role of Rho–ROCK activity is particularly evident in the case of amoeboid migration. Activation of Rho causes membrane blebs and Rho–ROCK activity is polarised in the direction of blebbing (Pinner and Sahai 2008). Myosin II and the upstream signalling involving Rho and ROCK are also required for retraction (Worthylake and Burridge 2003; Xu et al. 2003).

5.3 Orientation of the Front-to-Rear Polarity Axis During Directed Migration

The front-to-rear polarity axis is a prerequisite for random as well as directed cell migration. However, during directed migration, not only the establishment but also the orientation of the polarity axis is controlled by external stimuli. A wide variety of polarity cues steer cells to proper locations for tissue maintenance, restoration or remodelling. These cues correspond to gradients of soluble or non-soluble proteins, to anisotropic interactions with the extracellular matrix or with neighbouring cells or to local changes in the physical properties of the microenvironment (Fig. 5.2). These polarity cues can initiate cell polarisation or act on already polarised cells to maintain or redirect their front-to-rear axis (Van Haastert and Devreotes 2004; Iglesias and Devreotes 2008).

5.3.1 Polarisation in Response to Soluble Gradients

Chemotaxis is the directed migration of cells obtained in response to gradients of soluble factors which act as chemoattractants. One of the first and most characterised examples of gradient sensing is *Dictyostelium discoideum*. The social amoeba *D. discoideum* has a developmental cycle which heavily relies on chemotaxis. During development starving isolated cells secrete waves of cAMP (cyclic adenosine monophosphate) which acts as a chemoattractant for neighbouring cells. cAMP-induced directed migration leads to the aggregation of thousands of cells

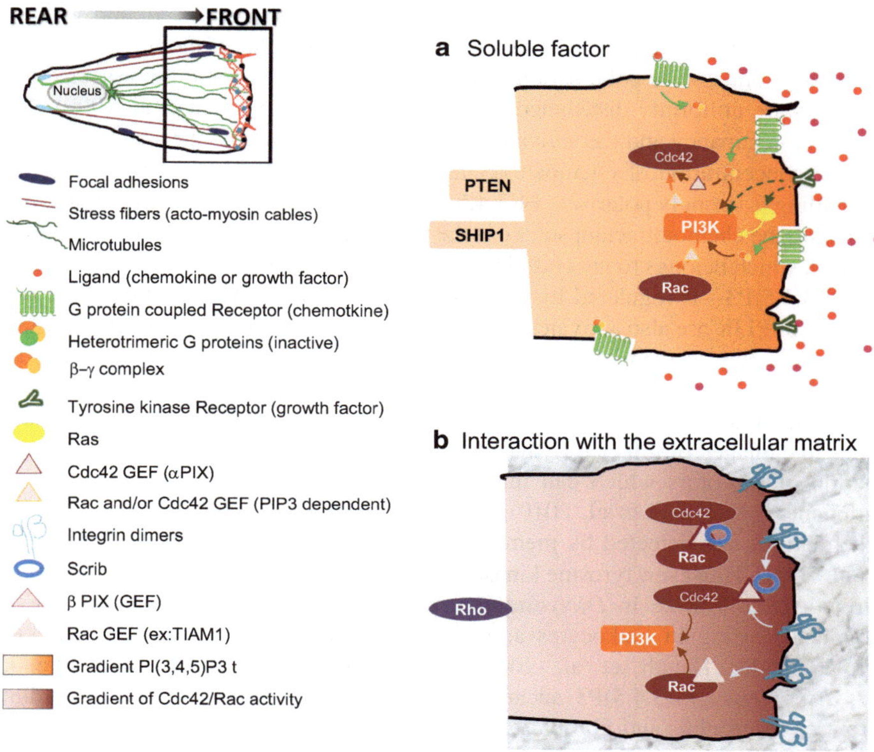

Fig. 5.2 Generating the front in response to polarity cues. The establishment and maintenance of the front-to-rear axis is controlled by a large variety of polarity cues including gradients of soluble proteins and chemical or physical properties of the extracellular matrix. This schematic depicts main signallings induced by these different polarity cues to promote the formation of a protrusive cell front. (**a**) Gradient of diffusible chemokines or growth factors promotes front-to-rear polarity mainly by generating a gradient of phosphoinositides. The example of a chemokine binding to a seven transmembrane receptors is shown here, but growth factor tyrosine kinase receptor would similarly lead to the recruitment and activation of PI3K by several parallel signals at the front (see text for details). In contrast, PTEN and SHIP recruited to the cell rear contribute to the generation of a PI(3,4,5)P3 front-to-rear gradient. In parallel, the small G proteins Rac and Cdc42 are activated at the front. Their regulation is mediated by GEF which are activated directly by G protein-coupled receptor signalling or by PI3K products. (**b**) Integrin engagement with the extracellular matrix at the cell front promotes a local polarity signal involved in the establishment of front-to-rear polarity. This signalling involves Scrib and βPIX which, when phosphorylated by Src acts as a GEF for Cdc42 (Tu et al. 2003), binds Rac and participates to Rac accumulation at the leading edge. Another GEF, like Tiam1, can (Tu et al. 2003). βPIX also activate Rac at the leading edge (Ellenbroek et al. 2012)

which eventually differentiate to form a fruiting body. These amoeboid cells use a G protein-coupled receptor to detect cAMP and transmit G protein-dependent intracellular signalling to promote directed migration (Manahan et al. 2004). There is only a few percent difference in concentration of cAMP between the cell edge facing the source of the chemoattractant and the opposite side; and yet, this induces totally different intracellular signals at the two poles of the cells in order to promote front-to-rear polarisation. The membrane receptors and the associated G proteins are uniformly distributed over the cell surface (Servant et al. 1999). The receptor occupancy and the activation of the G proteins follow the same variations as the concentration of chemoattractant (Jin et al. 2000). However, downstream signalling is strongly polarised. PI(3,4,5)P3 accumulation at the cell membrane that is facing the chemoattractant source is the first detectable polarised event (Fig. 5.2). PI3Kγ, which belongs to the class 1B of PI3K—the enzyme triggering the synthesis of PI(3,4,5)P3—is regulated by Gβγ subunits (Barber and Welch 2006). The two classes of PI3K are also activated by Ras (Parent et al. 1998; Kae et al. 2007). PI3K activation may also involve PLC and calcium signalling (Kortholt et al. 2007) (Fig. 5.2a).

In mammals, immunological responses require a rapid recruitment of immune cells to damaged or infected tissues. Lymphocytes or neutrophils express a combination of receptors which can interpret a complex network of chemoattractant gradients (McDonald et al. 2010). Most chemoattractants, like, for instance, the chemokines, are detected by members of G protein-coupled receptor family (Jin et al. 2008). However, tyrosine kinase receptors can also detect gradient of soluble growth factors. Like in *Dictyostelium*, phosphoinositides play a central role in gradient sensing in leukocytes and also in fibroblasts or cancer cells (Rickert et al. 2000; Haugh et al. 2000; Barber and Welch 2006) and like in *D. discoideum*, PI(3,4,5)P3 accumulation has been observed at the protruding front of neutrophils and fibroblasts in vitro and in vivo (Haugh et al. 2000; Parent et al. 1998; Yoo et al. 2010). PI3K is also involved in cell response to EGF (epidermal growth factor) or PDGF (platelet-derived growth factors) (Sossey-Alaoui et al. 2005; Yip et al. 2007). Class 1A PI3Ks (PI3Kα PI3Kβ and PI3Kδ) are directly regulated by tyrosine kinases (Cain and Ridley 2009). Leukocytes can also follow gradient of hydrogen peroxide in response to acute injury (Feng et al. 2010). Although the exact mechanisms by which hydrogen peroxide can induce directed migration are not fully understood, they seem to implicate a redox sensor such as Lyn, a Src kinase family member, which may then lead to PI3K activation (Yoo et al. 2011).

The polarised distribution of phosphoinositides, such as PI(3,4,5)P3 and PI(4,5)P2, is essential for defining front-to-rear polarity as well as for efficient migration (Fig. 5.1). In migrating *Dictyostelium*, PI(3,4,5)P3 and PI3K accumulate at the plasma membrane in front of the chemoattractant source (Funamoto et al. 2002; Merlot and Firtel 2003). The PI(3,4,5)P3 level is also regulated by two phosphatases: PTEN, a 3-phosphatase that converts PI(3,4,5)P3 to PI(4,5)P2, and SHIP1, a 5-phosphatase that converts PI(3,4,5)P3 to PI(3,4)P2. In wild-type mouse and zebrafish neutrophils, disruption of SHIP1 results in the accumulation of PI(3,4,5)P3 at the entire cell periphery (Mondal et al. 2012). PTEN is excluded from the front of migrating *D. discoideum* and prevents actin polymerisation on the lateral and rear sides of the cell (Iijima and Devreotes 2002). In neutrophils, the combined actions of SHIP1 and PTEN at the rear contribute to the polarised front-to-rear distribution of PI(3,4,5)P3, which is maintained at the leading edge and promotes cell polarisation and migration (Yoo et al. 2010) (Fig. 5.1 and 5.2a).

In amoeba and leukocytes, Rac activity parallels the PI(3,4,5)P3 gradient (Fig. 5.1). Rac is activated, in response to ligand binding, by the heterotrimeric G protein which directly associates with Rac guanine nucleotide-exchange factor (GEF), such as Elmo/DOCK (Yan et al. 2012). Rac is also activated in response to the accumulation of PI(3,4,5)P3 at the front via GEFs sensitive to phosphoinositides (Fig. 5.2a). G protein-coupled receptors also activate Cdc42 by a complex formed containing Gβγ subunits, the GEF PIXα and PAK (Li et al. 2003). However, Rac activation is not always sufficient to induce the actin polymerisation normally observed in all protrusions, suggesting that PI3K activates additional signalling cascades which also participate in the regulation of actin dynamics at the leading edge (Yoo et al. 2010). The Ras pathway, which controls TORC2 and PKB, also contributes to actin polymerisation (Charest et al. 2010; Cai et al. 2010). In general, inhibition of Ras or PI3K blocks directional sensing (Wang et al. 2002; Takeda et al. 2007; Loovers et al. 2006). However, in steep gradients, cell-directed cell migration can be induced independently of PI3K signalling, indicating that alternative pathways are at play (Hoeller and Kay 2007). The role of PLA2 in gradient sensing has been identified in screen performed in the presence of PI3K inhibitors and then confirmed using mutant cells and specific inhibitors (Chen et al. 2007; van Haastert et al. 2007).

5.3.2 Front-to-Rear Polarisation in Response to Cell Interaction with the Extracellular Matrix

Integrin engagement with the extracellular matrix at the leading edge of cell protrusion leads to cytoskeletal rearrangement resulting in two important features of front-to-rear polarisation (Vicente-Manzanares et al. 2009): the formation of protrusions in the direction of migration and the reorientation of the centrosome and the Golgi apparatus towards the direction of migration in some cells (Fig. 5.2b).

Each cell contains a specific integrin signature, and several integrin dimers have been involved in cell polarisation, depending on the nature of the substrate on which cells are migrating. For instance, α1β1 are crucial for the directed migration astrocytes on a laminin substrate (Peng et al. 2013), whereas αvβ3 have been implicated when astrocytes migrate on the neuronal protein Thy-1 (Kong et al. 2013). During migration, localised integrin signalling leads to the polarised recruitment of the small GTPases Cdc42 and Rac (Fukata et al. 1999; Del Pozo and Schwartz 2007). This pathway depends on the recruitment of the evolutionary conserved polarity protein Scrib at the front (Qin et al. 2005; Osmani et al. 2006). Scrib, in a complex with βPIX and GIT1 (G protein-coupled receptor interacting protein 1) (Nola et al. 2008; Audebert et al. 2004), controls Cdc42 activation and localisation of Rac (Osmani et al. 2006) (Fig. 5.2b). A similar mechanism is also observed in endothelial and mammary cells front (Qin et al. 2005).

In addition to its chemical nature, the physical properties of the extracellular matrix can influence directional migration. Contractile structures at the leading edge probe the substrate rigidity (Ghassemi et al. 2012). Focal adhesions are essential to mechanically couple the cell to the substratum and to induce intracellular signalling controlling cytoskeleton rearrangements and cell migration in response to the physical properties of the substrate (Choquet et al. 1997). The stiffness of the extracellular substrate is balanced by the actomyosin forces generated within the cell. The molecular mechanisms involved in the control of front-to-rear polarity by physical signals are not completely understood yet, but force-generating proteins like actin and myosin are central. Force detection is generally based on protein conformational changes resulting in the regulation of their activity. Ion channels, enzymes or scaffolding proteins can be involved. The rigidity of the extracellular matrix impacts on the maturation and the shape of focal adhesions, which are typically elongated in the direction of the stress fibres. Cells lying on soft elastic substrates develop elongated adhesions and stress fibres, whereas cells cultured on stiff plastic or glass substrate develop large focal adhesions (Pelham and Wang 1997). The complex formed between talin, vinculin and actin filaments can sense and translate mechanical tension into biochemical signals (Yu et al. 2009; Wang et al. 2012). Force unfolds talin and reveals cryptic vinculin binding domains (del Rio et al. 2009). Filamin conformation is also modified by external forces, which increase its binding to integrin β tails and reduces the binding of RhoGAP (Ehrlicher et al. 2011). A similar role for talin as link between integrins and actin cytoskeleton has been found in *C. elegans* and drosophila.

5.3.3 Front-to-Rear Polarisation in Response to Cellular Interactions

The role of cell–cell contacts in directional migration is perfectly illustrated in the collective migration of neural crest cells in Xenopus, zebrafish and chick embryos (Carmona-Fontaine et al. 2008; Carmona-Fontaine et al. 2011; Matthews et al. 2008; Becker et al. 2013). On one hand, cellular interactions promote the polarisation and the directed migration of neural crest cells. On the other hand, migrating neural crest cells undergo contact inhibition locomotion, a process that prevents cells to migrate over one another.

Adherens junctions (AJ) are essential structures for cell–cell adhesion, for the maintenance of tissue integrity, and for the cohesiveness of migrating cell groups (Etienne-Manneville 2011, 2012). They are mainly based on homophilic interaction between cadherin extracellular domains. Cadherin-mediated AJs are required for the migration of germ cells or of the lateral line migration in zebrafish (Kardash et al. 2010; Kerstetter et al. 2004). In mesendodermal cells in vivo, E-cadherin is involved in cell polarisation and for the correct orientation towards the animal pole (Dumortier et al. 2012). The anisotropic distribution of AJs is sufficient to induce front-to-rear polarisation (Fig. 5.3). The localisation of cadherin-mediated contacts impacts on nuclear position and promotes centrosome orientation towards the free cell edge (Dupin et al. 2009; Desai et al. 2009; Desai et al. 2013). The anisotropic distribution of AJs directs the actin-retrograde flow that is generated at the cell front towards the cell rear (Fig. 5.3). This flow of actin cable directly interacts with the nuclear envelop or acts on intermediate filaments to push the nucleus away from the cell-free edge (Dupin and Etienne-Manneville 2011; Dupin et al. 2011; Luxton et al. 2010). This rearward movement of the nucleus may facilitate centrosome reorientation and the polarisation of the microtubule network (Gomes et al. 2005; Palazzo et al. 2001b; Etienne-Manneville et al. 2005; Manneville et al. 2010). However, the molecular mechanisms underlying cadherin functions in the centrosome positioning are not fully understood and may involve a direct effect of cadherin-mediated junctions on the microtubule network. Microtubule regulation at cell–cell contacts, where Par3 interacts with dynein, contributes to the centrosome positioning (Schmoranzer et al. 2009). AJs may also have an indirect effect on front-to-rear polarity by controlling cell interactions with the extracellular matrix (Canel et al. 2013), and evidence point out an inhibitory effect of AJs on the formation of focal adhesions (Borghi et al. 2010; Dupin et al. 2009; Camand et al. 2012). Thus, the anisotropic distribution of AJs results in an asymmetric localisation of focal adhesions, which in turn can promote a polarisation signal by controlling microtubule anchoring and centrosome reorientation (see above) (Fig. 5.3). The restriction of focal adhesions at the cell front is also required for the generation of the polarised actin-retrograde flow which pushes the nucleus towards the rear (Borghi et al. 2010; Dupin et al. 2009; Camand et al. 2012).

Cumulative evidence point to AJs as mechanosensors of tissue rigidity relaying external forces to the actin cytoskeleton (Ladoux et al. 2010). Experiments in Xenopus and zebrafish show that force applied to C-cadherins induces cell polarisation and migration in the direction opposite to the applied force (Weber et al. 2012). Cells migrating together can generate and follow gradients of intercellular tension (Trepat et al. 2009; Tambe et al. 2011). Increased cell contractility induces AJ strengthening and resistance to mechanical stresses (Liu et al. 2010). α-catenin and vinculin are force-activated proteins in cadherin complexes. α-catenin is required for mechanical coupling between cadherin and actomyosin (Cavey et al. 2008; Desai et al. 2013; Gumbiner and McCrea 1993). Applied forces trigger the stretching α-catenin and consequently expose a cryptic vinculin-binding site in α-catenin that in turn recruits vinculin and actin to cadherin junctions. In fact, mechanosensitive vinculin/EPLIN recruitment at the junctions increased the binding between actin and cadherin (le Duc et al. 2010; Taguchi et al. 2011).

Interestingly, both the depletion and the overexpression of DE-cadherins impair the migration of drosophila border cells from the anterior part towards the posterior part of the egg chamber (Niewiadomska et al. 1999; Pacquelet et al. 2003; Pacquelet and Rorth 2005), highlighting the dual function of cadherin-mediated

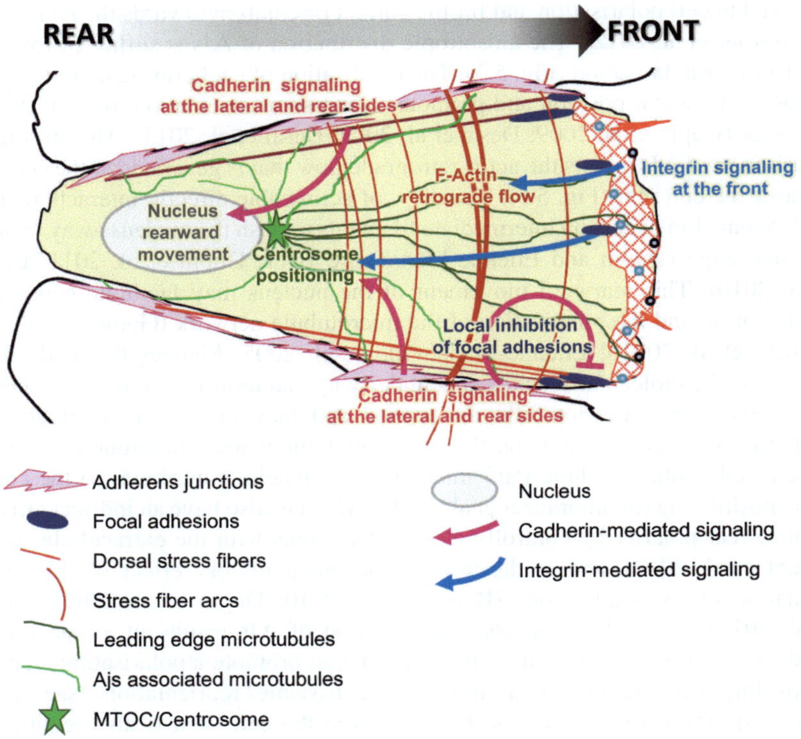

Fig. 5.3 Adherens junctions and front-to-rear polarity. During collective migration, the adherens junctions that support physical interactions between adjacent cells push the cells to migrate in the same direction by orienting the front-to-rear axis towards the free edge of the cells. Cadherin can either impact directly on the organisation of the cytoskeleton or act through the regulation of integrin engagement with the extracellular matrix. Cadherin-mediated signals are shown with *pink arrows*. The integrin-mediated signals which are restricted at the cell-free edge are shown as *blue arrows*

junction in the regulation of migration. When two cells collide, cadherins can form homophilic contacts which induce a very transient peak of Rac1 activity directly followed by an increase in RhoA activity (Carmona-Fontaine et al. 2008; Carmona-Fontaine et al. 2011; Theveneau et al. 2010; Etienne-Manneville 2011). This transient change participates in the repolarisation of the cell towards the opposite side and the emergence of a new protrusion at the free cell edge. In dense groups, contact inhibition promotes collective cell behaviour by driving cell movement towards regions devoid of intercellular contacts.

5.4 Maintenance of the Front-to-Rear Polarity Axis and Direction Persistence in Migration

Front-to-rear polarisation initiates migration, and the orientation of the front-to-rear axis by various extracellular cues induces directed migration. Once established, the front-to-rear polarity axis must be maintained in a stable direction. Thus, cells migrating in a complex 3D environment and submitted to constant minute changes in the concentration of soluble and insoluble ligands can keep migrating persistently in a constant direction. Several hypotheses have been formulated which generally combine positive feedback loops in signalling pathways (Fig. 5.4a) and microtubule-based cell organisation (Fig. 5.4b).

5.4.1 Positive Feedback Loops to Stabilise the Front

A positive feedback loop between phosphoinositide signalling and Rho GTPases activity is commonly thought to be a key element which maintains and amplifies actin-driven protrusion at the cell front (Weiner et al. 2002; Xu et al. 2003; Srinivasan et al. 2003) (Fig. 5.4a). The accumulation of PI(3,4,5)P3 triggers the activation of the small G proteins Rac and Cdc42, via the activation of PIP3-dependent exchange factors (Welch et al. 2002). The localised accumulation of PIP3 at the leading edge of zebrafish neutrophils promotes Rac activation and therefore the polarised protrusion (Yoo et al. 2010). PI3K directly interacts with Rac and the two proteins can activate each other (Janetopoulos and Firtel 2008). PI (3,4,5)P3 and PI(3,4)P2 either recruit or activate a number of Rac and/or Cdc42 GEFs such as p-Rex1, Vav, Tiam1 and PIX (Welch et al. 2003). PI(3,4,5)P3 also binds Rac and Cdc42 effectors like WASp and WAVE, as well as Rac itself (Suetsugu et al. 2006; Innocenti et al. 2003; Bokoch et al. 1996). Conversely, Rac-induced actin polymerisation is required to stabilise PI3K and reinforce the local accumulation of PI(3,4,5)P3 at the cell front (Weiner et al. 2002), while the accumulation of PI(3,4,5)P3 directly promotes localised actin polymerisation and pseudopod extension (Bolourani et al. 2006; Huang et al. 2003). Such a local coupling between receptor signalling and cell protrusion is sufficient to promote a biased random walk and chemotaxis (Arrieumerlou and Meyer 2005). In fact, disruption of the PI3K-Rac positive feedback loop in neutrophils alters the directionality of migration in response to chemoattractants (Wang et al. 2002; Weiner et al. 2002). Similarly, several feedback loops may be involved in maintaining active Cdc42 at the front, including phosphoinositides and PAR4/LKB1 (Raftopoulou et al. 2004; Zhang et al. 2008). Cdc42 has also been shown to contribute to integrin-mediated activation of PI3K (Schwartz and Shattil 2000). In fibroblasts, Cdc42 activation at the leading edge is increased by a positive feedback signal involving proton efflux at the leading edge (Frantz et al. 2007). Changes in the cytosolic pH may affect GEF binding to phosphoinositides and also

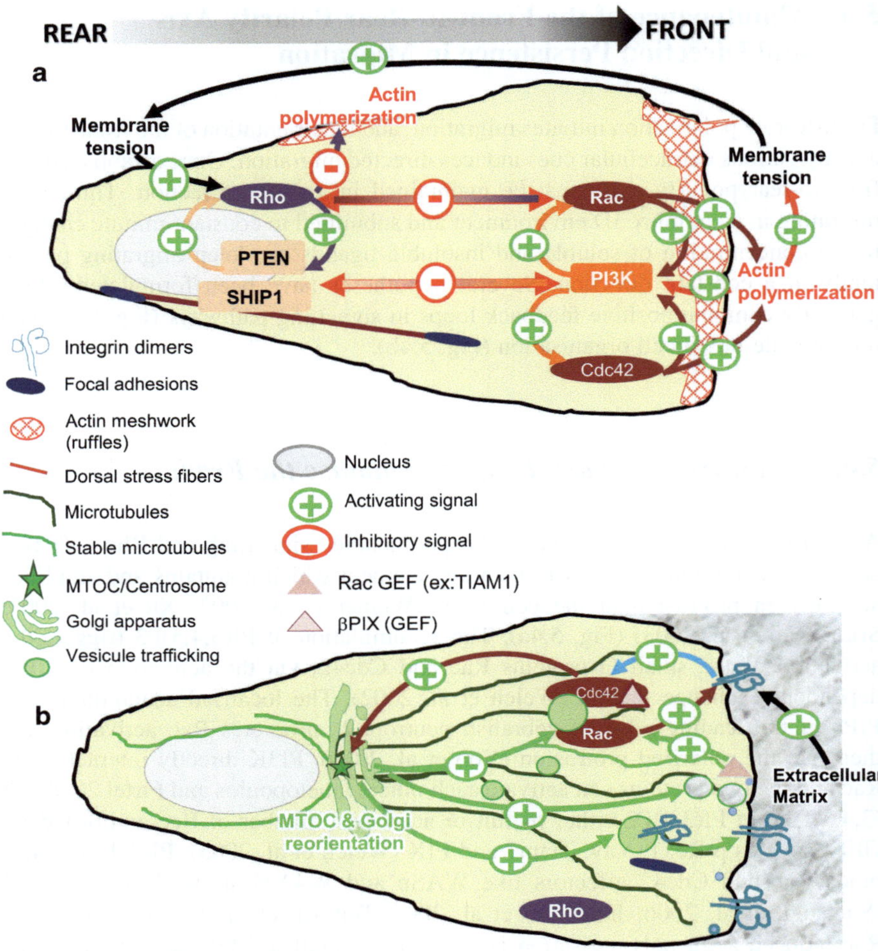

Fig. 5.4 Feedback loops and the maintenance of the polarity axis. (**a**) Positive feedback loops play a key role in the amplification and maintenance of front and rear signals. The cross talk between small G proteins of the Rho family and phosphoinositides is at the heart of these amplifying loops. Additional cross talk also involving cytoskeletal regulation and membrane tension contributes to the maintenance of the front-to-rear polarity axis. (**b**) The polarization of the microtubule network provides positive feedback loops mainly by favouring polarized vesicular traffic towards the leading edge and also by delivering key signalling molecules to specific location. Signalling pathways are symbolised by an *arrow the colour* of the initiating regulatory protein

the activity of myosin motors and the affinity of actin cross-linkers (Kohler et al. 2012). Inside–out signalling induced by small G proteins also creates a positive feedback loop at the front of adhesive migrating cells (Abram and Lowell 2009). The Rac-specific GEF Tiam1 interacts directly with Talin and activates the integrins, increasing integrin-induced cell polarity (Wang et al. 2012) (Fig. 5.4b).

5.4.2 Long-Range Signals to Maintain an Opposite Rear

How opposite molecular signals are initiated and maintained to function simultaneously at the front and at the rear of a unique cell whereas external stimuli only marginally vary across the cell diameter remains a matter of debate. Synchronisation of actin polymerisation at the cell front with myosin-dependent contractile forces at the rear requires numerous feedback mechanisms (Charest and Firtel 2006). The opposite distribution of Rac and Cdc42 on one hand and Rho on the other hand is paralleled by a polarised distribution of phosphoinositides.

Like at the front, the interplay between small G proteins and phosphoinositides tends to reinforce the specificity of the signalling (Fig. 5.4a). At the rear, Rho promotes the production of PI(4,5)P2 by activating the PI4P-5 kinase (Schwartz and Shattil 2000) and by stimulating PTEN activity through ROCK-dependent phosphorylation (Li et al. 2005; Sanchez et al. 2005). Complex models have been proposed to explain the induction of opposite signalling pathways at the front and the rear of migrating cells (for a review, Janetopoulos and Firtel 2008). Accumulation of PI3K and Rac signals at the front of the cell may result in the depletion of these molecules on the rest of the cell periphery and thus generate a rear zone by default of positive signals (Postma and Van Haastert 2001). Such a mechanism relies on the limited amount of some of the essential components of the "front" signal. This hypothesis is supported by the observation that PI3K recruitment to the front of migrating fibroblasts leads to the depletion of the cytosolic pool of the protein (Schneider and Haugh 2005). Alternatively, signals at the front may locally inhibit the signals specific of the cell rear, which would remain exclusively active at the cell rear to contribute to cell contractility and retraction. In neutrophils, Rac activation at the cell front counteracts Rho and Myosin activity and excludes Rho from this region (Pestonjamasp et al. 2006). Similarly, active Cdc42 is required to localise RhoA at the rear of leukocytes (Li et al. 2005). In collectively migrating cells, the presence of adherens junctions on the lateral and rear side of neighbouring cells generate signals that locally inhibit focal adhesion formation and signalling, restricting integrin-mediated protrusion at the front (Etienne-Manneville 2011, 2012; Borghi et al. 2010). Finally, a more general model of front-to-rear polarisation involves the local activation of pathways, promoting protrusion associated with a global and slower inhibitory signal (Janetopoulos and Firtel 2008; Jin 2013; Xiong et al. 2010). This "LEGI" (local excitation and global inhibition) model may rely on mechanical constrains due, for instance, to membrane tension, which would dampen actin dynamics (Houk et al. 2012). Actin-pushing forces at the leading edge significantly increase membrane tension (Lieber et al. 2013), which is immediately transmitted to the entire cell periphery (Fig. 5.4a). Membrane tension activates myosin contraction in fibroblasts and inhibits the recruitment of actin regulatory proteins in neutrophils (Gauthier et al. 2011) and acts as a global inhibitor of actin polymerisation confining protrusion at a single position (Houk et al. 2012). Membrane tension can also activate calcium influx via stretch-activated calcium channel and promote the retraction of motile keratocytes (Lee et al. 1999).

5.4.3 Microtubules: Guardians of the Polarity Axis

The polarised organisation of the microtubule network is crucial for persistent directed cell migration (Vasiliev et al. 1970). In addition to their direct role in the migration process, microtubules are essential in the maintenance of front-to-rear polarity in many cell types (Etienne-Manneville 2004a, b). Microtubule depolymerisation alters cell migration in vitro and in vivo (Redd et al. 2006). More precisely, the orientation of the microtubule network warrants the persistence of migration. In fact, alteration of microtubule dynamics by inhibition of microtubule +TIPs, like Clasp2, prevents persistent migration (Drabek et al. 2006). Microtubules may contribute their role in vesicular traffic to front-to-rear polarisation and cell migration. The position of the centrosome and the Golgi apparatus at the front of the nucleus, facing the leading edge and the stabilisation of microtubules elongating towards the leading edge, is likely to favour directed vesicle delivery to the leading edge plasma membrane (Etienne-Manneville 2013) (Fig. 5.4b). Localised vesicle delivery to the leading edge can reinforce the polarity cues by concentrating the relevant membrane receptors to specific membrane domain and thus reinforce the cell sensitivity to polarity signals and stabilise the direction of migration (Nieto et al. 1997). Migrating cells affect the composition and the organisation of the extracellular matrix, which in turn can modulate cell polarity and directional migration (Fig. 5.4b). Astrocytes produce laminin as they migrate and use laminin receptor to polarise (Peng et al. 2013). The rapid internalisation and recycling of integrins at the leading edge of migrating cells restrict integrin localisation to the front of the cell by outcompeting their slow lateral diffusion towards the cell body, suggesting that endocytic turnover of adhesion molecules at the leading edge serves to accumulate them there. The redistribution of integrins to the front of migrating cells is mediated by asymmetrical phosphorylation of Numb by aPKC (Nishimura and Kaibuchi 2007). In addition to essential membrane components like integrins which need to be recycled and renewed at the leading edge, the polarised traffic of molecules involved in polarity pathways such as Rac and Cdc42 contributes to the polarised localisation of these proteins and to the maintenance of their front activity (Osmani et al. 2010; Di Cesare et al. 2000) (Fig. 5.4b). The endocytic pathway is important in restricting the Cdc42 and Rac at the leading edge. In fact the recruitment and activation of Cdc42 at the leading edge is disturbed in Arf6 depleted cells (Osmani et al. 2010). In astrocyte, Arf6 which belongs to ARF family small promotes βPIX accumulation at the leading edge. This accumulation of βPIX is needed for Cdc42-restricted activation and for the polarised recruitment of Rac (Osmani et al. 2006; Cau and Hall 2005). The localisation of Rac is further facilitated at the leading edge by the delivery of lipid-raft-containing vesicles (Balasubramanian et al. 2007). Inhibition of the endocytic pathway is required for the maintenance of cell polarity (Wessels et al. 2000). Local regulation of membrane traffic may also affect membrane tension and indirectly impact on actin dynamics (Ji et al. 2008). In addition to their role in intracellular trafficking, microtubules can directly affect signalling.

The growth of microtubules at the cell front can activate Rac to further promote actin and microtubule polymerisation and cell protrusion (Waterman-Storer et al. 1999). The composition of the +TIP complex impacts on the signalling functions of microtubules. Growing leading edge microtubules which accumulate APC clusters can activate Asef, a Rac GEF interacting with APC (Kawasaki et al. 2003; Kawasaki et al. 2000). While growing microtubules can activate Rac, they can also inhibit Rho by sequestering and inactivating the RhoGEF-H1/Lfc (Krendel et al. 2002; Chang et al. 2008).

5.5 Conclusions

The conclusion/end point of front-to-rear polarisation is to functionally differentiate two opposite cell poles so that the protruding leading edge and the opposite contracting/retracting rear edge promote migration. Front-to-rear polarity is absolutely required for migration whether the cell migrates in a 2D, or a 3D, environment, in a random or a directed manner. Conserved polarised intracellular signals, involving small G proteins and phosphoinositide signalling, are essential regulator of front-to-rear polarity in all kinds of cell types and organisms. Migration in an in vivo environment is a tightly regulated process which requires the orientation of front-to-rear axis so that cells migrate in a directed manner following the multiple polarity cues present. How cells integrate and respond to the subtle and numerous variations of their microenvironment remains partially unclear. The maintenance of the front-to-rear polarity axis essentially relies on the continuous negative feedback between the leading edge and the side and rear edges of the cells. Beside its role in normal physiology, abnormal cell migration is involved in many pathological conditions, including inflammatory diseases and cancer invasion and metastasis. Loss of polarity is a hallmark of cancer cells. However, despite frequent alterations of GTPase activity and phosphoinositide signalling, the front-to-rear polarity and the underlying molecular mechanisms appear functional in invasive cancer cells. It contrast, it seems that, in this case, the ability to steer this polarity axis in response to external cues such intercellular interactions is perturbed, indicating that the molecular mechanisms responsible for the sensing and/or the interpretation of polarity cues in altered in these cells.

References

Abram CL, Lowell CA (2009) The ins and outs of leukocyte integrin signaling. Annu Rev Immunol 27:339–362

Affolter M, Weijer CJ (2005) Signaling to cytoskeletal dynamics during chemotaxis. Dev Cell 9(1):19–34. doi:10.1016/j.devcel.2005.06.003

Akhmanova A, Hoogenraad CC, Drabek K, Stepanova T, Dortland B, Verkerk T, Vermeulen W, Burgering BM, De Zeeuw CI, Grosveld F, Galjart N (2001) Clasps are CLIP-115 and -170 associating proteins involved in the regional regulation of microtubule dynamics in motile fibroblasts. Cell 104(6):923–935

Akhmanova A, Steinmetz MO (2008) Tracking the ends: a dynamic protein network controls the fate of microtubule tips. Nat Rev Mol Cell Biol 9(4):309–322

Alblas J, Ulfman L, Hordijk P, Koenderman L (2001) Activation of Rhoa and ROCK are essential for detachment of migrating leukocytes. Mol Biol Cell 12(7):2137–2145

Arber S, Barbayannis FA, Hanser H, Schneider C, Stanyon CA, Bernard O, Caroni P (1998) Regulation of actin dynamics through phosphorylation of cofilin by LIM-kinase. Nature 393 (6687):805–809. doi:10.1038/31729

Arias-Salgado EG, Lizano S, Shattil SJ, Ginsberg MH (2005) Specification of the direction of adhesive signaling by the integrin beta cytoplasmic domain. J Biol Chem 280(33):29699–29707

Arrieumerlou C, Meyer T (2005) A local coupling model and compass parameter for eukaryotic chemotaxis. Dev Cell 8(2):215–227. doi:10.1016/j.devcel.2004.12.007

Audebert S, Navarro C, Nourry C, Chasserot-Golaz S, Lecine P, Bellaiche Y, Dupont JL, Premont RT, Sempere C, Strub JM, Van Dorsselaer A, Vitale N, Borg JP (2004) Mammalian Scribble forms a tight complex with the betaPIX exchange factor. Curr Biol 14(11):987–995

Balasubramanian N, Scott DW, Castle JD, Casanova JE, Schwartz MA (2007) Arf6 and microtubules in adhesion-dependent trafficking of lipid rafts. Nat Cell Biol 9(12):1381–1391

Barber MA, Welch HC (2006) PI3K and RAC signalling in leukocyte and cancer cell migration. Bull Cancer 93(5):E44–E52

Becker SF, Mayor R, Kashef J (2013) Cadherin-11 mediates contact inhibition of locomotion during Xenopus neural crest cell migration. PLoS ONE 8(12):e85717. doi:10.1371/journal.pone.0085717

Beningo KA, Dembo M, Kaverina I, Small JV, Wang YL (2001) Nascent focal adhesions are responsible for the generation of strong propulsive forces in migrating fibroblasts. J Cell Biol 153(4):881–888

Bhatt A, Kaverina I, Otey C, Huttenlocher A (2002) Regulation of focal complex composition and disassembly by the calcium-dependent protease calpain. J Cell Sci 115(Pt 17):3415–3425

Bokoch GM, Vlahos CJ, Wang Y, Knaus UG, Traynor-Kaplan AE (1996) Rac GTPase interacts specifically with phosphatidylinositol 3-kinase. Biochem J 315(Pt 3):775–779

Bolourani P, Spiegelman GB, Weeks G (2006) Delineation of the roles played by RasG and RasC in cAMP-dependent signal transduction during the early development of Dictyostelium discoideum. Mol Biol Cell 17(10):4543–4550. doi:10.1091/mbc.E05-11-1019

Borghi N, Lowndes M, Maruthamuthu V, Gardel ML, Nelson WJ (2010) Regulation of cell motile behavior by crosstalk between cadherin- and integrin-mediated adhesions. Proc Natl Acad Sci U S A 107(30):13324–13329

Brandt DT, Grosse R (2007) Get to grips: steering local actin dynamics with IQGAPs. EMBO Rep 8(11):1019–1023. doi:10.1038/sj.embor.7401089

Cai H, Das S, Kamimura Y, Long Y, Parent CA, Devreotes PN (2010) Ras-mediated activation of the TORC2-PKB pathway is critical for chemotaxis. J Cell Biol 190(2):233–245. doi:10.1083/jcb.201001129

Cain RJ, Ridley AJ (2009) Phosphoinositide 3-kinases in cell migration. Biol Cell 101(1):13–29. doi:10.1042/BC20080079

Camand E, Peglion F, Osmani N, Sanson M, Etienne-Manneville S (2012) N-cadherin expression level modulates integrin-mediated polarity and strongly impacts on the speed and directionality of glial cell migration. J Cell Sci 125(Pt 4):844–857. doi:10.1242/jcs.087668

Canel M, Serrels A, Frame MC, Brunton VG (2013) E-cadherin-integrin crosstalk in cancer invasion and metastasis. J Cell Sci 126(Pt 2):393–401. doi:10.1242/jcs.100115

Carlier MF, Pantaloni D (2007) Control of actin assembly dynamics in cell motility. J Biol Chem 282(32):23005–23009. doi:10.1074/jbc.R700020200

Carmona-Fontaine C, Matthews HK, Kuriyama S, Moreno M, Dunn GA, Parsons M, Stern CD, Mayor R (2008) Contact inhibition of locomotion in vivo controls neural crest directional migration. Nature 456(7224):957–961. doi:10.1038/nature07441

Carmona-Fontaine C, Theveneau E, Tzekou A, Tada M, Woods M, Page KM, Parsons M, Lambris JD, Mayor R (2011) Complement fragment C3a controls mutual cell attraction during collective cell migration. Dev Cell 21(6):1026–1037. doi:10.1016/j.devcel.2011.10.012

Cau J, Hall A (2005) Cdc42 controls the polarity of the actin and microtubule cytoskeletons through two distinct signal transduction pathways. J Cell Sci 118(Pt 12):2579–2587

Cavey M, Rauzi M, Lenne PF, Lecuit T (2008) A two-tiered mechanism for stabilization and immobilization of E-cadherin. Nature 453(7196):751–756. doi:10.1038/nature06953

Chamberlain CE, Kraynov VS, Hahn KM (2000) Imaging spatiotemporal dynamics of Rac activation in vivo with FLAIR. Methods Enzymol 325:389–400

Chang YC, Nalbant P, Birkenfeld J, Chang ZF, Bokoch GM (2008) GEF-H1 couples nocodazole-induced microtubule disassembly to cell contractility via RhoA. Mol Biol Cell 19(5):2147–2153. doi:10.1091/mbc.E07-12-1269

Chao WT, Kunz J (2009) Focal adhesion disassembly requires clathrin-dependent endocytosis of integrins. FEBS Lett 583(8):1337–1343. doi:10.1016/j.febslet.2009.03.037

Charest PG, Firtel RA (2006) Feedback signaling controls leading-edge formation during chemotaxis. Curr Opin Genet Dev 16(4):339–347. doi:10.1016/j.gde.2006.06.016

Charest PG, Shen Z, Lakoduk A, Sasaki AT, Briggs SP, Firtel RA (2010) A Ras signaling complex controls the RasC-TORC2 pathway and directed cell migration. Dev Cell 18(5):737–749. doi:10.1016/j.devcel.2010.03.017

Charras GT, Hu CK, Coughlin M, Mitchison TJ (2006) Reassembly of contractile actin cortex in cell blebs. J Cell Biol 175(3):477–490. doi:10.1083/jcb.200602085

Charras GT, Yarrow JC, Horton MA, Mahadevan L, Mitchison TJ (2005) Non-equilibration of hydrostatic pressure in blebbing cells. Nature 435(7040):365–369. doi:10.1038/nature03550

Chen L, Iijima M, Tang M, Landree MA, Huang YE, Xiong Y, Iglesias PA, Devreotes PN (2007) PLA2 and PI3K/PTEN pathways act in parallel to mediate chemotaxis. Dev Cell 12(4):603–614. doi:10.1016/j.devcel.2007.03.005

Choquet D, Felsenfeld DP, Sheetz MP (1997) Extracellular matrix rigidity causes strengthening of integrin-cytoskeleton linkages. Cell 88(1):39–48

Daub H, Gevaert K, Vandekerckhove J, Sobel A, Hall A (2001) Rac/Cdc42 and p65PAK regulate the microtubule-destabilizing protein stathmin through phosphorylation at serine 16. J Biol Chem 276:1677–1680

Del Pozo MA, Schwartz MA (2007) Rac, membrane heterogeneity, caveolin and regulation of growth by integrins. Trends Cell Biol 17(5):246–250

del Rio A, Perez-Jimenez R, Liu R, Roca-Cusachs P, Fernandez JM, Sheetz MP (2009) Stretching single talin rod molecules activates vinculin binding. Science 323(5914):638–641

Desai RA, Gao L, Raghavan S, Liu WF, Chen CS (2009) Cell polarity triggered by cell-cell adhesion via E-cadherin. J Cell Sci 122(Pt 7):905–911

Desai RA, Gopal SB, Chen S, Chen CS (2013) Contact inhibition of locomotion probabilities drive solitary versus collective cell migration. J R Soc Interface 10(88):20130717. doi:10.1098/rsif.2013.0717

Di Cesare A, Paris S, Albertinazzi C, Dariozzi S, Andersen J, Mann M, Longhi R, de Curtis I (2000) p95-APP1 links membrane transport to Rac-mediated reorganization of actin. Nat Cell Biol 2(8):521–530

Drabek K, van Ham M, Stepanova T, Draegestein K, van Horssen R, Sayas CL, Akhmanova A, Ten Hagen T, Smits R, Fodde R, Grosveld F, Galjart N (2006) Role of CLASP2 in microtubule stabilization and the regulation of persistent motility. Curr Biol 16(22):2259–2264. doi:10.1016/j.cub.2006.09.065

Dujardin DL, Barnhart LE, Stehman SA, Gomes ER, Gundersen GG, Vallee RB (2003) A role for cytoplasmic dynein and LIS1 in directed cell movement. J Cell Biol 163(6):1205–1211

Dumortier JG, Martin S, Meyer D, Rosa FM, David NB (2012) Collective mesendoderm migration relies on an intrinsic directionality signal transmitted through cell contacts. Proc Natl Acad Sci U S A 109(42):16945–16950. doi:10.1073/pnas.1205870109

Dupin I, Camand E, Etienne-Manneville S (2009) Classical cadherins control nucleus and centrosome position and cell polarity. J Cell Biol 185(5):779–786

Dupin I, Etienne-Manneville S (2011) Nuclear positioning: mechanisms and functions. Int J Biochem Cell Biol 43(12):1698–1707

Dupin I, Sakamoto Y, Etienne-Manneville S (2011) Cytoplasmic intermediate filaments mediate actin-driven positioning of the nucleus. J Cell Sci 124(Pt 6):865–872. doi:10.1242/jcs.076356

Efimov A, Kharitonov A, Efimova N, Loncarek J, Miller PM, Andreyeva N, Gleeson P, Galjart N, Maia AR, McLeod IX, Yates JR 3rd, Maiato H, Khodjakov A, Akhmanova A, Kaverina I (2007) Asymmetric CLASP-dependent nucleation of noncentrosomal microtubules at the trans-Golgi network. Dev Cell 12(6):917–930. doi:10.1016/j.devcel.2007.04.002

Efimov A, Schiefermeier N, Grigoriev I, Ohi R, Brown MC, Turner CE, Small JV, Kaverina I (2008) Paxillin-dependent stimulation of microtubule catastrophes at focal adhesion sites. J Cell Sci 121(Pt 2):196–204. doi:10.1242/jcs.012666

Ehrlicher AJ, Nakamura F, Hartwig JH, Weitz DA, Stossel TP (2011) Mechanical strain in actin networks regulates FilGAP and integrin binding to filamin A. Nature 478(7368):260–263. doi:10.1038/nature10430

Eisenmann KM, Harris ES, Kitchen SM, Holman HA, Higgs HN, Alberts AS (2007) Dia-interacting protein modulates formin-mediated actin assembly at the cell cortex. Curr Biol 17(7):579–591. doi:10.1016/j.cub.2007.03.024

Ellenbroek SI, Iden S, Collard JG (2012) The Rac activator Tiam1 is required for polarized protrusional outgrowth of primary astrocytes by affecting the organization of the microtubule network. Small GTPases 3(1):4–14. doi:10.4161/sgtp.19379

Etienne-Manneville S (2004a) Actin and microtubules in cell motility: which one is in control? Traffic 5:470–477

Etienne-Manneville S (2004b) Cdc42–the centre of polarity. J Cell Sci 117(Pt 8):1291–1300

Etienne-Manneville S (2009) APC in cell migration. Adv Exp Med Biol 656:30–40

Etienne-Manneville S (2010) From signaling pathways to microtubule dynamics: the key players. Curr Opin Cell Biol 22(1):104–111

Etienne-Manneville S (2011) Control of polarized cell morphology and motility by adherens junctions. Semin Cell Dev Biol 22(8):850–857. doi:10.1016/j.semcdb.2011.07.023

Etienne-Manneville S (2012) Adherens junctions during cell migration. Subcell Biochem 60:225–249. doi:10.1007/978-94-007-4186-7_10

Etienne-Manneville S (2013) Microtubules in cell migration. Annu Rev Cell Dev Biol 29:471–499

Etienne-Manneville S, Hall A (2001) Integrin-mediated activation of Cdc42 controls cell polarity in migrating astrocytes through PKCzeta. Cell 106(4):489–498

Etienne-Manneville S, Hall A (2002) Rho GTPases in cell biology. Nature 420:629–635

Etienne-Manneville S, Hall A (2003a) Cdc42 regulates GSK3 and adenomatous polyposis coli (APC) to control cell polarity. Nature 421:753–756

Etienne-Manneville S, Hall A (2003b) Cell polarity: Par6, aPKC and cytoskeletal crosstalk. Curr Opin Cell Biol 15:67–72

Etienne-Manneville S, Manneville J, Nicholls S, Ferenczi MA, Hall A (2005) Cdc42 and Par6/PKCz regulate the spatially localized association of Dlg1 and APC to control cell polarization. J Cell Biol 170(6):895–901

Evangelista M, Zigmond S, Boone C (2003) Formins: signaling effectors for assembly and polarization of actin filaments. J Cell Sci 116:2603–2611

Ezratty EJ, Bertaux C, Marcantonio EE, Gundersen GG (2009) Clathrin mediates integrin endocytosis for focal adhesion disassembly in migrating cells. J Cell Biol 187(5):733–747. doi:10.1083/jcb.200904054

Fackler OT, Grosse R (2008) Cell motility through plasma membrane blebbing. J Cell Biol 181(6):879–884. doi:10.1083/jcb.200802081

Feng Y, Santoriello C, Mione M, Hurlstone A, Martin P (2010) Live imaging of innate immune cell sensing of transformed cells in zebrafish larvae: parallels between tumor initiation and wound inflammation. PLoS Biol 8(12):e1000562. doi:10.1371/journal.pbio.1000562

Franco SJ, Rodgers MA, Perrin BJ, Han J, Bennin DA, Critchley DR, Huttenlocher A (2004) Calpain-mediated proteolysis of talin regulates adhesion dynamics. Nat Cell Biol 6(10):977–983. doi:10.1038/ncb1175

Frantz C, Karydis A, Nalbant P, Hahn KM, Barber DL (2007) Positive feedback between Cdc42 activity and H+ efflux by the Na-H exchanger NHE1 for polarity of migrating cells. J Cell Biol 179(3):403–410. doi:10.1083/jcb.200704169

Friedl P (2004) Prespecification and plasticity: shifting mechanisms of cell migration. Curr Opin Cell Biol 16(1):14–23

Fukata M, Nakagawa M, Kuroda S, Kaibuchi K (1999) Cell adhesion and Rho small GTPases. J Cell Sci 112(Pt 24):4491–4500

Fukata M, Watanabe T, Noritake J, Nakagawa M, Yamaga M, Kuroda S, Matsuura Y, Iwamatsu A, Perez F, Kaibuchi K (2002) Rac1 and Cdc42 capture microtubules through IQGAP1 and CLIP-170. Cell 109(7):873–885

Funamoto S, Meili R, Lee S, Parry L, Firtel RA (2002) Spatial and temporal regulation of 3-phosphoinositides by PI 3-kinase and PTEN mediates chemotaxis. Cell 109(5):611–623

Gauthier NC, Fardin MA, Roca-Cusachs P, Sheetz MP (2011) Temporary increase in plasma membrane tension coordinates the activation of exocytosis and contraction during cell spreading. Proc Natl Acad Sci U S A 108(35):14467–14472. doi:10.1073/pnas.1105845108

Ghassemi S, Meacci G, Liu S, Gondarenko AA, Mathur A, Roca-Cusachs P, Sheetz MP, Hone J (2012) Cells test substrate rigidity by local contractions on submicrometer pillars. Proc Natl Acad Sci U S A 109(14):5328–5333. doi:10.1073/pnas.1119886109

Giannone G, Mege RM, Thoumine O (2009) Multi-level molecular clutches in motile cell processes. Trends Cell Biol 19(9):475–486

Goh WI, Ahmed S (2012) mDia1-3 in mammalian filopodia. Commun Integr Biol 5(4):340–344. doi:10.4161/cib.20214

Goh WI, Lim KB, Sudhaharan T, Sem KP, Bu W, Chou AM, Ahmed S (2012) mDia1 and WAVE2 proteins interact directly with IRSp53 in filopodia and are involved in filopodium formation. J Biol Chem 287(7):4702–4714. doi:10.1074/jbc.M111.305102

Goldstein B, Macara IG (2007) The PAR proteins: fundamental players in animal cell polarization. Dev Cell 13(5):609–622. doi:10.1016/j.devcel.2007.10.007

Gomes ER, Jani S, Gundersen GG (2005) Nuclear movement regulated by Cdc42, MRCK, myosin, and actin flow establishes MTOC polarization in migrating cells. Cell 121(3):451–463

Gumbiner BM, McCrea PD (1993) Catenins as mediators of the cytoplasmic functions of cadherins. J Cell Sci Suppl 17:155–158

Hahne P, Sechi A, Benesch S, Small JV (2001) Scar/WAVE is localised at the tips of protruding lamellipodia in living cells. FEBS Lett 492(3):215–220

Haugh JM, Codazzi F, Teruel M, Meyer T (2000) Spatial sensing in fibroblasts mediated by 3′ phosphoinositides. J Cell Biol 151(6):1269–1280

Hawkins RJ, Poincloux R, Benichou O, Piel M, Chavrier P, Voituriez R (2011) Spontaneous contractility-mediated cortical flow generates cell migration in three-dimensional environments. Biophys J 101(5):1041–1045. doi:10.1016/j.bpj.2011.07.038

Hoeller O, Kay RR (2007) Chemotaxis in the absence of PIP3 gradients. Curr Biol 17(9):813–817. doi:10.1016/j.cub.2007.04.004

Houk AR, Jilkine A, Mejean CO, Boltyanskiy R, Dufresne ER, Angenent SB, Altschuler SJ, Wu LF, Weiner OD (2012) Membrane tension maintains cell polarity by confining signals to the leading edge during neutrophil migration. Cell 148(1–2):175–188. doi:10.1016/j.cell.2011.10.050

Huang YE, Iijima M, Parent CA, Funamoto S, Firtel RA, Devreotes P (2003) Receptor-mediated regulation of PI3Ks confines PI(3,4,5)P3 to the leading edge of chemotaxing cells. Mol Biol Cell 14(5):1913–1922. doi:10.1091/mbc.E02-10-0703

Huttenlocher A, Ginsberg MH, Horwitz AF (1996) Modulation of cell migration by integrin-mediated cytoskeletal linkages and ligand-binding affinity. J Cell Biol 134(6):1551–1562

Huttenlocher A, Horwitz AR (2011) Integrins in cell migration. Cold Spring Harb Perspect Biol 3(9):a005074. doi:10.1101/cshperspect.a005074
Hynes RO (2002) A reevaluation of integrins as regulators of angiogenesis. Nat Med 8(9):918–921
Iden S, Collard JG (2008) Crosstalk between small GTPases and polarity proteins in cell polarization. Nat Rev Mol Cell Biol 9(11):846–859. doi:10.1038/nrm2521
Iglesias PA, Devreotes PN (2008) Navigating through models of chemotaxis. Curr Opin Cell Biol 20(1):35–40. doi:10.1016/j.ceb.2007.11.011
Iijima M, Devreotes P (2002) Tumor suppressor PTEN mediates sensing of chemoattractant gradients. Cell 109(5):599–610
Innocenti M, Frittoli E, Ponzanelli I, Falck JR, Brachmann SM, Di Fiore PP, Scita G (2003) Phosphoinositide 3-kinase activates Rac by entering in a complex with Eps8, Abi1, and Sos-1. J Cell Biol 160(1):17–23. doi:10.1083/jcb.200206079
Insall RH, Machesky LM (2009) Actin dynamics at the leading edge: from simple machinery to complex networks. Dev Cell 17(3):310–322. doi:10.1016/j.devcel.2009.08.012
Jaffe AB, Hall A (2005) Rho GTPases: biochemistry and biology. Annu Rev Cell Dev Biol 21:247–269
Janetopoulos C, Firtel RA (2008) Directional sensing during chemotaxis. FEBS Lett 582(14): 2075–2085. doi:10.1016/j.febslet.2008.04.035
Ji L, Lim J, Danuser G (2008) Fluctuations of intracellular forces during cell protrusion. Nat Cell Biol 10(12):1393–1400. doi:10.1038/ncb1797
Jin T (2013) Gradient sensing during chemotaxis. Curr Opin Cell Biol 25(5):532–537. doi:10.1016/j.ceb.2013.06.007
Jin T, Xu X, Hereld D (2008) Chemotaxis, chemokine receptors and human disease. Cytokine 44(1):1–8. doi:10.1016/j.cyto.2008.06.017
Jin T, Zhang N, Long Y, Parent CA, Devreotes PN (2000) Localization of the G protein betagamma complex in living cells during chemotaxis. Science 287(5455):1034–1036
Kae H, Kortholt A, Rehmann H, Insall RH, Van Haastert PJ, Spiegelman GB, Weeks G (2007) Cyclic AMP signalling in Dictyostelium: G-proteins activate separate Ras pathways using specific RasGEFs. EMBO Rep 8(5):477–482. doi:10.1038/sj.embor.7400936
Kardash E, Reichman-Fried M, Maitre JL, Boldajipour B, Papusheva E, Messerschmidt EM, Heisenberg CP, Raz E (2010) A role for Rho GTPases and cell-cell adhesion in single-cell motility in vivo. Nat Cell Biol 12(1):47–53. doi:10.1038/ncb2003, Sup pp 41–11
Kawano Y, Fukata Y, Oshiro N, Amano M, Nakamura T, Ito M, Matsumura F, Inagaki M, Kaibuchi K (1999) Phosphorylation of myosin-binding subunit (MBS) of myosin phosphatase by Rho-kinase in vivo. J Cell Biol 147(5):1023–1038
Kawasaki Y, Sato R, Akiyama T (2003) Mutated APC and Asef are involved in the migration of colorectal tumour cells. Nat Cell Biol 5(3):211–215
Kawasaki Y, Senda T, Ishidate T, Koyama R, Morishita T, Iwayama Y, Higuchi O, Akiyama T (2000) Asef, a link between the tumor suppressor APC and G-protein signaling. Science 289 (5482):1194–1197
Kerstetter AE, Azodi E, Marrs JA, Liu Q (2004) Cadherin-2 function in the cranial ganglia and lateral line system of developing zebrafish. Dev Dyn 230(1):137–143. doi:10.1002/dvdy.20021
Kohler S, Schmoller KM, Crevenna AH, Bausch AR (2012) Regulating contractility of the actomyosin cytoskeleton by pH. Cell Rep 2(3):433–439. doi:10.1016/j.celrep.2012.08.014
Kong M, Munoz N, Valdivia A, Alvarez A, Herrera-Molina R, Cardenas A, Schneider P, Burridge K, Quest AF, Leyton L (2013) Thy-1-mediated cell-cell contact induces astrocyte migration through the engagement of alphaVbeta3 integrin and syndecan-4. Biochim Biophys Acta 1833(6):1409–1420. doi:10.1016/j.bbamcr.2013.02.013
Kortholt A, King JS, Keizer-Gunnink I, Harwood AJ, Van Haastert PJ (2007) Phospholipase C regulation of phosphatidylinositol 3,4,5-trisphosphate-mediated chemotaxis. Mol Biol Cell 18(12):4772–4779. doi:10.1091/mbc.E07-05-0407
Kraynov VS, Chamberlain C, Bokoch GM, Schwartz MA, Slabaugh S, Hahn KM (2000) Localized Rac activation dynamics visualized in living cells. Science 290:333–337

Krendel M, Zenke FT, Bokoch GM (2002) Nucleotide exchange factor GEF-H1 mediates cross-talk between microtubules and the actin cytoskeleton. Nat Cell Biol 4(4):294–301

Kunda P, Craig G, Dominguez V, Baum B (2003) Abi, Sra1, and Kette control the stability and localization of SCAR/WAVE to regulate the formation of actin-based protrusions. Curr Biol 13(21):1867–1875

Kuntziger T, Gavet O, Manceau V, Sobel A, Bornens M (2001) Stathmin/Op18 phosphorylation is regulated by microtubule assembly. Mol Biol Cell 12(2):437–448

Ladoux B, Anon E, Lambert M, Rabodzey A, Hersen P, Buguin A, Silberzan P, Mege RM (2010) Strength dependence of cadherin-mediated adhesions. Biophys J 98(4):534–542

Lammermann T, Bader BL, Monkley SJ, Worbs T, Wedlich-Soldner R, Hirsch K, Keller M, Forster R, Critchley DR, Fassler R, Sixt M (2008) Rapid leukocyte migration by integrin-independent flowing and squeezing. Nature 453(7191):51–55. doi:10.1038/nature06887

Lammermann T, Renkawitz J, Wu X, Hirsch K, Brakebusch C, Sixt M (2009) Cdc42-dependent leading edge coordination is essential for interstitial dendritic cell migration. Blood 113(23): 5703–5710. doi:10.1182/blood-2008-11-191882

Lammermann T, Sixt M (2009) Mechanical modes of 'amoeboid' cell migration. Curr Opin Cell Biol 21(5):636–644. doi:10.1016/j.ceb.2009.05.003

Langridge PD, Kay RR (2006) Blebbing of Dictyostelium cells in response to chemoattractant. Exp Cell Res 312(11):2009–2017. doi:10.1016/j.yexcr.2006.03.007

Lansbergen G, Grigoriev I, Mimori-Kiyosue Y, Ohtsuka T, Higa S, Kitajima I, Demmers J, Galjart N, Houtsmuller AB, Grosveld F, Akhmanova A (2006) CLASPs attach microtubule plus ends to the cell cortex through a complex with LL5beta. Dev Cell 11(1):21–32

Lauffenburger DA, Horwitz AF (1996) Cell migration: a physically integrated molecular process. Cell 84(3):359–369

le Duc Q, Shi Q, Blonk I, Sonnenberg A, Wang N, Leckband D, de Rooij J (2010) Vinculin potentiates E-cadherin mechanosensing and is recruited to actin-anchored sites within adherens junctions in a myosin II-dependent manner. J Cell Biol 189(7):1107–1115

Lee J, Ishihara A, Oxford G, Johnson B, Jacobson K (1999) Regulation of cell movement is mediated by stretch-activated calcium channels. Nature 400(6742):382–386. doi:10.1038/22578

Li A, Ma Y, Yu X, Mort RL, Lindsay CR, Stevenson D, Strathdee D, Insall RH, Chernoff J, Snapper SB, Jackson IJ, Larue L, Sansom OJ, Machesky LM (2011) Rac1 drives melanoblast organization during mouse development by orchestrating pseudopod-driven motility and cell-cycle progression. Dev Cell 21(4):722–734. doi:10.1016/j.devcel.2011.07.008

Li Z, Dong X, Wang Z, Liu W, Deng N, Ding Y, Tang L, Hla T, Zeng R, Li L, Wu D (2005) Regulation of PTEN by Rho small GTPases. Nat Cell Biol 7(4):399–404. doi:10.1038/ncb1236

Li Z, Hannigan M, Mo Z, Liu B, Lu W, Wu Y, Smrcka AV, Wu G, Li L, Liu M, Huang CK, Wu D (2003) Directional sensing requires G beta gamma-mediated PAK1 and PIX alpha-dependent activation of Cdc42. Cell 114(2):215–227

Lieber AD, Yehudai-Resheff S, Barnhart EL, Theriot JA, Keren K (2013) Membrane tension in rapidly moving cells is determined by cytoskeletal forces. Curr Biol 23(15):1409–1417. doi:10.1016/j.cub.2013.05.063

Liu Z, Tan JL, Cohen DM, Yang MT, Sniadecki NJ, Ruiz SA, Nelson CM, Chen CS (2010) Mechanical tugging force regulates the size of cell-cell junctions. Proc Natl Acad Sci U S A 107(22):9944–9949

Loovers HM, Postma M, Keizer-Gunnink I, Huang YE, Devreotes PN, van Haastert PJ (2006) Distinct roles of PI(3,4,5)P3 during chemoattractant signaling in Dictyostelium: a quantitative in vivo analysis by inhibition of PI3-kinase. Mol Biol Cell 17(4):1503–1513. doi:10.1091/mbc.E05-09-0825

Luo BH, Carman CV, Springer TA (2007) Structural basis of integrin regulation and signaling. Annu Rev Immunol 25:619–647

Luo BH, Springer TA (2006) Integrin structures and conformational signaling. Curr Opin Cell Biol 18(5):579–586

Luxton GW, Gomes ER, Folker ES, Vintinner E, Gundersen GG (2010) Linear arrays of nuclear envelope proteins harness retrograde actin flow for nuclear movement. Science 329(5994): 956–959

Machacek M, Hodgson L, Welch C, Elliott H, Pertz O, Nalbant P, Abell A, Johnson GL, Hahn KM, Danuser G (2009) Coordination of Rho GTPase activities during cell protrusion. Nature 461(7260):99–103. doi:10.1038/nature08242

Machesky LM, Atkinson SJ, Ampe C, Vandekerckhove J, Pollard TD (1994) Purification of a cortical complex containing two unconventional actins from Acanthamoeba by affinity chromatography on profilin-agarose. J Cell Biol 127(1):107–115

Machesky LM, Insall RH (1998) Scar1 and the related Wiskott-Aldrich syndrome protein, WASP, regulate the actin cytoskeleton through the Arp2/3 complex. Curr Biol 8(25):1347–1356

Maekawa M, Ishizaki T, Boku S, Watanabe N, Fujita A, Iwamatsu A, Obinata T, Ohashi K, Mizuno K, Narumiya S (1999) Signaling from Rho to the actin cytoskeleton through protein kinases ROCK and LIM-kinase. Science 285(5429):895–898

Manabe R, Kovalenko M, Webb DJ, Horwitz AR (2002) GIT1 functions in a motile, multimolecular signaling complex that regulates protrusive activity and cell migration. J Cell Sci 115(Pt 7):1497–1510

Manahan CL, Iglesias PA, Long Y, Devreotes PN (2004) Chemoattractant signaling in dictyostelium discoideum. Annu Rev Cell Dev Biol 20:223–253. doi:10.1146/annurev.cellbio.20.011303.132633

Manneville JB, Etienne-Manneville S (2006) Positioning centrosomes and spindle poles: looking at the periphery to find the centre. Biol Cell 98(9):557–565

Manneville JB, Jehanno M, Etienne-Manneville S (2010) Dlg1 binds GKAP to control dynein association with microtubules, centrosome positioning, and cell polarity. J Cell Biol 191(3): 585–598

Matthews HK, Marchant L, Carmona-Fontaine C, Kuriyama S, Larrain J, Holt MR, Parsons M, Mayor R (2008) Directional migration of neural crest cells in vivo is regulated by Syndecan-4/Rac1 and non-canonical Wnt signaling/RhoA. Development 135(10):1771–1780. doi:10.1242/dev.017350

McDonald B, Pittman K, Menezes GB, Hirota SA, Slaba I, Waterhouse CC, Beck PL, Muruve DA, Kubes P (2010) Intravascular danger signals guide neutrophils to sites of sterile inflammation. Science 330(6002):362–366. doi:10.1126/science.1195491

Merlot S, Firtel RA (2003) Leading the way: Directional sensing through phosphatidylinositol 3-kinase and other signaling pathways. J Cell Sci 116(Pt 17):3471–3478

Mimori-Kiyosue Y, Grigoriev I, Lansbergen G, Sasaki H, Matsui C, Severin F, Galjart N, Grosveld F, Vorobjev I, Tsukita S, Akhmanova A (2005) CLASP1 and CLASP2 bind to EB1 and regulate microtubule plus-end dynamics at the cell cortex. J Cell Biol 168(1):141–153

Mondal S, Subramanian KK, Sakai J, Bajrami B, Luo HR (2012) Phosphoinositide lipid phosphatase SHIP1 and PTEN coordinate to regulate cell migration and adhesion. Mol Biol Cell 23(7):1219–1230

Nalbant P, Hodgson L, Kraynov V, Toutchkine A, Hahn KM (2004) Activation of endogenous Cdc42 visualized in living cells. Science 305(5690):1615–1619

Narumiya S, Ishizaki T, Watanabe N (1997) Rho effectors and reorganization of actin cytoskeleton. FEBS Lett 410(1):68–72

Nieto M, Frade JM, Sancho D, Mellado M, Martinez AC, Sanchez-Madrid F (1997) Polarization of chemokine receptors to the leading edge during lymphocyte chemotaxis. J Exp Med 186(1): 153–158

Niewiadomska P, Godt D, Tepass U (1999) DE-Cadherin is required for intercellular motility during Drosophila oogenesis. J Cell Biol 144(3):533–547

Nimnual AS, Taylor LJ, Bar-Sagi D (2003) Redox-dependent downregulation of Rho by Rac. Nat Cell Biol 5(3):236–241. doi:10.1038/ncb938

Nishimura T, Kaibuchi K (2007) Numb controls integrin endocytosis for directional cell migration with aPKC and PAR-3. Dev Cell 13(1):15–28

Niwa R, Nagata-Ohashi K, Takeichi M, Mizuno K, Uemura T (2002) Control of actin reorganization by Slingshot, a family of phosphatases that dephosphorylate ADF/cofilin. Cell 108(2): 233–246

Nobes CD, Hall A (1995) Rho, Rac and Cdc42 regulate the assembly of multimolecular focal complexes associated with actin stress fibers, lamellipodia and filopodia. Cell 81:53–62

Nola S, Sebbagh M, Marchetto S, Osmani N, Nourry C, Audebert S, Navarro C, Rachel R, Montcouquiol M, Sans N, Etienne-Manneville S, Borg JP, Santoni MJ (2008) Scrib regulates PAK activity during the cell migration process. Hum Mol Genet 17(22):3552–3565

Obermeier A, Ahmed S, Manser E, Yen SC, Hall C, Lim L (1998) PAK promotes morphological changes by acting upstream of Rac. EMBO J 17(15):4328–4339. doi:10.1093/emboj/17.15.4328

Osmani N, Peglion F, Chavrier P, Etienne-Manneville S (2010) Cdc42 localization and cell polarity depend on membrane traffic. J Cell Biol 191(7):1261–1269

Osmani N, Vitale N, Borg JP, Etienne-Manneville S (2006) Scrib controls Cdc42 localization and activity to promote cell polarization during astrocyte migration. Curr Biol 16(24):2395–2405

Pacquelet A, Lin L, Rorth P (2003) Binding site for p120/delta-catenin is not required for Drosophila E-cadherin function in vivo. J Cell Biol 160(3):313–319. doi:10.1083/jcb.200207160

Pacquelet A, Rorth P (2005) Regulatory mechanisms required for DE-cadherin function in cell migration and other types of adhesion. J Cell Biol 170(5):803–812. doi:10.1083/jcb.200506131

Palazzo AF, Cook TA, Alberts AS, Gundersen GG (2001a) mDia mediates Rho-regulated formation and orientation of stable microtubules. Nat Cell Biol 3:723–729

Palazzo AF, Eng CH, Schlaepfer DD, Marcantonio EE, Gundersen GG (2004) Localized stabilization of microtubules by integrin- and FAK-facilitated Rho signaling. Science 303(5659): 836–839

Palazzo AF, Joseph HL, Chen YJ, Dujardin DL, Alberts AS, Pfister KK, Vallee RB, Gundersen GG (2001b) Cdc42, dynein, and dynactin regulate MTOC reorientation independent of Rho-regulated microtubule stabilization. Curr Biol 11:1536–1541

Paluch E, Sykes C, Prost J, Bornens M (2006) Dynamic modes of the cortical actomyosin gel during cell locomotion and division. Trends Cell Biol 16(1):5–10. doi:10.1016/j.tcb.2005.11.003

Parent CA, Blacklock BJ, Froehlich WM, Murphy DB, Devreotes PN (1998) G protein signaling events are activated at the leading edge of chemotactic cells. Cell 95(1):81–91

Pelham RJ Jr, Wang Y (1997) Cell locomotion and focal adhesions are regulated by substrate flexibility. Proc Natl Acad Sci U S A 94(25):13661–13665

Peng H, Ong YM, Shah WA, Holland PC, Carbonetto S (2013) Integrins regulate centrosome integrity and astrocyte polarization following a wound. Dev Neurobiol 73(5):333–353. doi:10.1002/dneu.22055

Pertz O, Hodgson L, Klemke RL, Hahn KM (2006) Spatiotemporal dynamics of RhoA activity in migrating cells. Nature 440(7087):1069–1072. doi:10.1038/nature04665

Pestonjamasp KN, Forster C, Sun C, Gardiner EM, Bohl B, Weiner O, Bokoch GM, Glogauer M (2006) Rac1 links leading edge and uropod events through Rho and myosin activation during chemotaxis. Blood 108(8):2814–2820. doi:10.1182/blood-2006-01-010363

Pinner S, Sahai E (2008) Imaging amoeboid cancer cell motility in vivo. J Microsc 231(3):441–445. doi:10.1111/j.1365-2818.2008.02056.x

Poincloux R, Collin O, Lizarraga F, Romao M, Debray M, Piel M, Chavrier P (2011) Contractility of the cell rear drives invasion of breast tumor cells in 3D Matrigel. Proc Natl Acad Sci U S A 108(5):1943–1948. doi:10.1073/pnas.1010396108

Postma M, Van Haastert PJ (2001) A diffusion-translocation model for gradient sensing by chemotactic cells. Biophys J 81(3):1314–1323. doi:10.1016/S0006-3495(01)75788-8

Pruyne D, Evangelista M, Yang C, Bi E, Zigmond S, Bretscher A, Boone C (2002) Role of formins in actin assembly: nucleation and barbed-end association. Science 297(5581):612–615. doi:10.1126/science.1072309

Qin Y, Capaldo C, Gumbiner BM, Macara IG (2005) The mammalian Scribble polarity protein regulates epithelial cell adhesion and migration through E-cadherin. J Cell Biol 171(6):1061–1071. doi:10.1083/jcb.200506094

Raftopoulou M, Etienne-Manneville S, Self A, Nicholls S, Hall A (2004) Regulation of cell migration by the C2 domain of the tumor suppressor PTEN. Science 303(5661):1179–1181

Redd MJ, Kelly G, Dunn G, Way M, Martin P (2006) Imaging macrophage chemotaxis in vivo: studies of microtubule function in zebrafish wound inflammation. Cell Motil Cytoskeleton 63(7):415–422. doi:10.1002/cm.20133

Rickert P, Weiner OD, Wang F, Bourne HR, Servant G (2000) Leukocytes navigate by compass: roles of PI3Kgamma and its lipid products. Trends Cell Biol 10(11):466–473

Ridley AJ, Schwartz MA, Burridge K, Firtel RA, Ginsberg MH, Borisy G, Parsons JT, Horwitz AR (2003) Cell migration: integrating signals from front to back. Science 302(5651):1704–1709

Rivero S, Cardenas J, Bornens M, Rios RM (2009) Microtubule nucleation at the cis-side of the Golgi apparatus requires AKAP450 and GM130. EMBO J 28(8):1016–1028. doi:10.1038/emboj.2009.47

Sagot I, Klee SK, Pellman D (2002) Yeast formins regulate cell polarity by controlling the assembly of actin cables. Nat Cell Biol 4(1):42–50. doi:10.1038/ncb719

Sahai E (2005) Mechanisms of cancer cell invasion. Curr Opin Genet Dev 15(1):87–96. doi:10.1016/j.gde.2004.12.002

Sanchez C, Perez M, Avila J (2000) GSK3beta-mediated phosphorylation of the microtubule-associated protein 2C (MAP2C) prevents microtubule bundling. Eur J Cell Biol 79(4):252–260

Sanchez C, Tompa P, Szucs K, Friedrich P, Avila J (1996) Phosphorylation and dephosphorylation in the proline-rich C-terminal domain of microtubule-associated protein 2. Eur J Biochem 241(3):765–771

Sanchez T, Thangada S, Wu MT, Kontos CD, Wu D, Wu H, Hla T (2005) PTEN as an effector in the signaling of antimigratory G protein-coupled receptor. Proc Natl Acad Sci U S A 102(12):4312–4317. doi:10.1073/pnas.0409784102

Schmoranzer J, Fawcett JP, Segura M, Tan S, Vallee RB, Pawson T, Gundersen GG (2009) Par3 and dynein associate to regulate local microtubule dynamics and centrosome orientation during migration. Curr Biol 19(13):1065–1074

Schneider IC, Haugh JM (2005) Quantitative elucidation of a distinct spatial gradient-sensing mechanism in fibroblasts. J Cell Biol 171(5):883–892. doi:10.1083/jcb.200509028

Schwartz MA, Shattil SJ (2000) Signaling networks linking integrins and Rho family GTPases. Trends Biochem Sci 25:388–391

Servant G, Weiner OD, Neptune ER, Sedat JW, Bourne HR (1999) Dynamics of a chemoattractant receptor in living neutrophils during chemotaxis. Mol Biol Cell 10(4):1163–1178

Sossey-Alaoui K, Li X, Ranalli TA, Cowell JK (2005) WAVE3-mediated cell migration and lamellipodia formation are regulated downstream of phosphatidylinositol 3-kinase. J Biol Chem 280(23):21748–21755. doi:10.1074/jbc.M500503200

Srinivasan S, Wang F, Glavas S, Ott A, Hofmann F, Aktories K, Kalman D, Bourne HR (2003) Rac and Cdc42 play distinct roles in regulating PI(3,4,5)P3 and polarity during neutrophil chemotaxis. J Cell Biol 160(3):375–385

Stehbens S, Wittmann T (2012) Targeting and transport: how microtubules control focal adhesion dynamics. J Cell Biol 198(4):481–489. doi:10.1083/jcb.201206050

Stradal T, Courtney KD, Rottner K, Hahne P, Small JV, Pendergast AM (2001) The Abl interactor proteins localize to sites of actin polymerization at the tips of lamellipodia and filopodia. Curr Biol 11(11):891–895

Suetsugu S, Kurisu S, Oikawa T, Yamazaki D, Oda A, Takenawa T (2006) Optimization of WAVE2 complex-induced actin polymerization by membrane-bound IRSp53, PIP(3), and Rac. J Cell Biol 173(4):571–585. doi:10.1083/jcb.200509067

Szczur K, Zheng Y, Filippi MD (2009) The small Rho GTPase Cdc42 regulates neutrophil polarity via CD11b integrin signaling. Blood 114(20):4527–4537. doi:10.1182/blood-2008-12-195164

Taguchi K, Ishiuchi T, Takeichi M (2011) Mechanosensitive EPLIN-dependent remodeling of adherens junctions regulates epithelial reshaping. J Cell Biol 194(4):643–656. doi:10.1083/jcb.201104124

Takeda K, Sasaki AT, Ha H, Seung HA, Firtel RA (2007) Role of phosphatidylinositol 3-kinases in chemotaxis in Dictyostelium. J Biol Chem 282(16):11874–11884. doi:10.1074/jbc.M610984200

Tambe DT, Hardin CC, Angelini TE, Rajendran K, Park CY, Serra-Picamal X, Zhou EH, Zaman MH, Butler JP, Weitz DA, Fredberg JJ, Trepat X (2011) Collective cell guidance by cooperative intercellular forces. Nat Mater 10(6):469–475. doi:10.1038/nmat3025

Theveneau E, Marchant L, Kuriyama S, Gull M, Moepps B, Parsons M, Mayor R (2010) Collective chemotaxis requires contact-dependent cell polarity. Dev Cell 19(1):39–53

Trepat X, Wasserman MR, Angelini TE, Millet E, Weitz DA, Butler JP, Fredberg JJ (2009) Physical forces during collective cell migration. Nat Phys 5:426–430

Tu S, Wu WJ, Wang J, Cerione RA (2003) Epidermal growth factor-dependent regulation of Cdc42 is mediated by the Src tyrosine kinase. J Biol Chem 278(49):49293–49300

Uehata M, Ishizaki T, Satoh H, Ono T, Kawahara T, Morishita T, Tamakawa H, Yamagami K, Inui J, Maekawa M, Narumiya S (1997) Calcium sensitization of smooth muscle mediated by a Rho-associated protein kinase in hypertension. Nature 389(6654):990–994. doi:10.1038/40187

Van Haastert PJ, Devreotes PN (2004) Chemotaxis: signalling the way forward. Nat Rev Mol Cell Biol 5(8):626–634. doi:10.1038/nrm1435

van Haastert PJ, Keizer-Gunnink I, Kortholt A (2007) Essential role of PI3-kinase and phospholipase A2 in Dictyostelium discoideum chemotaxis. J Cell Biol 177(5):809–816. doi:10.1083/jcb.200701134

Vasiliev JM, Gelfand IM, Domnina LV, Ivanova OY, Komm SG, Olshevskaja LV (1970) Effect of colcemid on the locomotory behaviour of fibroblasts. J Embryol Exp Morphol 24(3):625–640

Vicente-Manzanares M, Choi CK, Horwitz AR (2009) Integrins in cell migration–the actin connection. J Cell Sci 122(Pt 2):199–206. doi:10.1242/jcs.018564

Vidali L, Chen F, Cicchetti G, Ohta Y, Kwiatkowski DJ (2006) Rac1-null mouse embryonic fibroblasts are motile and respond to platelet-derived growth factor. Mol Biol Cell 17(5):2377–2390. doi:10.1091/mbc.E05-10-0955

Wang F, Herzmark P, Weiner OD, Srinivasan S, Servant G, Bourne HR (2002) Lipid products of PI(3)Ks maintain persistent cell polarity and directed motility in neutrophils. Nat Cell Biol 4:513–518

Wang S, Watanabe T, Matsuzawa K, Katsumi A, Kakeno M, Matsui T, Ye F, Sato K, Murase K, Sugiyama I, Kimura K, Mizoguchi A, Ginsberg MH, Collard JG, Kaibuchi K (2012) Tiam1 interaction with the PAR complex promotes talin-mediated Rac1 activation during polarized cell migration. J Cell Biol 199(2):331–345. doi:10.1083/jcb.201202041

Watanabe T, Noritake J, Kaibuchi K (2005) Regulation of microtubules in cell migration. Trends Cell Biol 15(2):76–83

Watanabe T, Noritake J, Kakeno M, Matsui T, Harada T, Wang S, Itoh N, Sato K, Matsuzawa K, Iwamatsu A, Galjart N, Kaibuchi K (2009) Phosphorylation of CLASP2 by GSK-3beta regulates its interaction with IQGAP1, EB1 and microtubules. J Cell Sci 122(Pt 16):2969–2979. doi:10.1242/jcs.046649

Waterman-Storer C, Desai A, Salmon ED (1999) Fluorescent speckle microscopy of spindle microtubule assembly and motility in living cells. Methods Cell Biol 61:155–173

Webb DJ, Parsons JT, Horwitz AF (2002) Adhesion assembly, disassembly and turnover in migrating cells – over and over and over again. Nat Cell Biol 4(4):E97–E100. doi:10.1038/ncb0402-e97

Weber GF, Bjerke MA, DeSimone DW (2012) A mechanoresponsive cadherin-keratin complex directs polarized protrusive behavior and collective cell migration. Dev Cell 22(1):104–115. doi:10.1016/j.devcel.2011.10.013

Weiner OD, Neilsen PO, Prestwich GD, Kirschner MW, Cantley LC, Bourne HR (2002) A PtdInsP(3) and Rho GTPase-mediated positive feedback loop regulates neutrophil polarity. Nat Cell Biol 4:509–513

Welch HC, Coadwell WJ, Ellson CD, Ferguson GJ, Andrews SR, Erdjument-Bromage H, Tempst P, Hawkins PT, Stephens LR (2002) P-Rex1, a PtdIns(3,4,5)P3- and Gbetagamma-regulated guanine-nucleotide exchange factor for Rac. Cell 108(6):809–821

Welch HC, Coadwell WJ, Stephens LR, Hawkins PT (2003) Phosphoinositide 3-kinase-dependent activation of Rac. FEBS Lett 546(1):93–97

Wessels D, Reynolds J, Johnson O, Voss E, Burns R, Daniels K, Garrard E, O'Halloran TJ, Soll DR (2000) Clathrin plays a novel role in the regulation of cell polarity, pseudopod formation, uropod stability and motility in Dictyostelium. J Cell Sci 113(Pt 1):21–36

Wheeler AP, Wells CM, Smith SD, Vega FM, Henderson RB, Tybulewicz VL, Ridley AJ (2006) Rac1 and Rac2 regulate macrophage morphology but are not essential for migration. J Cell Sci 119(Pt 13):2749–2757. doi:10.1242/jcs.03024

Wickstrom SA, Lange A, Hess MW, Polleux J, Spatz JP, Kruger M, Pfaller K, Lambacher A, Bloch W, Mann M, Huber LA, Fassler R (2010) Integrin-linked kinase controls microtubule dynamics required for plasma membrane targeting of caveolae. Dev Cell 19(4):574–588. doi:10.1016/j.devcel.2010.09.007

Worthylake RA, Burridge K (2003) RhoA and ROCK promote migration by limiting membrane protrusions. J Biol Chem 278(15):13578–13584. doi:10.1074/jbc.M211584200

Xiong Y, Huang CH, Iglesias PA, Devreotes PN (2010) Cells navigate with a local-excitation, global-inhibition-biased excitable network. Proc Natl Acad Sci U S A 107(40):17079–17086. doi:10.1073/pnas.1011271107

Xu J, Wang F, Van Keymeulen A, Herzmark P, Straight A, Kelly K, Takuwa Y, Sugimoto N, Mitchison T, Bourne HR (2003) Divergent signals and cytoskeletal assemblies regulate self-organizing polarity in neutrophils. Cell 114(2):201–214

Yamana N, Arakawa Y, Nishino T, Kurokawa K, Tanji M, Itoh RE, Monypenny J, Ishizaki T, Bito H, Nozaki K, Hashimoto N, Matsuda M, Narumiya S (2006) The Rho-mDia1 pathway regulates cell polarity and focal adhesion turnover in migrating cells through mobilizing Apc and c-Src. Mol Cell Biol 26(18):6844–6858. doi:10.1128/MCB. 00283-06

Yan J, Mihaylov V, Xu X, Brzostowski JA, Li H, Liu L, Veenstra TD, Parent CA, Jin T (2012) A Gbetagamma effector, ElmoE, transduces GPCR signaling to the actin network during chemotaxis. Dev Cell 22(1):92–103. doi:10.1016/j.devcel.2011.11.007

Yip SC, El-Sibai M, Coniglio SJ, Mouneimne G, Eddy RJ, Drees BE, Neilsen PO, Goswami S, Symons M, Condeelis JS, Backer JM (2007) The distinct roles of Ras and Rac in PI 3-kinase-dependent protrusion during EGF-stimulated cell migration. J Cell Sci 120(Pt 17):3138–3146. doi:10.1242/jcs.005298

Yoo SK, Deng Q, Cavnar PJ, Wu YI, Hahn KM, Huttenlocher A (2010) Differential regulation of protrusion and polarity by PI3K during neutrophil motility in live zebrafish. Dev Cell 18(2): 226–236. doi:10.1016/j.devcel.2009.11.015

Yoo SK, Starnes TW, Deng Q, Huttenlocher A (2011) Lyn is a redox sensor that mediates leukocyte wound attraction in vivo. Nature 480(7375):109–112. doi:10.1038/nature10632

Yu JA, Deakin NO, Turner CE (2009) Paxillin-kinase-linker tyrosine phosphorylation regulates directional cell migration. Mol Biol Cell 20(22):4706–4719. doi:10.1091/mbc.E09-07-0548

Zhang S, Schafer-Hales K, Khuri FR, Zhou W, Vertino PM, Marcus AI (2008) The tumor suppressor LKB1 regulates lung cancer cell polarity by mediating cdc42 recruitment and activity. Cancer Res 68(3):740–748

Zhou FQ, Zhou J, Dedhar S, Wu YH, Snider WD (2004) NGF-induced axon growth is mediated by localized inactivation of GSK-3beta and functions of the microtubule plus end binding protein APC. Neuron 42(6):897–912. doi:10.1016/j.neuron.2004.05.011

Chapter 6
Neuronal Polarity

Bhavin Shah, Alejandro Lopez Tobon, and Andreas W. Püschel

Abstract Neurons are highly polarized cells that form distinct axonal and somatodendritic compartments. The establishment of this neuronal polarity, i.e., the specification of an axon and multiple dendrites, is essential for the normal structure and function of the nervous system. During embryonic development, proliferation, asymmetric division, and migration transform a single layer of highly polarized neuronal precursors into a structure with six distinct layers that is characteristic for the mammalian neocortex. The neuronal progenitor cells in the ventricular zone of the brain serve as the major source of pyramidal neurons in the telencephalon. Postmitotic neurons become polarized during their migration from the ventricular zone to the cortical plate by extending a leading process and a trailing axon. However, neuronal development is difficult to analyze in situ and requires advanced microscopy setups for imaging. Therefore, cultures of dissociated neurons have been instrumental in identifying the pathways that direct the establishment of neuronal polarity. These cultures allow to observe neuronal differentiation in an accessible and homogeneous environment with reduced complexity. In this chapter, we will discuss factors required for the establishment of neuronal polarity that were identified using cultured neurons and the extent to which their physiological function has been confirmed by the analysis of knockout mice.

Keywords Cortical development • Hippocampal neurons • Knockout mouse • Neuronal migration • Radial glia

Neurons are highly polarized cells that form morphologically, molecularly, and functionally distinct axonal and somatodendritic compartments, typically a single axon and multiple dendrites, that facilitate the directional flow of information in the nervous system. This chapter will discuss the molecules required for the establishment and maintenance of this neuronal polarity during the development of the nervous system. We will mainly focus on the neocortex because most of the

B. Shah • A.L. Tobon • A.W. Püschel (✉)
Institut für Molekulare Zellbiologie, Westfälische Wilhelms-Universität, Schloßplatz 5, 48149 Münster, Germany
e-mail: apuschel@uni-muenster.de

K. Ebnet (ed.), *Cell Polarity 1*, DOI 10.1007/978-3-319-14463-4_6

molecules shown to be important for neuronal polarity have been identified in cortical or hippocampal neurons.

During embryonic development, the mammalian nervous system develops from the neuroepithelium that consists of a single layer of highly polarized cells. These cells mature into radial glial cells (RGCs) that line the ventricular zone and serve as the major source of pyramidal neurons in the dorsal telencephalon (Götz and Huttner 2005). Cell proliferation, asymmetric cell division, and cell migration transform the pseudostratified neuroepithelium into a structure with six distinct layers characteristic for the mammalian neocortex (Gupta et al. 2002; Gao et al. 2013; Götz and Huttner 2005). RGCs undergo symmetrical divisions to expand the pool of neuronal progenitors and asymmetrical divisions that generate postmitotic neurons or intermediate progenitors (IPs). IP cells reside in the subventricular region where they undergo further divisions to give rise to neurons (Götz and Huttner 2005). Postmitotic neurons migrate along the basal processes of RGCs to form the cortical plate (CP). The CP expands as sequential waves of postmitotic neurons are born and form the different cortical layers (layers 1–6) in an inside-out pattern by migrating past older neurons. The earliest born neurons form the deepest layers, and the latest born neurons occupy the most superficial layer. Layer 1 is mainly occupied by the dendrites of the superficial layer neurons, radial glia endfeet, and Cajal-Retzius cells that secrete the extracellular matrix protein reelin. Projection neurons from the CP extend their axons through the intermediate zone (IZ) that is located below the CP and is relatively free of cell somata.

Live cell imaging of cortical slices from embryonic rodent brains has allowed to investigate the behavior of migrating neurons in situ and revealed how they polarize during their radial migration (Noctor et al. 2004; Sakakibara et al. 2014). These studies have shown that neurons undergo distinct morphological changes during their migration coinciding with their polarization to extend a leading process that will become a dendrite and a trailing process that is the axon (Barnes and Polleux 2009). Neurons born in the VZ initially move into the SVZ where they assume a multipolar morphology by extending multiple processes of similar length that dynamically extend and retract (Tabata and Nakajima 2003; Nadarajah and Parnavelas 2002). Upon polarization in the IZ, these cells become bipolar with a leading and a trailing process and start their migration along the radial glia into the CP. When approaching the marginal zone (MZ), neurons attach their leading processes to the MZ and switch to a glial-independent mode of migration referred to as terminal or somal translocation (Nadarajah et al. 2001).

6.1 Studying the Specification of Axons and Dendrites

The development of neuronal polarity during embryonic development depends on the interaction of newly born neurons with RGCs and is influenced not only by different intrinsic processes but also extrinsic factors produced by other cells (Lathia et al. 2007; Lakoma et al. 2011; Arimura and Kaibuchi 2007). Because neurons are difficult to access in situ and require advanced microscopy setups for imaging, several culture systems have been established that allow to observe the development of neuronal polarity in an accessible environment with reduced complexity. Most of our current knowledge about the mechanism underlying neuronal polarization has come from these studies. These cultures allow to study neurons in a homogeneous environment in which the distinct morphological changes leading to the differentiation of axons and dendrites depend only on intrinsic factors.

The most widely used model to study neuronal polarity is cultures of hippocampal neurons isolated from rat or mouse embryos (Barnes and Polleux 2009). The morphological changes that occur until a mature neuron is visible can be divided into five stages (Fig. 6.1) (Dotti et al. 1988). Freshly dissociated neurons that have attached to the culture substrate first develop lamellipodia (stage 1) and subsequently several immature neurites (stage 2). These short neurites have highly active growth cones at their tip and repeatedly extend and retract (Arimura and Kaibuchi 2007). Stage 2 neurons may correspond to the multipolar phase of neuronal migration (Barnes and Polleux 2009). All immature neurites have the potential to become an axon, but only one of them is specified as the axon and extends rapidly (stage 3), while the other neurites begin to mature into dendrites (stage 4) and eventually form dendritic spines and synapses (stage 5).

A widely accepted model for neuronal differentiation explains the establishment of neuronal polarity by the interaction of local positive and global negative feedback loops (Andersen and Bi 2000) (Fig. 6.1). In this model, each neurite produces both positive and negative signals that promote its own extension and suppress the growth of other neurites. Initially, these signals are balanced in stage 2 neurons. Eventually, either extracellular stimuli or stochastic fluctuations of polarity factors lead to an asymmetry, favoring the extension of one neurite by an increase in its growth promoting signal. In addition, also the negative signals that prevent the remaining neurites from also turning into an axon become stronger. This event is called "break of symmetry" (Andersen and Bi 2000; Goslin and Banker 1989). The coupling of positive and negative feedback loops amplifies small initial differences in signal strength and eventually leads to the selection of a single process as the axon. Once the axon has been selected, the remaining neurites differentiate into dendrites but retain the ability to convert into axons for some time (Gomis-Ruth et al. 2008). The signaling cascades linked to the break of symmetry specify the axon and trigger its rapid extension by affecting processes like membrane fusion, actin dynamics, and microtubule polymerization and stabilization (Barnes and Polleux 2009). However, the molecular identity of the factors that initiate the break of symmetry and that mediate these feedback loops is still largely unknown. Several candidates have been proposed, but it remains to be shown that they are essential for polarity in vivo (Toriyama et al. 2010; Arimura and Kaibuchi 2007). Studies focused on the first asymmetries displayed by neurons in culture suggest that axonal fate may be defined already quite early when the first two neurites emerge at opposite poles of the neuron (Calderon de Anda et al. 2008). The break of symmetry may depend on signals linked to remnants of the cytokinesis that gave

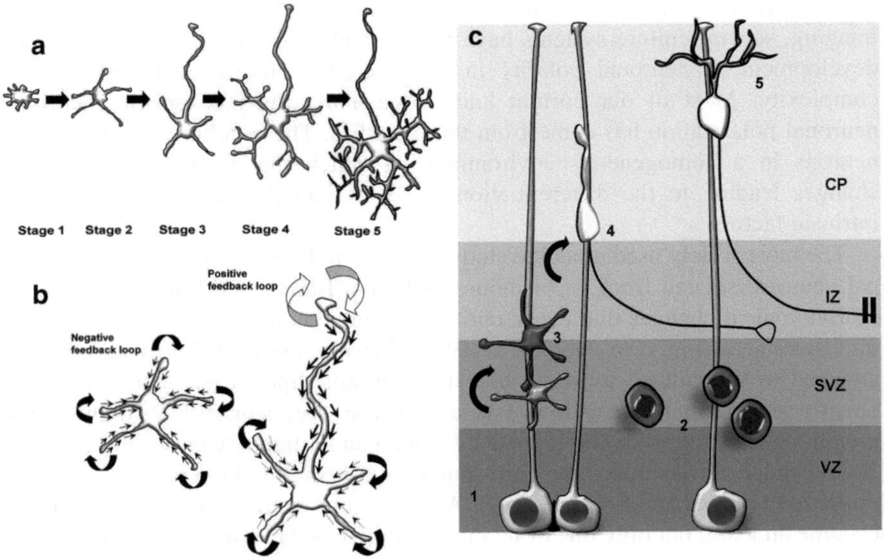

Fig. 6.1 Polarization of neurons in culture and in vivo. (**a**) Upon attachment to the culture substrate, dissociated hippocampal neurons initially form lamellipodia and filopodia (stage 1) and subsequently multiple short protrusions (minor neurites), which are similar in length (stage 2). At some point between 12 and 24 h of culture, one of the neurites starts to accumulate specific polarity factors that allow this neurite to extend rapidly as it becomes the axon (stage 3). Once the neuron is polarized, further maturation occurs between 5 and 12 days in vitro in which the remaining neurites differentiate into dendrites (stage 4) and mature to form dendritic spines and synaptic contacts (stage 5). (**b**) At stage 2, negative feedback signals inhibit the differentiation of neurites into an axon, while positive signals promote their extension. At the transition from stage 2 to 3, this balance is broken and one neurite specified as the axon to extend rapidly. (**c**) Radial glial cells (1, RGCs) give rise to neurons by asymmetric division. These cells have long basal processes that adhere to the pial surface, while their apical cell bodies remain in the ventricular zone (VZ) and are linked by adherens junctions. RGCs can also give rise to intermediate (basal) progenitors (2), which are another source for neurons during cortical development and mainly occupy the subventricular zone (SVZ). After their terminal division, postmitotic neuronal daughter cells initially assume a multipolar morphology with multiple processes (3). Upon polarization, neurons become bipolar with a radially directed leading process and an axon as the trailing process in the intermediate zone (IZ) that elongates tangentially (4). During their final stages of migration, neurons switch from glial-dependent locomotion to glial-independent somal translocation to assume their final positions in the cortical plate (5 and onwards)

rise to the postmitotic neuron (Pollarolo et al. 2011). The cell adhesion molecule N-cadherin was shown to accumulate asymmetrically in neurons and could initiate the break of symmetry (Gärtner et al. 2012). However, it remains to be shown that cell division or N-cadherin is indeed the signals that initiate the break of symmetry in newly born neurons in the developing cortex.

Numerous molecules have been identified that are involved in axon specification and extension (Arimura and Kaibuchi 2007). In the remaining chapter, we will discuss the most important factors shown to be required for the specification of axons. Neuronal differentiation is intrinsically controlled by transcriptional programs that determine the expression of polarity factors. Recent progress in this area is summarized in excellent reviews and will not be discussed here (Itoh et al. 2013; de la Torre-Ubieta and Bonni 2011).

6.2 Symmetry Breaking and the Role of Extrinsic Signals

One of the earliest events during axon specification is the activation of phosphatidylinositol-4,5-bisphosphate 3-kinase (PI3K) and accumulation of phosphatidylinositol-3,4,5-trisphosphate (PIP_3) (Shi et al. 2003). Blocking the activity of PI3K by pharmacological inhibitors prevents the formation of the axon in hippocampal neurons. Activation of PI3K leads to the local accumulation of PIP_3 in the axon and activates the kinase Akt/PKB (Menager et al. 2004; Shi et al. 2004; Schwamborn and Püschel 2004). The factors that act upstream of PI3K in vivo still remain to be identified. In neuronal cultures, different growth factors or adhesion molecules can stimulate PI3K activity directly through their cell receptors or through the activation of Ras GTPases (Arimura and Kaibuchi 2007). In addition to the local production of PIP_3 by PI3K, its polarized transport is also important for axon elongation (Horiguchi et al. 2006). Early enrichment of PIP_3 at the tip of the growing axon requires specific vesicles transported by kinesin family member 13B (KIF13B) (also known as GAKIN). It has been proposed that this initial polarized delivery of PIP_3 initiates axon formation while the local level of PIP_3 is maintained by the activation of PI3K. PI3K itself is transported into the axon by its interaction with the kinesin-1 subunit kinesin light chain 2 (Amato et al. 2011). Another factor linked to PI3K is shootin1 that is enriched in the axon and indirectly interacts with PI3K (Toriyama et al. 2010). Shootin1 appears to be required for the accumulation of PI3K in the axonal growth cone by an unknown mechanism. A quantitative model suggests that the length-dependent accumulation of shootin1, together with shootin1-dependent neurite growth, constitutes a positive feedback loop that amplifies stochastic fluctuations of shootin1 to facilitate a break of symmetry (Toriyama et al. 2010). Shootin1 interacts with the motor protein KIF20B that transports it into the developing axon in cultured neurons to promote PIP_3 accumulation in the growth cone. In vivo, KIF20B promotes the transition from the multipolar to the bipolar stage (Sapir et al. 2013). The asymmetric distribution of Akt to the axon at stage 3 depends on its local degradation in dendrites by the ubiquitin-proteasome system (UPS) (Yan et al. 2006). The activation of PI3K promotes axon formation through several pathways. The most important ones include the GTPase Rap1B, mTOR pathway, aPKC, and GSK3β (Fig. 6.2).

In vivo, axon formation probably depends on extracellular signals (Barnes and Polleux 2009). Growth factors like insulin-like growth factor 1 (IGF1) or transforming growth factor β (TGFβ) and adhesion molecules like laminin and N-cadherin can initiate axon formation in culture (Sosa et al. 2006; Lei et al. 2012; Gärtner et al. 2012; Yi et al. 2010). Cortical neurons lacking the type II TGFβ

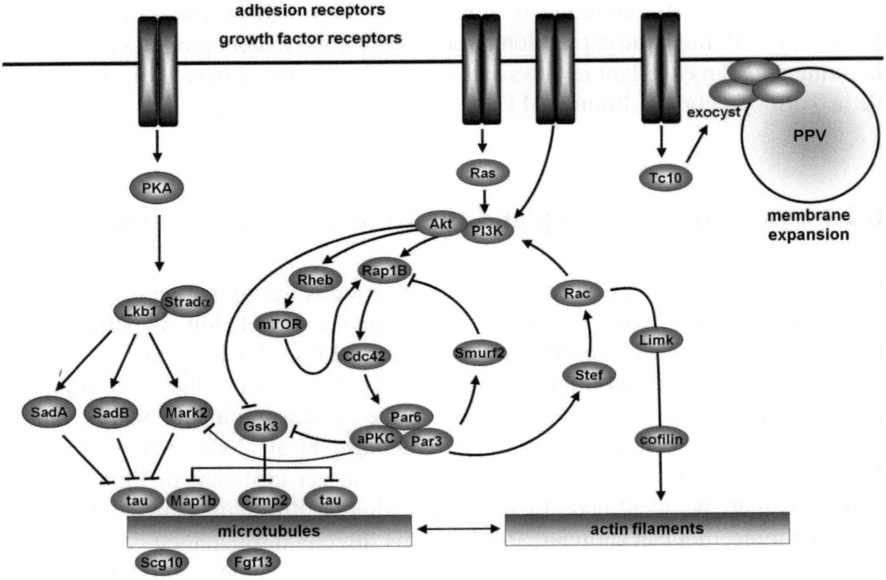

Fig. 6.2 Molecular mechanisms regulating axonal specification. Binding of extrinsic factors like BDNF to their membrane receptors leads to the activation of Ras GTPases and activation of PI3K. Among the targets of PI3K are Rheb/mTOR, Rap1B, and GSK3s. Rap1B recruits and activates Cdc42 to the axon through a yet unknown guanine-nucleotide exchange factor (GEF) at the transition from stage 2 to stage 3. Cdc42 binds to the Par polarity complex and triggers the activation of the Rac GEF Tiam2. Rac1 regulates actin dynamics by increasing actin polymerization through cofilin. It can also initiate a positive feedback loop by activating PI3K. PI3K stimulate the phosphorylation and inhibition of GSK3 through Akt. This inhibition of GSK3 promotes microtubule stabilization through microtubule-binding proteins like Crmp2 together with additional factors such as Fgf13 and Scg10. Another pathway that promotes microtubule stabilization by phosphorylating microtubule-binding proteins like tau includes the kinases Lkb1, SadA/B, and Mark2. Together these pathways promote microtubule polymerization and stability that is required for the elongation of axons. The activation of membrane receptors also increases polarized transport into the axon and the fusion of plasmalemmal precursor vesicles with the membrane

receptor show impaired axon formation (Yi et al. 2010). Another growth factor implicated in axon formation is BDNF. The asymmetric presentation of BDNF promotes axon formation by triggering a local mechanism involving its receptor TrkB, PI3K, PKA, and local translation of BDNF (Cheng et al. 2011a). This suggests that BDNF acts as a paracrine self-amplifying factor that accelerates axonal development through the induction of a positive feedback loop. Sema3A regulates axon formation by inhibiting axon and promoting dendrite formation through increasing local cGMP production and reducing cAMP (Shelly et al. 2010, 2011). The suppression of axon formation by Sema3A was attributed to an antagonizing effect on the PKA-dependent phosphorylation of Lkb1 and GSK3β. N-Cadherin accumulates asymmetrically in neurons and initiate axon formation by a PI3K-dependent mechanism (Gärtner et al. 2012).

6.3 Downstream of PI3K

One of the steps downstream of PI3K is the accumulation of Rap1B in a single neurite (Schwamborn and Püschel 2004). Initially, Rap1 is present in all neurites of stage 2 neurons but becomes restricted to a single neurite at late stage 2 and eventually to the axon. Rap1B is necessary and sufficient for axon formation. Overexpression of constitutively active Rap1B rescues the loss of polarity after treatment with a PI3K inhibitor, indicating that Rap1B is one of the main targets of this pathway. The restriction of Rap1B to a single neurite is mediated by the UPS (Schwamborn et al. 2007a). Inactive, GDP-bound Rap1B is ubiquitinated by the ubiquitin E3 ligase Smurf2, which initiates its destruction through the proteasome. Only in the future axon, Rap1 is activated strongly enough to protect it from degradation, while it is removed from the remaining neurites. The degradation of Rap1B is balanced by its translation that is regulated by the mTOR pathway. In this pathway, the GTPase Rheb acts downstream of PI3K to stimulate the translation of Rap1 through the kinase mTOR (Li et al. 2008). PI3K activation leads to the phosphorylation and inhibition of the Tsc1/2 complex, a GAP for Rheb, by Akt. Rheb activates mTOR that is a central regulator of translation and phosphorylates eukaryotic translation initiation factor 4E-binding proteins (4EBP1s) and S6 kinases (Hay and Sonenberg 2004). PI3K stimulates the translation of Rap1B through Rheb and suppression of Rheb or mTOR blocks axon formation. The balance of Rap1B degradation and translation together with the activation of Rap1B through an as yet unidentified GEF restricts Rap1B to a single neurite.

Rap1B acts upstream of Cdc42 and the Par polarity complex to specify the axon (Schwamborn and Püschel 2004). The Par complex consists of Par3 and Par6 and atypical protein kinase C (aPKC: aPKCλ and aPKMζ) and interacts with Cdc42 through Par6 (Insolera et al. 2011; Parker et al. 2013). Par3, Par6, and aPKC are detectable in all neurites at stage 2 but become enriched in the axon during polarization (Shi et al. 2003). The components of the Par complex interact with several additional factors that together regulate the formation of axons (Insolera et al. 2011). Par3 also links Smurf2 to kinesin-2 to mediate its transport (Schwamborn et al. 2007b). In addition, Par3 acts as a microtubule-binding protein that promotes microtubule stability and organization. In addition to Cdc42, several kinases regulate the function of the Par complex during axon formation. These include ROCK, Aurora A, aPKC, Mark2, and aPKC (Khazaei and Püschel 2009; Funahashi et al. 2013; Chen et al. 2006b; Nakayama et al. 2008). The level of Par6 also depends on its ubiquitination by the ubiquitin ligase Smurf1 (Cheng et al. 2011b). Phosphorylation by PKA switches the substrate preference of Smurf1 from Par6 to the growth inhibitory GTPase RhoA. Decreasing the levels of RhoA by degradation promotes axon extension. The establishment of the Par complex also requires dishevelled that activates and stabilizes aPKC to promote axon differentiation (Zhang et al. 2007).

The Par complex appears to have multiple outputs. One is mediated by aPKC that inhibits glycogen synthase kinase 3 alpha and beta (GSK3α and β) and

MAP/microtubule affinity-regulating kinase 2 (Mark2) (Chen et al. 2006b). Suppression of Mark2 induces the extension on multiple axons, which can be rescued by overexpression of Par3/Par6/aPKC (Chen et al. 2006b). Par6 also links Cdc42 to the Rac GEF Tiam1/STEF that forms a complex with Par3 (Kunda et al. 2001; Nishimura et al. 2005). The GTPase Rac1 stimulates the growth of both axons and dendrites. Its selective activation in the axon favors extension through its effectors in the WAVE complex (Tahirovic et al. 2010).

The highly homologous GSK3α and β are constitutively active kinases that phosphorylate multiple microtubule-associated proteins (MAPs) like collapsin response mediator protein 2 (Crmp2) (Jiang et al. 2005; Shi et al. 2004; Yoshimura et al. 2005). GSK3β has also been suggested to regulate the subcellular distribution Par3 to the tip of the axon (Shi et al. 2004). Phosphorylation of GSK3α or GSK3β at Ser9 or Ser21, respectively, inhibits their kinase activity. Axon formation requires the phosphorylation and inhibition of GSK3β by Akt or aPKC specifically at the tip of the axon. Active GSK3β blocks axon formation, while pharmacological inhibitors of GSK3s induce the formation of multiple axons (Jiang et al. 2005; Shi et al. 2004; Yoshimura et al. 2005). However, the role of GSK3β in neuronal polarity is controversial. Neurons from knock-in mice carrying mutations of GSK3β Ser9 and GSK3α Ser21 that prevent their phosphorylation establish normal neuronal polarity (Gärtner et al. 2006). Conditional knockouts for GSK3α and β revealed that they have redundant functions in neuronal development (Kim et al. 2006). A careful study using conditional knockout mice and a pharmacological inhibitor selective for GSK3 suggests that the contradictory conclusions about the function of GSK3s could result from incomplete inhibition of these kinases and low selectivity of some inhibitors. Knockdown of GSK3α and GSK3β leads to a defect in axon outgrowth in dissociated neurons and brain slices. This analysis of GSK3 function showed that they are required for axon extension and branching. Local inhibition at the axon tip prevents the phosphorylation of a specific subset of substrates including Crmp2 that are negatively regulated by GSK3s. Another subset of substrates that are positively regulated by GSK3s remains unaffected. GSK3β inhibition also requires the scaffolding protein axin, which is required for axon formation (Fang et al. 2011). Cdk5-dependent phosphorylation of axin facilitates its interaction with GSK3β. This interaction inhibits its kinase activity, thereby decreasing the phosphorylation of its substrate Crmp2. Together these results suggest that several pathways that regulate axon formation converge on GSK3s to regulate microtubule stability.

Other scaffolding proteins involved in neuronal polarity are the Jnk-interacting proteins 1 (Jip1) and 3 (Jip3) that mediate c-Jun N-terminal kinase (Jnk) signaling. Jip1 and phosphorylated Jnk accumulate specifically in the developing axon (Dajas-Bailador et al. 2008; Oliva et al. 2006). The kinesin-1-dependent transport of Jip3 is a prerequisite for enhancing Jnk phosphorylation in the axon and promotes axon elongation (Sun et al. 2013). Inhibition of Jnk activity prevents the formation of the axon but not dendrites. Jnk activation modulates actin filaments through cofilin activity and microtubule dynamics through Scg10 (also called stathmin-2) (Westerlund et al. 2011; Tararuk et al. 2006; Sun et al. 2013).

6.4 The Function of Cytoskeletal Dynamics in Neuronal Polarity

The extension of axons is driven mainly by the polymerization of microtubules, which is restrained by their interaction with actin filaments in the growth cone (Stiess and Bradke 2011; Witte and Bradke 2008). Before an axon extends, its growth cone enlarges and actin filaments become more dynamic and less stable, whereas they stay relatively static in the remaining neurites (Kunda et al. 2001; Stiess and Bradke 2011). The important role of the cytoskeleton was revealed by the effect of drugs that modulate actin or microtubule polymerization. Depolymerization of actin filaments by cytochalasin D or mild stabilization of microtubules by low doses of taxol can induce minor neurites to become an axon (Witte et al. 2008; Bradke and Dotti 2000).

Among the Rho GTPase effectors that control actin dynamics in the axonal growth cone, actin-depolymerizing factor (ADF) and cofilin were shown to regulate axon formation (Sumi et al. 1999). Cofilin is enriched in the growth cone where it severs and depolymerizes actin (Pak et al. 2008). Its activity is regulated by inhibitory phosphorylation at Ser9. Neurons from Cdc42 knockout mice show a specific increase in the inhibitory phosphorylation of cofilin by Limk (Garvalov et al. 2007). Knockdown of cofilin results in polarity defects similar to those caused by Cdc42 ablation, and depletion of cofilin impairs axon formation.

In addition to actin dynamics, the local regulation of microtubules is essential for axon extension. The relative content of tyrosinated and acetylated α-tubulin reflects differences in their dynamic properties (Conde and Caceres 2009). In growing axons, dynamic microtubules are highly concentrated in the axonal tip, whereas the more stable polymer predominates in the shaft and base of the axon (Witte et al. 2008). Microtubule stability depends on numerous proteins, whose interaction with microtubules is regulated by kinases like SadA/B, Mark2, or GSK3. One of the microtubule-binding proteins involved in axon formation is Crmp2 (Inagaki et al. 2001). Crmp2 is enriched in the axon, and its overexpression induces the formation of multiple axons. During axon formation, the translation of Crmp2 and the microtubule-associated protein tau is regulated by the mTOR pathway (Morita and Sobue 2009). Other factors that regulate microtubule dynamics during axonal extension are Fgf13 and Scg10. Fgf13 acts as a microtubule-stabilizing protein that is required for axon formation and neuronal migration in the embryonic cortex (Wu et al. 2012). Scg10 is a microtubule-destabilizing protein that is required for axon extension (Li et al. 2009). Interaction with the GTPase Rnd1 stimulates its activity, and a knockdown of either protein impairs axon formation. Phosphorylation by Jnk1 also regulates Scg10 activity and is required for the exit from the multipolar stage (Westerlund et al. 2011).

6.5 The Role of Polarized Transport

The initiation and maintenance of polarity require the transport of specific proteins and vesicles into axons and dendrites to achieve the polarized localization of proteins that are characteristic for these compartments. Kinesins are molecular motors that mediate the anterograde transport of various cargos into the axon (Nakata and Hirokawa 2007). A constitutively activated Kif5C accumulates at the tip of the axon but not in the dendrites (Nakata and Hirokawa 2003; Jacobson et al. 2006). This accumulation can be seen already at stage 2, suggesting that the polarization of transport precedes axon specification (Jacobson et al. 2006). Key components that are delivered by kinesins include Par3 that interacts with the kinesin-2 subunit Kif3a (Shi et al. 2004) and Crmp2 that is transported by kinesin-1 (Kawano et al. 2005). Blocking the transport of these factors interferes with the normal establishment of neuronal polarity. After the establishment of polarity, various mechanisms assure that specific cargos are transported into axons and dendrites (Nakata and Hirokawa 2007). How kinesins manage to distinguish between the different processes is still not well understood. Important mechanisms to maintain neuronal polarity involve the covalent modification of microtubules and the formation of a barrier (Rasband 2010). During the maturation of neurons, this barrier is established between the axonal and the somatodendritic compartment at the axon initial segment to prevent the uncontrolled exchange of molecules between them. It is essential to maintain the polarized structure of the neuron and depends on the actin cytoskeleton (Song et al. 2009) and ankyrin G (Hedstrom et al. 2008; Rasband 2010). Suppression of ankyrin G by RNAi disrupts the axon initial segment and leads to a loss of neuronal polarity (Hedstrom et al. 2008).

The extension of the axon also requires the addition of new membranes in the axonal growth that is enabled by the selective transport of a specific vesicle population (Pfenninger 2009; Lockerbie et al. 1991). These plasmalemmal precursor vesicles (PPVs) are characterized by their ultrastructural morphology and their lipid and protein composition (Wang et al. 2011; Pfenninger 2009). Three types of PPVs carrying diverse sets of membrane proteins are transported into the axon by different motors of the kinesin family: KIF2, KIF4, and kinesin-1 (Peretti et al. 2000). The trafficking of vesicles is controlled by Rab GTPases (Rodman and Wandinger-Ness 2000). The mammalian homologue of *Drosophila* lethal giant larvae 1 Lgl1 controls membrane trafficking by directly interacting with Rab10 and releasing it from GDP dissociation inhibitor (Wang et al. 2011). Rab10 interacts with myosin Vb and regulates the formation of post-Golgi vesicles that are important for axon development (Liu et al. 2013). Thereby, Lgl1 and Rab10 promote the insertion of new membranes in the axon. The activation of IGF1R leads to the regulated fusion of PPVs with the membrane in the axon growth cone. This fusion requires the exocyst complex, which is regulated by the small GTPases TC10 and RalA (Dupraz et al. 2009; Lalli 2009). RalA participates in axon formation by linking Par3 and aPKC to the exocyst complex (Lalli 2009). Expression of constitutively active RalA inhibits axon formation, while a RalA mutant that is unable to interact with the exocyst complex has no effect. The important function of the exocyst complex is confirmed by the loss of neuronal polarity after depletion of the exocyst subunits Sec6, Sec8, or Exo84. How PPVs are linked to the other signaling pathways described above remains to be investigated.

6.6 Genetic Analysis of Neuronal Polarity

As summarized in the previous paragraphs, many factors have been identified that are required for polarity in cultured neurons. However, only a few of these molecules have been investigated genetically for their in vivo role in establishing and maintaining neuronal polarity so far. In the developing nervous system, another layer of complexity is added by the interaction of neurons with RGCs that is essential for their polarization and radial migration (Nadarajah and Parnavelas 2002; Nadarajah et al. 2001). These interactions are involved in the polarization and orientation of neurons during their transition from the multipolar to the bipolar morphology (Kawauchi et al. 2010). Several brain-specific transgenic lines are available to generate conditional knockouts by expressing the Cre recombinase at different time points of neuronal development. Emx1-Cre expression starts at around E9.5, Foxg1-Cre at about E9, and Nestin-Cre around E10.5 (Liang et al. 2012). Nex-Cre-mediated recombination is restricted to postmitotic neurons. The use of these Cre lines has allowed to dissect the functions of different genes in neurons and RGCs with respect to polarity and migration. From the large number of signaling molecules implicated in the establishment and maintenance of neuronal polarity, only the GTPase Cdc42, the kinases Lkb1 and Sad, and the growth factor TGFβ (Garvalov et al. 2007; Kishi et al. 2005; Barnes et al. 2007; Yi et al. 2010) have been confirmed for their role in polarity during neocortical development genetically.

6.7 Lkb1 and Sad Kinases Regulate Axon Formation During Neuronal Development

LKB1 is a serine-threonine kinase that regulates polarity in many cell types and is the master regulator of adenosine monophosphate-activated protein kinase (AMPK) and AMPK-related kinases like SadA (Brsk2), SadB (Brsk1), and Mark2 (Williams and Brenman 2008; Lizcano et al. 2004). In addition, Lkb1 regulates Golgi morphology and neuronal polarization through Stk25 and Golgi matrix protein 130 (GM130) (Matsuki et al. 2010). Complex formation with the pseudokinase Stradα activates Lkb1 resulting in phosphorylation of both partners and initiates neuronal polarization (Baas et al. 2003; Barnes et al. 2007; Shelly et al. 2007). Activated Lkb1 phosphorylates and activates the highly related kinases

SadA and SadB as well as Mark2. Active Mark2 and Sad kinases in turn regulate microtubule-associated proteins like tau, which stabilize microtubules. Both kinases are probably activated in response to extracellular cues, thus linking this intracellular molecular cascade with extracellular cues from the developing cortical environment (Arimura and Kaibuchi 2007).

The conditional inactivation of *Lkb1* using the *Emx1-Cre* line leads to an almost complete loss of axons in the cortex, while cortical lamination is not affected (Barnes et al. 2007). Deletion of *Lkb1* in postmitotic neurons using *Nex-Cre*, however, does not show any polarity defects, indicating an early role of *Lkb1* in establishing polarity (Courchet et al. 2013). Downstream of Lkb1, the Sad kinases are important for determining neuronal polarity in vivo (Kishi et al. 2005). $Sada^{-/-}$; $Sadb^{-/-}$ double mutants show a loss of axons in the cortex. Cultured neurons from the $Sada^{-/-}$;$Sadb^{-/-}$ hippocampus extend several neurites that are positive for both dendritic and axonal markers. However, the downstream targets of Lkb1 as well as the kinases activating SadA/B appear to be cell type specific (Lilley et al. 2013). Lkb1 and Sad kinases are required at early stages of axon formation only in cortical pyramidal neurons but not in the spinal cord or brain stem. Sad kinases are involved in axon branching in some types of sensory neurons at a later stage independent of Lkb1. Another AMPK-related kinase, Mark2, has also been implicated in neuronal polarity (Chen et al. 2006b; Biernat et al. 2002). However, a *Mark2* knockout does not show obvious developmental defects in the nervous system probably because of redundancy with other Mark kinases (Segu et al. 2008).

6.8 Loss of Cdc42 Leads to Defects in Neuronal Polarity In Vivo

Cdc42 was identified as a central regulator of neuronal polarity initially by knockdown experiments in cultured neurons (Schwamborn and Püschel 2004). So far, Cdc42 is the only GTPase shown to be required for neuronal polarity in vivo. Cdc42 has been conditionally inactivated at different time points of development revealing its important role in the establishment and maintenance of polarity in the developing nervous system (Garvalov et al. 2007; Peng et al. 2013; Chen et al. 2006a; Cappello et al. 2006). Loss of Cdc42 at E9.5 by *Emx1-Cre*-mediated recombination leads to a disruption of adherens junctions and affects the neurogenic fate of the progenitors, indicating a role for Cdc42 in the maintenance of the apicobasal polarity in neuronal progenitors. Deletion of Cdc42 using FoxG1-Cre leads a similar result and abolishes the apical localization of Par6, aPKC, E-cadherin, β-catenin, and Numb in RGCs (Chen et al. 2006a). It also impairs the extension of nestin-positive radial fibers.

A *Nestin-Cre*-mediated deletion of Cdc42 confirmed its function in neuronal polarity. Neurons cultured from Cdc42 mutant embryos fail to form axons in culture (Garvalov et al. 2007). Analysis by DiI tracing revealed a severe reduction of axon formation in the embryonic cortex at E16.5. However, some short axonal bundles remain detectable in this mutant, which may result from an incomplete inactivation of the gene by *Nestin-Cre*. The *Nestin-Cre* conditional knockout of Cdc42 also showed a considerable reduction in the thickness of the cortex because of increased apoptosis during cortical development (Peng et al. 2013). Cdc42 knockout neurons showed severely altered cytoskeletal dynamics, and growth cones were immobile due to a misregulation of the actin-depolymerizing protein cofilin. These results show that the regulation of microtubule and actin dynamics in response to external cues is an important factor for polarity during cortical development.

6.9 The Role of TGFβ in Determining Neuronal Polarity In Vivo

One of the most important questions that still remains to be answered concerns the extrinsic signals that initiate axon formation. Several growth factors initiate axon formation in neuronal cultures (see above). So far, only one of them (TGFβ) has been shown to be required for axon formation in vivo (Yi et al. 2010). TGFβ acts through a heterodimer of type II and type I TGF receptor serine/threonine kinases (TβR1 and TβR2). TGFβs are restricted to the SVZ and VZ of the embryonic cortex in the region where axon formation is initiated, and TβR2 is detectable at E14.5 in axons. Loss of TβR2 by Cre-mediated recombination results in the loss of axons but does not affect the formation of a leading process and the migration of neurons indicating a specific role restricted to axogenesis. The role of TβR2 in axon specification depends on the phosphorylation of Par6 at S345 by TβR2.

6.10 Conclusions and Perspectives

A large number of factors have been implicated in the establishment of neuronal polarity in cultures. It is likely that several polarity pathways cooperate during the establishment of neuronal polarity in the complex environment of the developing nervous system. The current challenge is to confirm their role in vivo by generating knockout mutants. The availability of different Cre lines will allow to dissect the roles of different factors in polarization and their function in RGCs and neurons. Not only axon specification but also the directed migration of neurons along RGC processes depends on the polarization of neurons. One question that remains to be addressed is the link between the extensions of a leading process by multipolar neurons that is required for neuronal migration and axon formation. Live cell imaging showed that neuronal polarization in the cortex occurs during radial migration. The analysis of knockout mutants with defects in axon formation indicates that axon specification can occur independently of migration, uncoupling the formation of leading and trailing process. The role of the interaction of multipolar neurons with RGCs requires further studies and may be more important for migration than for axon formation. This idea is supported by the observation that defects in radial glia polarity affect neuronal migration and result in the arrest of neurons in the IZ (Cappello et al. 2012). Advances in high resolution live cell imaging in combination with genetic approaches will allow to address these issues in the future.

References

Amato S, Liu X, Zheng B, Cantley L, Rakic P, Man HY (2011) AMP-activated protein kinase regulates neuronal polarization by interfering with PI 3-kinase localization. Science 332 (6026):247–251

Andersen S, Bi G (2000) Axon formation: a molecular model for the generation of neuronal polarity. Bioessays 22(2):172–179

Arimura N, Kaibuchi K (2007) Neuronal polarity: from extracellular signals to intracellular mechanisms. Nat Rev Neurosci 8(3):194–205

Baas A, Boudeau J, Sapkota G, Smit L, Medema R, Morrice N, Alessi D, Clevers H (2003) Activation of the tumour suppressor kinase LKB1 by the STE20-like pseudokinase STRAD. EMBO J 22(12):3062–3072

Barnes AP, Lilley BN, Pan YA, Plummer LJ, Powell AW, Raines AN, Sanes JR, Polleux F (2007) LKB1 and SAD kinases define a pathway required for the polarization of cortical neurons. Cell 129(3):549–563

Barnes AP, Polleux F (2009) Establishment of axon-dendrite polarity in developing neurons. Annu Rev Neurosci 32:347–381

Biernat J, Wu YZ, Timm T, Zheng-Fischhofer Q, Mandelkow E, Meijer L, Mandelkow EM (2002) Protein kinase MARK/PAR-1 is required for neurite outgrowth and establishment of neuronal polarity. Mol Biol Cell 13(11):4013–4028

Bradke F, Dotti C (2000) Differentiated neurons retain the capacity to generate axons from dendrites. Curr Biol 10(22):1467–1470

Calderon de Anda F, Gärtner A, Tsai LH, Dotti CG (2008) Pyramidal neuron polarity axis is defined at the bipolar stage. J Cell Sci 121(Pt2):178–185

Cappello S, Attardo A, Wu X, Iwasato T, Itohara S, Wilsch-Bräuninger M, Eilken H, Rieger M, Schroeder T, Huttner W, Brakebusch C, Götz M (2006) The Rho-GTPase cdc42 regulates neural progenitor fate at the apical surface. Nat Neurosci 9(9):1099–1107

Cappello S, Bohringer CR, Bergami M, Conzelmann KK, Ghanem A, Tomassy GS, Arlotta P, Mainardi M, Allegra M, Caleo M, van Hengel J, Brakebusch C, Götz M (2012) A radial glia-specific role of RhoA in double cortex formation. Neuron 73(5):911–924

Chen L, Liao G, Yang L, Campbell K, Nakafuku M, Kuan CY, Zheng Y (2006a) Cdc42 deficiency causes Sonic hedgehog-independent holoprosencephaly. Proc Natl Acad Sci U S A 103 (44):16520–16525

Chen YM, Wang QJ, Hu HS, Yu PC, Zhu J, Drewes G, Piwnica-Worms H, Luo ZG (2006b) Microtubule affinity-regulating kinase 2 functions downstream of the PAR-3/PAR-6/atypical PKC complex in regulating hippocampal neuronal polarity. Proc Natl Acad Sci U S A 103 (22):8534–8539

Cheng P-L, Song A-H, Wong Y-H, Wang S, Zhang X, Poo M-M (2011a) Self-amplifying autocrine actions of BDNF in axon development. Proc Natl Acad Sci U S A 108 (45):18430–18435

Cheng PL, Lu H, Shelly M, Gao H, Poo MM (2011b) Phosphorylation of E3 ligase Smurf1 switches its substrate preference in support of axon development. Neuron 69(2):231–243
Conde C, Caceres A (2009) Microtubule assembly, organization and dynamics in axons and dendrites. Nat Rev Neurosci 10(5):319–332
Courchet J, Lewis TL Jr, Lee S, Courchet V, Liou DY, Aizawa S, Polleux F (2013) Terminal axon branching is regulated by the LKB1-NUAK1 kinase pathway via presynaptic mitochondrial capture. Cell 153(7):1510–1525
Dajas-Bailador F, Jones EV, Whitmarsh AJ (2008) The JIP1 scaffold protein regulates axonal development in cortical neurons. Curr Biol 18(3):221–226
de la Torre-Ubieta L, Bonni A (2011) Transcriptional regulation of neuronal polarity and morphogenesis in the mammalian brain. Neuron 72(1):22–40
Dotti CG, Sullivan CA, Banker GA (1988) The establishment of polarity by hippocampal neurons in culture. J Neurosci 8(4):1454–1468
Dupraz S, Grassi D, Bernis ME, Sosa L, Bisbal M, Gastaldi L, Jausoro I, Caceres A, Pfenninger KH, Quiroga S (2009) The TC10-Exo70 complex is essential for membrane expansion and axonal specification in developing neurons. J Neurosci 29(42):13292–13301
Fang WQ, Ip JP, Li R, Ng YP, Lin SC, Chen Y, Fu AK, Ip NY (2011) Cdk5-mediated phosphorylation of Axin directs axon formation during cerebral cortex development. J Neurosci 31(38):13613–13624
Funahashi Y, Namba T, Fujisue S, Itoh N, Nakamuta S, Kato K, Shimada A, Xu C, Shan W, Nishioka T, Kaibuchi K (2013) ERK2-mediated phosphorylation of Par3 regulates neuronal polarization. J Neurosci 33(33):13270–13285
Gao P, Sultan KT, Zhang XJ, Shi SH (2013) Lineage-dependent circuit assembly in the neocortex. Development 140(13):2645–2655
Gärtner A, Fornasiero EF, Munck S, Vennekens K, Seuntjens E, Huttner WB, Valtorta F, Dotti CG (2012) N-cadherin specifies first asymmetry in developing neurons. EMBO J 31(8):1893–1903
Gärtner A, Huang X, Hall A (2006) Neuronal polarity is regulated by glycogen synthase kinase-3 (GSK-3beta) independently of Akt/PKB serine phosphorylation. J Cell Sci 119 (Pt 19):3927–3934
Garvalov BK, Flynn KC, Neukirchen D, Meyn L, Teusch N, Wu X, Brakebusch C, Bamburg JR, Bradke F (2007) Cdc42 regulates cofilin during the establishment of neuronal polarity. J Neurosci 27(48):13117–13129
Gomis-Ruth S, Wierenga CJ, Bradke F (2008) Plasticity of polarization: changing dendrites into axons in neurons integrated in neuronal circuits. Curr Biol 18(13):992–1000
Goslin K, Banker G (1989) Experimental observations on the development of polarity by hippocampal neurons in culture. J Cell Biol 108(4):1507–1516
Götz M, Huttner WB (2005) The cell biology of neurogenesis. Nat Rev Mol Cell Biol 6 (10):777–788
Gupta A, Tsai LH, Wynshaw-Boris A (2002) Life is a journey: a genetic look at neocortical development. Nat Rev Genet 3(5):342–355
Hay N, Sonenberg N (2004) Upstream and downstream of mTOR. Genes Dev 18(16):1926–1945
Hedstrom KL, Ogawa Y, Rasband MN (2008) AnkyrinG is required for maintenance of the axon initial segment and neuronal polarity. J Cell Biol 183(4):635–640
Horiguchi K, Hanada T, Fukui Y, Chishti AH (2006) Transport of PIP3 by GAKIN, a kinesin-3 family protein, regulates neuronal cell polarity. J Cell Biol 174(3):425–436
Inagaki N, Chihara K, Arimura N, Menager C, Kawano Y, Matsuo N, Nishimura T, Amano M, Kaibuchi K (2001) CRMP-2 induces axons in cultured hippocampal neurons. Nat Neurosci 4 (8):781–782
Insolera R, Chen S, Shi SH (2011) Par proteins and neuronal polarity. Dev Neurobiol 71 (6):483–494
Itoh Y, Tyssowski K, Gotoh Y (2013) Transcriptional coupling of neuronal fate commitment and the onset of migration. Curr Opin Neurobiol 23(6):957–964

Jacobson C, Schnapp B, Banker GA (2006) A change in the selective translocation of the Kinesin-1 motor domain marks the initial specification of the axon. Neuron 49(6):797–804

Jiang H, Guo W, Liang X, Rao Y (2005) Both the establishment and the maintenance of neuronal polarity require active mechanisms: critical roles of GSK-3beta and its upstream regulators. Cell 120(1):123–135

Kawano Y, Yoshimura T, Tsuboi D, Kawabata S, Kaneko-Kawano T, Shirataki H, Takenawa T, Kaibuchi K (2005) CRMP-2 is involved in kinesin-1-dependent transport of the Sra-1/WAVE1 complex and axon formation. Mol Cell Biol 25(22):9920–9935

Kawauchi T, Sekine K, Shikanai M, Chihama K, Tomita K, K-i K, Nakajima K, Nabeshima Y-I, Hoshino M (2010) Rab GTPases-dependent endocytic pathways regulate neuronal migration and maturation through N-cadherin trafficking. Neuron 67(4):588–602

Khazaei MR, Püschel AW (2009) Phosphorylation of the par polarity complex protein Par3 at serine 962 is mediated by aurora a and regulates its function in neuronal polarity. J Biol Chem 284(48):33571–33579

Kim WY, Zhou FQ, Zhou J, Yokota Y, Wang YM, Yoshimura T, Kaibuchi K, Woodgett JR, Anton ES, Snider WD (2006) Essential roles for GSK-3s and GSK-3-primed substrates in neurotrophin-induced and hippocampal axon growth. Neuron 52(6):981–996

Kishi M, Pan YA, Crump JG, Sanes JR (2005) Mammalian SAD kinases are required for neuronal polarization. Science 307(5711):929–932

Kunda P, Paglini G, Quiroga S, Kosik K, Caceres A (2001) Evidence for the involvement of Tiam1 in axon formation. J Neurosci 21(7):2361–2372

Lakoma J, Garcia-Alonso L, Luque JM (2011) Reelin sets the pace of neocortical neurogenesis. Development 138(23):5223–5234

Lalli G (2009) RalA and the exocyst complex influence neuronal polarity through PAR-3 and aPKC. J Cell Sci 122(Pt 10):1499–1506

Lathia JD, Patton B, Eckley DM, Magnus T, Mughal MR, Sasaki T, Caldwell MA, Rao MS, Mattson MP, ffrench-Constant C (2007) Patterns of laminins and integrins in the embryonic ventricular zone of the CNS. J Comp Neurol 505(6):630–643

Lei W-L, Xing S-G, Deng C-Y, Ju X-C, Jiang X-Y, Luo Z-G (2012) Laminin/β1 integrin signal triggers axon formation by promoting microtubule assembly and stabilization. Cell Res 22 (6):954–972

Li Y-H, Ghavampur S, Bondallaz P, Will L, Grenningloh G, Püschel A (2009) Rnd1 regulates axon extension by enhancing the microtubule destabilizing activity of SCG10. J Biol Chem 284(1):363–371

Li YH, Werner H, Püschel AW (2008) Rheb and mTOR regulate neuronal polarity through Rap1B. J Biol Chem 283(48):33784–33792

Liang H, Hippenmeyer S, Ghashghaei HT (2012) A Nestin-cre transgenic mouse is insufficient for recombination in early embryonic neural progenitors. Biol Open 1(12):1200–1203

Lilley BN, Pan YA, Sanes JR (2013) SAD kinases sculpt axonal arbors of sensory neurons through long- and short-term responses to neurotrophin signals. Neuron 79(1):39–53

Liu Y, Xu XH, Chen Q, Wang T, Deng CY, Song BL, Du JL, Luo ZG (2013) Myosin Vb controls biogenesis of post-Golgi Rab10 carriers during axon development. Nat Commun 4:2005

Lizcano J, Göransson O, Toth R, Deak M, Morrice N, Boudeau J, Hawley S, Udd L, Mäkelä T, Hardie D, Alessi D (2004) LKB1 is a master kinase that activates 13 kinases of the AMPK subfamily, including MARK/PAR-1. EMBO J 23(4):833–843

Lockerbie R, Miller V, Pfenninger K (1991) Regulated plasmalemmal expansion in nerve growth cones. J Cell Biol 112(6):1215–1227

Matsuki T, Matthews RT, Cooper JA, van der Brug MP, Cookson MR, Hardy JA, Olson EC, Howell BW (2010) Reelin and stk25 have opposing roles in neuronal polarization and dendritic Golgi deployment. Cell 143(5):826–836

Menager C, Arimura N, Fukata Y, Kaibuchi K (2004) PIP3 is involved in neuronal polarization and axon formation. J Neurochem 89(1):109–118

Morita T, Sobue K (2009) Specification of neuronal polarity regulated by local translation of CRMP2 and Tau via the mTOR-p70S6K pathway. J Biol Chem 284(40):27734–27745
Nadarajah B, Brunstrom J, Grutzendler J, Wong R, Pearlman A (2001) Two modes of radial migration in early development of the cerebral cortex. Nat Neurosci 4(2):143–150
Nadarajah B, Parnavelas JG (2002) Modes of neuronal migration in the developing cerebral cortex. Nat Rev Neurosci 3(6):423–432
Nakata T, Hirokawa N (2003) Microtubules provide directional cues for polarized axonal transport through interaction with kinesin motor head. J Cell Biol 162(6):1045–1055
Nakata T, Hirokawa N (2007) Neuronal polarity and the kinesin superfamily proteins. Sci STKE 2007(372):pe6
Nakayama M, Goto TM, Sugimoto M, Nishimura T, Shinagawa T, Ohno S, Amano M, Kaibuchi K (2008) Rho-kinase phosphorylates PAR-3 and disrupts PAR complex formation. Dev Cell 14 (2):205–215
Nishimura T, Yamaguchi T, Kato K, Yoshizawa M, Nabeshima Y, Ohno S, Hoshino M, Kaibuchi K (2005) PAR-6-PAR-3 mediates Cdc42-induced Rac activation through the Rac GEFs STEF/ Tiam1. Nat Cell Biol 7(3):270–277
Noctor SC, Martinez-Cerdeno V, Ivic L, Kriegstein AR (2004) Cortical neurons arise in symmetric and asymmetric division zones and migrate through specific phases. Nat Neurosci 7 (2):136–144
Oliva AA Jr, Atkins CM, Copenagle L, Banker GA (2006) Activated c-Jun N-terminal kinase is required for axon formation. J Neurosci 26(37):9462–9470
Pak CW, Flynn KC, Bamburg JR (2008) Actin-binding proteins take the reins in growth cones. Nat Rev Neurosci 9(2):136–147
Parker S, Mandell E, Hapak S, Maskaykina I, Kusne Y, Kim J-Y, Moy J, St John P, Wilson J, Gothard K, Price T, Ghosh S (2013) Competing molecular interactions of aPKC isoforms regulate neuronal polarity. Proc Natl Acad Sci U S A 110(35):14450–14455
Peng X, Lin Q, Liu Y, Jin Y, Druso JE, Antonyak MA, Guan JL, Cerione RA (2013) Inactivation of Cdc42 in embryonic brain results in hydrocephalus with ependymal cell defects in mice. Protein Cell 4(3):231–242
Peretti D, Peris L, Rosso S, Quiroga S, Cáceres A (2000) Evidence for the involvement of KIF4 in the anterograde transport of L1-containing vesicles. J Cell Biol 149(1):141–152
Pfenninger KH (2009) Plasma membrane expansion: a neuron's Herculean task. Nat Rev Neurosci 10(4):251–261
Pollarolo G, Schulz JG, Munck S, Dotti CG (2011) Cytokinesis remnants define first neuronal asymmetry *in vivo*. Nat Neurosci 14(12):1525–1533
Rasband MN (2010) The axon initial segment and the maintenance of neuronal polarity. Nat Rev Neurosci 11(8):552–562
Rodman JS, Wandinger-Ness A (2000) Rab GTPases coordinate endocytosis—commentary. J Cell Sci 113(2):183–192
Sakakibara A, Sato T, Ando R, Noguchi N, Masaoka M, Miyata T (2014) Dynamics of centrosome translocation and microtubule organization in neocortical neurons during distinct modes of polarization. Cereb Cortex 24(5):1301–1310
Sapir T, Levy T, Sakakibara A, Rabinkov A, Miyata T, Reiner O (2013) Shootin1 acts in concert with KIF20B to promote polarization of migrating neurons. J Neurosci 33(29):11932–11948
Schwamborn J, Müller M, Becker A, Püschel A (2007a) Ubiquitination of the GTPase Rap1B by the ubiquitin ligase Smurf2 is required for the establishment of neuronal polarity. EMBO J 26 (5):1410–1422
Schwamborn JC, Khazaei MR, Püschel AW (2007b) The interaction of mPar3 with the ubiquitin ligase Smurf2 is required for the establishment of neuronal polarity. J Biol Chem 282 (48):35259–35268
Schwamborn JC, Püschel AW (2004) The sequential activity of the GTPases Rap1B and Cdc42 determines neuronal polarity. Nat Neurosci 7(9):923–929

Segu L, Pascaud A, Costet P, Darmon M, Buhot MC (2008) Impairment of spatial learning and memory in ELKL Motif Kinase1 (EMK1/MARK2) knockout mice. Neurobiol Aging 29 (2):231–240

Shelly M, Cancedda L, Heilshorn S, Sumbre G, Poo MM (2007) LKB1/STRAD promotes axon initiation during neuronal polarization. Cell 129(3):565–577

Shelly M, Cancedda L, Lim BK, Popescu AT, Cheng PL, Gao H, Poo MM (2011) Semaphorin3A regulates neuronal polarization by suppressing axon formation and promoting dendrite growth. Neuron 71(3):433–446

Shelly M, Lim BK, Cancedda L, Heilshorn SC, Gao H, Poo MM (2010) Local and long-range reciprocal regulation of cAMP and cGMP in axon/dendrite formation. Science 327 (5965):547–552

Shi S, Jan L, Jan Y (2003) Hippocampal neuronal polarity specified by spatially localized mPar3/ mPar6 and PI 3-kinase activity. Cell 112:63–75

Shi SH, Cheng T, Jan LY, Jan YN (2004) APC and GSK-3beta are involved in mPar3 targeting to the nascent axon and establishment of neuronal polarity. Curr Biol 14(22):2025–2032

Song AH, Wang D, Chen G, Li Y, Luo J, Duan S, Poo MM (2009) A selective filter for cytoplasmic transport at the axon initial segment. Cell 136(6):1148–1160

Sosa L, Dupraz S, Laurino L, Bollati F, Bisbal M, Caceres A, Pfenninger KH, Quiroga S (2006) IGF-1 receptor is essential for the establishment of hippocampal neuronal polarity. Nat Neurosci 9(8):993–995

Stiess M, Bradke F (2011) Neuronal polarization: the cytoskeleton leads the way. Dev Neurobiol 71(6):430–444

Sumi T, Matsumoto K, Takai Y, Nakamura T (1999) Cofilin phosphorylation and actin cytoskeletal dynamics regulated by Rho- and Cdc42-activated LIM-kinase 2. J Cell Biol 147 (7):1519–1532

Sun T, Yu N, Zhai LK, Li N, Zhang C, Zhou L, Huang Z, Jiang XY, Shen Y, Chen ZY (2013) c-Jun NH2-terminal kinase (JNK)-interacting protein-3 (JIP3) regulates neuronal axon elongation in a kinesin- and JNK-dependent manner. J Biol Chem 288(20):14531–14543

Tabata H, Nakajima K (2003) Multipolar migration: the third mode of radial neuronal migration in the developing cerebral cortex. J Neurosci 23(31):9996–10001

Tahirovic S, Hellal F, Neukirchen D, Hindges R, Garvalov BK, Flynn KC, Stradal TE, Chrostek-Grashoff A, Brakebusch C, Bradke F (2010) Rac1 regulates neuronal polarization through the WAVE complex. J Neurosci 30(20):6930–6943

Tararuk T, Ostman N, Li W, Bjorkblom B, Padzik A, Zdrojewska J, Hongisto V, Herdegen T, Konopka W, Courtney MJ, Coffey ET (2006) JNK1 phosphorylation of SCG10 determines microtubule dynamics and axodendritic length. J Cell Biol 173(2):265–277

Toriyama M, Sakumura Y, Shimada T, Ishii S, Inagaki N (2010) A diffusion-based neurite length-sensing mechanism involved in neuronal symmetry breaking. Mol Syst Biol 6:394

Wang T, Liu Y, Xu XH, Deng CY, Wu KY, Zhu J, Fu XQ, He M, Luo ZG (2011) Lgl1 activation of rab10 promotes axonal membrane trafficking underlying neuronal polarization. Dev Cell 21 (3):431–444

Westerlund N, Zdrojewska J, Padzik A, Komulainen E, Bjorkblom B, Rannikko E, Tararuk T, Garcia-Frigola C, Sandholm J, Nguyen L, Kallunki T, Courtney MJ, Coffey ET (2011) Phosphorylation of SCG10/stathmin-2 determines multipolar stage exit and neuronal migration rate. Nat Neurosci 14(3):305–313

Williams T, Brenman JE (2008) LKB1 and AMPK in cell polarity and division. Trends Cell Biol 18(4):193–198

Witte H, Bradke F (2008) The role of the cytoskeleton during neuronal polarization. Curr Opin Neurobiol 18(5):479–487

Witte H, Neukirchen D, Bradke F (2008) Microtubule stabilization specifies initial neuronal polarization. J Cell Biol 180(3):619–632

Wu QF, Yang L, Li S, Wang Q, Yuan XB, Gao X, Bao L, Zhang X (2012) Fibroblast growth factor 13 is a microtubule-stabilizing protein regulating neuronal polarization and migration. Cell 149 (7):1549–1564

Yan D, Guo L, Wang Y (2006) Requirement of dendritic Akt degradation by the ubiquitin-proteasome system for neuronal polarity. J Cell Biol 174(3):415–424

Yi JJ, Barnes AP, Hand R, Polleux F, Ehlers MD (2010) TGF-beta signaling specifies axons during brain development. Cell 142(1):144–157

Yoshimura T, Kawano Y, Arimura N, Kawabata S, Kikuchi A, Kaibuchi K (2005) GSK-3beta regulates phosphorylation of CRMP-2 and neuronal polarity. Cell 120(1):137–149

Zhang X, Zhu J, Yang GY, Wang QJ, Qian L, Chen YM, Chen F, Tao Y, Hu HS, Wang T, Luo ZG (2007) Dishevelled promotes axon differentiation by regulating atypical protein kinase C. Nat Cell Biol 9(7):743–754

Chapter 7
Epithelial Apicobasal Polarity in the *Drosophila* Embryo

Tao Jiang, Daryl J.V. David, and Tony J.C. Harris

Abstract Epithelia are comprised sheets of interconnected cells that function in tissue morphogenesis, tissue protection, selective absorption, and other essential processes. Epithelial cells harbor different molecular properties in their apical versus basolateral domains. This apicobasal polarity is critical for normal epithelial functions and its loss is linked with cancer progression. Because of its experimental accessibility by genetic, molecular, and imaging approaches, *Drosophila melanogaster* has become a major model for understanding how conserved epithelial polarity circuitries function during animal development. In this chapter, we first review the components and logic of molecular circuits formed from the Par complex; the Crb complex; the basolateral kinase Par-1; the basolateral proteins Scribble, Discs Large, and Lethal Giant Larvae; and the Yurt/Cora group (Fig. 7.1). We focus on local positive feedback loops for forming polarized domains of the cell, long-range negative feedback loops between the apical and basolateral domains of the cell, as well as local negative feedback loops that regulate specific domains. Next, we discuss how these circuits are organized into pathways that establish and elaborate epithelial apicobasal polarity in the *Drosophila* embryo. Finally, we outline examples of cross talk between epithelial apicobasal polarity and morphogenesis.

Keywords Apical complexes • Basolateral complexes • Feedback loops • Polarity • Epithelia • Morphogenesis • Drosophila

---

T. Jiang • D.J.V. David • T.J.C. Harris (✉)
Department of Cell & Systems Biology, University of Toronto, 25 Harbord Street, Toronto, ON, Canada M5S 3G5
e-mail: tony.harris@utoronto.ca

K. Ebnet (ed.), *Cell Polarity 1*, DOI 10.1007/978-3-319-14463-4_7

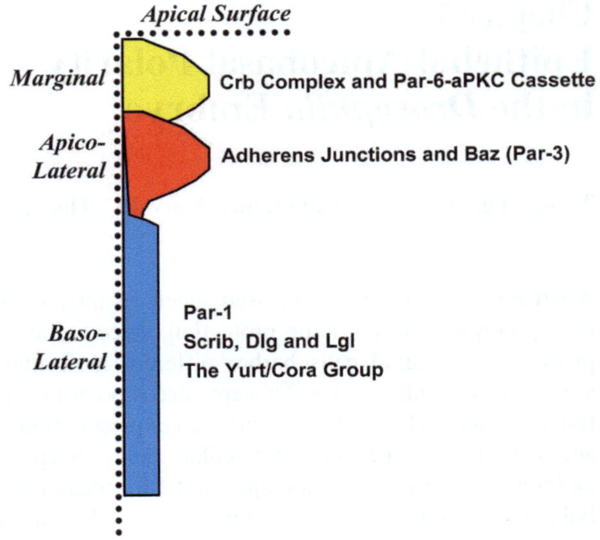

Fig. 7.1 The main domains of a polarized epithelial cell are compared with the distributions of key polarity proteins and complexes. One corner of an epithelial cell is represented. See Sects. 7.2 and 7.3 for details

7.1 Local Positive Feedback Loops for Plasma Membrane Complex Assembly

Positive feedback mechanisms can establish cell polarity spontaneously without external cues. In yeast, for example, expression of a constitutively active form of Cdc42 can drive cell polarization autonomously, and mathematical simulations argue that local positive feedback circuits are sufficient for such polarization. In vivo, this positive feedback emerges from activated Cdc42 inducing local actin-based Cdc42 recruitment and local Cdc42-GEF activity for polarized accumulation of more and more activated Cdc42 (Slaughter et al. 2009). Similarly, local positive feedback loops appear to function in the apicobasal polarization of *Drosophila* epithelial cells (Fig. 7.2).

7.1.1 The Par Complex

The Par complex is composed of Bazooka (Baz; *Drosophila* Par-3), Par-6, and atypical protein kinase C (aPKC). It is evolutionarily conserved and is important for epithelial polarization across animals (Goldstein and Macara 2007; St Johnston and Ahringer 2010; Tepass 2012). Although specific interactions among its components have been identified, its components appear to localize and function separately in the apical domain of *Drosophila* epithelial cells. Specifically, Baz predominantly localizes to apicolateral adherens junctions (AJs), whereas Par-6 and aPKC localize with the Crumbs (Crb) complex in the apical domain just above (Harris and Peifer

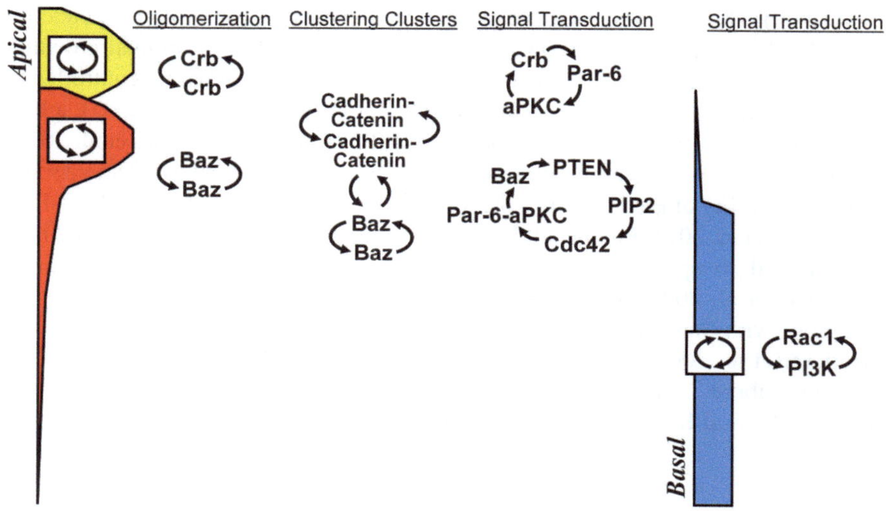

Fig. 7.2 Examples of local positive feedback among polarity regulators. See Sect. 7.2 for details

2005). Moreover, the functions of Baz are distinguishable from those of Par-6 and aPKC during polarity establishment and during later epithelial regulation in the *Drosophila* embryo (David et al. 2010; Harris and Peifer 2004, 2007; Muller and Wieschaus 1996). Thus, it may be more accurate to view the Par complex as a transient assembly involving interaction between the scaffold protein Baz and the Par-6-aPKC cassette. Despite these complexities, positive feedback mechanisms are evident among the Par complex proteins.

One potential positive feedback mechanism is Baz oligomerization. A conserved domain at the N-terminus of Baz mediates its self-oligomerization, important for Baz localization and function (Benton and St Johnston 2003a). Structural analyses of the corresponding mammalian domain show that this oligomerization occurs through front-to-back interactions between positively and negatively charged faces of the domain (Feng et al. 2007). Such oligomerization has the potential to assemble large and thus less mobile complexes in one part of the cell. As is apparent for the polarization of the MEX-5 RNA-binding protein in *C. elegans* (Griffin et al. 2011), less mobile complexes in one part of the cell could be fed by diffusion of more mobile single proteins from other parts of the cell, and in this way protein polarization can be promoted.

Higher-order clustering may also drive positive feedback. As polarity is established in the *Drosophila* embryo, Baz oligomerization is evident from the formation of discrete puncta of Baz around the apicolateral domain. These puncta normally colocalize with early AJs (so-called spot AJs). Strikingly, Baz puncta can form with undetectable AJ assembly but then fail to properly mature around the apical domain (Harris and Peifer 2004). Similarly, small cadherin-catenin clusters can form with loss of Baz but then fail to mature into spot or belt-like AJs (McGill et al. 2009). Thus, Baz puncta and cadherin-catenin clusters can form independently of each other, but appear to engage one another to promote their mutual growth.

Another source of positive feedback may arise from lipid signaling controlled by Baz. Baz directly binds and colocalizes with PTEN (von Stein et al. 2005), a lipid phosphatase that promotes the production of PIP2 from PIP3 (Shewan et al. 2011). Analyses of *Drosophila* embryonic and postembryonic epithelia have demonstrated the potent ability of Baz to control the localization of PTEN and thus PIP2 and PIP3 (Pickering et al. 2013; Pinal et al. 2006). In mammalian cells, PTEN and PIP2 have been linked to apical domain assembly through activation of Cdc42 (Martin-Belmonte et al. 2007). Cdc42 is known to activate the Par-6-aPKC cassette and functions with these players to establish polarity in the early *Drosophila* embryo (Hutterer et al. 2004). Although the mechanism of Cdc42 activation in the *Drosophila* embryo is unknown, these data frame a model in which Baz may help activate Par-6-aPKC, which in turn support Baz.

7.1.2 The Crb Complex

The Crb complex is centered around the Crb transmembrane protein (Pocha and Knust 2013; Tepass 2012). Cytoplasmically, Crb forms a complex with a number of proteins including Stardust (Sdt) and Patj. As discussed, Crb also associates with the Par-6-aPKC cassette. Together these proteins are polarized to the apical surface and its margins (the marginal zone) just above the AJs. Molecular interactions on the extracellular and intracellular sides of the Crb complex appear to form positive feedback loops for its polarized assembly.

One positive feedback mechanism involves extracellular Crb oligomerization. Crb has a large extracellular domain containing epidermal growth factor-like repeats (Tepass et al. 1990). Recently, this domain has been shown to promote Crb apical localization through its homophilic interactions. For example, Crb lacking its intracellular domain can be recruited to the apical domain of *Drosophila* epithelia through a mechanism requiring the presence of full-length Crb (Fletcher et al. 2012). Moreover, interactions between Crb extracellular domains are sufficient for mediating adhesion between *Drosophila* S2 cells and for recruiting full-length Crb to opposing cell-cell contacts within *Drosophila* epithelia (Letizia et al. 2013). Together, these studies implicate *cis*- and *trans*-interactions between Crb extracellular domains in the polarized assembly of Crb complexes. Similar to the oligomerization of Baz and cadherin-catenin complexes, Crb oligomerization may provide positive feedback for Crb complex assembly.

Other evidence for positive feedback is apparent on the cytoplasmic face of the Crb complex. Specifically, binding of Par-6 to Crb aids in drawing the Par-6-aPKC cassette away from Baz (Morais-de-Sa et al. 2010), and, in turn, aPKC phosphorylation of Crb stabilizes the Crb complex (Sotillos et al. 2004). Because of the multiple interactions involved in assembling the Crb complex, additional intracellular positive feedback loops may soon be discovered. Overall, these examples highlight the potential for positive feedback in organizing the apical domain of *Drosophila* epithelial cells.

7.1.3 Basolateral Positive Feedback

Positive feedback among basolateral polarity regulators is less clear. However, one example has recently come to light. Specifically, basolateral PI3K activity has been shown to promote Rac1 activation, which reciprocally activates PI3K signaling (Chartier et al. 2011).

7.2 Long-Range Inhibitory Mechanisms for Displacing Plasma Membrane Proteins

Although positive feedback loops can be sufficient for generating polarity, combining them with long-range inhibitory mechanisms from other domains of the cell can make polarity establishment more robust (Chau et al. 2012; Fletcher et al. 2012). There are multiple basolateral players that antagonize apical proteins to robustly define apicobasal polarity in epithelial cells, and the repulsion is mutual (Fig. 7.3).

7.2.1 Dynein-Mediated Transport

As polarity is first established in the *Drosophila* embryo, microtubules (MTs) provide tracks for long-range transport along the apicobasal axis of epithelial

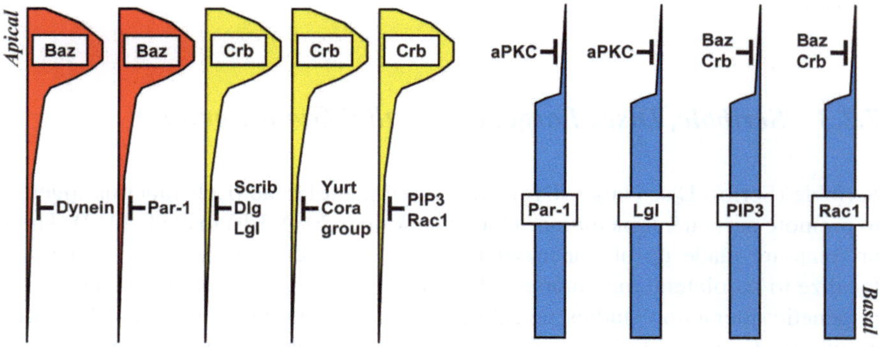

Fig. 7.3 Examples of long-range inhibition among polarity regulators. See Sect. 7.3 for details

cells. The MTs are oriented with their minus ends at the apical domain and their plus ends at the basolateral domain. This alignment of MT polarity with apicobasal polarity is used to position mRNA species (Wilkie and Davis 2001) and intercellular organelles (Papoulas et al. 2005) through long-range transport by Dynein, the minus end-directed motor. At the cell cortex, Dynein also functions to displace Baz from the basolateral domain (Harris and Peifer 2005). This long-range transport contributes to the polarization of Baz protein to the apical domain. Intriguingly, Dynein can also affect the polarization of apical factors by transporting their mRNA transcripts. In other *Drosophila* epithelia, Dynein transports the mRNAs for crb and sdt to the apical domain and thereby promotes Crb complex localization (Horne-Badovinac and Bilder 2008; Li et al. 2008).

7.2.2 Par-1 Kinase Activity

The localization patterns of antagonistic kinases are also coupled to apicobasal polarization. Specifically, the conserved kinase Par-1 localizes to the basolateral domain (Bayraktar et al. 2006; Benton and St Johnston 2003b; McKinley and Harris 2012). There, it inhibits assembly of Baz complexes by phosphorylating Baz at two sites (Benton and St Johnston 2003b). These phosphorylation events recruit 14-3-3 (also known as Par-5) to Baz, and as a result, other Baz protein-protein interactions are inhibited. One phosphorylation event occurs near the oligomerization domain of Baz and antagonizes the Baz-Baz interactions. The other phosphorylation event occurs near an aPKC binding site in Baz and inhibits the Baz-aPKC interactions. Thus, Par-1 blocks the formation of Baz complexes in the basolateral domain, and Baz can only form complexes after displacement to the apical domain (Benton and St Johnston 2003b). Importantly, the phosphorylation of Baz by Par-1 is reversible. For example, protein phosphatase 2A functions to dephosphorylate the Par-1 phosphorylation sites on Baz (Krahn et al. 2009; Nam et al. 2007). Also, Par-1 can also be phosphorylated and inhibited by aPKC in *Drosophila* cells (Doerflinger et al. 2010), suggesting a mutual antagonism between apical and basolateral kinases.

7.2.3 Scribble, Discs Large, and Lethal Giant Larvae

Scribble (Scrib), Discs large (Dlg), and Lethal giant larvae (Lgl) function together to promote basolateral membrane character (Bilder 2004; Elsum et al. 2012). These proteins are made up of nonenzymatic, protein-protein interaction domains and localize to basolateral membranes. Their roles in apicobasal polarity became clear in genetic interaction studies revealing a mutual antagonism with apical polarity regulators. Loss of individual apical or basal protein function leads to an expansion of one domain at the expense of the other and a loss of polarized epithelial structure.

However, with combined loss of both apical and basal protein functions, the system can be rebalanced, with a restoration of properly sized apical and basolateral domains (Bilder et al. 2003; Tanentzapf and Tepass 2003). The specific mechanisms by which Scrib, Dlg, or Lgl antagonize apical proteins remain unclear. However, Lgl is a phosphorylation target of aPKC (Betschinger et al. 2003), and this phosphorylation is important for excluding Lgl from the apical domain of epithelial cells in the *Drosophila* embryo (Hutterer et al. 2004). The mutual antagonism between these apical and basolateral proteins provides another example of long-range inhibition important for epithelial polarity.

7.2.4 The Yurt/Cora Group

The FERM domain-containing proteins Yurt and Coracle (Cora) also promote the basolateral domain in conjunction with a number of other proteins, making up the Yurt/Cora group (Laprise et al. 2009). Expanded apical membranes arise with loss of either Yurt or Cora, and the expansion is intensified with loss of both, suggesting they function in separate pathways. Through double mutant analyses, Crb activity was shown to be responsible for these abnormal apical expansions, indicating that Yurt and Cora normally antagonize Crb (Laprise et al. 2009).

7.2.5 Interplay with Lipids and Small GTPases

Phosphoinositide and Rac small GTPase signaling also contribute to epithelial polarity in *Drosophila*. Specifically, apicobasal polarity is balanced by antagonistic signaling between the Crb complex and the basolateral Rac1-PI3K positive feedback loop (Chartier et al. 2011). Connections between Baz and these basolateral players are also significant. For example, the apical recruitment of PTEN by Baz promotes the polarization of PIP2 and PIP3 to the apical and basolateral domains, respectively (Pinal et al. 2006). Additionally, Baz acts upstream of the Rac-GEF Still life (*Drosophila* Tiam1) to restrict Rac1 activity to the basolateral domain (Georgiou and Baum 2010).

7.3 Local Inhibitory Mechanisms for Tuning Complex Localization and Turnover

Besides long-range inhibitory mechanisms between apical and basolateral proteins, local inhibitory mechanisms also exist. As we expand upon below, such local inhibition appears to have two major effects. First, it promotes a wider distribution of polarity proteins. The combination of local positive feedback and long-range inhibition can provide robust polarity, but this polarity can be too extreme. In epithelial cells, the polarization of apical proteins must be coupled with a mechanism for distributing the proteins around the apical circumference. Second, polarized complexes are rarely static. They must remain dynamic for tissue morphogenesis to occur.

7.3.1 Segregating and Distributing Proteins of the Par and Crb Complexes

As discussed, proteins of the Par and Crb complexes are somewhat interchangeable, and specific mechanisms control their segregation and distribution around the apical domain. Specifically, although Baz, Par-6, and aPKC can interact and function as the Par complex, they also localize and function separately. Baz localizes to AJs, whereas Par-6 and aPKC localize to the marginal zone with Crb. Crb plays a key role in this segregation as evident in *crb* mutants in which Baz, Par-6, and aPKC co-aggregate in large puncta segregated from basolateral proteins (Harris and Peifer 2005). Biochemically, Crb is able to out-compete Baz for binding to Par-6 (Morais-de-Sa et al. 2010). aPKC plays another major role in the segregation. Specifically, after aPKC phosphorylates Baz, it appears to dissociate from Baz and localize with Crb. This model is based on the behavior of a non-phosphorylatable Baz construct which both binds aPKC strongly and promotes the aggregation of all apparent apical proteins into large single puncta in the apical domain (Morais-de-Sa et al. 2010; Walther and Pichaud 2010). As discussed above and in more detail below, the apical proteins promote associations among each other. Here, however, we see evidence for them inhibiting interactions among each other. Thus, negative feedback or incoherent feed-forward loops may be at work (Hart and Alon 2013; Lim et al. 2013). These loops can act as buffers. When an inhibitory/buffering effect is removed from the system, such as when aPKC dissociation from Baz is prevented, polarization becomes extreme—all apparent apical proteins become focused into a single enlarged puncta instead of being distributed more evenly around the apical domain in two layers. Thus, local inhibitory mechanisms appear to counteract the polarity generated by local positive feedback loops and long-range inhibition for a moderated form of polarity (Fig. 7.4).

7.3.2 Endocytosis of Apical Complexes

It is also clear that apical complexes are subject to endocytosis. For example, a syntaxin and Rab5 restrict the size of the apical domain of *Drosophila* epithelial cells by promoting the endocytosis of Crb (Lu and Bilder 2005). AJs, as well, are

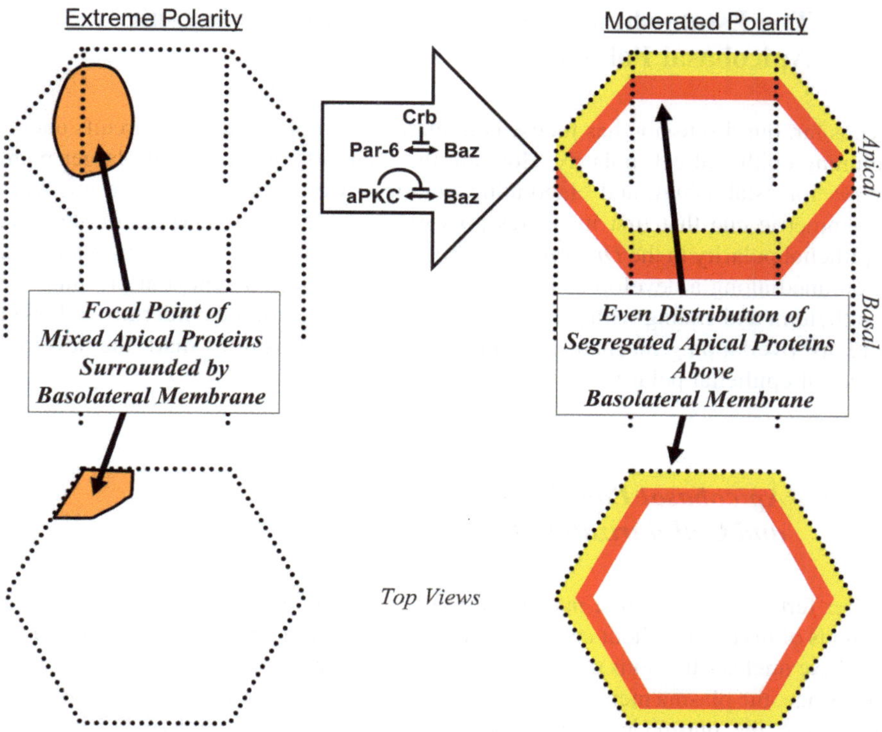

Fig. 7.4 Segregating and distributing proteins of the Par and Crb complexes around the apical domain. See Sect. 7.4 for details

subject to endocytic regulation in *Drosophila*, and this local inhibition is important for tissue morphogenesis during a number of processes (Harris 2012). Strikingly, the Par proteins regulate the endocytosis and trafficking of these complexes. For example, abnormal endocytic delays and tubulation have been observed during AJ endocytosis with the loss of the Par-6-aPKC cassette, suggesting a positive role for the Par proteins in the endocytic mechanism (Georgiou et al. 2008; Leibfried et al. 2008). However, aPKC has also been reported to suppress endocytosis of Crb in both the *Drosophila* embryo (Harris and Tepass 2008) and the follicular epithelium (Fletcher et al. 2012). These apparent discrepancies are consistent with a model in which apical polarity proteins both inhibit and support one another to tune the structure of the apical domain. The balance of such effects may be context dependent. Further work is needed to test these hypotheses.

7.4 The Establishment and Elaboration of Embryonic Apicobasal Polarity

Thus far, our discussion has focused on relatively simple regulatory circuits underpinning epithelial cell polarity. How are these circuits organized into larger pathways for establishing and elaborating polarity? Moreover, what is the upstream asymmetric cue that first instigates polarity? One of the advantages of studying epithelial polarity in the *Drosophila* embryo is that polarization mechanisms can be examined along a developmental timeline. Starting from a syncytial presumptive epithelium and ending with fully differentiated epithelia of the epidermis and other organs, overlapping mechanisms and stage-specific transitions mark the development of epithelial polarity in the embryo.

7.4.1 Apicobasal Polarity of Pseudocleavage and Cellularization Furrows

After fertilization, the first nucleus of the *Drosophila* embryo undergoes multiple rounds of division without cell division, creating a syncytium. The first nine rounds of these nuclear divisions occur centrally in the embryo (Foe and Alberts 1983). At this stage, the plasma membrane surrounding syncytium appears relatively uniform. However, the plasma membrane is coated with a non-AJ pool of DE-cadherin (DE-cad) which contributes to the anchoring of actin cytoskeletal networks (Abreu-Blanco et al. 2011). When the nuclei migrate to the embryo periphery, they undergo another four rounds of division, generating ~6, 000 evenly distributed nuclei (Foe and Alberts 1983). This migration provides the first major cue for apicobasal polarity—each cell compartment exists on an axis running from the surface to the interior of the embryo. Recruitment of nuclei to the embryo surface depends on microtubules (Baker et al. 1993; Zalokar and Erk 1976), although the mechanism of microtubule attachment to the outer plasma membrane remains unknown. The plasma membrane first invaginates between nuclei to separate them during mitotic divisions (Foe and Alberts 1983). These pseudocleavage (or metaphase) furrows form transiently around each nucleus along the apicobasal axis of each cell compartment. Strikingly, they display polarization of core apicobasal polarity machinery (Mavrakis et al. 2009). Specifically, DE-cad and the septin Peanut localize to the apical regions of the furrow, whereas Patj (the Crb complex component) localizes to the basal tips of the furrows.

One conspicuous aspect of this early plasma membrane polarity is that Patj localizes to the basal rather than the apical end of membranes. Remarkably, Patj retains its localization to the basal tips of furrows into cellularization (Bhat et al. 1999). Moreover, other proteins with later apical enrichment also accumulate at or near the base of the membranes, specifically AJ proteins (Hunter and Wieschaus 2000). One major correlation with these inverted localization patterns, and their later apical shifts, is the similar behavior of actomyosin networks (networks formed between actin filaments and non-muscle myosin II—myosin hereafter). These networks form at the basal tips of cellularization furrows, where they support furrow growth and ultimately assemble cytokinetic rings to separate the epithelial cells from the yolk below. With cellularization completed, the networks largely shift to the apicolateral and apical domain to function in epithelial morphogenesis at gastrulation (Harris et al. 2009). Thus, the inverted polarity of some apicobasal polarity proteins along pseudocleavage and cellularization furrows may reflect their associations with actomyosin networks (Mavrakis et al. 2009; Sokac and Wieschaus 2008).

During cellularization, other apical polarity proteins are recruited apically. Perhaps best understood is the positioning of Baz by three partially overlapping mechanisms. One mechanism involves an apical anchor based around apical actin networks (Harris and Peifer 2005). These apical networks also associate with AJ proteins and, at this stage, apical microvilli appear to sweep AJ proteins into clusters (McGill et al. 2009). Baz shows minimal association with these initial apical-most cadherin-catenin clusters, but when they move basally, Baz clusters engage and retain them around the apical circumference (McGill et al. 2009)—the "clustering of clusters" positive feedback loop proposed above. The actin organization in this circumferential zone is poorly understood, but would presumably contribute to clustering of complexes as well. Indeed, a recent study demonstrated that the small GTPase Rap1 and the actin- and AJ-associated protein Canoe (*Drosophila* afadin) are required for effective positioning of cadherin-catenin and Baz clusters to this region (Choi et al. 2013). The second Baz positioning mechanism is also spatially associated with this apicolateral zone (Harris and Peifer 2005). Specifically, the Baz clusters are found at the same apicobasal position as centrosomes. At this stage, centrosomal microtubule arrays extend basally from the apical domain, placing microtubule minus ends next to the sites of Baz accumulation on the plasma membrane. Strikingly, when the apical actin-based anchor is saturated by Baz overexpression, discrete Baz clusters are displaced and first flow basally with the forming plasma membrane. Later, however, they reverse course and display directed transport to the apical domain next to the centrosomes. This transport is dependent on the minus end-directed motor Dynein (Harris and Peifer 2005) as well as the three PDZ domains of Baz (McKinley and Harris 2012). The third positioning mechanism is evident when the transport mechanism is compromised. In mutants affecting Dynein (Harris and Peifer 2005) or with deletion of the Baz PDZ domains (McKinley and Harris 2012), Baz remains basal for a more extended period of time, but at the end of cellularization, the basally mislocalized puncta disappear. Knockdown of Par-1 and mutation of the Par-1 phosphorylation sites in Baz revealed that this later mechanism is based on phosphorylation of Baz by Par-1 (McKinley and Harris 2012). Overall, these results implicate a robust Baz polarization system possibly involving multiple local positive feedback loops and two long-range redistribution mechanisms.

Baz is important for positioning AJs (Harris and Peifer 2004; McGill et al. 2009; Muller and Wieschaus 1996) and also for positioning the other components of the

Par complex (Harris and Peifer 2005). The positioning of aPKC and PAR-6 at the end of cellularization is, however, poorly understood. Baz is essential for apical aPKC localization, but Par-6 can localize apically when Baz is perturbed (Harris and Peifer 2005). The localization of aPKC, however, must also involve other players since aPKC accumulates in the marginal zone above Baz, and in contrast to Baz, aPKC accumulates normally when Dynein is compromised (Harris and Peifer 2005). Another major player is Cdc42, which promotes the apical localization of aPKC and Par-6 at this stage (Hutterer et al. 2004). How local Cdc42 activity is controlled and how the Cdc42 pathway is coupled with Baz remain to be determined.

During these early stages, basal polarity proteins such as Lgl, Dlg, and Par-1 are not yet excluded from the apical domain (Harris and Peifer 2004; Hutterer et al. 2004; McKinley and Harris 2012). Thus, their role in the apicobasal polarity of pseudocleavage and cellularization furrows is not entirely clear. They have, however, been shown to function at these stages. Dlg plays a role in membrane trafficking (Lee et al. 2003), and Par-1 promotes the actomyosin networks at the base of cellularization furrows (McKinley and Harris 2012).

7.4.2 The Transition to Gastrulation and Polarity Elaboration Thereafter

As embryos enter gastrulation, the properties of the ectodermal epithelium change dramatically. One major change is the accumulation of apicolateral and apicomedial actomyosin networks. In the ventral furrow, a mechanistic understanding of apical actomyosin recruitment involving localized activation of Rho1 is becoming clearer (Harris et al. 2009), but in other regions of the embryo, this recruitment is poorly understood. The recruitment is coupled with an increase in the continuity of cadherin-catenin clusters and a focusing of AJs into a sharp plane across all cells in the tissue (Harris 2012). AJs and Baz are not passively affected by the cytoskeleton networks but impact the networks as well, through proteins such as α-catenin (Cavey et al. 2008), Canoe (Sawyer et al. 2009), and Bitesize (Pilot et al. 2006). Gaining a fuller understanding of how these apical trans-cellular AJ-cytoskeletal networks form and function is critical for our understanding of the tissue morphogenesis that drives gastrulation.

Another major change to the cytoskeleton at gastrulation involves microtubule networks. Specifically, the centrosomal microtubule networks of cellularization are converted into non-centrosomal networks typical of more mature epithelia. aPKC is required for this full transition (Harris and Peifer 2007). Without aPKC, centrosomal microtubule asters are maintained in the apical domain into gastrulation and, strikingly, are each linked with a strong single puncta of Baz and a local accumulation of AJ proteins. These defects coincide with significant disruptions to gastrulation. By inhibiting microtubule-Baz interactions, aPKC appears critical for this stage-specific polarity transition. How aPKC inhibits this earlier polarity generating mechanism is unknown.

In addition to the cytoskeletal networks and their effects on polarity complexes, other substantial changes to polarity complexes also occur at gastrulation. One is the accumulation of the apical Crb complex. A major mechanism dictating this assembly has recently come to light (Krahn et al. 2010). It also centers on aPKC. Specifically, the Crb complex component Sdt binds Baz at its aPKC phosphorylation site. The phosphorylation of Baz by aPKC displaces Sdt and permits Sdt recruitment to the Crb complex. Closely coupled with this mechanism are those that promote the segregation of Baz and the Par-6-aPKC cassette to AJs and the Crb complex, respectively (as discussed above) (Morais-de-Sa et al. 2010).

Long-range inhibition of the basolateral players Dlg, Lgl, and Par-1 also becomes apparent during the transition into gastrulation (Harris and Peifer 2004; Hutterer et al. 2004; McKinley and Harris 2012). For Lgl, aPKC phosphorylation has been shown to displace the protein from the apical domain (Hutterer et al. 2004), and since aPKC functions to displace Par-1 in other contexts (Doerflinger et al. 2010), it is tempting to speculate that it may displace it during gastrulation as well. Thus, aPKC activity at the onset of gastrulation may be critical for many of the polarity transitions that occur at this stage. One major question is how aPKC activity is specifically triggered or elevated at this stage.

The Scrib complex becomes critical for maintaining polarity following gastrulation, but remarkably loss of polarity in Scrib complex mutants can be restored during later development. As the organ systems of the embryo are developing, the Yrt/Cora group functions to maintain basolateral polarity, redundantly with the Scrib complex (Laprise et al. 2006, 2009). Further highlighting the robustness of the system, when both the Scrib complex and Yrt/Cora group are compromised, polarity is lost for a more extended period of time, but in late embryos, epithelial polarity is restored significantly, suggesting that there is a later unknown mechanism that contributes to apicobasal polarity (Laprise et al. 2009). Overall, epithelial apicobasal polarity appears to be established and maintained through partially overlapping layers of regulatory circuits, as well as transitions tied to specific developmental stages.

7.5 The Coupling of Apicobasal Polarity Proteins to Epithelial Morphogenesis

Another major advantage of studying apicobasal polarity in the *Drosophila* embryo is the ability to probe how apicobasal polarity proteins contribute to epithelial morphogenesis. Developing epithelia are not static, and thus neither are apicobasal polarity complexes. As animals develop, cells change shape and exchange neighbors to drive tissue morphogenesis. How are such events influenced by cell polarity?

7.5.1 *Permissive Roles*

Polarity proteins can serve a permissive role during epithelial morphogenesis by maintaining overall tissue integrity while individual cells rearrange. This permissive role first became apparent in mutants in which DE-cad levels are substantially reduced (Tepass et al. 1996; Uemura et al. 1996). The tissues most susceptible to breakdown are highly morphogenetically active—e.g., the neuroectoderm which requires constant cell-cell contact establishments following single-cell delaminations. The critical role of DE-cad in maintaining morphogenetically active tissues was confirmed by genetically blocking the tissue dynamics—with less active tissues, low levels of DE-cad are sufficient for maintaining tissue structure. More recently, similar roles have been attributed to proteins of the Par and Crb complexes (Campbell et al. 2009; Harris and Tepass 2008). Thus, normal levels of AJs and apicobasal polarity proteins appear essential for controlling local losses and reestablishments of cell-cell adhesion for maintaining overall tissue integrity with morphogenesis.

7.5.2 *Instructive Roles*

Apicobasal polarity proteins can also serve instructive roles during epithelial morphogenesis. In some cases, the proteins display an even circumferential or radial distribution and regulate an isotropic change to the cells. In other cases, the proteins become planar polarized and regulate anisotropic changes to the cells. Here, we use the terms isotropic and anisotropic to describe the planar polarity of epithelial cells. By isotropic, we mean an even property around the circumference of an epithelial cell, keeping in mind that the property is polarized along the apicobasal axis. By anisotropic, we mean an uneven property around the circumference of an epithelial cell, again recognizing that the property is polarized along the apicobasal axis.

7.5.2.1 Actions of Isotropic Apicobasal Polarity Proteins

There are several examples in which isotropically localized apicobasal polarity proteins affect isotropic behaviors of epithelial cells in the *Drosophila* embryo. One recently documented example is the formation of folds in the ectoderm on the dorsal side of the embryo just after cellularization (Wang et al. 2012). These dorsal folds are formed through specific basal shifts of AJs in rows of cells in which Par-1 levels are normally reduced. Here, a modified balance between basolateral Par-1 activity and apical aPKC activity regulates the sizes of the apical and basolateral domains and in turn the positioning of AJs around the interface between the domains. Thus, when Par-1 activity is normally reduced in specific rows of cells, the AJs of these cells shift basally, and folds in the tissue are induced. This example illustrates how a regulated balance between the apical and basolateral domains can direct tissue morphogenesis.

A balance between apical and basolateral proteins also controls the size of epithelial tubes in the embryonic tracheal system (Laprise et al. 2010). Here, loss of Yrt or Cora activity leads to enlarged tubes. Indicating that this defect is due to abnormally elevated Crb activity, it can be suppressed by additional loss of Crb. Moreover, overexpression of Crb alone can also enlarge the tubes. Thus, a balance between activities of the Crb complex and the Yurt/Cora group appears to dictate the size of the apical domain of each epithelial cell in the tube, and cumulatively the size of all apical domains helps to determine the full size of the tube.

Apical domains can also change their size through the contractility of apical actomyosin networks. Polarity proteins can regulate such networks, as illustrated during apical constriction of amnioserosa cells during embryo dorsal closure. During this process, squamous epithelial amnioserosa cells initially undergo pulsatile apical constrictions and then shift to a final phase of persistent apical constriction that helps drive their internalization (Blanchard et al. 2010; Solon et al. 2009). These constrictions are mediated by apicomedial actomyosin networks and their engagements with circumferential AJs (Blanchard et al. 2010; David et al. 2010, 2013). Baz, Par-6, and aPKC initially localize around the apical circumference of these cells but are progressively recruited to the apicomedial domain through the influence of the actomyosin networks and positive interactions among the Par proteins themselves (David et al. 2013). The Par-6-aPKC cassette inhibits the actomyosin networks and thus may function in a negative feedback loop for promoting network disassembly and the pulsatile behavior of the cells. Strikingly, Baz has the opposite effect, promoting actomyosin networks (David et al. 2010). Baz appears to do so by acting as a competitive inhibitor of aPKC kinase activity (David et al. 2013). During early dorsal closure, this inhibition may delay the antagonistic effects of Par-6-aPKC on the actomyosin networks to promote oscillatory behavior. As the apical levels of the Par proteins build by late dorsal closure, enhanced inhibition of Par-6-aPKC by Baz may lift their antagonism of actomyosin to promote network persistence. This example illustrates how polarity proteins can tune the behavior of apical actomyosin networks to reshape cells and how the nature of the regulatory circuits involved can change over developmental time.

7.5.2.2 Actions of Anisotropic Apicobasal Polarity Proteins

In the clearest examples during which anisotropically localized apicobasal polarity proteins affect anisotropic behaviors of epithelial cells in the *Drosophila* embryo, there is also interplay between the polarity proteins and actin-based networks.

One example involves the assembly of actomyosin cables and actin-based protrusions at the leading edges of cells bordering groups of invaginating cells. During dorsal closure, the epidermal cells surrounding the amnioserosa form a supracellular actomyosin cable as well as numerous filopodial protrusions at their leading edges and use these assemblies to promote closure. One contributor to this anisotropy is the cell adhesion molecule Echinoid, which is expressed in the epidermis but not in the amnioserosa (Laplante and Nilson 2011). This differential expression leads to a loss of Echinoid at the leading edge contact of the epidermal cells bordering the amnioserosa. In turn, this normal loss has been linked to assembly of the actomyosin cable and to the loss of Baz from the leading edge cell-cell contact. Thus, Baz becomes anisotropically localized in the leading edge cells (Laplante and Nilson 2011). This Baz planar polarity restricts the distribution pattern of the Baz binding partner PTEN to cell-cell contacts away from the leading edge, and as a result, PIP3 accumulates at the leading edge contact and promotes filopodia formation (Pickering et al. 2013). This inverse planar polarization of Baz-PTEN and PIP3-actin also promotes filopodia formation at the leading edges of cells surrounding wounds in the embryo (Pickering et al. 2013). In an analogous context, the *Drosophila* embryonic salivary gland placode, a planar polarized depletion of Crb and aPKC from the placode-epidermis boundary, is important for assembling an actomyosin belt at the interface (Roper 2012). These examples illustrate how anisotropically localized polarity proteins can control the assembly of cytoskeletal networks to help specific regions of an epithelium internalize.

Planar polarization of Baz also contributes to convergent extension of the germ band. Here, planar polarized actomyosin cables also form at specific cell-cell contacts, but in this tissue, the networks instigate loss of the contacts and cell-cell intercalation for extending the body axis (Lecuit and Lenne 2007; Zallen 2007). Similar to the leading edge of epidermal cells at dorsal closure or around wounds, Baz accumulates at cell-cell contacts where actomyosin is weakest (Baz and actomyosin have reciprocal planar polarity) (Zallen and Wieschaus 2004). In the germ band, the planar polarity of Baz is dictated by an actomyosin assembly pathway (Simoes Sde et al. 2010). Specifically, Rho kinase localizes to the contacts where actomyosin networks are to be assembled and induces their assembly. An apparently simultaneous effect is Rho kinase phosphorylation of Baz and the resulting loss of Baz from the contacts with elevated Rho kinase. Intriguingly, however, Baz also supports the planar polarity of myosin, suggesting mutual antagonism between the complexes at the two contact types (Simoes Sde et al. 2010). How Baz antagonizes myosin remains to be determined.

Planar polarization of apicobasal polarity proteins is additionally important for defining precise boundaries between segments of the later epidermis. At this stage, cell-cell contacts become amazingly aligned along the dorsal-ventral axis of the embryo through the activity of actomyosin cables running along them. Baz also has reciprocal planar polarity with these actomyosin cables (Simone and DiNardo 2010). In this context, microtubules run across the apical domain along the dorsal-ventral axis such that their plus ends terminate at the Baz-enriched contacts. By genetically manipulating the microtubules, they were found to both promote the planar polarity of Baz and locally inhibit the RhoGEF2-Rho1-Rho kinase pathway. This mechanism helps to planar polarize the cells and to align them across the epidermal sheet (Bulgakova et al. 2013). Remarkably, the reciprocal contacts enriched with myosin are not themselves homogeneous. Typically, for a single strip of cell-cell contacts along the dorsal-ventral axis, actomyosin cables form first at two contacts per cell. However, this arrangement gives the overall strip of contacts a zigzag appearance. To straighten the strip, specific contacts are lost so that only one contact retains an actomyosin cable per cell. For this regulation, a distinctive planar polarization of Dlg apparently opposes actomyosin activity at the contact to be retained (Simone and DiNardo 2010). It will be interesting to see how Dlg becomes planar polarized and how it elicits its effects on myosin.

7.6 Final Comment

This chapter has reviewed how local positive feedback loops, long-range negative feedback loops, and local negative feedback loops function together to establish and elaborate epithelial apicobasal polarity in the *Drosophila* embryo and to couple this polarity with epithelial morphogenesis. Although numerous specific polarity mechanisms have been clearly defined, it is always challenging to connect them into larger models to provide a system-level view of the cell. Thus, many of the syntheses we have presented remain working hypotheses that require more rigorous evaluation in the years to come.

Acknowledgments Tony Harris is supported by a Tier 2 Canada Research Chair; the lab's research is supported by operating grants from the Natural Sciences and Engineering Research Council of Canada and the Canadian Institutes of Health Research.

References

Abreu-Blanco MT, Verboon JM, Parkhurst SM (2011) Cell wound repair in Drosophila occurs through three distinct phases of membrane and cytoskeletal remodeling. J Cell Biol 193:455–464

Baker J, Theurkauf WE, Schubiger G (1993) Dynamic changes in microtubule configuration correlate with nuclear migration in the preblastoderm Drosophila embryo. J Cell Biol 122:113–121

Bayraktar J, Zygmunt D, Carthew RW (2006) Par-1 kinase establishes cell polarity and functions in Notch signaling in the Drosophila embryo. J Cell Sci 119:711–721

Benton R, St Johnston D (2003a) A conserved oligomerization domain in drosophila Bazooka/PAR-3 is important for apical localization and epithelial polarity. Curr Biol 13:1330–1334

Benton R, St Johnston D (2003b) Drosophila PAR-1 and 14-3-3 inhibit Bazooka/PAR-3 to establish complementary cortical domains in polarized cells. Cell 115:691–704

Betschinger J, Mechtler K, Knoblich JA (2003) The Par complex directs asymmetric cell division by phosphorylating the cytoskeletal protein Lgl. Nature 422:326–330

Bhat MA, Izaddoost S, Lu Y, Cho KO, Choi KW, Bellen HJ (1999) Discs Lost, a novel multi-PDZ domain protein, establishes and maintains epithelial polarity. Cell 96:833–845

Bilder D (2004) Epithelial polarity and proliferation control: links from the Drosophila neoplastic tumor suppressors. Genes Dev 18:1909–1925

Bilder D, Schober M, Perrimon N (2003) Integrated activity of PDZ protein complexes regulates epithelial polarity. Nat Cell Biol 5:53–58

Blanchard GB, Murugesu S, Adams RJ, Martinez-Arias A, Gorfinkiel N (2010) Cytoskeletal dynamics and supracellular organisation of cell shape fluctuations during dorsal closure. Development 137:2743–2752

Bulgakova NA, Grigoriev I, Yap AS, Akhmanova A, Brown NH (2013) Dynamic microtubules produce an asymmetric E-cadherin-Bazooka complex to maintain segment boundaries. J Cell Biol 201:887–901

Campbell K, Knust E, Skaer H (2009) Crumbs stabilises epithelial polarity during tissue remodelling. J Cell Sci 122:2604–2612

Cavey M, Rauzi M, Lenne PF, Lecuit T (2008) A two-tiered mechanism for stabilization and immobilization of E-cadherin. Nature 453:751–756

Chartier FJ, Hardy EJ, Laprise P (2011) Crumbs controls epithelial integrity by inhibiting Rac1 and PI3K. J Cell Sci 124:3393–3398

Chau AH, Walter JM, Gerardin J, Tang C, Lim WA (2012) Designing synthetic regulatory networks capable of self-organizing cell polarization. Cell 151:320–332

Choi W, Harris NJ, Sumigray KD, Peifer M (2013) Rap1 and Canoe/afadin are essential for establishment of apical-basal polarity in the Drosophila embryo. Mol Biol Cell 24:945–963

David DJ, Tishkina A, Harris TJ (2010) The PAR complex regulates pulsed actomyosin contractions during amnioserosa apical constriction in Drosophila. Development 137:1645–1655

David DJV, Wang Q, Feng JJ, Harris TJC (2013) Bazooka inhibits aPKC to limit antagonism of actomyosin networks during amnioserosa apical constriction. Development 140:4719. doi:10.1242/dev.098491

Doerflinger H, Vogt N, Torres IL, Mirouse V, Koch I, Nusslein-Volhard C, St Johnston D (2010) Bazooka is required for polarisation of the Drosophila anterior-posterior axis. Development 137:1765–1773

Elsum I, Yates L, Humbert PO, Richardson HE (2012) The Scribble-Dlg-Lgl polarity module in development and cancer: from flies to man. Essays Biochem 53:141–168

Feng W, Wu H, Chan LN, Zhang M (2007) The Par-3 NTD adopts a PB1-like structure required for Par-3 oligomerization and membrane localization. EMBO J 26:2786–2796

Fletcher GC, Lucas EP, Brain R, Tournier A, Thompson BJ (2012) Positive feedback and mutual antagonism combine to polarize Crumbs in the Drosophila follicle cell epithelium. Curr Biol 22:1116–1122

Foe VE, Alberts BM (1983) Studies of nuclear and cytoplasmic behaviour during the five mitotic cycles that precede gastrulation in Drosophila embryogenesis. J Cell Sci 61:31–70

Georgiou M, Baum B (2010) Polarity proteins and Rho GTPases cooperate to spatially organise epithelial actin-based protrusions. J Cell Sci 123:1089–1098

Georgiou M, Marinari E, Burden J, Baum B (2008) Cdc42, Par6, and aPKC regulate Arp2/3-mediated endocytosis to control local adherens junction stability. Curr Biol 18:1631–1638

Goldstein B, Macara IG (2007) The PAR proteins: fundamental players in animal cell polarization. Dev Cell 13:609–622

Griffin EE, Odde DJ, Seydoux G (2011) Regulation of the MEX-5 gradient by a spatially segregated kinase/phosphatase cycle. Cell 146:955–968

Harris KP, Tepass U (2008) Cdc42 and Par proteins stabilize dynamic adherens junctions in the Drosophila neuroectoderm through regulation of apical endocytosis. J Cell Biol 183:1129–1143

Harris TJ (2012) Adherens junction assembly and function in the Drosophila embryo. Int Rev Cell Mol Biol 293:45–83

Harris TJ, Peifer M (2004) Adherens junction-dependent and -independent steps in the establishment of epithelial cell polarity in Drosophila. J Cell Biol 167:135–147

Harris TJ, Peifer M (2005) The positioning and segregation of apical cues during epithelial polarity establishment in Drosophila. J Cell Biol 170:813–823

Harris TJ, Peifer M (2007) aPKC controls microtubule organization to balance adherens junction symmetry and planar polarity during development. Dev Cell 12:727–738

Harris TJ, Sawyer JK, Peifer M (2009) How the cytoskeleton helps build the embryonic body plan: models of morphogenesis from Drosophila. Curr Top Dev Biol 89:55–85

Hart Y, Alon U (2013) The utility of paradoxical components in biological circuits. Mol Cell 49:213–221

Horne-Badovinac S, Bilder D (2008) Dynein regulates epithelial polarity and the apical localization of stardust A mRNA. PLoS Genet 4:e8

Hunter C, Wieschaus E (2000) Regulated expression of nullo is required for the formation of distinct apical and basal adherens junctions in the Drosophila blastoderm. J Cell Biol 150:391–401

Hutterer A, Betschinger J, Petronczki M, Knoblich JA (2004) Sequential roles of Cdc42, Par-6, aPKC, and Lgl in the establishment of epithelial polarity during Drosophila embryogenesis. Dev Cell 6:845–854

Krahn MP, Buckers J, Kastrup L, Wodarz A (2010) Formation of a Bazooka-Stardust complex is essential for plasma membrane polarity in epithelia. J Cell Biol 190:751–760

Krahn MP, Egger-Adam D, Wodarz A (2009) PP2A antagonizes phosphorylation of Bazooka by PAR-1 to control apical-basal polarity in dividing embryonic neuroblasts. Dev Cell 16:901–908

Laplante C, Nilson LA (2011) Asymmetric distribution of Echinoid defines the epidermal leading edge during Drosophila dorsal closure. J Cell Biol 192:335–348

Laprise P, Beronja S, Silva-Gagliardi NF, Pellikka M, Jensen AM, McGlade CJ, Tepass U (2006) The FERM protein Yurt is a negative regulatory component of the Crumbs complex that controls epithelial polarity and apical membrane size. Dev Cell 11:363–374

Laprise P, Lau KM, Harris KP, Silva-Gagliardi NF, Paul SM, Beronja S, Beitel GJ, McGlade CJ, Tepass U (2009) Yurt, Coracle, Neurexin IV and the Na(+), K(+)-ATPase form a novel group of epithelial polarity proteins. Nature 459:1141–1145

Laprise P, Paul SM, Boulanger J, Robbins RM, Beitel GJ, Tepass U (2010) Epithelial polarity proteins regulate Drosophila tracheal tube size in parallel to the luminal matrix pathway. Curr Biol 20:55–61

Lecuit T, Lenne PF (2007) Cell surface mechanics and the control of cell shape, tissue patterns and morphogenesis. Nat Rev Mol Cell Biol 8:633–644

Lee OK, Frese KK, James JS, Chadda D, Chen ZH, Javier RT, Cho KO (2003) Discs-Large and Strabismus are functionally linked to plasma membrane formation. Nat Cell Biol 5:987–993

Leibfried A, Fricke R, Morgan MJ, Bogdan S, Bellaiche Y (2008) Drosophila Cip4 and WASp define a branch of the Cdc42-Par6-aPKC pathway regulating E-cadherin endocytosis. Curr Biol 18:1639–1648

Letizia A, Ricardo S, Moussian B, Martin N, Llimargas M (2013) A functional role of the extracellular domain of Crumbs in cell architecture and apicobasal polarity. J Cell Sci 126:2157–2163

Li Z, Wang L, Hays TS, Cai Y (2008) Dynein-mediated apical localization of crumbs transcripts is required for Crumbs activity in epithelial polarity. J Cell Biol 180:31–38

Lim WA, Lee CM, Tang C (2013) Design principles of regulatory networks: searching for the molecular algorithms of the cell. Mol Cell 49:202–212

Lu H, Bilder D (2005) Endocytic control of epithelial polarity and proliferation in Drosophila. Nat Cell Biol 7:1232–1239

Martin-Belmonte F, Gassama A, Datta A, Yu W, Rescher U, Gerke V, Mostov K (2007) PTEN-mediated apical segregation of phosphoinositides controls epithelial morphogenesis through Cdc42. Cell 128:383–397

Mavrakis M, Rikhy R, Lippincott-Schwartz J (2009) Plasma membrane polarity and compartmentalization are established before cellularization in the fly embryo. Dev Cell 16:93–104

McGill MA, McKinley RF, Harris TJ (2009) Independent cadherin-catenin and Bazooka clusters interact to assemble adherens junctions. J Cell Biol 185:787–796

McKinley RF, Harris TJ (2012) Displacement of basolateral Bazooka/PAR-3 by regulated transport and dispersion during epithelial polarization in Drosophila. Mol Biol Cell 23:4465–4471
Morais-de-Sa E, Mirouse V, St Johnston D (2010) aPKC phosphorylation of Bazooka defines the apical/lateral border in Drosophila epithelial cells. Cell 141:509–523
Muller HA, Wieschaus E (1996) armadillo, bazooka, and stardust are critical for early stages in formation of the zonula adherens and maintenance of the polarized blastoderm epithelium in Drosophila. J Cell Biol 134:149–163
Nam SC, Mukhopadhyay B, Choi KW (2007) Antagonistic functions of Par-1 kinase and protein phosphatase 2A are required for localization of Bazooka and photoreceptor morphogenesis in Drosophila. Dev Biol 306:624–635
Papoulas O, Hays TS, Sisson JC (2005) The golgin Lava lamp mediates dynein-based Golgi movements during Drosophila cellularization. Nat Cell Biol 7:612–618
Pickering K, Alves-Silva J, Goberdhan D, Millard TH (2013) Par3/Bazooka and phosphoinositides regulate actin protrusion formation during Drosophila dorsal closure and wound healing. Development 140:800–809
Pilot F, Philippe JM, Lemmers C, Lecuit T (2006) Spatial control of actin organization at adherens junctions by a synaptotagmin-like protein Btsz. Nature 442:580–584
Pinal N, Goberdhan DC, Collinson L, Fujita Y, Cox IM, Wilson C, Pichaud F (2006) Regulated and polarized PtdIns(3,4,5)P3 accumulation is essential for apical membrane morphogenesis in photoreceptor epithelial cells. Curr Biol 16:140–149
Pocha SM, Knust E (2013) Complexities of Crumbs function and regulation in tissue morphogenesis. Curr Biol 23:R289–R293
Roper K (2012) Anisotropy of Crumbs and aPKC drives myosin cable assembly during tube formation. Dev Cell 23:939–953
Sawyer JK, Harris NJ, Slep KC, Gaul U, Peifer M (2009) The Drosophila afadin homologue Canoe regulates linkage of the actin cytoskeleton to adherens junctions during apical constriction. J Cell Biol 186:57–73
Shewan A, Eastburn DJ, Mostov K (2011) Phosphoinositides in cell architecture. Cold Spring Harb Perspect Biol 3:a004796
Simoes Sde M, Blankenship JT, Weitz O, Farrell DL, Tamada M, Fernandez-Gonzalez R, Zallen JA (2010) Rho-kinase directs Bazooka/Par-3 planar polarity during Drosophila axis elongation. Dev Cell 19:377–388
Simone RP, DiNardo S (2010) Actomyosin contractility and Discs large contribute to junctional conversion in guiding cell alignment within the Drosophila embryonic epithelium. Development 137:1385–1394
Slaughter BD, Smith SE, Li R (2009) Symmetry breaking in the life cycle of the budding yeast. Cold Spring Harb Perspect Biol 1:a003384
Sokac AM, Wieschaus E (2008) Zygotically controlled F-actin establishes cortical compartments to stabilize furrows during Drosophila cellularization. J Cell Sci 121:1815–1824
Solon J, Kaya-Copur A, Colombelli J, Brunner D (2009) Pulsed forces timed by a ratchet-like mechanism drive directed tissue movement during dorsal closure. Cell 137:1331–1342
Sotillos S, Diaz-Meco MT, Caminero E, Moscat J, Campuzano S (2004) DaPKC-dependent phosphorylation of Crumbs is required for epithelial cell polarity in Drosophila. J Cell Biol 166:549–557
St Johnston D, Ahringer J (2010) Cell polarity in eggs and epithelia: parallels and diversity. Cell 141:757–774
Tanentzapf G, Tepass U (2003) Interactions between the crumbs, lethal giant larvae and bazooka pathways in epithelial polarization. Nat Cell Biol 5:46–52
Tepass U (2012) The apical polarity protein network in Drosophila epithelial cells: regulation of polarity, junctions, morphogenesis, cell growth, and survival. Annu Rev Cell Dev Biol 28:655–685

Tepass U, Gruszynski-DeFeo E, Haag TA, Omatyar L, Torok T, Hartenstein V (1996) shotgun encodes Drosophila E-cadherin and is preferentially required during cell rearrangement in the neurectoderm and other morphogenetically active epithelia. Genes Dev 10:672–685

Tepass U, Theres C, Knust E (1990) crumbs encodes an EGF-like protein expressed on apical membranes of Drosophila epithelial cells and required for organization of epithelia. Cell 61:787–799

Uemura T, Oda H, Kraut R, Hayashi S, Kotaoka Y, Takeichi M (1996) Zygotic Drosophila E-cadherin expression is required for processes of dynamic epithelial cell rearrangement in the Drosophila embryo. Genes Dev 10:659–671

von Stein W, Ramrath A, Grimm A, Muller-Borg M, Wodarz A (2005) Direct association of Bazooka/PAR-3 with the lipid phosphatase PTEN reveals a link between the PAR/aPKC complex and phosphoinositide signaling. Development 132:1675–1686

Walther RF, Pichaud F (2010) Crumbs/DaPKC-dependent apical exclusion of bazooka promotes photoreceptor polarity remodeling. Curr Biol 20:1065

Wang YC, Khan Z, Kaschube M, Wieschaus EF (2012) Differential positioning of adherens junctions is associated with initiation of epithelial folding. Nature 484:390–393

Wilkie GS, Davis I (2001) Drosophila wingless and pair-rule transcripts localize apically by dynein-mediated transport of RNA particles. Cell 105:209–219

Zallen JA (2007) Planar polarity and tissue morphogenesis. Cell 129:1051–1063

Zallen JA, Wieschaus E (2004) Patterned gene expression directs bipolar planar polarity in Drosophila. Dev Cell 6:343–355

Zalokar M, Erk I (1976) Division and migration of nuclei during early embryogenesis of Drosophila melanogaster. J Mic Biol Cell 23:97–106

Chapter 8
The Polarized Distribution of the Na^+,K^+-ATPase

M. Cereijido, R.G. Contreras, M.I. Larre, and L. Shoshani

Abstract The life of a cell depends on the perennial inflow of metabolites and outflow of catabolites, ultimately driven by membrane pumps or by the electrochemical potential gradients that these pumps generate. Metazoans have cells in which pumps have an additional function: they accumulate in a certain domain of the membrane to induce *polarity*. Surprisingly, the polarized distribution of Na^+, K^+-ATPase does not arise only from canonical signals or classical mechanisms but also from the peculiar affinities between its own subunits. For example, subunits α and β have an affinity for each other that binds them together right after synthesis, and they then migrate through the endoplasmic reticulum and the Golgi apparatus and are delivered to the plasma membrane. In keeping with this role of subunit affinities, we have shown that the polarized distribution of the whole enzyme at the plasma membrane facing the intercellular space arises from the very specific affinity of one β subunit for another. In addition to being distributed in a polarized manner, Na^+,K^+-ATPase participates in cell polarization by acting as a receptor for the ouabain hormone, thereby promoting ciliogenesis; obviously, the enzyme can act as a receptor because this is polarized toward the blood side where hormones come from. In this chapter, we review the polarized distribution of Na^+,K^+-ATPase and suggest that the very existence of higher metazoans depends on this polarized expression of pumps.

Keywords Claudin • c-Src • Hormone • Na,K-ATPase • Ouabain

8.1 Introduction

The information gathered through more than a century of research on the permeability of the plasma membrane required several parallel efforts, including those needed to elucidate the movement of ions and molecules, because the information could not be understood on only the basis of the electrochemical potential gradient

M. Cereijido (✉) • R.G. Contreras • M.I. Larre • L. Shoshani
Department of Physiology, Biophysics & Neurosciences, Center for Research & Advanced Studies (Cinvestav), Av. I.P.N. 2508, Col. Zacatenco, 07360 México, D.F., Mexico
e-mail: cereijido@fisio.cinvestav.mx

K. Ebnet (ed.), *Cell Polarity 1*, DOI 10.1007/978-3-319-14463-4_8

between the two sides of the plasma membrane. This challenged the imagination of physiologists, who conceived theoretical mechanisms such as pores, carriers, and pumps and assigned to them properties that frequently appeared to be in violation of the fundamental laws of physics. A case in point was the use of chemical reactions (in those times thought to be scalar phenomena) to produce the vectorial movement of ions, which was clearly demonstrated once J. Ch. Skou discovered the presence of Na^+,K^+-ATPase in crab nerves (Skou 1957; Onsager 1967). Another more recent example became evident after we introduced an experimental model system to study the development of the two fundamental features of the transporting epithelial phenotype: tight junctions (TJs) and polarity (de Donder and Van Rysselberghe 1936; Cereijido et al. 1978; Boulan and Sabatini 1978). The knowledge we gained on the signals and mechanisms that are accounted for by the polarized distribution of most membrane proteins was insufficient to explain the polarized distribution of Na^+,K^+-ATPase. In a way, this was ironic because this enzyme is responsible not only for its own polarity but, together with the polarized expression of other transporter proteins, for the polarized (net) flux of glucose, amino acids, and ions across plasma membranes and epithelial membranes (Csaky and Thale 1960; Crane et al. 1961, 1965; Kedem and Essig 1965; Schultz and Curran 1970). Fortunately, the polarized distribution of this enzyme was recently elucidated, and the data demonstrated that the distribution is due to mechanisms that seem to be specific to this pump (Cereijido and Rotunno 1970; Shoshani et al. 2005; Padilla-Benavides et al. 2010).

8.2 Incongruences, Stumbling Blocks, and Apparent Violations of Thermodynamics (Cereijido et al. 1971)

In the second half of the nineteenth century, du Bois Reymond (Bois-Reymond 1848; Berridge et al. 2003) demonstrated that frog skin maintains an electrical potential difference ($\Delta\Psi$) between the inner and outer sides of the epithelium (Fig. 8.1a), a feature no inert membrane can show when mounted between identical saline solutions. Half a century later, Galeotti (Galeotti 1904; Griffiths 2007) attributed this asymmetry to a higher Na^+ permeability in the inward versus outward direction (Fig. 8.1b). This attribution was disregarded on the basis that it would violate the first principle of thermodynamics and create a *perpetuum mobile*: Na^+ would forever rotate counterclockwise in the *gedankenexperiment* illustrated in Fig. 8.1b. Yet, after the Second World War, radioisotopes became available to measure unidirectional fluxes, and it was easy to demonstrate that the inward flux of sodium across frog skin is up to 20 times higher than the outward flux, a feature that many took as solid evidence that life would never be explained on the basis of pure physical laws. Later on, biologists demonstrated that $\Delta\Psi$ is by no means perpetual and lasts as long as the epithelium is alive and proposed that the net flux of Na^+ (influx *minus* outflux) across an epithelium depends on the energy generated by

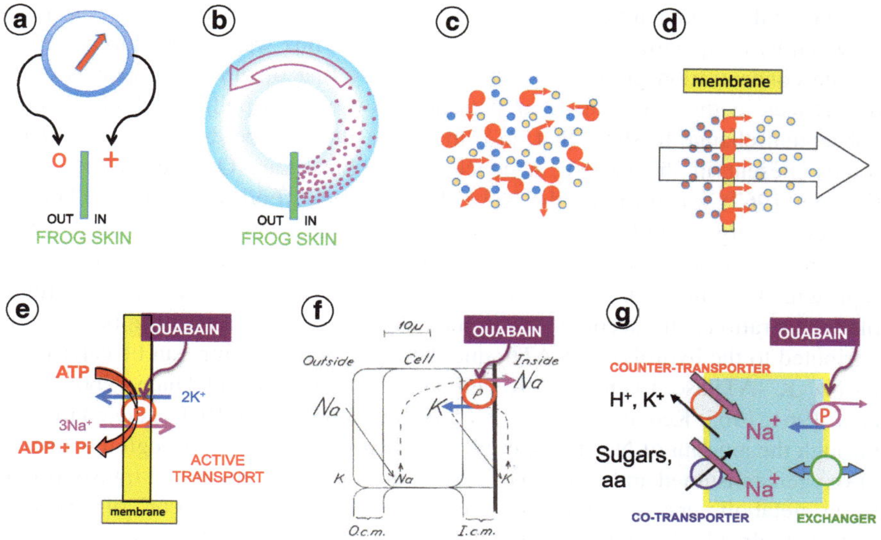

Fig. 8.1 Historical cornerstones toward the polarity of the Na^+,K^+-ATPase. (**a**) Emile du Bois Reymond discovered that frog skin maintains an electrical potential difference between its outer and inner surfaces. (**b**) Galeotti proposed that such electrical potential could be explained by assuming that frog skin has a higher permeability to Na^+ in the inward than in the outward direction. His proposal was discarded on the basis that this would increase the concentration of this ion (*magenta dots*) on the inner side and the ensuing diffusion would originate a counter-clockwise *perpetuum mobile*. (**c**) Avoiding a skirmish with Curie's principle, pumping is vectorial at the microscopic level (i.e., each pump works in a given direction). Yet, a study at the macroscopic level in a homogenized preparation where pumps point in all directions masked their intrinsic vectoriality. (**d**) In a cell membrane where all Na^+,K^+-ATPase pumps are aligned, vectoriality is recovered. (**e**) J.Ch Skou found Na^+,K^+-ATPase in the plasma membrane of an extract of crab nerve. The enzyme simultaneously pumps three Na^+ and $2K^+$ (albeit in different directions) per molecule of ATP hydrolyzed. (**f**) Koefoed-Johnsen and Ussing (1958) proposed that the Na^+,K^+-ATPase proposed by Skou is distributed in a polarized manner in the epithelial cells of frog skin. O.c.m., outer cell membrane, which is permeable to Na^+. I.c.m., inner cell membrane, impermeable to Na^+ but passively permeable to K^+. *P* place where the Na^+,K^+-ATPase is needed to act as an ion pump. (**g**) The pumping of Na^+ out of the cell and toward the extracellular space creates an electrochemical potential difference that drives most co- and counter-transporters, which act as "secondary pumps;" however, these are not considered to be active transporters because they are not driven directly by the chemical energy generated by ATP hydrolysis

metabolism. This suggestion was also rejected on the basis that it would be in violation of Curie's principle: *phenomena of different tensorial order cannot be coupled*. In plain words, metabolism is the sum of chemical processes, which were assumed to be *scalar* phenomena (no particular direction in space), implying that metabolism would never sustain net fluxes, which are clearly *vectorial* processes. Eventually, it was argued that chemical phenomena, such as the splitting of ATP by Na^+,K^+-ATPase, are vectorial at the microscopic level but lose vectoriality at macroscopic levels. Figure 8.1c illustrates this paradox as well as the solution provided (Fig. 8.1d).

Classic thermodynamics assumed that each flow (e.g., of Na^+, K^+, H^+, etc.) was powered by its specific chemical potential, i.e., Na^+ flux was assumed to be driven by the concentration gradient of Na^+ but not by that of other ion species or by differences in the temperature or pressure between two points. It took a major conceptual change to show that, in principle, a flux can be driven by *any* of the forces present in the system, a concept for which Lars Onsager (Skou 1957; Onsager 1967) was awarded the Nobel Prize in 1968. Following this, de Donder and Rysselberghe (de Donder and Van Rysselberghe 1936; Cereijido et al. 1978; Boulan and Sabatini 1978) demonstrated that even chemical reactions can be represented as fluxes driven by chemical affinity. Therefore, taking both groups of considerations into account, fluxes such as those of Na^+ and K^+ can be formally attributed to the hydrolysis of ATP, and therefore, ion pumping can be carried out by Na^+,K^+-ATPase. Eventually, Kedem and Essig (Csaky and Thale 1960; Crane et al. 1961, 1965; Kedem and Essig 1965; Schultz and Curran 1970) gauged ion flux through the amount of Na^+ transported and chemical reactions through the amount of oxygen consumed and were able to show that epithelia can in fact transport ions using metabolic energy. Ironically, this was a "permission to occur" of a process that had already been experimentally demonstrated. Nevertheless, this is the essence of science: to rest assured that the body of knowledge is self-consistent, theories and experimental observations should not disagree.

8.3 Why Do Cells Need Ion Pumps?

Pumps create an asymmetric distribution of ions between the cytoplasm and the external bathing solution that offers some physiological advantages to the cell. (1) The asymmetric distribution enables the plasma membrane to act as an electric capacitor that can, in a given moment, spark an action potential used by muscle fibers to contract and neurons to communicate with each other. (2) The principle of electroneutrality requires that the amount of negative charges in a given volume exactly matches the amount of positive ones, implying that mobile ions should penetrate into the cell to neutralize fixed electric charges from proteins and other molecules trapped inside the cytoplasm. However, simple coulombic considerations (*see* Cereijido and Rotunno 1970; Shoshani et al. 2005; Padilla-Benavides et al. 2010) show that these charges will be perennially neutralized by protons (H^+) because their negligible radius enables their positively charged nuclei to closely approach and bind to fixed negative charges in the cytoplasm with a force far stronger than between the negative charge and any other mobile cation, thereby provoking a molecular debacle in the cell. Hence, cells protect themselves from this disaster by expressing pumps and counter-transporters in their plasma membrane, which, by extruding H^+ toward the extracellular space, maintain such low concentrations of H^+ (in the nanomolar range) that protons are unable to compete with other mobile cations (in the millimolar range). Ca^{2+} faces the same situation because even though its radius compares to that of Na^+ and K^+, calcium has two positive charges instead of one. Again, the combined work of the Na^+,K^+-ATPase and Na^+/Ca^{2+} co-transporters, as well as the endoplasmic reticulum's pumps and chelating proteins, keeps the activity of Ca^{2+} very low in the cell water (Bois-Reymond 1848; Berridge et al. 2003). Therefore, in spite of the fact that hydrogen and calcium are two strong binders to fixed negative charges, the work of the Na^+, K^+-ATPase keeps them at sufficiently low concentrations to insure physiological exchanges among protons, calcium, and monovalent cations. (3) Counter-transporters present in the plasma membrane, such as those of Na^+/glucose, have an affinity for glucose proportional to the Na^+ concentration. Therefore, when facing the extracellular solution, the high Na^+ concentration makes the counter-transporter develop a high affinity for glucose and load itself with this sugar; however, due to the low concentration of Na^+ in the cytoplasm, as soon as the carrier faces the inside of the cell, it releases its Na^+, loses its affinity for glucose, does not combine with the sugar on this side, and returns empty to the outer bathing solution. Notice that this "secondary transport" generates a *net* transport of the sugar and works as long as Na^+,K^+-ATPase maintains the Na^+ asymmetry across the plasma membrane. An analogous mechanism is observed in Na^+-amino acid co-transporters.

It is generally assumed that the water/ion composition of the cytoplasm reflects the composition that the primeval ocean had the day a droplet was trapped inside a vesicle surrounded by a lipid bilayer (Galeotti 1904; Griffiths 2007). Without discarding this possibility, the considerations made in the previous paragraph indicate that the nature of cellular content is the consequence of having a plasma membrane studded with proteins, i.e., synthesized and installed with mechanisms coded in the genome.

8.4 An "Ocean" Less Than a Micron Thick

The intense exchange of a myriad of different substances across the cell membrane of a unicellular organism does not exhaust the nutrients or pollute the ocean (*external milieu*); rather, this external milieu acts as a reservoir (Fig. 8.2a). However, when the cell belongs to a metazoan, say a neuron in the brain (Fig. 8.2b), the "ocean" is replaced by an extremely narrow extracellular space (Fig. 8.2c, *black*), less than a micron thick, that would be quickly exhausted and spoiled were it not for a circulatory apparatus that shuttles its extracellular fluid to and from "transporting epithelia," where the exchange with the outer milieu takes place. To illustrate the amount of epithelia involved in this exchange with the environment, the area of several of them is scaled with the silhouette of a man (Fig. 8.2d). In turn, the exchange between blood and cells proceeds across endothelia, which have basically the same cellular phenotype as epithelia but occupy an even larger area (Fig. 8.2e, *pink*). Figure 8.2f represents the transporting epithelial phenotype of a kidney tube, which is a continuous layer of cells surrounding the lumen, showing its two fundamental features: TJs and apical/basolateral polarity. Interestingly, in spite of

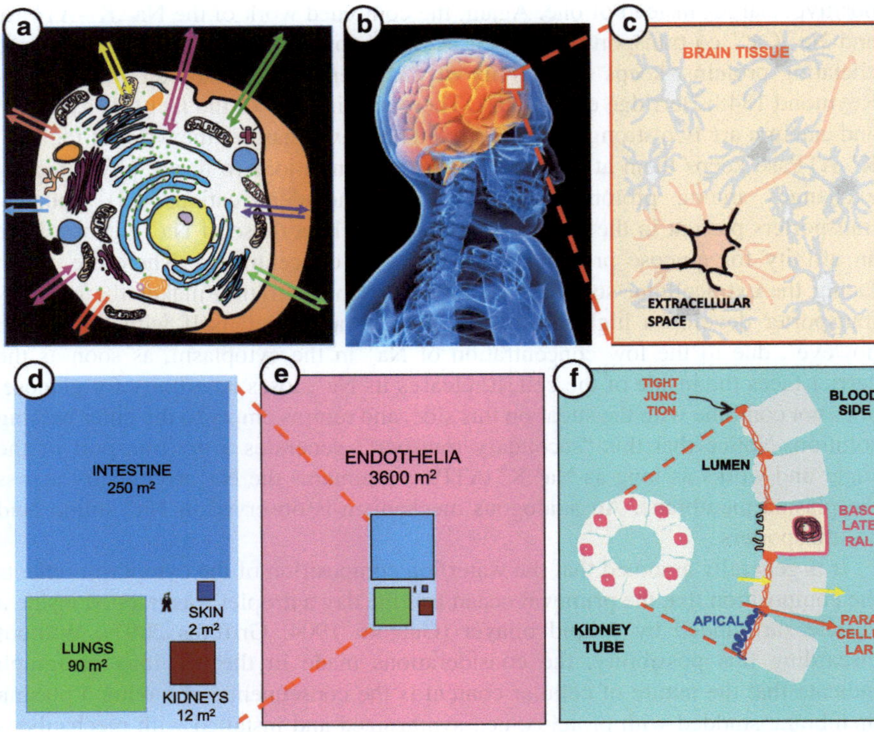

Fig. 8.2 Transporting epithelial phenotype. (**a**) A unicellular organism in the primitive ocean that acts as a reservoir; hence, it cannot be exhausted nor spoiled as a consequence of exchanges with the cell. (**b**) A cell in a metazoan has the "ocean" reduced to a narrow extracellular space (**c**, *black*) that, nevertheless, is able to act as a reservoir due to a circulatory apparatus that takes its fluid back and forth to and from transporting epithelia where the exchange with the environment takes place. (**d**) A few transporting epithelia scaled with the figure of a man to give an idea of the large area devoted to the exchange of substances with the environment. (**e**) Represents (**d**) at a smaller scale to compare the area of epithelia with those of endothelia (*pink*). (**f**) Schematic representation of five epithelial cells forming a segment of the wall of a kidney tube, showing the two fundamental differentiated features: *tight junctions* (*red*) that seal the intercellular pace and *apical/basolateral polarity*. The apical domain of the cell membrane (*blue*) is in contact with the lumen and the basolateral one (*magenta*) with the interstitial fluid on the blood side

this lumen running at the central axis of the tube, the tube is considered to be full of *external* milieu because it connects with the urinary tract, which interacts with the external milieu bathing the animal.

Therefore, the phenotype composed of a sealing element, the TJs, and an apical/basolateral polarity is the preferred structure implemented by higher metazoans to seclude a large portion of their bodies, e.g., the gastrointestinal space, the renal space, or the space within the genitalia, blood, and lymphatic fluids, or even to create a "castle-within-the-castle" effect as seen with the brain and its surrounding extracellular fluid. This suggests that polarization became an important requirement in the evolution from single cells to metazoans. Let's consider this suggestion.

Shortly after J. Ch. Skou introduced Na^+,K^+-ATPase (Skou 1957) (Fig. 8.1e), V. Koefoed-Johnsen and H. H. Ussing (Koefoed-Johnsen and Ussing 1958) (KJU) used this pump to develop a fertile working model that acted as a blueprint to explain active transport across most epithelia (Fig. 8.1f). Interestingly, KJU transformed a common cell into one of the transporting epithelia by proposing that the pump is expressed in a polarized manner on the basal domain of the plasma membrane (Fig. 8.1f). Significantly, further adaptations of the KJU model to other epithelia proposed that co- and counter-transporters are also expressed in a polarized manner (Fig. 8.1g).

8.5 The Polarized Distribution of Proteins in the Plasma Membrane

The generation of cell surface polarity for most membrane proteins involves sorting signals encoded in their amino acid sequence, trafficking routes that include apical or basolateral recycling endosomes, and interactions with epithelial-specific protein complexes such as AP-1B and clathrin, which can be regulated by small GTPases (Cereijido et al. 2003; Duffield et al. 2008; Bryant and Mostov 2008; Mellman and Nelson 2008). Early studies demonstrated that the Na^+-K^+-ATPase, composed of a catalytic subunit (α) and an accessory subunit (β), is assembled in the endoplasmic reticulum, sorted in the trans-Golgi network, and delivered directly to the basolateral membrane of epithelial cells (Caplan et al. 1986; Gottardi and Caplan 1993; Zurzolo and Rodríguez-Boulan 1993). Therefore, a basolateral signal was assumed to exist in the α-subunit of the Na^+-K^+-ATPase (Muth et al. 1998). Na^+-K^+-ATPase and H^+-K^+-ATPase are highly homologous ion pumps; yet in LLC-PK1 cells, the former is polarized to the basolateral domain, whereas the latter is localized to the apical plasma membrane. To identify the sorting signals of these ion pumps, the polarized expression of chimeric constructs of the α-subunit of the H^+-K^+-ATPase and the Na^+-K^+-ATPase was studied (Muth et al. 1998). Apical sorting information was recognized within the fourth transmembrane domain of the α-subunit of the H^+-K^+-ATPase that is sufficient to redirect the Na^+-K^+-ATPase from the basolateral to the apical surface of these cells (Dunbar et al. 2000). However, it remains unclear whether basolateral sorting information exists in the fourth transmembrane domain of the α-subunit of the Na^+-K^+-ATPase; thus, a nonconventional signal could be involved in the basolateral targeting of this pump (Dunbar and Caplan 2001). Moreover, efforts to elucidate the trafficking mechanism of newly synthesized Na^+-K^+-ATPase revealed that it is independent of AP-1B because the pump localizes to the basolateral surface in the μ1B-deficient cell line LLC-PK1 (Duffield et al. 2004) and in MDCK cells in which μ1B expression is suppressed via RNA interference (Gravotta et al. 2007).

Although the manner in which the Na^+-K^+-ATPase achieves polarized distribution remains mysterious, several clues have emerged on how this is accomplished. The first clue came from the observation that in MDCK cells, the pump is not expressed in the apical or basal domains but only at the lateral membrane (Fig. 8.3a) (Hammerton et al. 1991; Contreras et al. 1995a; Shoshani et al. 2005). The second clue was obtained when monolayers of MDCK cells were treated with EDTA to chelate Ca^{2+}. The cells detached from each other and took their own Na^+-K^+-ATPase pumps with them (Fig. 8.3c) (Contreras et al. 1995b), indicating that the fluorescent mark observed under control conditions (Fig. 8.3b) was formed by pumps on both neighboring cells (Cereijido et al. 2000). The third clue was that monolayers formed with cells belonging to different epithelia and even different animal species formed sealed TJs but did not express Na^+-K^+-ATPase at a given lateral border, unless both epithelial cells belonged to the same animal species (Fig. 8.3d) (Contreras et al. 1995a; Shoshani et al. 2005). The fourth clue was that the β-subunit of the Na^+-K^+-ATPase resembles an adhesion molecule: it has a short cytoplasmic domain, a single transmembrane domain, and a long and heavily glycosylated extracellular domain. Likewise, experiments implemented by the Schachner group (Antonicek et al. 1987; Antonicek and Schachner 1988) revealed that the adhesion molecule on glia (AMOG) is an isoform of the β-subunit of the Na^+-K^+-ATPase (Gloor et al. 1990). Finally, the structure of the Na^+-K^+-ATPase, with its three subunits, has been resolved by X-ray crystallography (Fig. 8.3e) (Morth et al. 2007; Ogawa et al. 2009). These clues indicated that the β-subunit is an adhesion molecule. In accordance with this attribute of the β-subunit, we have demonstrated that Chinese hamster ovary cells transfected with the canine β_1-subunit of Na^+-K^+-ATPase (CHO-β) increase their tendency to form aggregates (Fig. 8.3f) (Shoshani et al. 2005). Using cocultures of MDCK and CHO-β cells, we showed that the Na^+-K^+-ATPase of MDCK cells was now polarized to the lateral border even when the adjacent cell was of another species (Fig. 8.3g) (Shoshani et al. 2005). We also showed by a pulldown assay that the dog β_1-subunit could specifically bind to the soluble extracellular domain of the β_1-subunit of the same animal species and that β_1-subunits of neighboring epithelial cells interact directly with each other in vivo (FRET and Co-IP essays). In the crystal structure of the Na^+, K^+-ATPase, the β-subunit is mostly exposed toward the intercellular space (Morth et al. 2007; Ogawa et al. 2009). This position of the β-subunit would favor the β-β association between Na^+-K^+-ATPases of adjoining cells at the intercellular space (Fig. 8.3h).

In this respect, several studies in mammals have shown that polarized targeting of the Na^+-K^+-ATPase in transporting epithelial cells is related to the expression of specific β-isoforms. Basolateral targeting is related to the expression of the β_1- and β_3-isoforms, while apical targeting is related to the expression of the β_2-isoform (Burrow et al. 1999). Accordingly, it has been shown that in the human gastric adenocarcinoma cell line (HGT-1), the pump is localized to the apical membrane domain and constitutes the β_2-isoform. When the β_1-isoform is expressed in this cell line, the pump is delivered to the basolateral domain (Vagin et al. 2005). Furthermore, there is a wealth of information suggesting that the N-glycans of the

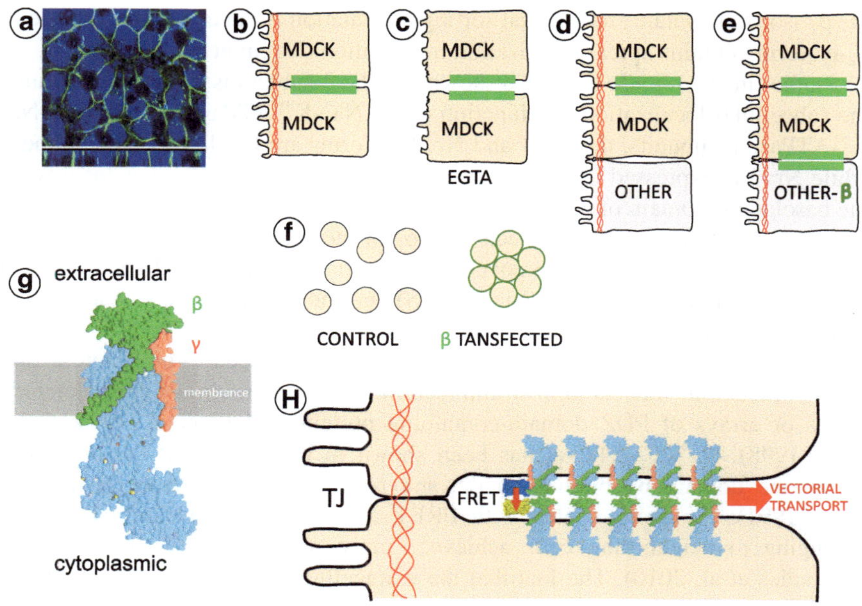

Fig. 8.3 Role of the β-subunit in the polarized distribution of Na^+,K^+-ATPase. (**a**) Normal (*upper panel*) and lateral (*lower panel*) view of a monolayer of MDCK cells, with the β-subunit stained in *green* and nuclei in *blue*, showing that the enzyme only occupies the lateral membrane. (**b**) Two neighboring cells expressing lateral Na^+,K^+-ATPase (*green*). (**c**) Ca^{2+} removal shows that detached cells separate, hauling their own enzyme. (**d**) A monolayer formed with a mixed population of MDCK cells plus "other" cell types (Ma104, LLC-PK_1, or CHO) expresses Na^+, K^+-ATPase on homotypic but not heterotypic lateral borders. (**e**) Mixed monolayers of MDCK and CHO cells transfected with the dog β-subunit (other-β) express this subunit in homotypic and heterotypic contacts. (**f**) The β-subunit of Na^+,K^+-ATPase has the characteristics of an adhesion molecule. Transfecting this subunit in CHO fibroblasts (β-transfected) confers adhesion properties. (**g**). Crystallography indicates that the enzyme on the plasma membrane exposes its α- (*light blue*), β- (*green*), and γ- (*orange*) subunits as depicted. The C-terminal lobe of the β-subunit is exposed to the intercellular space and shares structural similarities with proteins from the IgG-like superfamily. (**h**) Monolayers prepared with a mixed population of MDCK cells transfected with a β-subunit fused to a cyan fluorescent protein (*blue silhouette*) and MDCK cells transfected with a β-subunit fused to yellow fluorescent protein (*yellow silhouette*), as depicted in the first pair of Na^+,K^+-ATPases, show in a FRET assay that energy can be transferred from one β-subunit to the other, indicating that two β-subunits can interact directly from a distance less than 10 nm, thereby anchoring the whole enzyme at the cell membrane facing the intercellular space. Combining the crystal silhouette in G with the position of the enzyme in epifluorescence (**a**) shows that Na^+,K^+-ATPases can pump Na^+ toward the intercellular space. Given that this space is closed by TJs at its outermost end, Na^+ can only diffuse toward the basal (blood) side, in a net (vectorial) manner (*red arrow*)

β-subunit play an important role in the polarized sorting and trafficking of Na^+ pumps. Studies in which N-glycosylation of the Na^+,K^+-ATPase β-subunit was modified by either pharmacological or site-directed mutagenesis have shown that the β_1-isoform contains basolateral sorting information (Lian et al. 2006), while the β_2-isoform contains apical polarization information (Vagin et al. 2005, 2007).

Studies in *Drosophila* have also shown that the β-subunit is a key determinant of the subcellular localization and function of the Na^+-K^+-ATPase. Of the three Na^+-K^+-ATPase β-subunits, the Nrv1 and Nrv2 isoforms are localized to the epithelia, while Nrv3 is expressed in the nervous system. Interestingly, Nrv1 is localized to the basolateral domain of almost all epithelial cells; by contrast, Nrv2 is expressed at septate junctions (SJs) (the insects' analog of the vertebrate TJ) and co-localizes with the SJ marker coracle. Moreover, it has been shown that Nrv2 controls, by its extracellular domain, the functionality of SJ and the tracheal tube size in a pump-independent function (Paul and Palladino 2007).

The basic mechanism of lateral polarization is reinforced through interactions with extracellular ligands or with intracellular scaffolds such as cytoskeletal elements or arrays of PDZ domain-containing proteins (Mays et al. 1995; Cohen et al. 1998). Na^+-K^+-ATPase has been shown to be retained at the basolateral membrane domain by binding to the ankyrin-fodrin cytoskeleton (Hammerton et al. 1991). As demonstrated with FRET analysis, the two external moieties belonging to neighboring cells achieve a close proximity (<10 nm) (Padilla-Benavides et al. 2010). The fact that the extracellular moieties of β-subunits and that the α- and β-subunits have a high affinity for each other results in the anchoring of the Na^+-K^+-ATPase of both neighboring cells to the lateral border. This has profound functional consequences for the overall transport of Na^+ in the inward direction because Na^+ pumped into the intercellular space can only diffuse toward the blood side because the outermost end of this space is sealed by a TJ.

8.6 Apical Distribution of the Na^+,K^+-ATPase

In contrast to most epithelia, in the choroid plexus and the retinal pigment epithelium (RPE), both with neuroepithelial origins, the Na+,K(+)-ATPase is localized to the apical plasma membrane domain. The retinal pigment epithelium (RPE) separates the photoreceptors (rods and cones) from the choroid generating a proper ionic environment for the photoreceptor's function in the subretinal space. The apical domain of RPE is in intimate contact with the distal segments of rods and cones, generating a subretinal space. The subretinal K^+ and Na^+ concentrations that generate the dark current necessary for vision are under the control of the RPE cells. This crucial task requires that the Na^+,K^+-ATPase localize to the apical surface (Bok 1982; Frambach and Misfeldt 1983). In addition, RPE also transports a net amount of fluid in the apical to basal direction, creating a negative pressure that helps attach the neural retina to the RPE (Adijanto et al. 2009) in coordination with apical and basolateral bicarbonates and Cl^- transporters. Cl^- and Na^+ transport from the apical to the basolateral compartments drives fluid transport, which can be facilitated by aquaporin 1 (Strauss 2005). The retinal activity drastically changes the levels of Na^+, K^+, and CO_2 in the subretinal space, which are balanced by changes in the activity of various Na^+, K^+, Cl^-, and bicarbonate transporters, including the Na^+,K^+-ATPase. The retina produces a large amount of lactate that RPE cells transport through proton-coupled monocarboxylate transporters (MCTs). Thus, lactate is captured through the apical MCT1 and excluded from the basolateral MCT3 (Philp et al. 1998), resulting in the swelling of RPE cells (Philp et al. 1998; Hamann et al. 2003; Adijanto et al. 2009).

The choroid plexus epithelium (CPE) produces 600 ml of cerebrospinal fluid per day (Wright 1978), which is required for mechanical support, for communication, and as a pathway for waste removal and nutrient supply to the brain. The apical membrane of the CPE contacts the ventricle space and the basolateral surface, a highly vascularized compartment that provides a high blood supply. Na^+, Cl^- and HCO_3^- transports drive apical fluid secretion from the blood to the ventricles (Brown et al. 2004). This transport is possible due to the exquisite polarized structure of the choroid plexus epithelial cells that includes the expression of Na^+,K^+-ATPase in the apical membrane (Masuzawa et al. 1984; Siegel et al. 1984), apical and basolateral bicarbonate transporters, and active intracellular carbonic anhydrases that promote the intracellular formation of bicarbonate (Johanson et al. 2011), as in the kidney proximal tubule, Na^+, and bicarbonate gradients that constitute the driving force for the movement of fluid, which is facilitated by the aquaporin AQP1 (Wolburg and Paulus 2010).

The cellular mechanism responsible for the apical polarization of the Na^+,K^+-ATPase in both RPE and CPE is far from being elucidated. Nevertheless, as neuroepithelial cells, both tissues are expected to express the β_2-isoform that contains an apical signal and therefore would deliver the $\alpha\beta_2$ dimer to this domain. Evidently, the Na^+,K^+-ATPase does not carry a simple or classic basolateral sorting determinant. Moreover, plasma membrane proteins composed of two or more subunits, such as Na^+,K^+-ATPase, are interesting to study in terms of sorting mechanisms, as sorting signals could be present in one or more subunits that act hierarchically. Which subunit dominates the sorting of the Na^+,K^+-ATPase heterodimer is still not well understood. Remarkably, sorting signals can be interpreted depending on the cell type and be recognized by different components of the cellular sorting machinery (Philp et al. 2011; Castorino et al. 2011).

8.7 Na^+,K^+-ATPase Acts as a Receptor for the Hormone Ouabain Due to Its Polarized Distribution in the Plasma Membrane

It has recently been demonstrated that Na^+,K^+-ATPase is a receptor for the hormone ouabain. This hormone modulates signaling routes, regulating the survival of the cell in stressful situations, proliferation, and differentiation (Aizman and Aperia 2003; Xie 2006; Aperia 2007; Liu and Xie 2010).

We have recently demonstrated that ouabain regulates the two basic features of transporting epithelia: TJs and polarity (Larre et al. 2010, 2011). This was tested in monolayers of MDCK cells where the Na^+,K^+-ATPase is expressed at the plasma membrane in contact with the intercellular space. We observed that ouabain only acts when added from the basal side, i.e., the side that a hormone will reach in a living animal. Keep in mind that this is also the side where the enzyme is needed as a pump. Once it reaches the binding site, ouabain can elicit the following effects as a consequence of the polarized expression of the enzyme:

(a) Hormonal effects of ouabain on the TJ.
The effect of ouabain consists of an increase in the degree of tightness as gauged by TEER (transepithelial electrical resistance) and is mediated by the individual expression of specific claudin isoforms through specific signaling pathways. Thus, while the cell content of cln-1 is modulated through a route involving c-Src and ERK1/2, cln-4 is regulated through ERK1/2 but not c-Src. This specificity is also reflected in the modulation of specific permeabilities. Thus, while the ion flux through the TJ is controlled by c-Src and partially through ERK1/2, the flux of neutral 3 kDa dextran is regulated through ERK1/2 but not c-Src (Larre et al. 2010).

(b) Hormonal effects of ouabain on polarity.
The development of polarity was gauged through the development of a cilium in the middle of the apical domain of MDCK cells. Ciliogenesis is stimulated by ouabain through ERK1/2, provided cell proliferation is arrested. This can even be observed in single cells whose proliferation is prevented by plating a mixture of 5 % MDCK cells and 95 % NRK cells at saturating densities. Under this condition, single MDCK cells can be found completely surrounded by NRK cells. Because the expression of E-cadherin occurs at homotypic MDCK/MDCK contacts, but not in heterotypic ones, this molecule is absent when an MDCK cell is completely surrounded by NRK cells. Interestingly, in spite of the fact that under this situation these MDCK cells do not express E-cadherin, they nevertheless undergo ciliogenesis (Larre et al. 2011).

Another interesting case of ouabain-controlled polarity is the reduction in the expression of the apical Na^+/H^+ exchanger isoform 3 in proximal tubular epithelial cells (Cai et al. 2008). The hormone also reduces the basolateral expression of the pump itself and therefore the transepithelial transport of Na^+. This mechanism seems to operate in the proximal tubular epithelial cells and is impaired in salt-sensitive hypertension (Liu et al. 2011).

(c) Ciliar expression of claudin-2.
This isoform of claudin is an important component of the TJ, where it confers permeability to Na^+. However, we have observed that MDCK cells express claudin-2 at the cilium as well, where it cannot possibly play a role in permeation. Given that claudin-2 has an exquisite sensitivity to Na^+, it is conceivable that it could act as a sensor of Na^+ concentration in the fluid bathing the apical domain of the cell. The expression of claudin isoforms at the TJ and at the cilium follows independent kinetics, but both are modulated through ERK1/2 (Larre et al. 2010).

In summary, ouabain modulates the two fundamental differentiated features of transporting epithelial cells, TJs and polarity, thereby playing a crucial role in metazoan life. Ouabain binds to Na^+,K^+-ATPase once the cell stops proliferation and engages in differentiation, and therefore, it accesses its receptor from the basolateral domain of epithelial cells.

8.8 Na^+,K^+-ATPase Polarity and the Emergence of Metazoan Life

The ocean and the internal milieu act as reservoirs, the first because of its enormous size and the second because, in spite of having a very small size, its composition is kept constant by the extremely efficient quickness of the circulatory apparatus and the multitude of organs that participate in maintaining its homeostasis. All cells have to devote part of their efforts to housekeeping, a task that is enormously simplified when the internal milieu is constant (same pH, availability of nutrients, clearance of catabolites, etc.). That is why the basolateral domain of the plasma membrane of all cells in multicellular organisms is virtually identical. The apical domain, however, has to adapt to wildly different external milieus (gastric juice, bile, intestinal flora, tears, sea water, and glomerular filtrate that is progressively becoming urine), whose composition, pH, content of hormones, and other properties vary drastically throughout the day, with the diet, and has to coordinate peristalsis and movement of flagella. One may say that, having the house secured (i.e., the constancy of the internal milieu), epithelial cells can indulge in differentiation in much the same way as gene duplication, when one gene fulfills an indispensable fixed requirement and the other is "free" to diverge and explore other possibilities. This promoted the expression of diverse cell phenotypes, which enabled metazoans to progress in a mindboggling range of environments.

Acknowledgments This work was supported by research grants from CONACYT (National Research Council of México) and Secretaría de Ciencia, Tecnología e Inovación del Distrito Federal (SECITIDF). I. Larre had a postdoctoral fellowship from SECITIDF. We wish to acknowledge the efficient technical assistance of A. Castillo, L. Hinojosa, M. L. Roldán and C. Flores-Maldonado and Miss E. del Oso, Y. de Lorenz, E. Estrada, E. Méndez, and Javier Soriano.

References

Adijanto J, Banzon T, Jalickee S et al (2009) CO2-induced ion and fluid transport in human retinal pigment epithelium. J Gen Physiol 133:603–622. doi:10.1085/jgp.200810169

Aizman O, Aperia A (2003) Na, K-ATPase as a signal transducer. Ann N Y Acad Sci 986:489–496

Antonicek H, Persohn E, Schachner M (1987) Biochemical and functional characterization of a novel neuron-glia adhesion molecule that is involved in neuronal migration. J Cell Biol 104:1587–1595

Antonicek H, Schachner M (1988) The adhesion molecule on glia (AMOG) incorporated into lipid vesicles binds to subpopulations of neurons. J Neurosci 8:2961–2966

Aperia A (2007) New roles for an old enzyme: Na, K-ATPase emerges as an interesting drug target. J Intern Med 261:44–52. doi:10.1111/j.1365-2796.2006.01745.x

Berridge MJ, Bootman MD, Roderick HL (2003) Calcium: calcium signalling: dynamics, homeostasis and remodelling. Nat Rev Mol Cell Biol 4:517–529. doi:10.1038/nrm1155

Du Bois-Reymond E (1848) Untersuchungen Uber Thierische Elektricitat, 1st edn. G. Reimer, Berlin

Bok D (1982) Autoradiographic studies on the polarity of plasma membrane receptors in retinal pigment epithelial cells. In: Hollyfield JG (ed) The structure of the eye. North Holland, Elsevier, pp 247–256

Boulan ER, Sabatini DD (1978) Asymmetric budding of viruses in epithelial monolayers: a model system for study of epithelial polarity. Proc Natl Acad Sci U S A 75:5071–5075

Brown PD, Davies SL, Speake T, Millar ID (2004) Molecular mechanisms of cerebrospinal fluid production. Neuroscience 129:957–970. doi:10.1016/j.neuroscience.2004.07.003

Bryant DM, Mostov KE (2008) From cells to organs: building polarized tissue. Nat Rev Mol Cell Biol 9:887–901. doi:10.1038/nrm2523

Burrow CR, Devuyst O, Li X et al (1999) Expression of the beta2-subunit and apical localization of Na+-K+-ATPase in metanephric kidney. Am J Physiol 277:F391–F403

Cai H, Wu L, Qu W et al (2008) Regulation of apical NHE3 trafficking by ouabain-induced activation of the basolateral Na+-K+-ATPase receptor complex. Am J Physiol Cell Physiol 294:C555–C563. doi:10.1152/ajpcell.00475.2007

Caplan MJ, Anderson HC, Palade GE, Jamieson JD (1986) Intracellular sorting and polarized cell surface delivery of (Na+, K+)ATPase, an endogenous component of MDCK cell basolateral plasma membranes. Cell 46:623–631

Castorino JJ, Deborde S, Deora A et al (2011) Basolateral sorting signals regulating tissue-specific polarity of heteromeric monocarboxylate transporters in epithelia. Traffic 12:483–498. doi:10.1111/j.1600-0854.2010.01155.x

Cereijido M, Contreras RG, Shoshani L, García-Villegas MR (2003) Membrane targeting. Prog Biophys Mol Biol 81:81–115

Cereijido M, Robbins ES, Dolan WJ et al (1978) Polarized monolayers formed by epithelial cells on a permeable and translucent support. J Cell Biol 77:853–880

Cereijido M, Rotunno CA (1970) Introduction to the study of biological membranes. Gordon and Breach, New York, NY

Cereijido M, Rotunno CA (1971) Introduction to the study of biological membranes. Gordon and Breach, New York, NY

Cereijido M, Shoshani L, Contreras RG (2000) Molecular physiology and pathophysiology of tight junctions. I. Biogenesis of tight junctions and epithelial polarity. Am J Physiol Gastrointest Liver Physiol 279:G477–G482

Cohen AR, Woods DF, Marfatia SM et al (1998) Human CASK/LIN-2 binds syndecan-2 and protein 4.1 and localizes to the basolateral membrane of epithelial cells. J Cell Biol 142:129–138

Contreras RG, Lázaro A, Bolivar JJ et al (1995a) A novel type of cell-cell cooperation between epithelial cells. J Membr Biol 145:305–310

Contreras RG, Lázaro A, Mújica A et al (1995b) Ouabain resistance of the epithelial cell line (Ma104) is not due to lack of affinity of its pumps for the drug. J Membr Biol 145:295–300

Crane R, Miller D, Bihler I (1961) The restrictions on possible mechanisms of intestinal active transport of sugars. In: Kleinzeller A, Kotyk A (eds) Membrane transport and metabolism, 1st edn. Academic, London, pp 439–449

Crane RK, Forstner G, Eichholz A (1965) Studies on the mechanism of the intestinal absorption of sugars X. An effect of Na+ concentration on the apparent michaelis constants for intestinal sugar transport, in vitro. Biochim Biophys Acta 109:467–477. doi:10.1016/0926-6585(65)90172-X

Csaky TZ, Thale M (1960) Effect of ionic environment on intestinal sugar transport. J Physiol 151:59–65

de Donder T, Van Rysselberghe P (1936) L'affinité, 1st edn. Gauthier-Villars & cie, Paris

Duffield A, Caplan MJ, Muth TR (2008) Protein trafficking in polarized cells. Int Rev Cell Mol Biol 270:145–179. doi:10.1016/S1937-6448(08)01404-4

Duffield A, Fölsch H, Mellman I, Caplan MJ (2004) Sorting of H, K-ATPase beta-subunit in MDCK and LLC-PK cells is independent of mu 1B adaptin expression. Traffic 5:449–461. doi:10.1111/j.1398-9219.2004.00192.x

Dunbar LA, Aronson P, Caplan MJ (2000) A transmembrane segment determines the steady-state localization of an ion-transporting adenosine triphosphatase. J Cell Biol 148:769–778

Dunbar LA, Caplan MJ (2001) Ion pumps in polarized cells: sorting and regulation of the Na+, K+- and H+, K+-ATPases. J Biol Chem 276:29617–29620. doi:10.1074/jbc.R100023200

Frambach DA, Misfeldt DS (1983) Furosemide-sensitive Cl transport in embryonic chicken retinal pigment epithelium. Am J Physiol 244:F679–F685

Galeotti G (1904) Über die electromotorichen Kräfte, welch an der Oberfläche tierischer membranen bei der Berührung mit verschiedenen Elektrolyten Zustande kommen. Zellforsch Phys Chem 49:542–562

Gloor S, Antonicek H, Sweadner KJ et al (1990) The adhesion molecule on glia (AMOG) is a homologue of the beta subunit of the Na, K-ATPase. J Cell Biol 110:165–174

Gottardi CJ, Caplan MJ (1993) An ion-transporting ATPase encodes multiple apical localization signals. J Cell Biol 121:283–293

Gravotta D, Deora A, Perret E et al (2007) AP1B sorts basolateral proteins in recycling and biosynthetic routes of MDCK cells. Proc Natl Acad Sci U S A 104:1564–1569. doi:10.1073/pnas.0610700104

Griffiths G (2007) Cell evolution and the problem of membrane topology. Nat Rev Mol Cell Biol 8:1018. doi:10.1038/nrm2287

Hamann S, Kiilgaard JF, la Cour M et al (2003) Cotransport of H+, lactate, and H2O in porcine retinal pigment epithelial cells. Exp Eye Res 76:493–504

Hammerton RW, Krzeminski KA, Mays RW et al (1991) Mechanism for regulating cell surface distribution of Na+, K(+)-ATPase in polarized epithelial cells. Science 254:847–850

Johanson CE, Stopa EG, McMillan PN (2011) The blood-cerebrospinal fluid barrier: structure and functional significance. Methods Mol Biol 686:101–131. doi:10.1007/978-1-60761-938-3_4

Kedem O, Essig A (1965) Isotope flows and flux ratios in biological membranes. J Gen Physiol 48:1047–1070

Koefoed-Johnsen V, Ussing HH (1958) The nature of the frog skin potential. Acta Physiol Scand 42:298–308

Larre I, Lazaro A, Contreras RG, Balda MS, Matter K, Flores-Maldonado C, Ponce A, Flores-Benitez D, Rincon-Heredia R, Padilla-Benavides T, Castillo A, Shoshani L, Cereijido M (2010) Ouabain modulates epithelial cell tight junction. Proc Natl Acad Sci U S A 107 (25):11387–11392

Larre I, Castillo A, Flores-Maldonado C, Contreras RG, Galvan I, Muñoz-Estrada J, Cereijido M (2011) Ouabain modulates ciliogenesis in epithelial cells. Proc Natl Acad Sci U S A 108 (51):20591–20596

Lian W-N, Wu T-W, Dao R-L et al (2006) Deglycosylation of Na+/K+-ATPase causes the basolateral protein to undergo apical targeting in polarized hepatic cells. J Cell Sci 119:11–22. doi:10.1242/jcs.02706

Liu J, Xie Z-J (2010) The sodium pump and cardiotonic steroids-induced signal transduction protein kinases and calcium-signaling microdomain in regulation of transporter trafficking. Biochim Biophys Acta 1802:1237–1245. doi:10.1016/j.bbadis.2010.01.013

Liu J, Yan Y, Liu L et al (2011) Impairment of Na/K-ATPase signaling in renal proximal tubule contributes to Dahl salt-sensitive hypertension. J Biol Chem 286:22806–22813. doi:10.1074/jbc.M111.246249

Masuzawa T, Ohta T, Kawamura M, Nakahara N (1984) Immunohistochemical localization of Na +,K+-ATPase in the choroid plexus. Brain Res 302:357

Mays RW, Siemers KA, Fritz BA et al (1995) Hierarchy of mechanisms involved in generating Na/K-ATPase polarity in MDCK epithelial cells. J Cell Biol 130:1105–1115

Mellman I, Nelson WJ (2008) Coordinated protein sorting, targeting and distribution in polarized cells. Nat Rev Mol Cell Biol 9:833–845. doi:10.1038/nrm2525

Morth JP, Pedersen BP, Toustrup-Jensen MS et al (2007) Crystal structure of the sodium-potassium pump. Nature 450:1043–1049. doi:10.1038/nature06419

Muth TR, Gottardi CJ, Roush DL, Caplan MJ (1998) A basolateral sorting signal is encoded in the alpha-subunit of Na-K-ATPase. Am J Physiol 274:C688–C696

Ogawa H, Shinoda T, Cornelius F, Toyoshima C (2009) Crystal structure of the sodium-potassium pump (Na+, K+-ATPase) with bound potassium and ouabain. Proc Natl Acad Sci U S A 106:13742–13747. doi:10.1073/pnas.0907054106

Onsager L (1967) Thermodynamics and some molecular aspects of biology. In: Quarton GC, Melinchick T, Schmitt FO (eds) The neurosciences. Rockefeller University Press, New York, NY, p 75

Padilla-Benavides T, Roldán ML et al (2010) The polarized distribution of Na+, K+-ATPase: role of the interaction between {beta} subunits. Mol Biol Cell 21:2217. doi:10.1091/mbc.E10-01-0081

Paul S, Palladino M (2007) A pump-independent function of the Na, K-ATPase is required for epithelial junction function and tracheal tube-size control. Development 134:147

Philp N, Shoshani L, Cereijido M, Rodriguez-Boulan E (2011) Epithelial domains. In: Nabi IR (ed) Cellular domains. John Wiley & Sons, Hoboken, NJ, pp 1–31

Philp NJ, Yoon H, Grollman EF (1998) Monocarboxylate transporter MCT1 is located in the apical membrane and MCT3 in the basal membrane of rat RPE. Am J Physiol 274:R1824–R1828

Schultz SG, Curran PF (1970) Coupled transport of sodium and organic solutes. Physiol Rev 50:637–718

Shoshani L, Contreras RG, Roldán ML et al (2005) The polarized expression of Na+, K+-ATPase in epithelia depends on the association between beta-subunits located in neighboring cells. Mol Biol Cell 16:1071–1081. doi:10.1091/mbc.E04-03-0267

Siegel GJ, Holm C, Schreiber JH et al (1984) Purification of mouse brain (Na+ + K+)-ATPase catalytic unit, characterization of antiserum, and immunocytochemical localization in cerebellum, choroid plexus, and kidney. J Histochem Cytochem 32:1309–1318

Skou JC (1957) The influence of some cations on an adenosine triphosphatase from peripheral nerves. Biochim Biophys Acta 23:394–401

Strauss O (2005) The retinal pigment epithelium in visual function. Physiol Rev 85:845–881. doi:10.1152/physrev.00021.2004

Vagin O, Turdikulova S, Sachs G (2005) Recombinant addition of N-glycosylation sites to the basolateral Na, K-ATPase beta1 subunit results in its clustering in caveolae and apical sorting in HGT-1 cells. J Biol Chem 280:43159–43167. doi:10.1074/jbc.M508262200

Vagin O, Turdikulova S, Tokhtaeva E (2007) Polarized membrane distribution of potassium-dependent ion pumps in epithelial cells: different roles of the N-glycans of their beta subunits. Cell Biochem Biophys 47:376–391

Wolburg H, Paulus W (2010) Choroid plexus: biology and pathology. Acta Neuropathol 119:75–88. doi:10.1007/s00401-009-0627-8

Wright EM (1978) Transport processes in the formation of the cerebrospinal fluid. Rev Physiol Biochem Pharmacol 83:3–34

Xie Z (2006) Membrane transporters and signal transduction. Cell Mol Biol (Noisy-le-Grand) 52:1–2

Zurzolo C, Rodríguez-Boulan E (1993) Delivery of Na+, K(+)-ATPase in polarized epithelial cells. Science 260:550–552, author reply 554–6

Chapter 9
Endothelial Cell Polarization During Lumen Formation, Tubulogenesis, and Vessel Maturation in 3D Extracellular Matrices

George E. Davis, Katherine R. Speichinger, Pieter R. Norden, Dae Joong Kim, and Stephanie L.K. Bowers

Abstract A fundamental event during vascular morphogenesis is the formation of endothelial cell (EC) lumens and tube networks in 3D extracellular matrices (ECM). This morphogenic step leads to rapid EC polarization in that apical and basal membrane domains are created, with fluid contacting the developing apical membrane surface, while ECM is in contact with the basal membrane surface. Interestingly, it has been difficult to identify specific apical vs. basal markers in ECs. This suggests that their ability to polarize is distinct from other cell types such as epithelial cells which have more extensive cell–cell junctional contacts and more specialized apical domains with exocrine abilities where substantial fluids can be generated and released apically. One key difference is that ECs in contact with blood must be flat in order to properly function and respond to fluid flow forces. Interestingly, our recent studies reveal that tubulins and, in particular, modified tubulins such as acetylated tubulin are distributed in a subapical domain that supports the developing apical membrane surface during EC lumen formation and tube maintenance. Disruption of the tubulin cytoskeleton can lead to failure of tube formation and rapid tube collapse suggesting that the polarized shape of EC-lined tubes in 3D extracellular matrices is strongly dependent on the tubulin cytoskeleton rather than the actin cytoskeleton (although both are clearly important). EC tube polarization is also strongly affected by the recruitment of mural cells (i.e., pericytes) which, in conjunction with ECs, leads to critical extracellular matrix remodeling including deposition of the vascular basement membrane matrix which is laid down on the basal surface in between the basally distributed pericytes which co-contribute this matrix along with ECs. In conclusion, the mechanisms that stimulate EC tubulogenesis establish apical–basal cytoskeletal and membrane polarity during morphogenic events in 3D matrices. EC polarization is further enhanced by the recruitment of mural cells such as pericytes along the abluminal

G.E. Davis (✉) • K.R. Speichinger • P.R. Norden • D.J. Kim • S.L.K. Bowers
Department of Medical Pharmacology and Physiology, Dalton Cardiovascular Sciences Center, University of Missouri School of Medicine, MA415 Medical Sciences Building, Columbia, MO 65212, USA
e-mail: davisgeo@health.missouri.edu

K. Ebnet (ed.), *Cell Polarity 1*, DOI 10.1007/978-3-319-14463-4_9

EC surface which stimulates vascular basement membrane assembly along this basal surface to further contribute to EC polarity and maintenance of EC-lined tubes. At this point, flow forces and components of blood together will provide additional polarizing information to the EC apical surface and contribute to vascular tube stabilization and the development of EC quiescence in the mature vasculature.

Keywords Endothelial cells • Extracellular matrix • Lumen formation • Pericytes • Polarization • Tubulogenesis • Vascular basement membrane matrix • Vascular morphogenesis

9.1 Introduction

In this chapter, we will discuss the development of EC polarity and its key relationship to vascular morphogenesis (Davis et al. 2011; Sacharidou et al. 2012; Xu and Cleaver 2011; Lizama and Zovein 2013), where EC-lined tubes are created in 3D extracellular matrix (ECM) environments (Senger and Davis 2011) (Figs. 9.1 and 9.2). We will address what is known concerning how EC polarity is established and maintained during these events and discuss how mural cells and, in particular, pericytes affect these processes (Fig. 9.2). One of the most important functions of ECs is to form tube networks (Davis et al. 2011; Sacharidou et al. 2012). When this occurs, EC polarization is established with an apical surface exposed to fluid and the basal surface exposed to ECM (Figs. 9.1 and 9.2). Other cell types such as mural cells do not form tubes but migrate and invade in 3D matrices as single cells (Stratman and Davis 2012; Armulik et al. 2011). In contrast, ECs form lumens and tubes and assemble together as multicellular structures in tubular networks which then attract other cell types such as pericytes (Davis et al. 2011) (Figs. 9.1, 9.2, and 9.3). In fact, during EC–pericyte tube co-assembly, pericytes recruit as single cells to developing EC tubes in 3D matrices (Stratman and Davis 2012). We recently demonstrated that pericyte motility in 3D matrices under serum-free defined conditions depends on the presence of ECs and that this occurs due to EC-derived PDGF-BB and HB-EGF, which affects both motility and proliferation of pericytes during these events (Stratman and Davis 2012; Stratman et al. 2010). Pericyte recruitment along EC tubes occurs exclusively along the EC abluminal surface, and this recruitment results in vascular basement membrane assembly along this polarized EC basal surface (Stratman and Davis 2012; Stratman et al. 2009a) (Fig. 9.4). Thus, the intrinsic polarity of capillary tubes is enhanced and perhaps maintained by this polarized recruitment of pericytes (Figs. 9.2, 9.3, and 9.4). This is followed by the polarized deposition of vascular basement membrane matrix proteins along this abluminal surface and in between the ECs and recruited pericytes (Stratman et al. 2009a; Stratman and Davis 2012) (Fig. 9.4).

We have demonstrated using multiple systems in our laboratory in either collagen or fibrin matrices that ECs create tube networks and pericytes recruit to these tubes on the abluminal surface where basement membrane matrices form (Davis

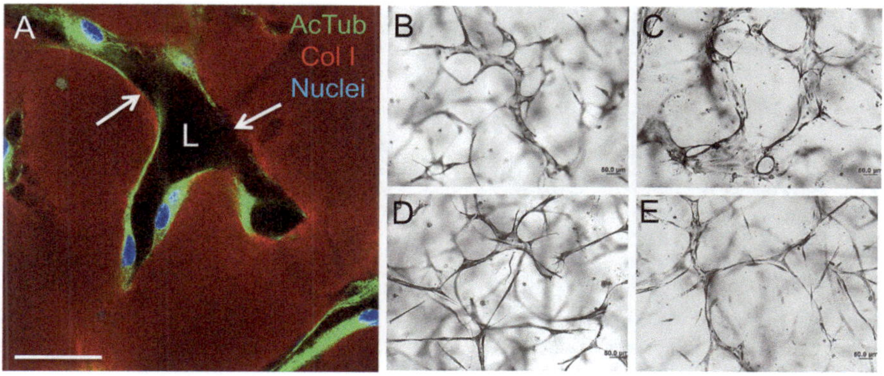

Fig. 9.1 Critical functional ability of endothelial cells to form cell-lined tubes in the absence or presence of pericytes in 3D extracellular matrices. (**A**) Human ECs were seeded as single cells in 3D collagen matrices and after 24 h were fixed and stained with the indicated antigens including acetylated tubulin (*AcTub*) and collagen type I (*Col I*) or were stained with a nuclear dye. *Arrows* indicate the border of the polarized EC tube within a vascular guidance tunnel space created through proteolysis of collagen matrices during the morphogenic process. *L* indicates EC lumen. Bar equals 50 μm. (**B–E**) ECs were seeded as single cells and form multicellular networks of tubes in 3D collagen matrices in the absence (**B** and **C**) or presence of pericytes (**D** and **E**) over 3 days (**B** and **D**), or 5 days (**c** and **e**) of culture. The hematopoietic stem cell cytokines, SCF, IL-3, and SDF-1α, in conjunction with FGF-2 were added to stimulate the tube morphogenic process. Note that the EC tube diameter progressively becomes wider from 3 to 5 days in EC-only cultures while they remain narrow in the EC–pericyte co-cultures. Bar equals 50 μm

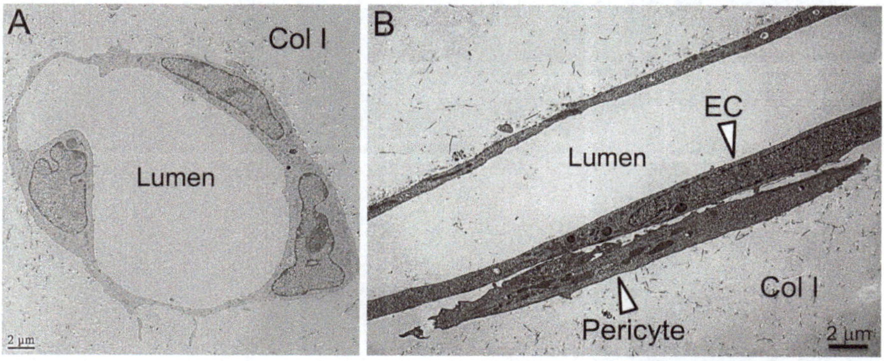

Fig. 9.2 Transmission electron microscopy of EC tubulogenesis and EC–pericyte tube co-assembly in 3D collagen matrices reveals polarized EC-lined tubes with an apical domain exposed to fluid and a basal surface in contact with ECM or recruited pericytes. EC-only (**A**) or EC–pericyte co-cultures (**B**) were established in 3D collagen matrices using hematopoietic stem cell cytokines with FGF-2 and were processed for transmission electron microscopy. The EC lumen space is indicated and *arrowheads* indicate ECs or pericytes in the co-culture. Collagen type I matrix is indicated as Col I. Bar equals 2 μm

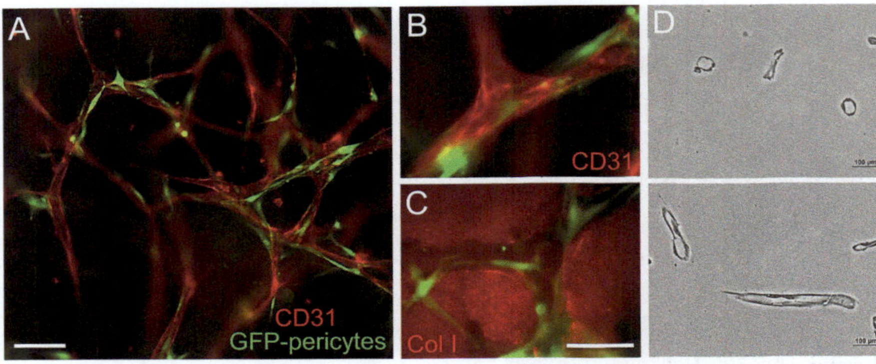

Fig. 9.3 Polarized recruitment of pericytes to the abluminal surface of EC-lined tubes within vascular guidance tunnels. EC–pericyte co-cultures were established and allowed to assemble in collagen matrices over 3 days using hematopoietic stem cell cytokines and FGF-2. After this time, cultures were fixed and stained with CD31 to label ECs while pericytes express GFP (**A** and **B**). (**C**) Cultures were stained with antibodies to collagen type I (*Col I*) to reveal vascular guidance tunnels which are created during EC tube formation and pericytes are recruited to the abluminal surface within these tunnel spaces. (**D**) Cross sections from plastic embedded cultures reveal lumen and tube structures throughout the 3D collagen matrices. Bar equals 50 μm

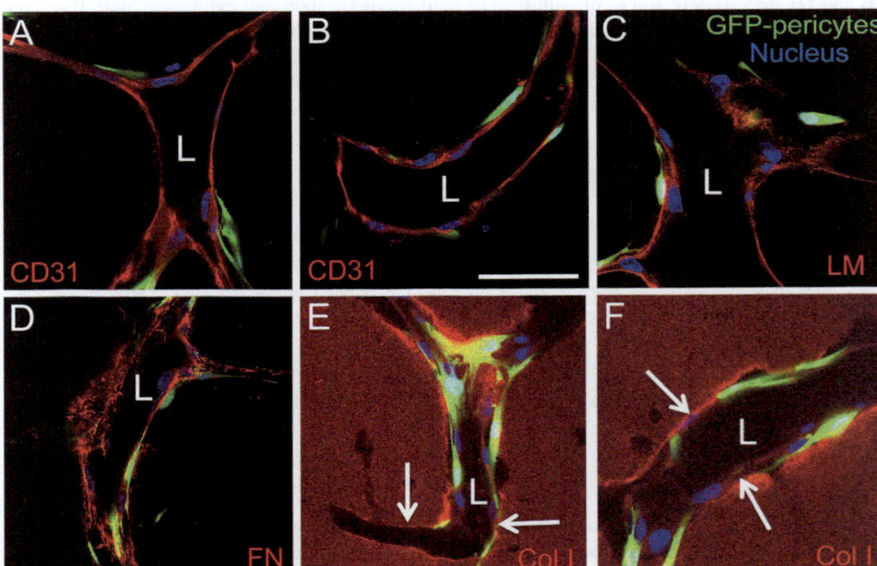

Fig. 9.4 Confocal microscopy of EC–pericyte tube co-assembly reveals polarized vascular basement membrane deposition along the EC tube abluminal surface in between ECs and recruited pericytes. EC–pericyte co-cultures were fixed and immunostained with the indicated antibodies. (**A** and **B**) CD31; (**C**) laminin (*LM*); (**D**) fibronectin (*FN*); (**E** and **F**) collagen type I (*Col I*) which reveals the vascular guidance tunnels in which EC tubes and the vascular basement membrane are embedded. Pericytes were labeled with GFP. *L* indicates EC lumen. *Arrows* indicate the border of vascular guidance tunnels in which EC tubes are embedded. Bar equals 50 μm

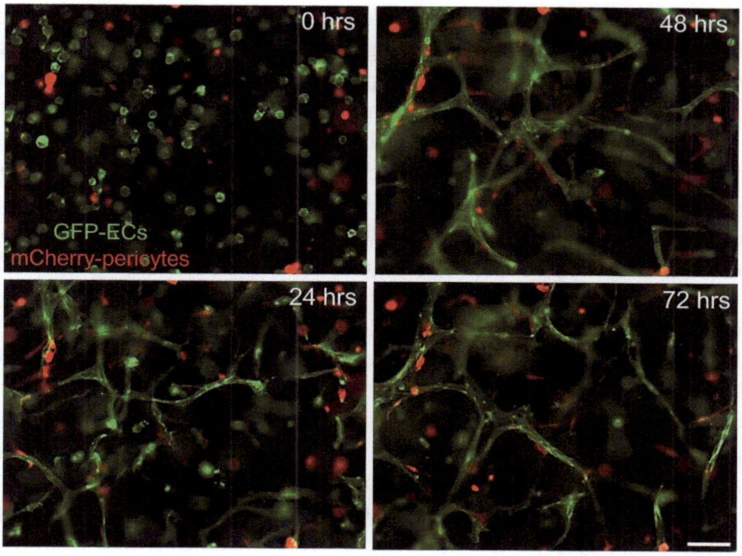

Fig. 9.5 Time-lapse microscopy of EC–pericyte tube co-assembly in 3D fibrin matrices in response to recombinant hematopoietic stem cell cytokines and FGF-2. The ECs are labeled with membrane GFP while the pericytes were labeled with mCherry. A representative field is shown from 0 to 72 h of culture, which reveals marked EC tube morphogenesis and recruitment of pericytes to the EC tube abluminal surface. Bar equals 100 μm

et al. 2011; Stratman et al. 2009a, 2010; Smith et al. 2013) (Figs. 9.1, 9.2, 9.3, 9.4, and 9.5). In each case, EC tubes are polarized apically with their apical membrane in contact with fluid (and in preparation for the onset of flow forces), and their basal surfaces are in contact with provisional ECM (while they are forming) and then in contact with remodeled ECM when ECs and recruited pericytes work together to form polarized basement membranes underlying the abluminal surface (Stratman and Davis 2012; Senger and Davis 2011). Importantly, both ECs and mural cells are polarized with respect to each other and the ECM that they are exposed to during vascular morphogenesis and maturation events. Many important questions remain concerning the molecular basis for these processes, and it is clear that a combination of in vitro and in vivo models will be necessary to elucidate this critical biology, which is fundamental to our ability to appropriately develop therapeutic strategies to regulate the vasculature to treat human disease.

9.2 Development of Polarity During EC Tube Formation in 3D Matrices

The ability of ECs to form tubes creates cell polarization during vascular morphogenesis in 3D extracellular matrices (Davis et al. 2011; Senger and Davis 2011; Xu and Cleaver 2011; Lizama and Zovein 2013). This process rapidly creates both apical and basal membrane surfaces which are necessary for this intrinsic polarity (Bryant and Mostov 2008). A fascinating aspect of the tubulogenic process is that ECs locally degrade the ECM to create physical spaces in the matrix in which the tubes reside to facilitate the polarization process (Saunders et al. 2006; Stratman et al. 2009b). We have termed these physical spaces vascular guidance tunnels (Stratman et al. 2009b), which support EC tubulogenesis and allow for rapid tube remodeling events once the tunnels have been formed (Figs. 9.1 and 9.4). Our work has shown that these tunnels are created through proteolysis by the cell surface-expressed membrane type 1-matrix metalloproteinase (MT1-MMP) (or MMP-14) (Stratman et al. 2009b; Saunders et al. 2006) which acts in conjunction with integrin signaling and downstream Cdc42- and Rac-dependent events to create EC lumens and tube networks. Blockade of any of these molecules leads to absence of EC tubes as well as vascular guidance tunnels showing that the creation of both structures occurs in a coordinated fashion (Stratman et al. 2009b). Importantly, the tunnel spaces allow for cell adhesion events on the EC basal surface to occur during tube formation, which directly contributes to polarization of the apical vs. basal surfaces while tubes are remodeled and stabilized. It is also very interesting that these processes occur very rapidly (within hours to a few days); this is necessary in order to establish the vasculature for embryonic viability. The highly dynamic nature of these processes has been observed using real-time movies in vitro and in vivo, suggesting that dramatic remodeling events are central to the process (Saunders et al. 2006; Kamei et al. 2006; Iruela-Arispe and Davis 2009; Stratman et al. 2009b) (Fig. 9.5). Thus, many views of tube morphogenesis, such as the role of stable junctional contacts that are observed during epithelial tubulogenesis (Bryant and Mostov 2008; Bryant et al. 2010; Goldstein and Macara 2007; Martin-Belmonte and Mostov 2008), might not be operative during this much more rapid EC process.

There appear to be two major ways that apical membrane surfaces are generated during EC tubulogenesis (i.e., intracellular vacuolation within single cells or membrane invagination between two cells) (Davis and Camarillo 1996; Davis et al. 2002; Kamei et al. 2006; Bayless and Davis 2002; Bayless et al. 2000), but in both cases, a transfer of basal membranes to an apical position is occurring (Sacharidou et al. 2012) (Fig. 9.6). This has been discussed in detail in previous reviews from our laboratory (Sacharidou et al. 2012; Davis et al. 2011). Recent data suggests that the signaling molecules and pathways that control EC lumen formation appear to be controlling this membrane transfer process in either case. We have previously reported that Cdc42 and Rac1 are expressed in apical membranes and that membranes which appear to be enriched in these GTPases can be shown to be targeting apically during EC lumen formation in 3D matrices (Bayless and Davis 2002) (Figs. 9.6 and 9.7). Both of these GTPases are critical for the EC tubulogenic process (Bayless and Davis 2002; Koh et al. 2008a, 2009; Sacharidou et al. 2010). Furthermore, our data shows a fundamental role for integrins, MT1-MMP, protein kinase C epsilon, Src family kinases (i.e., Src, Yes, Fyn), the Cdc42 effectors, Pak-2 and Pak-4, as well as downstream kinases such as Raf, Mek, and Erk (Koh et al. 2008a, 2009; Davis et al. 2011). In addition, we have demonstrated a

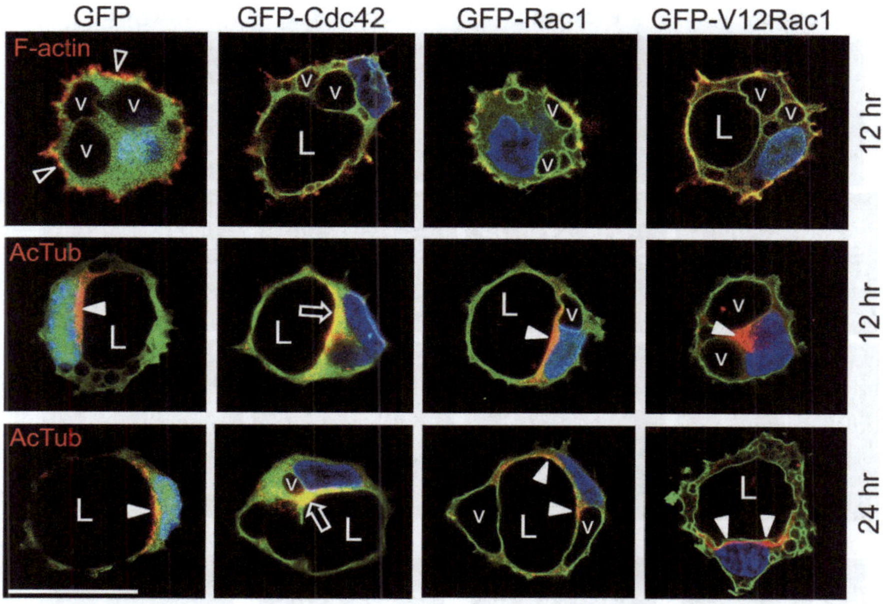

Fig. 9.6 Polarized trafficking of Rac1- and Cdc42-enriched membrane vacuoles to the developing apical surface along subapically polarized microtubules containing acetylated tubulin during EC lumen formation in 3D matrices. ECs were induced to express GFP, GFP-Cdc42, GFP-Rac1, or GFP-V12Rac1 (constitutively active Rac1). Cultures were established in 3D collagen matrices and at 12 or 24 h were fixed and stained for acetylated tubulin (*AcTub*) or for F-actin with phalloidin. Note that F-actin is concentrated in a basal position (*black arrowhead*) while acetylated tubulin shows a strong subapical polarization (*white arrowheads*). There is strong co-localization of acetylated tubulin with GFP-Cdc42 (*black arrows*). *L* indicates EC lumen and *v* equals intracellular vacuoles. Bar equals 50 μm

concurrent requirement for polarity signaling involving Par6b, Par3 (Koh et al. 2008a), and the Par3 binding proteins, JamB and JamC (Sacharidou et al. 2010; Ebnet et al. 2003). Par6b is also a direct Cdc42 effector (Joberty et al. 2000) which plays a role along with Pak-2 and Pak-4 during EC lumen formation (Koh et al. 2008a). Thus, there is a direct link between Cdc42 and polarity signaling pathways during EC lumen formation (Koh et al. 2008a; Zovein et al. 2010; Sacharidou et al. 2010, 2012; Davis et al. 2011). In addition, work by others has revealed a role for VE-cadherin during these processes, and this may occur through its ability to also interact with Par3 (Lampugnani et al. 2010). Interestingly, in unpublished work, we have shown that siRNA suppression leads to blockade of EC lumen formation even at a single cell level prior to them assembling with other ECs, showing a non-junctional role for VE-cadherin. We reported similar findings for JamB and JamC (Sacharidou et al. 2010), so it is likely that these three adhesion molecules coordinate with integrins, MT1-MMP, polarity proteins, Rho GTPase signaling, and activation of kinase cascades to control EC tube formation (Davis et al. 2011; Sacharidou et al. 2012). In support of these

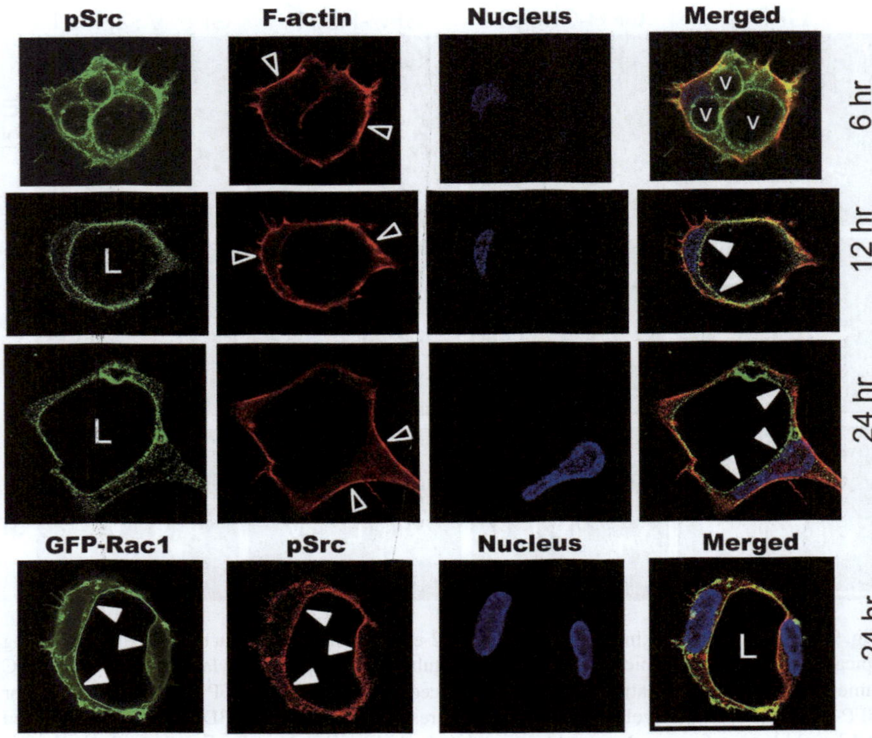

Fig. 9.7 Activated Src family kinases target the developing EC apical surface to control lumen and tube formation in 3D matrices. ECs were seeded in 3D collagen matrices and at the indicated times were stained with phospho-Src antibodies or with phalloidin to label F-actin. Note that pSrc staining shows a strong localization to intracellular vacuoles (observed at 6 h) and the developing apical surface (observed at 12 and 24 h) (*white arrowheads*), while F-actin is localized in a predominant basal position (*black arrowheads*). The *lower panels* show ECs that were induced to express GFP-Rac1 and then were stained with pSrc which demonstrates a partial overlap of GFP-Rac1 with apically targeted pSrc during EC lumen formation in 3D matrices. *L* indicates EC lumen and *v* indicates intracellular vacuoles. Bar equals 50 μm

concepts, previous work revealed a role for JamA in the directional guidance of single neutrophils during inflammatory responses (Cera et al. 2009).

9.3 Cytoskeletal Polarization During EC Lumen and Tube Formation in 3D Matrices

Our previous studies address which major cytoskeletal structures are critical for the shape and branching pattern of EC-lined tubes in 3D matrices (Bayless and Davis 2004; Kim et al. 2013). When networks of EC tubes were treated with cytoskeletal disrupting agents including those that caused actin disassembly vs. microtubule disassembly, it became apparent that only microtubule disruption led to rapid EC tube collapse (Bayless and Davis 2004). This important data suggested that the major cytoskeletal structure that supports what we observe as EC-lined tubes are secondary to microtubules (Bayless and Davis 2004; Kim et al. 2013). If these agents were added at the onset of EC morphogenesis, both types of agents completely disrupted the ability of ECs to undergo lumen and tube formation (Davis and Camarillo 1996).

To address these issues in more molecular detail, we recently investigated how microtubules affected the ability of ECs to form tubes and then maintain these structures once they have formed. We identified the microtubule tip complex proteins, EB1, p150glued, and Clasp1, to be critical in EC lumen formation as they control EC microtubule assembly and posttranslational modifications such as tubulin acetylation (Kim et al. 2013). siRNA suppression of these molecules blocked tube formation and interfered with tubulin acetylation, a modification that is strongly increased during lumen formation but which also directly correlated with the time course of EC tube formation (Kim et al. 2013). Their influence is likely mediated in part through modulating the activity of the tubulin deacetylases, Sirt2 and HDAC6 (a known binding partner of EB1). siRNA suppression of these deacetylases increases lumen formation, while increasing their expression blocks the process (Kim et al. 2013).

9.4 Polarized Trafficking of Intracellular Vacuoles Along Microtubule Tracks with Acetylated Tubulin to Direct and Support the Developing Apical Membrane Surface During EC Lumen Formation

Immunofluorescent staining and confocal microscopy during a time course of EC lumen formation suggest that acetylated tubulin is oriented in a subapical location which supports the developing apical membrane surface during lumen formation (Kim et al. 2013) (Fig. 9.6). A key point is that this modified tubulin is observed to surround intracellular vacuoles (i.e., which are similar to macropinosomes) to facilitate membrane transport from a basal to apical position during this process (Fig. 9.6). Thus, our data suggests that microtubule arrays (or tracks) which emanate from the centrosome are playing a critical role in vacuole trafficking to create the apical membrane surface which is required to create EC polarization secondary to lumen formation. We have previously reported that GFP-Rac1 or GFP-Cdc42 could be shown to target to intracellular vacuole membranes and that these vacuoles move toward the apical membrane and fuse to stimulate lumen expansion (Bayless and Davis 2002; Davis et al. 2002; Kamei et al. 2006) (Fig. 9.6). Our new data suggests that microtubules and, in particular, modified tubulins such as acetylated tubulin appear to be critical for directing this membrane transport event (Fig. 9.6). Cdc42 is known to affect centrosome and nuclear positioning as well as directional motility and cell invasion of 3D matrices (Cau and Hall 2005; Etienne-Manneville and Hall 2001; Gomes et al. 2005; Fisher et al. 2009) and is required for EC lumen formation which we first demonstrated more than a decade ago (Bayless and Davis 2002). GFP-Cdc42 appears in particular to cup or support vacuoles (Davis et al. 2007) as well as the perinuclear subapical membrane area and is strongly co-localized with acetylated tubulin in this polarized region of ECs during lumen formation (Fig. 9.6).

Studies from our laboratory have demonstrated that EC lumen formation and invasion of single cells in 3D matrices require co-dependent Cdc42 and MT1-MMP signaling events that allows for polarized morphogenic and invasive cellular events (Sacharidou et al. 2010; Fisher et al. 2009; Stratman et al. 2009b). In both of these distinct cases, the Cdc42 and MT1-MMP co-dependent processes require a concomitant kinase activation cascade (Sacharidou et al. 2010; Fisher et al. 2009). During EC lumen formation, key required kinases include PKC epsilon, Src family kinases, Pak-2, Pak-4, Raf, Mek, and Erk (Koh et al. 2008a, 2009). Interestingly, increased expression of PKC epsilon leads to marked Src phosphorylation which then activates Pak kinases, followed by activation of Raf kinases (B-Raf and C-Raf) (Koh et al. 2009). Raf activation further leads to downstream Mek and Erk phosphorylation events which are necessary for EC lumen formation (Koh et al. 2009). To address whether there was evidence for polarized kinase activation during this process, immunostaining was performed which identified a strong apical targeting of activated Src kinases. Activated phospho-Src is observed in early intracellular vacuoles as they pinocytose from the basal surface and then is highly localized in the apical membranes as lumen structures develop (Fig. 9.7). Blockade of Src kinases using chemical inhibitors or the protein inhibitor, Csk, markedly interferes with EC lumen formation in 3D matrices (Koh et al. 2009). Overall, this suggests that Cdc42-, Rac1-, MT1-MMP-, and integrin-dependent signaling leads to Src activation which stimulates membrane pinocytosis (from the basal surface) and trafficking of intracellular vacuole membranes along microtubules (Figs. 9.6 and 9.7). These vacuoles then fuse to the EC apical membrane surface to expand the lumen space during the tubulogenic process (Fig. 9.7).

9.5 Fundamental Role for Pericytes in Regulating and Maintaining EC Luminal Polarity in Capillary Tube Networks in 3D Matrices

For many years, pericytes were defined by their anatomical presence around small capillary vessels and by the presence of immunomarkers such as NG2 proteoglycan, 3G5 antigen, or PDGFRβ (Armulik et al. 2011; Stratman and Davis 2012). Past work suggested that small vessels were more susceptible to regression when they were devoid of pericytes, raising the possibility that they promoted the stability of vessels (Benjamin et al. 1998, 1999). Work from our laboratory has begun to provide a functional definition for pericytes based on their behaviors in defined in vitro model systems (Saunders et al. 2006; Stratman et al. 2009a, 2010; Stratman and Davis 2012; Smith et al. 2013). We have shown that pericytes (1) invade and recruit to small blood vessel networks in 3D matrices in an EC-dependent manner secondary to EC production of PDGF-BB and HB-EGF (Stratman et al. 2010), (2) proliferate in response to EC-derived PDGF-BB and HB-EGF during these events (in contrast to ECs which do not proliferate during tube assembly or pericyte-induced tube maturation) (Stratman et al. 2010), (3) recruit to the abluminal surface of EC-lined tubes leading to vascular basement membrane matrix assembly which results in more narrowed and extended tubes in 3D matrices (Stratman et al. 2009a; Smith et al. 2013), (4) actively migrate along the abluminal EC tube surface to work in conjunction with actively motile ECs to assemble the vascular basement membrane (Stratman et al. 2009a; Smith et al. 2013), and (5) protect EC tubes from pro-regressive proteinases including MMP-1 and MMP-10 by their production of the MMP and ADAM proteinase inhibitor, TIMP-3 (Saunders et al. 2006). Thus, pericytes have many important functions that can now be investigated in detail using in vitro models of EC–pericyte tube co-assembly in 3D matrices under serum-free defined conditions (Stratman and Davis 2012) (Figs. 9.1, 9.2, 9.3, 9.4, and 9.5).

Pericytes modulate EC polarity during tube morphogenesis by affecting the outcome of tubulogenesis in that tubes continue to widen and become less branched and unstable in EC only compared to EC–pericyte co-cultures, where tubes are narrow, highly elongated, and protected against pro-regressive stimuli (Stratman and Davis 2012) (Fig. 9.1). During tubulogenesis, EC-only tubes show polarity as indicated in that the EC apical surface is exposed to fluid and the basal surface is exposed to ECM such as collagen or fibrin. Under these circumstances, morphogenesis can continue because collagen and fibrin exposure promotes tubulogenesis and sprouting (Senger and Davis 2011), while pericyte recruitment leads to ECM remodeling and deposition of vascular basement membrane matrix (Fig. 9.4) which also contains TIMP-3, an MMP, and MT1-MMP inhibitor, which reduces and prevents further morphogenic processes (Stratman and Davis 2012; Senger and Davis 2011; Saunders et al. 2006). Furthermore, the vascular basement membrane matrix appears to have additional inhibitory signals such as from laminin isoforms (Davis and Senger 2005) which results in changes in differential integrin signaling (activating integrins that bind basement membrane proteins such as $\alpha 3\beta 1$, $\alpha 6\beta 1$, $\alpha 1\beta 1$, $\alpha 5\beta 1$) to promote a more mature EC tube phenotype which likely also facilitates maturation of pericytes as well (Stratman et al. 2009a). Importantly, we demonstrated that these basement membrane binding integrins were important selectively during EC–pericyte tube co-assembly in 3D collagen matrices, but not during tube formation in EC-only cultures ($\alpha 2\beta 1$-dependent, a collagen-binding integrin) (Stratman et al. 2009a). Thus, pericyte recruitment narrows EC tube diameter allowing for more elongated and branched tubes that have an abluminal surface with deposited basement membrane and pericytes, which are selectively localized along this surface (Stratman et al. 2009a; Stratman and Davis 2012)

(Fig. 9.1). How this recruitment alters the overall polarity of ECs with regard to their distribution of apical or basally oriented proteins is currently not known; however, models are now in place that can directly address this question at a molecular level.

We hypothesize that unique signaling events occur between the two cell types due to growth factor and other molecule cross talks. These cell–cell interactions are further coordinated through unique signals delivered by the basement membrane matrix (via integrins, other ECM receptors, and additional co-associated receptors) which is uniquely deposited during EC–pericyte tube co-assembly. This specialized ECM can be decorated by growth factors (EC and pericyte derived) and molecules such as TIMP-3 (pericyte derived) which adsorb to this matrix on the abluminal tube surface. The unique matrix–growth factor interactions and other molecules that associate with vascular basement membrane proteins and glycosaminoglycans allow for unique signaling opportunities for the recipient ECs and pericytes. These signals will play a major role in controlling the behavior of ECs (shape, gene profile, polarity) or pericytes in different vascular beds. Functional EC behavioral and structural differences (including polarity) in distinct tissues may be explained by alterations in the signals delivered by variations in the vascular basement membrane matrices in these tissues since the unique cells that are adjacent to small vessels will contribute differing combinations of molecules that decorate these matrices. Thus, the polarized vascular basement membrane will contribute unique basal signals that could act in conjunction with unique apical signals delivered to blood ECs depending on the arterial, capillary, or venous origins of the ECs (which are exposed to different shear forces). This discussion reveals at least one way that the known functional diversity of ECs could be achieved through differential signaling related to EC apical–basal polarization.

9.6 Growth Factor Requirements for the Establishment of EC Tube Networks and Luminal Polarity in 3D Extracellular Matrices

A fundamental question that underlies how ECs establish polarity in 3D matrices during vascular tube morphogenesis is what growth factors are necessary for this process to occur. Our laboratory has established over many years the first serum-free defined assays of human EC tubulogenesis and sprouting that allow us to define the actual growth factor requirements for tube formation in 3D matrices (Koh et al. 2008b; Stratman et al. 2011; Smith et al. 2013). A central assumption for many years is that isoforms of vascular endothelial growth factor (VEGF) are the primary growth factors that stimulate vascular morphogenesis (Adams and Alitalo 2007). Using our serum-free defined models adding VEGF-A or VEGF-C even in combination with fibroblast growth factor (FGF)-2 were unable to stimulate or sustain human EC tube formation. For many years, we added phorbol ester to stimulate tubulogenesis, and it was the major agent that affected these processes (Davis and Camarillo 1996; Koh et al. 2008b). Several years ago, we screened hundreds of combinations of recombinant growth factors and identified a combination that supported both tube assemblies when ECs were seeded as single cells within the 3D ECM or when they were seeded on a monolayer surface and allowed to sprout to form tubes (Stratman et al. 2011). Four factors in combination allow human ECs to undergo tubulogenesis and sprouting and they are stem cell factor (SCF), interleukin-3 (IL-3), stroma-derived factor-1α (SDF-1α), and FGF-2 (Stratman et al. 2011; Smith et al. 2013) (Figs. 9.1, 9.2, 9.3, 9.4, and 9.5). VEGF isoforms do not substitute for either SCF or FGF-2 in these systems (Stratman et al. 2011; Smith et al. 2013). After many additional years of screening other factors, we have yet to find another combination of recombinant growth factors that can support human EC tubulogenesis and sprouting under serum-free defined conditions. Importantly, these factors work very well under serum-free defined conditions in collagen or fibrin matrices and also when pericytes are added (Stratman et al. 2011; Smith et al. 2013). As described above, this serum-free defined system with hematopoietic stem cell cytokines and FGF-2 allows for dramatic pericyte recruitment and vascular basement membrane assembly around the human EC tubes in 3D collagen or fibrin matrices (Stratman et al. 2009a; Smith et al. 2013; Stratman and Davis 2012). The growth factors work best when they are added into the matrices prior to polymerization, suggesting a relationship between the ECM and growth factor presentation as an important regulator of vascular morphogenesis. A key question that is under investigation is how this combination of recombinant growth factors is able to coordinate the human EC tubulogenic signaling pathway in conjunction with the other regulators of cell polarity and lumen formation that are discussed above. The general approach that we have taken is critical to understand both the growth factor and downstream molecular signaling requirements for vascular morphogenesis in 3D matrices. In this way, many important applications of this knowledge could be forwarded that have implications for the successful therapeutics of human diseases where the vasculature plays a pathogenic role.

9.7 Conclusions

In this chapter, we have discussed the functional relationship between the ability of ECs to form tube networks in 3D matrices and the establishment of polarization at the level of apical vs. basal membranes but also the cytoskeleton where tubulins such as acetylated tubulin are distributed subapically, while F-actin is predominantly oriented in a basal fashion. It is important to state that although other cells make tubes such as epithelial cells, EC-lined tubes are unique in that they are exposed to fluid and shear forces, particularly EC tubes exposed to blood. ECs are very flat compared to epithelial cells in tube structures which is necessitated due to these shear forces, but it also implies that the polarity mechanisms will likely show distinct features. It should also be stated that much work needs to be done to understand how this unique EC polarity is established and how it varies depending on which vascular bed is examined. This chapter focused on the fact that the EC tubulogenic pathway, for which considerable information now exists, is fundamentally required for the establishment of EC polarity in a 3D matrix environment. Furthermore, the recruitment of pericytes around capillary tubes, the major vessels that perfuse and nourish tissues, adds another element of polarity since this recruitment is necessary for vascular basement membrane assembly, an event which selectively occurs on the abluminal tube surface.

Thus, the major determinants of EC polarity are tubulogenesis which creates apical and basal membrane surfaces which become the surfaces that allow for stimuli such as flow forces on the apical membrane as well as basement membrane matrices and pericytes on the basal membrane to provide further EC polarity specialization to developing and maturing EC tubes. This sequence of events also implies that EC polarization will change over time during the processes of EC tubulogenesis, maturation, and eventual tube stabilization with EC quiescence (processes that can be altered in disease states). Recognizing EC polarity as a dynamic process, and determining the key proteins and signals involved in these dynamics, may reveal more appropriately targeted vascular therapeutics for the treatment of a wide range of human diseases.

Acknowledgments The authors would like to thank the contributions of Dr. Amber Stratman toward our understanding of these processes as well as the technical contributions of Annie Smith and Matthew Davis. This work was supported by NIH grant, HL105606, to G.E.D.

References

Adams RH, Alitalo K (2007) Molecular regulation of angiogenesis and lymphangiogenesis. Nat Rev Mol Cell Biol 8(6):464–478

Armulik A, Genove G, Betsholtz C (2011) Pericytes: developmental, physiological, and pathological perspectives, problems, and promises. Dev Cell 21(2):193–215. doi:10.1016/j.devcel.2011.07.001, S1534-5807(11)00269-3 [pii]

Bayless KJ, Davis GE (2002) The Cdc42 and Rac1 GTPases are required for capillary lumen formation in three-dimensional extracellular matrices. J Cell Sci 115(Pt 6):1123–1136

Bayless KJ, Davis GE (2004) Microtubule depolymerization rapidly collapses capillary tube networks in vitro and angiogenic vessels in vivo through the small GTPase Rho. J Biol Chem 279(12):11686–11695. doi:10.1074/jbc.M308373200, M308373200 [pii]

Bayless KJ, Salazar R, Davis GE (2000) RGD-dependent vacuolation and lumen formation observed during endothelial cell morphogenesis in three-dimensional fibrin matrices involves the alpha(v)beta(3) and alpha(5)beta(1) integrins. Am J Pathol 156(5):1673–1683

Benjamin LE, Hemo I, Keshet E (1998) A plasticity window for blood vessel remodelling is defined by pericyte coverage of the preformed endothelial network and is regulated by PDGF-B and VEGF. Development 125(9):1591–1598

Benjamin LE, Golijanin D, Itin A, Pode D, Keshet E (1999) Selective ablation of immature blood vessels in established human tumors follows vascular endothelial growth factor withdrawal. J Clin Invest 103(2):159–165

Bryant DM, Mostov KE (2008) From cells to organs: building polarized tissue. Nat Rev Mol Cell Biol 9(11):887–901. doi:10.1038/nrm2523, nrm2523 [pii]

Bryant DM, Datta A, Rodriguez-Fraticelli AE, Peranen J, Martin-Belmonte F, Mostov KE (2010) A molecular network for de novo generation of the apical surface and lumen. Nat Cell Biol 12(11):1035–1045. doi:10.1038/ncb2106, ncb2106 [pii]

Cau J, Hall A (2005) Cdc42 controls the polarity of the actin and microtubule cytoskeletons through two distinct signal transduction pathways. J Cell Sci 118(Pt 12):2579–2587

Cera MR, Fabbri M, Molendini C, Corada M, Orsenigo F, Rehberg M, Reichel CA, Krombach F, Pardi R, Dejana E (2009) JAM-A promotes neutrophil chemotaxis by controlling integrin internalization and recycling. J Cell Sci 122(Pt 2):268–277. doi:10.1242/jcs.037127, 122/2/268 [pii]

Davis GE, Camarillo CW (1996) An alpha 2 beta 1 integrin-dependent pinocytic mechanism involving intracellular vacuole formation and coalescence regulates capillary lumen and tube formation in three-dimensional collagen matrix. Exp Cell Res 224(1):39–51

Davis GE, Senger DR (2005) Endothelial extracellular matrix: biosynthesis, remodeling, and functions during vascular morphogenesis and neovessel stabilization. Circ Res 97(11): 1093–1107

Davis GE, Bayless KJ, Mavila A (2002) Molecular basis of endothelial cell morphogenesis in three-dimensional extracellular matrices. Anat Rec 268(3):252–275

Davis GE, Koh W, Stratman AN (2007) Mechanisms controlling human endothelial lumen formation and tube assembly in three-dimensional extracellular matrices. Birth Defects Res C Embryo Today 81(4):270–285

Davis GE, Stratman AN, Sacharidou A, Koh W (2011) Molecular basis for endothelial lumen formation and tubulogenesis during vasculogenesis and angiogenic sprouting. Int Rev Cell Mol Biol 288:101–165

Ebnet K, Aurrand-Lions M, Kuhn A, Kiefer F, Butz S, Zander K, Meyer zu Brickwedde MK, Suzuki A, Imhof BA, Vestweber D (2003) The junctional adhesion molecule (JAM) family members JAM-2 and JAM-3 associate with the cell polarity protein PAR-3: a possible role for JAMs in endothelial cell polarity. J Cell Sci 116(Pt 19):3879–3891

Etienne-Manneville S, Hall A (2001) Integrin-mediated activation of Cdc42 controls cell polarity in migrating astrocytes through PKCzeta. Cell 106(4):489–498

Fisher KE, Sacharidou A, Stratman AN, Mayo AM, Fisher SB, Mahan RD, Davis MJ, Davis GE (2009) MT1-MMP- and Cdc42-dependent signaling co-regulate cell invasion and tunnel formation in 3D collagen matrices. J Cell Sci 122(Pt 24):4558–4569

Goldstein B, Macara IG (2007) The PAR proteins: fundamental players in animal cell polarization. Dev Cell 13(5):609–622. doi:10.1016/j.devcel.2007.10.007, S1534-5807(07)00385-1 [pii]

Gomes ER, Jani S, Gundersen GG (2005) Nuclear movement regulated by Cdc42, MRCK, myosin, and actin flow establishes MTOC polarization in migrating cells. Cell 121(3):451–463

Iruela-Arispe ML, Davis GE (2009) Cellular and molecular mechanisms of vascular lumen formation. Dev Cell 16(2):222–231

Joberty G, Petersen C, Gao L, Macara IG (2000) The cell-polarity protein Par6 links Par3 and atypical protein kinase C to Cdc42. Nat Cell Biol 2(8):531–539

Kamei M, Saunders WB, Bayless KJ, Dye L, Davis GE, Weinstein BM (2006) Endothelial tubes assemble from intracellular vacuoles in vivo. Nature 442(7101):453–456. doi:10.1038/nature04923, nature04923 [pii]

Kim DJ, Martinez-Lemus LA, Davis GE (2013) EB1, p150Glued, and Clasp1 control endothelial tubulogenesis through microtubule assembly, acetylation, and apical polarization. Blood 121(17):3521–3530. doi:10.1182/blood-2012-11-470179, blood-2012-11-470179 [pii]

Koh W, Mahan RD, Davis GE (2008a) Cdc42- and Rac1-mediated endothelial lumen formation requires Pak2, Pak4 and Par3, and PKC-dependent signaling. J Cell Sci 121(Pt 7):989–1001

Koh W, Stratman AN, Sacharidou A, Davis GE (2008b) In vitro three dimensional collagen matrix models of endothelial lumen formation during vasculogenesis and angiogenesis. Methods Enzymol 443:83–101

Koh W, Sachidanandam K, Stratman AN, Sacharidou A, Mayo AM, Murphy EA, Cheresh DA, Davis GE (2009) Formation of endothelial lumens requires a coordinated PKC{epsilon}-, Src-, Pak- and Raf-kinase-dependent signaling cascade downstream of Cdc42 activation. J Cell Sci 122(Pt 11):1812–1822

Lampugnani MG, Orsenigo F, Rudini N, Maddaluno L, Boulday G, Chapon F, Dejana E (2010) CCM1 regulates vascular-lumen organization by inducing endothelial polarity. J Cell Sci 123(Pt 7):1073–1080. doi:10.1242/jcs.059329, 123/7/1073 [pii]

Lizama CO, Zovein AC (2013) Polarizing pathways: balancing endothelial polarity, permeability, and lumen formation. Exp Cell Res 319(9):1247–1254. doi:10.1016/j.yexcr.2013.03.028, S0014-4827(13)00151-1 [pii]

Martin-Belmonte F, Mostov K (2008) Regulation of cell polarity during epithelial morphogenesis. Curr Opin Cell Biol 20(2):227–234. doi:10.1016/j.ceb.2008.01.001, S0955-0674(08)00003-3 [pii]

Sacharidou A, Koh W, Stratman AN, Mayo AM, Fisher KE, Davis GE (2010) Endothelial lumen signaling complexes control 3D matrix-specific tubulogenesis through interdependent Cdc42- and MT1-MMP-mediated events. Blood 115(25):5259–5269. doi:10.1182/blood-2009-11-252692, blood-2009-11-252692 [pii]

Sacharidou A, Stratman AN, Davis GE (2012) Molecular mechanisms controlling vascular lumen formation in three-dimensional extracellular matrices. Cells Tissues Organs 195(1–2): 122–143. doi:10.1159/000331410, 000331410 [pii]

Saunders WB, Bohnsack BL, Faske JB, Anthis NJ, Bayless KJ, Hirschi KK, Davis GE (2006) Coregulation of vascular tube stabilization by endothelial cell TIMP-2 and pericyte TIMP-3. J Cell Biol 175(1):179–191

Senger DR, Davis GE (2011) Angiogenesis. Cold Spring Harb Perspect Biol 3(8):a005090. doi:10.1101/cshperspect.a005090, a005090 [pii], cshperspect.a005090 [pii]

Smith AO, Bowers SL, Stratman AN, Davis GE (2013) Hematopoietic stem cell cytokines and fibroblast growth factor-2 stimulate human endothelial cell-pericyte tube co-assembly in 3D fibrin matrices under serum-free defined conditions. PLoS One 8(12):e85147. doi:10.1371/journal.pone.0085147, PONE-D-13-31310 [pii]

Stratman AN, Davis GE (2012) Endothelial cell-pericyte interactions stimulate basement membrane matrix assembly: influence on vascular tube remodeling, maturation, and stabilization. Microsc Microanal 18(1):68–80. doi:10.1017/S1431927611012402, S1431927611012402 [pii]

Stratman AN, Malotte KM, Mahan RD, Davis MJ, Davis GE (2009a) Pericyte recruitment during vasculogenic tube assembly stimulates endothelial basement membrane matrix formation. Blood 114(24):5091–5101

Stratman AN, Saunders WB, Sacharidou A, Koh W, Fisher KE, Zawieja DC, Davis MJ, Davis GE (2009b) Endothelial cell lumen and vascular guidance tunnel formation requires MT1-MMP-dependent proteolysis in 3-dimensional collagen matrices. Blood 114(2):237–247

Stratman AN, Schwindt AE, Malotte KM, Davis GE (2010) Endothelial-derived PDGF-BB and HB-EGF coordinately regulate pericyte recruitment during vasculogenic tube assembly and stabilization. Blood 116(22):4720–4730

Stratman AN, Davis MJ, Davis GE (2011) VEGF and FGF prime vascular tube morphogenesis and sprouting directed by hematopoietic stem cell cytokines. Blood 117(14):3709–3719

Xu K, Cleaver O (2011) Tubulogenesis during blood vessel formation. Semin Cell Dev Biol 22(9): 993–1004. doi:10.1016/j.semcdb.2011.05.001, S1084-9521(11)00076-0 [pii]

Zovein AC, Luque A, Turlo KA, Hofmann JJ, Yee KM, Becker MS, Fassler R, Mellman I, Lane TF, Iruela-Arispe ML (2010) Beta1 integrin establishes endothelial cell polarity and arteriolar lumen formation via a Par3-dependent mechanism. Dev Cell 18(1):39–51. doi:10.1016/j.devcel.2009.12.006, S1534-5807(09)00488-2 [pii]

Chapter 10
Phosphoinositides as Determinants of Membrane Identity, Apicobasal Polarity, and Lumen Formation

Annette M. Shewan, Aline Awad, Juan Peng, and Ama Gassama-Diagne

Abstract Epithelial cells sense their surrounding environment via mechanical and chemical stimuli, responding to multiple sources of tension and force, including those generated by cell–cell and cell–extracellular matrix (ECM) interaction: they respond to these cues by developing an apicobasal axis of polarity. Phosphoinositides (PIs) are structural components of biological membranes that control a diverse array of signaling pathways through spatiotemporal recruitment of effectors containing PI-specific binding domain(s). Thus they have been shown to modulate a plethora of cellular processes including actin polymerization, cell migration, proliferation, differentiation, and vesicular trafficking. PIs are enriched in different membranes and their levels are tightly regulated by specific PI kinases and phosphatases. During the past decade, PIs have come to the fore as specific markers that define membrane identity, acting as critical regulators of the cell polarization process. In this review, we have examined how PIs are able to assign identity to polarized epithelial cell plasma membrane domains and integrate in space and time complex signaling pathways to trigger appropriate cellular responses to environmental cues. PIs are implicated in a vast array of cellular responses that are central for morphogenesis such as, but not limited to, cytoskeletal changes, cytokinesis, and recruitment of downstream effectors to govern mechanisms involved in polarization and lumen formation. Subversion by pathogens of PI metabolism and

A.M. Shewan
School of Chemistry and Molecular Biosciences, University of Queensland, St. Lucia, QLD 4072, Australia

A. Awad • J. Peng
Univ Paris-Sud, UMR-S 785, Villejuif 94800, France

Inserm Unité 785, Villejuif 94800, France

A. Gassama-Diagne (✉)
Univ Paris-Sud, UMR-S 785, Villejuif 94800, France

Inserm Unité 785, Villejuif 94800, France

Centre Hépato-Biliaire, Hôpital Paul Brousse, 12-14 Avenue Paul Vaillant Couturier, Villejuif 94800, France
e-mail: ama.gassama@inserm.fr

K. Ebnet (ed.), *Cell Polarity 1*, DOI 10.1007/978-3-319-14463-4_10

plasma membrane identity in polarized cells and the clinical relevance of research on PIs were also discussed.

Keywords Epithelial polarity • Phosphoinositide • PI3K • SHIP2 • PTEN • Membrane identity • Lumen formation

10.1 Introduction

Epithelia are coherent sheets of cells that form a barrier between the interior of the body and the outside. These sheets can be one cell thick in the case of simple epithelia or many cells thick for stratified epithelia. Some epithelia cover the outside of the organism, whereas others are a spherical monolayer of cells that line internal organs and enclose a central lumen. Timely and proper establishment of intrinsic cellular polarity is crucial to tissue form and function and follows precise spatial and temporal parameters. Epithelial cells differentiate and polarize along their apicobasal axis, providing the basic building blocks that support tissue architecture and that enabled the tissues to achieve their various functions, including barrier formation (Gibson and Perrimon 2003). The study of the molecular mechanisms of epithelial polarity is important for understanding a number of disease processes where polarity is disrupted, such as cancer and infectious diseases.

Polarity is a multistage and dynamic process requiring cues from cell–cell and cell–extracellular matrix (ECM) that leads to the asymmetric organization of the plasma membrane, intracellular organelles, and the cytoskeleton (Nelson 2003; Nelson and Bissell 2006; Bryant and Mostov 2008; Guillot and Lecuit 2013). Analysis of the polarization of unicellular eukaryotes has yielded enormous insights into the mechanisms that underlie cell polarity (Goldstein and Macara 2007). During the last decade further studies have revealed the role of mechanotransduction as a main driver of cell differentiation and polarity during morphogenesis (Houk et al. 2012; Nelson 2013; Rodríguez-Fraticelli and Martín-Belmonte 2013).

The polarized cell phenotype is established and maintained by the segregation and retention of specific proteins and lipids in distinct apical and basolateral plasma membrane domains. The biological membranes are constituted of bilayer of lipids mainly composed of phospholipids that move freely and make the membrane a fluid area. Fluid organization of the membrane facilitates redistribution of lipids to ensure membrane remodeling and curvature required for vesicle formation and trafficking are achievable.

Phosphoinositides (PIs) represent only a small percentage of the total cellular phospholipids, but they have a well-established, important role in the activation and regulation of signaling pathways (Berridge and Irvine 1989; Martin 1998). Major advances have been made toward understanding their role in an increasing number of biological processes, including proliferation, migration, differentiation, and polarization (Gassama-Diagne and Payraste 2009 and Comer 2007). A comprehensive review of the role of PIs in all of these important processes is beyond the scope of this review; however the contribution of PIs to these processes has been extensively reviewed recently (Balla 2013). In this chapter we present some important features of PIs and discuss how this minor class of lipids shape and determine membrane identity and integrate cascades of signaling pathways to govern apicobasal polarization and lumen formation in epithelial cells.

10.2 Phosphoinositides, Basic Features

Phosphatidylinositol (PtdIns) is the most abundant member of the family of PIs and represents up to 10 % of total phospholipids. Spatiotemporal regulation of production and turnover of PIs by enzymes that metabolize them, coupled with the recruitment of specific effectors that control PI function and their specific presence within the different cellular membranes, act in concert to define and reinforce membrane identity (Fig. 10.1) (Cho and Stahelin 2005; Lemmon 2008). The complexity of these reactions is such that 18 PI interconversion reactions mediated by 19 PI kinases and 28 PI phosphatases have been identified in mammals (Payrastre 2004; Sasaki et al. 2009). PIs relay signals through recruitment and binding to cytosolic protein effectors that are able to recognize the arrangement of phosphate groups around the inositol ring (Kutateladze 2010). The pleckstrin homology (PH) domain was the first effector found to associate with PIs (Harlan et al. 1994). The number of PI-binding domains has since grown rapidly and these modules display a wide range of affinities and selectivities for lipid membranes as depicted in Fig. 10.2 (Lemmon 2008). The number is still increasing with the recent characterization of a pseudo C2 domain in IQGAP proteins (Dixon et al. 2012).

The function of PIs as determinants of internal membrane identity and as crucial regulators of membrane trafficking has been reviewed elsewhere (Di Paolo and De Camilli 2006; Jean and Kiger 2012; Balla 2013). Here we will consider the role of PIs in plasma membrane identity in polarized epithelial cells.

10.3 Phosphoinositides and Plasma Membrane Identity in Polarized Cells

Phosphoinositide 3-kinases (PI3Ks) phosphorylate the hydroxyl group on 3-position of the inositol ring of PtdIns. The PI3K family of enzymes comports three different classes (class I, II, and III), based on their substrate specificity and molecular structure (Vanhaesebroeck et al. 2012). Phosphatidylinositol-3,4,5-trisphosphate (PtdIns(3,4,5)P3) is the major product of class I PI3Ks and a key mediator in diverse intracellular signaling pathways including cell proliferation, differentiation, and migration and is targeted for research in inflammation,

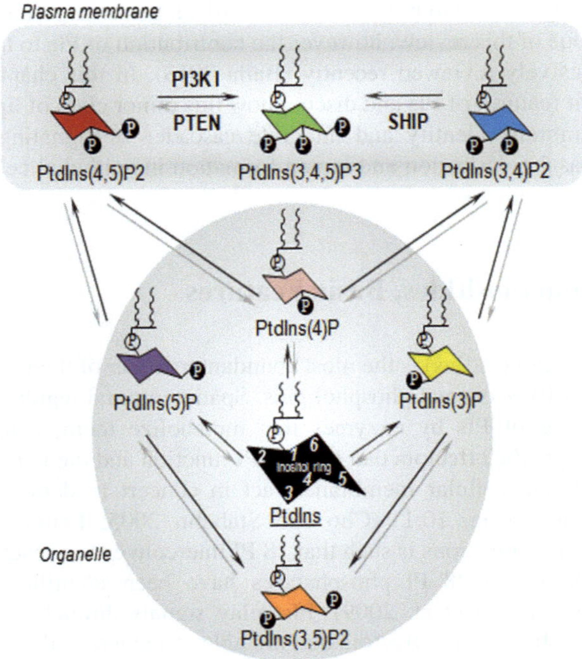

Fig. 10.1 The network of interconversions between phosphoinositides. The inositol ring (in *black* and *numbered*) of phosphatidylinositol (PtdIns) can be phosphorylated at three hydroxyl positions (positions 3, 4, and 5), generating seven phosphoinositides (PIs): one PI triphosphate (PtdIns (3,4,5)P3 in *green*), three PI biphosphates (PtdIns(3,4)P2 in *blue*, PtdIns(4,5)P2 in *red*, and PtdIns (3,5)P2 in *orange*), and finally three PI monophosphates (PtdIns(3)P in *yellow*, PtdIns(4)P in *pink*, and PtdIns(5)P in *purple*). These PIs are produced by kinases (*black arrows*) and phosphatases (*gray arrows*) that catalyze these phosphorylation and dephosphorylation events, respectively (Krauss and Haucke 2007; Sasaki et al. 2009). Indicated in the figure are the relevant enzymes of PtdIns(3,4,5)P3 signaling participating in the reactions discussed in this review (PI3K, PTEN, and SHIP). PtdIns(3,4,5)P3 is generated at the plasma membrane by the class I PI3K that phosphorylate PtdIns(4,5)P2 at the 3 position of the inositol ring (Cantley 2002). PtdIns(3,4,5) P3 is dephosphorylated on the 3 position by the phosphatase and tensin homolog (PTEN) and by SHIPs (SH2-cotaining inositol phosphatase) on the 5 position to produce PtdIns(4,5)P2 and PtdIns (3,4)P2, respectively (Balla 2013). PtdIns(3,4,5)P3, PtdIns(3,4)P2, and PtdIns(4,5)P2 are enriched at the plasma membrane, and all other PIs are mainly present at the membrane of subcellular organelles

immunity, cancer, and metabolic diseases (Cantley 2002; Vanhaesebroeck et al. 2001; Wymann and Marone 2005; Wymann et al. 2008). This signaling is mediated in part by recruiting proteins that carry a specialized lipid-binding domain, such as a PH domain (Fig. 10.2). These PH domain-containing proteins include guanine nucleotide exchange factors (GEFs) that activate Rho family GTPases and also the serine/threonine protein kinase B (also known as Akt) which is one of the best-studied targets of the PtdIns(3,4,5)P3. Besides the PI3Ks, several inositol polyphosphatases such as PTEN and SHIP (Backers et al. 2003;

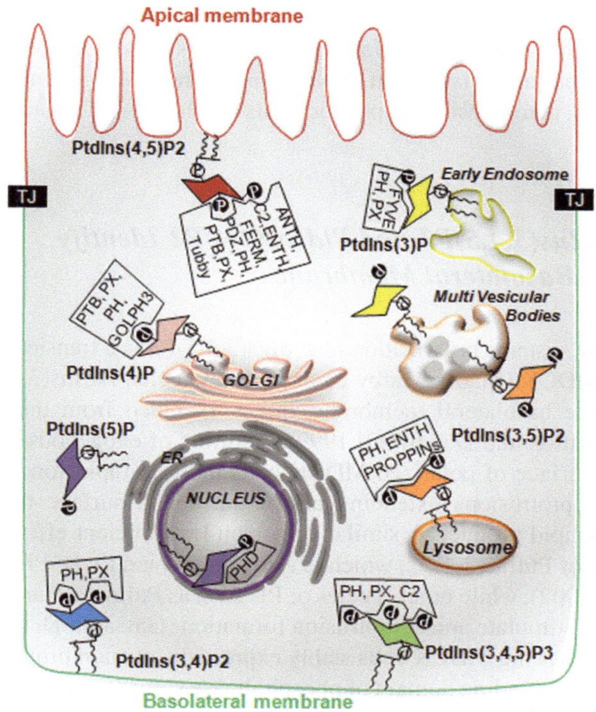

Fig. 10.2 Subcellular distribution of PIs and their recognized effectors binding domains in polarized epithelial cells. PIs have specific subcellular positions and are polarized. PtdIns(4,5)P2 in *red* is enriched at the apical membrane while PtdIns(3,4)P2 in *blue* and PtdIns(3,4,5)P3 in *green* are enriched at the basolateral membrane. PtdIns(3)P in *yellow* is concentrated on early endosomal membranes and in multivesicular bodies. PtdIns(4)P in *pink* is localized at the Golgi membrane. PtdIns(5)P in *purple* is found in the nucleus and is also associated with the ER, although there is some debate about its exact distribution. PtdIns(3,5)P2 in *orange* is localized in lysosomes and multivesicular bodies. Each PI binds specific domains with unique design features that play a critical role in their cellular localization and function (Lemmon 2008; Kutateladze 2010). The different PI-binding domains and their target PIs are presented according to their predominant subcellular distribution. ANTH (AP180 amino-terminal homology), C2 (conserved region 2 of protein kinase C), ENTH (epsin amino-terminal homology), FERM (band 4.1, ezrin, radixin, moesin), FYVE (Fab1, YOTB, Vac1, and EEA1), GOLPH3 (Golgi phosphoprotein 3), PDZ (postsynaptic density protein (PSD95)), *Drosophila* disc large tumor suppressor (Dlg1), and zonula occludens-1 protein (ZO-1), PH (pleckstrin homology), PTB (phosphotyrosine-binding domain), PX (phox homology), PROPPINs (B-propellers that bind PIs), PHD (plant homeodomain). *TJ* tight junctions

Elong Edimo et al. 2014) actively control the dephosphorylation of the PI3K products (Fig. 10.1).

The segregation of PIs in the plasma membrane is a crucial event for the cell polarization process across different systems. Among them, the role of PtdIns (3,4,5)P3 in chemotaxic cell migration using *Dictyostelium discoideum* or neutrophils as a model system is among the most thoroughly studied examples of PI function in plasma membrane polarization (Kölsch et al. 2008; Cai and Devreotes 2011; Weiger and Parent 2012; Wang et al. 2002; Weiner et al. 2002). In addition, multiple cell types use polarization of phosphatidylinositol signaling components to drive migration, cytokinesis, and phagocytosis (Balla 2013; Cain and Ridley 2009).

10.3.1 PtdIns(3,4,5)P3 and PtdIns(3,4)P2 Identify the Basolateral Membrane

While in many systems polarization of PtdIns(3,4,5)P3 is a transient phenomenon, in polarized MDCK (Madin–Darby canine kidney) cells, PtdIns(3,4,5)P3 is stably localized at the basolateral membrane and is excluded from the apical plasma membrane (Watton and Downward 1999). Addition of exogenous PtdIns(3,4,5)P3 to the apical surface of polarized MDCK cells induced rapid formation of PtdIns (3,4,5)P3-rich protrusions extending above the apical surface that is normally devoid of this lipid product. A similar but much less efficient effect was observed upon addition of PtdIns(3,4)P2, which is also able to bind to the PH domain of Akt (Scheid et al. 2002), while other species of PIs such as PtdIns(4)P and PtdIns(4,5)P2 were unable to stimulate apical protrusion formation (Gassama-Diagne et al. 2006). Previous studies using MDCK cells stably expressing a fusion protein consisting of the PtdIns(4,5)P2-binding domain of phospholipase C fused to GFP (GFP-PH-PLC) (Balla and Várnai 2009; Várnai et al. 2002) indicated that apical addition of PtdIns (3,4,5)P3 did not result in GFP-PH-PLC labeling of the protrusions. These studies demonstrate that PtdIns(4,5)P2 is not concentrated in the protrusions from the apical surface, thus confirming the specificity of the PI3K lipid products for the basolateral membrane. Together, these results indicate that PtdIns(3,4,5)P3 is able to rapidly form transient membranous protrusions in the absence of growth factors or other stimuli.

The Ras and Rho family GTPases are key regulators of cell polarization, chemotaxis, and cell migration (Fukata et al. 2003; Sasaki and Firtel 2006; Heasman and Ridley 2008). Complex direct and indirect feedback interactions dictate how these proteins influence each other's activity. In chemotaxing neutrophils a positive feedback loop between PtdIns(3,4,5)P3, the small GTPases, and actin polymerization contributes to the sharp, front–rear polarization as well as the secondary recruitment and activation of PI3K (Wang et al. 2002; Weiner et al. 2002; Srinivasan et al. 2003; Marée et al. 2012). In polarized epithelial cells, it has been suggested that *de novo* synthesis of endogenous PtdIns(3,4,5)P3 by PI3K is required to amplify the initial signal provided by exogenous PtdIns (3,4,5)P3 (Gassama-Diagne et al. 2006), suggesting a positive feedback loop similar to that operating in the generation of polarity during chemotaxis (Weiner et al. 2002). Another study suggested that PI3-kinase exerts a Rac-dependent morphogenetic effect in cultured epithelial cells through regulation of cell height (Jeanes et al. 2009). A recent study by Yang and colleagues has further unraveled the complexities of Ras and Rho cooperativity in migrating cells. These authors demonstrate cooperative activation of PI3K by Ras and Rho family small GTPases (Yang et al. 2012). Conditionally specific activation of endogenous small GTPases failed to activate PI3K, while the broad-spectrum GEF, Vav2, induced PI3K activation. Using a combination of *in vitro* data and mathematical modeling, this study demonstrates that PI3K activity can be amplified by its downstream activators through cooperative, positive feedback loops in migrating cells. The authors suggest that simultaneous activation of multiple downstream activators of PI3K (Rac1, RhoG, and Cdc42) is required to maintain cell polarity during cell migration. It is possible that this circumstance contributes to spatiotemporally controlled PI3K activation in polarized epithelial cells and tissues.

A recent study highlighted the role of PtdIns(3,4)P2 and SHIP2 as additional determinants of basolateral membrane formation (Awad et al. 2013). In non-polarized and unstimulated cells, SHIP2 has a perinuclear and cytoplasmic localization, while in serum-stimulated cells SHIP2 can be localized at the plasma membrane and at focal contacts (Elong Edimo et al. 2014). Of particular interest, and as has been reported for PtdIns(3,4,5)P3 (Yu et al. 2008) in MDCK cells grown as cysts in 3D Matrigel culture, SHIP2 was enriched in the basolateral membrane. The lipid product of SHIP2, PtdIns(3,4)P2, was also detected at the basolateral domain using either a lipid antibody or the PH domain of TAPP (Wullschleger et al. 2011) as a probe (Awad et al. 2013). The presence of PtdIns(3,4)P2 at the basal membrane is regulated by SHIP2 enzymatic activity. Indeed overexpression of SHIP2 in MDCK cysts led to an increase of both SHIP2 and PtdIns(3,4)P2 at the basal membrane, while the PtdIns(3,4)P2 signal was strongly reduced and dispersed in the cysts expressing the catalytic mutant of SHIP2 (D607A) (Pesesse et al. 2001; Zhang et al. 2007). Thus, these data argue in favor of the presence of PtdIns(3,4) P2 at the basal membrane and its production by SHIP2.

Together these data indicate that both PtdIns(3,4,5)P3 and PtdIns(3,4)P2 are present at the basolateral membrane of polarized MDCK cysts and are supportive of a comprehensive interrogation of their specific functional contributions to apicobasal polarity. The function of SHIP2 may not be limited to the downregulation of PtdIns(3,4,5)P3 levels, since SHIP2 is involved in signaling at focal adhesions (Elong Edimo et al. 2014) and could play a role in the establishment and stabilization of the basement membrane. It is worth noting that the majority of studies across different systems of polarization make use of a probe to detect PtdIns (3,4,5)P3 that consists of the PH domain of Akt, which recognizes both PtdIns (3,4,5)P3 and PtdIns(3,4)P2. So the observed cellular phenotypes could be attributed to either of these particular PIs; thus further studies using specific tools and cutting edge approaches (Lippincott-Schwartz and Manley 2009) will be required to determine the spatiotemporal localization of these two lipids during cell polarization.

10.3.2 PtdIns(4,5)P2 Identifies the Apical Membrane

MDCK cells grown in three-dimensional (3D) culture form polarized cysts with a central lumen located at the apical surface (O'Brien et al. 2001). During the early stages of cyst formation, PtdIns(3,4,5)P3 and PtdIns(4,5)P2 are colocalized at the plasma membrane of non-polarized cells, while PtdIns(4,5)P2 became concentrated at the apical surface of polarized cells (Martin-Belmonte et al. 2007). Importantly, the knockdown of PTEN caused defects in the segregation of PtdIns(3,4,5)P3 and PtdIns(4,5)P2 and disruption of lumen formation. Analogous to the change in membrane identity following addition of exogenous PtdIns(3,4,5)P3 to the apical surface (Gassama-Diagne et al. 2006), the delivery of exogenous PtdIns(4,5)P2 to the basolateral surface of cysts caused lumen reduction and ectopic recruitment of apical membrane proteins, indicative of a change in membrane identity (Martin-Belmonte et al. 2007).

The localization of PTEN and PtdIns(4,5)P2 to the apical membrane is also shown by the work on *Drosophila* embryonic epithelia (Von Stein et al. 2005; Pilot et al. 2006). However, PtdIns(3,4,5)P3 may not always be strictly localized to the basolateral domain. There are likely to be PI microdomains found within the greater apical and basolateral surfaces to carry out specific functions. In a specialized *Drosophila* photoreceptor cell, PTEN is localized to cell–cell junctions where it functions to restrict PtdIns(3,4,5)P3 to the apical membrane domain (Pinal et al. 2006). This photoreceptor domain is a modified cilium which is a specialized organelle distinct from the bulk of the apical membrane (Reiter and Mostov 2006).

As suggested previously, the function of PTEN at the apical membrane could be restriction of PtdIns(3,4,5)P3 generated by PI3K to the basolateral membrane (Martin-Belmonte et al. 2007; Shewan et al. 2011). If this is the case, then the question arises: how are PtdIns(4,5)P2 levels generated and maintained at the apical pole? A common mechanism for PI(4,5)P2 synthesis at the plasma membrane is the sequential phosphorylation of the large pool of PtdIns by PI4-kinases to form PtdIns4P, the latter being phosphorylated in turn by type I phosphatidylinositol-4-phosphate 5-kinases (PIP5Ks) at the 5′ position of its inositol group (De Matteis and Godi 2004; Krauss and Haucke 2007). Perhaps this scenario occurs at the apical membrane of polarized epithelial cells? Indeed recent publications report the role of PIP5K in apical membrane trafficking (Rousso et al. 2013; Szalinski et al. 2013). Nevertheless, an unresolved issue concerns the origin and the exact pool of PtdIns and PtdIns(4)P and the associated kinases present at the apical membrane. Indeed it was recently demonstrated that the presence of PtdIns(4)P at the plasma membrane of non-polarized cells is not the only source of PtdIns(4)P required for PtdIns(4,5) P2 synthesis (Hammond et al. 2012).

An increasing number of studies highlight the critical function of PIs as an identity code that label specific membranes, including both intracellular membranes (Di Paolo and De Camilli 2006; Krauss and Haucke 2007; Jean and Kiger 2012) or plasma membranes, as described here. It is important to take into account that tight regulation of local levels of the different PIs occurs due to the action of cascades of kinases and phosphatases: their regulation in time and space requires further study. Insights into the molecular mechanisms at play in governing the spatial and temporal formation and distribution of PIs will uncover events that drive establishment of polarity.

10.4 PIs, Apicobasal Polarity, and Lumen Formation

In polarized epithelial cells, the plasma membrane is separated into apical and basolateral domains by tight junctions (Gibson and Perrimon 2003). Work by many labs has led to a general model for the acquisition of epithelial polarity via signaling generated through cell interactions with neighboring cells and with the surrounding extracellular matrix (ECM) (O'Brien et al. 2002). β1-Integrin is a critical component of this signaling pathway in MDCK cells and is a known modulator of PI(3)-kinase and PtdIns(3,4,5)P3 levels (Parise et al. 2000; Yu et al. 2008). Consequently, the specification of basolateral membrane identity through the generation of integrin-mediated PtdIns(3,4,5)P3 production could be involved in the earliest steps in epithelial polarization. Furthermore, E-cadherin-mediated PI(3)-kinase activation following cell–cell contact formation provides a significant source of PtdIns(3,4,5)P3 during initial polarization (Pece et al. 1999; Woodfield et al. 2001; Yap and Kovacs 2003). The activation of PI3K at sites of cell–cell and cell–ECM contacts could in turn contribute to increased activity of Rac1 (Noren et al. 2001). This cascade subsequently directs assembly of the cytoskeleton that serves as a scaffold for recruitment and binding of cytosolic signaling proteins and polarity complexes to the membrane that further define and reinforce the identity of the respective membrane domain (Jacobson and Mostov 2007; Miyoshi and Takai 2008).

Cortical domains are defined by a complex that incorporates Par3, Par6, aPKC, and Cdc42 (Goldstein and Macara 2007). PTEN regulated the apical recruitment of Par3, Par6, and Cdc42 and annexin 2 (Anx2) and is required for lumen development (Martin-Belmonte et al. 2007). Interestingly, it was shown that Par3 membrane targeting is dependent of the binding of its PDZ domain to PtdIns(4,5)P2, the product of PTEN (Wu et al. 2007; Krahn et al. 2010; Krahn and Wodarz 2012). Likewise, as outlined below, SHIP2 enzymatic activity is required to maintain discs large 1 (Dlg1) at the basolateral membrane. These studies highlight the function of PI3K signaling in the establishment of apicobasal polarity (Fig. 10.3).

In order to form epithelial organs, the component cells must polarize and construct an apical cell surface and lumen (Fig. 10.3). Coordinated interactions of specific Rabs and PIs are coming to the fore as regulated mechanisms to control the fidelity of membrane trafficking (Jean and Kiger 2012). The Rab GTPases are the largest family of proteins that regulate membrane traffic. Recently, a molecular mechanism was reported that couples membrane trafficking and cortical polarity machinery in generation of the apical surface *de novo* (Bryant, Datta et al. 2010). This mechanism is driven by Rab11a that regulates lumen formation and apical

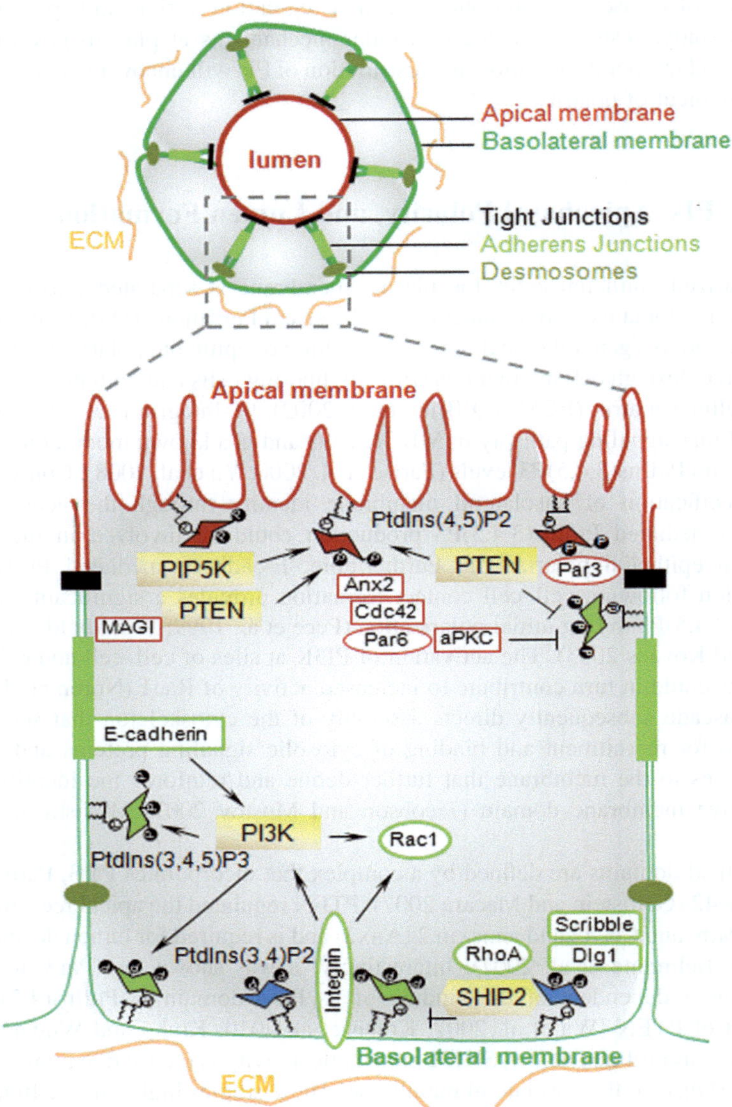

Fig. 10.3 PIs in epithelial cell polarity. Polarized epithelial cells form a cyst whereby a monolayer of tightly associated cells encloses a central lumen. The plasma membrane of polarized epithelial cell consists of basolateral membrane that contains adherens junctions and desmosomes (involved in cell–cell adhesion) and that also interacts with underlying basement membrane and extracellular matrix (ECM) components. The apical membranes are facing the lumen and are physically separated from basolateral membrane by tight junctions (TJs; O'Brien et al. 2002). At the initial steps of epithelial cell polarization, signals from cell–cell and cell–ECM contacts mediated through E-cadherin and integrin, respectively, induce PI3K activation leading to accumulation of PtdIns(3,4,5)P3 to define the apicobasal axis and lumen initiation (Bryant and Mostov 2008). Indeed PtdIns(3,4,5)P3 acts as a compass for orientation of the cell polarity axis (Gassama-Diagne et al. 2006). Later, phosphatases are activated to downregulate PtdIns(3,4,5)P3. PTEN is activated at the apical pole to generate PtdIn(4,5)P2 and has a critical role in lumen formation by recruiting

traffic via association with Rabin8, the GEF for Rab8a. Par3 was targeted to the cortex via the action of Rab8a and Rab11a through the exocyst, which in turn controlled Cdc42 activation at the apical pole. Here we consider the apically implicated Rab proteins and their interactions with PIs and accessory proteins in lumen development.

Mammalian phosphatidylinositol 4-kinase beta (PI4KB) interacts with GTP-bound Rab11a, and inhibition of this interaction blocked Rab11 recruitment to the Golgi in kidney fibroblasts (De Graaf et al. 2004). In *Drosophil*a ovarian follicle cells, phosphatidylinositol 4-kinase III alpha (PI4KIIIα) is required to control apical protein localization and oocyte polarization (Yan et al. 2011). Furthermore, in polarized MDCK cells, PI4KB regulated Golgi to basolateral membrane traffic (Bruns et al. 2002), and exit from the recycling endosome is facilitated by the association of Rab11 with Myosin V (Hales et al. 2002; Lapierre et al. 2001). Myo2p (the yeast homolog of Myosin V) binds Ypt31/32p (the yeast homolog of Rab11) and Sec4p (the yeast homolog of Rab8a), although the structural requirements for Myo2p binding are distinct (Santiago-Tirado et al. 2011). This is analogous to the interaction of Myosin V with Rab11a and Rab8a (Roland et al. 2007). As suggested, distinct structural requirements for Rab binding to Myosin V could facilitate compartment-specific regulation. Thus, the ternary complex consisting of the Rab protein, PtdIns(4)P, and Myosin V serves as a coincidence detection mechanism, to allow the motor to discern between secretory compartments and the cell surface. Importantly, in polarized epithelial cells, Myosin V is required for apical membrane trafficking and lumen formation and was dependent upon association with both Rab8a and Rab11a, although whether this requires simultaneous or sequential interactions is unclear (Roland, Bryant et al. 2011). In addition, targeting of Myosin V to recycling vesicles required Rab11a association. Whether this association requires PtdIns(4)P is not known, but it is plausible to suggest that this mechanism could contribute to apical membrane traffic in polarized epithelial cells. Indeed, in yeast, it has been shown that the secretory membranes transported by Myo2p, namely, the late Golgi, the TGN Ypt31–32p compartment, and Sec4p secretory vesicles, are enriched in PtdIns(4)P (Santiago-Tirado et al. 2011). The arrival of PtdIns(4)P-decorated secretory vesicles at the cortex could provide a source of PtdIns(4)P for PtdIns(4,5)P2 synthesis to generate the apical membrane as discussed in the precedent section.

Fig. 10.3 (continued) TJ complex proteins including Annexin2 (Anx2), Cdc42, Par3, Par6, and atypical protein kinase C (aPKC) (Martin-Belmonte et al. 2007). The aPKC restricts PtdIns(3,4,5)P3 to the basolateral membrane (Takahama et al. 2008). PTEN is recruited at adherens junction through MAGI (membrane-associated guanylate kinase-inverted proteins) (Kotelevets et al. 2005). Phosphatidylinositol 5-kinase that produces PtdIns(4,5)P2 from PtdIns(4)P is also present at the apical domain and involved in apical vesicle trafficking (Rousso et al. 2013; Szalinski et al. 2013). SHIP2 is activated at the basal pole to produce PtdIns(3,4)P2 and maintain cell polarity by stabilization of polarity complex proteins like Dlg1/Scribble which supports epithelial architecture (Awad et al. 2013)

Type Iγ phosphatidylinositol-4-phosphate 5-kinase (PIPKIγ) associates with the exocyst via a direct interaction with Exo70 (Xiong et al. 2012). Exo70 is the subunit that targets the polarized delivery of the exocyst to the cortex. This interaction mediates the association of E-cadherin with Exo70, targeting Exo70 to adherens junctions. The Exo70 subunit of the exocyst associates with the plasma membrane via its association with PtdIns(4,5)P2 where it is required for docking and fusion of post-Golgi secretory vesicles (Liu et al. 2007). Interestingly, accumulation of PtdIns(4,5)P2 was observed as nascent E-cadherin adhesions were assembling. PIPKIγ-generated PtdIns(4,5)P2 is necessary to recruit Exo70 to newly formed E-cadherin adhesions, where it is required for the polarized targeting and clustering of E-cadherin at the cell surface. This is potentially an important point of convergence between the axis of polarity pathway and the pathway controlling *de novo* generation of the apical domain, whereby PIPKIγ acts upon PtdIns(4)P delivered via polarized vesicular trafficking to nascent E-cadherin adhesions. This mechanism could reinforce the initial symmetry-breaking cue, acting to define the apical pole via spatiotemporally localized PtdIns(4,5)P2 production. Together these data and those indicating the association of Myo2p-transported Sec4p vesicles with PtdIns(4)P reinforced the presumed role of PtdIns(4)P from vesicles delivered to the apical pole as the main precursors of apical membrane PtdIns(4,5)P2.

Furthermore, microlumens have been observed at the basolateral surface of polarized MDCK cysts, possibly forming after cell division, and these coalesce to form the lumen (Cerruti et al. 2013). How PIs contribute to this process is not understood. It is reasonable to suggest that E-cadherin dynamism is a key factor that enables microlumen sliding along the cell–cell contact, suggesting a role for PIs and their metabolizing enzymes in this process.

Apical secretion in epithelial tubules in the *Drosophila* embryo requires the Formin family protein, Diaphanous (Dia) (Massarwa et al. 2009). Dia is apically localized and is activated by Rho1 and a pair of apically localized GEFs (Gef2 and Gef64C). Myosin V has been linked to polarized secretion, and the simultaneous disruption of apical actin organization and apical secretion in *Dia* mutant tissues suggested that Dia builds actin filaments to serve as highways for Myosin V-based polarized apical secretion (Massarwa et al. 2009). Importantly, PtdIns(4,5)P2 levels at the cortex regulate Dia localization in polarized MDCK cysts and in *Drosophila* tubular epithelia (Rousso et al. 2013). The interaction with Rho triggers a conformation switch in Dia that facilitates interaction with PtdIns(4,5)P2. The N-terminal basic domain of Dia is crucial for apical localization, implying that Dia binds directly to PtdIns(4,5)P2 (Rousso et al. 2013). Dia family actin nucleators function in apical secretion in the secretory organs of mice (pancreas) and flies (Geron et al. 2013). Interestingly, the *in vivo* distribution of PtdIns(4,5)P2 in the rat pancreas appears to be somewhat different to the distribution observed in cultured cells (Hicks et al. 2006). Fujimoto and colleagues (Ozato-Sakurai et al. 2011) show that instead of being enriched apically, PtdIns(4,5)P2 exhibits its highest concentration at the gap junction, possibly via association with connexin proteins (Van Zeijl et al. 2007). The significance and function of the differential PtdIns(4,5)P2 distribution for Myosin V-polarized secretion and tissue organization remains to be fully elucidated.

In addition to the 3-phosphatase action of PTEN, cellular PtdIns(3,4,5)P3 concentrations are also regulated by the activity of the inositol 5-phosphatase SHIP2, producing PtdIns(3,4)P2. SHIP2 is mainly located to the cell–ECM contact in polarized cells (Awad et al. 2013) and is reflective of reports of the role of SHIP2 as a docking protein for a large number of cytoskeletal, focal adhesion proteins and tyrosine kinase-associated receptors (Dyson et al. 2001; Prasad et al. 2001; Erneux et al. 2011; Elong Edimo et al. 2014). Accordingly SHIP2 could regulate polarity via the stabilization of both cell to cell and cell to ECM contacts. Let us consider this possibility in more detail.

The specific function of SHIP family of phosphatases in cell polarization has been reported in migrating cells. In murine neutrophils, deletion of the PTEN gene had little or no effect on chemotaxis or polarization in response to fMLP or C5a, whereas deletion of SHIP1 caused a major failure in polarization and cell motility (Nishio et al. 2007). Additional data supporting a significant role for PtdIns(3,4)P2 and 5-phosphatase action in regulating cellular polarity comes from studies of neuronal polarization, wherein aberrant polarization was caused by mutation of PI3K/age-1 or PTEN/daf-18 or the selective PtdIns(3,4)P2-binding target lamellipodin/MIG10 (Adler et al. 2006).

What do we know of the interactions of the basolateral PIs, SHIP2, and the basolateral polarity machinery? The Scribble polarity module defines the basolateral surface and is comprised of Scrib (Scribble), Discs large (Dlg), and Lgl (Lethal giant larvae; Elsum et al. 2012). Awad et al. (2013) recently reported that SHIP2 regulates apicobasal polarity. Repression of SHIP2 via RNAi led to the formation of MDCK cysts with multilumens, indicating that SHIP2 contributes to lumen formation. This phenotype was dependent of SHIP2 lipid phosphatase activity since its catalytically inactive mutant disrupted apicobasal polarity with formation of multilumen cysts. In SHIP2-depleted cysts, the basolateral polarity proteins β-catenin, Scribble, and Dlg1 were delocalized from cell contacts, and their expression was significantly reduced, suggesting that SHIP2 is required for the localization and stability of basolateral complex proteins to maintain epithelial morphogenesis. Furthermore, Dlg1 (SAP97) bound mainly to PtdIns(3,4)P2 and to a lesser extent to PtdIns(3,4,5)P3, the two PIs localized at the basolateral membrane and depletion of Dlg1 led to the formation of cysts with filled lumens and multilumen cysts. In contrast to the reduction in Scrib levels observed upon SHIP2 depletion, no change was observed in Scrib expression levels in Dlg1 RNAi-treated cysts; however an important mislocalization of Scrib was observed in these cells. Mutual dependence of the basolateral polarity proteins for appropriate subcellular localization has been reported previously (Bilder et al. 2000). Collectively, these data indicate that SHIP2 contributes to Dlg1 and Scrib localization in mammalian cells and supports the model whereby the SHIP2/Scrib/Dlg1 axis contributes to epithelial architecture.

Furthermore, Dlg1 is an important regulator of myelination where it acts as part of the myelin "brake" (Macklin 2010). Akt activity is an important driver of myelination (Flores et al. 2008; Narayanan et al. 2009; Tyler et al. 2009; Goebbels et al. 2010). The Dlg1–PTEN interaction in Schwann cells stabilizes PTEN, resulting in reduced Akt activity and transiently braking myelination (Cotter et al. 2010; Noseda et al. 2013). Moreover, abnormal distribution of Dlg1 and PTEN has been observed in premalignant lesions and invasive cervical cancer where abnormal distribution or expression of PTEN correlates with mislocalization of Dlg1 (Vázquez-Ulloa et al. 2011). In light of the above-mentioned reports and considering that Dlg interacts with the p85 subunit of PI3K (Laprise et al. 2004), it can be assumed that basolaterally localized PIs co-operate to recruit polarity proteins at E-cadherin adhesions; thereby building an adhesion dependent PI3K signaling platform at the surface to reinforce basolateral identity. This pathway acts to maintain and support epithelial architecture and suppress cancer progression. However, more studies are required to identify the network of binding partners, how these interactions are regulated, and the domain(s) of Dlg1 that are involved in its binding to PtdIns(3,4)P2 and other accessory proteins.

The interaction of polarity proteins with specific PIs to stabilize signaling networks at the membrane could be a common mechanism in epithelial morphogenesis. Support for this premise is drawn from studies of the multi-PDZ domain protein FRMPD2 (FERM and PDZ-domain-containing 2). FRMPD2 is localized at the basolateral surface in polarized epithelial cells, where the binding of FRMPD2 to PtdIns(3,4P)2 is dependent of its PDZ2 domain and is sufficient for its basolateral localization (Stenzel et al. 2009). Furthermore PTEN binding to MAGI (membrane-associated guanylate kinase-inverted proteins) stabilizes complexes containing E-cadherin at the apical domain of adherens junctions (Kotelevets et al. 2005; Subauste et al. 2005). These data strongly suggest that a similar mechanism, governed by the specific interaction of the PDZ domain of the polarity proteins to PIs, could be responsible for the polarization process in epithelial cells.

Moreover SHIP2 induced increased activation of RhoA in epithelial cells (Awad et al. 2013) consistent with a report of enhanced RhoA activation in SHIP2-dependent polarization and migration of glioma cells (Kato et al. 2012). Indeed the Rho family of GTPases, in particular RhoA and Rac1 that, respectively, regulate the formation of stress fibers and lamellipodia (Burridge and Wennerberg 2004), play a pivotal role in cell polarity (O'Brien et al. 2002; Datta et al. 2011).

To conclude, these insights into SHIP2 function have increased our understanding of the role of PIs in the apicobasal polarization process. It will be important to determine what determines the spatiotemporal regulation of the cellular pools of PtdIns(3,4,5)P3, PtdIns(3,4)P2, and PtdIns(4,5)P2 during cell polarization and in particular to define the precise involvement of PTEN and SHIP2 at the apical and basolateral membranes, respectively. For instance, PtdIns(3,4,5)P3 is absent from, or is a minor component of, the PI pool at the apical membrane while it is present in higher proportions at the basolateral membrane (Gassama-Diagne et al. 2006; Martin-Belmonte et al. 2007). Hence, it is reasonable to suggest then that SHIP2 may act through distinct signals of PI3K to perform complementary roles that are required for basolateral identity and epithelial polarization.

10.5 Subversion of PIs and Apicobasal Polarity by Pathogens

Polarized epithelia are the most fundamental component of the innate immune system, protecting organisms from the vast environmental microbiota: indeed >90 % of infectious agents enter through the epithelia. Thus numerous pathogens including viruses and bacteria acquired strategies to subvert or inactivate host defense mechanisms, leading to colonization, invasion, and disease (Kazmierczak et al. 2001; Bomsel and Alfsen 2003). Given the importance of PI metabolism for cellular signaling and trafficking events (Di Paolo and De Camilli 2006), many intracellular pathogens modulate and exploit PIs to ensure survival and efficient intracellular replication (Behnia and Munro 2005), and an important literature addresses these aspects (Pizarro-Cerdá and Cossart 2004; Chukkapalli et al. 2008; Weber et al. 2009; Hsu et al. 2010; Alvisi et al. 2011; Inlora et al. 2011; Altan-Bonnet and Balla 2012; Bishé et al. 2012). Here we provide a brief oversight of the literature around subversion of PIs by pathogens.

Bacteria such as *Salmonella* and *Shigella* are enteric pathogens that inject toxins into the host cell that subvert host signal transduction pathways and manipulate the cell cytoskeleton and membrane trafficking machineries in order to allow entry of the pathogen into the apical surface of the mucosal barrier (Cossart and Sansonetti 2004; Muthuswamy and Xue 2012). SigD, a virulence factor of *Salmonella*, is a 4-phosphatase that dephosphorylates PtdIns(4,5)P2 to form PI(5)P. As a consequence of SigD microinjection, depletion of PtdIns(4,5)P2 at the apical domain was observed and was associated with modifications of cytoskeletal architecture and loss of tight junctions (Mason et al. 2007).

Pseudomonas aeruginosa (*P. aeruginosa*), a gram-negative pathogen, is a leading cause of nosocomial infections in hospitalized patients (Salyers and Whitt 2002; Engel 2002). In tissue culture models, *P. aeruginosa* is observed to preferentially bind to and enter the cells at the edge and at basolateral surfaces (Geiser et al. 2001). Other studies, using MDCK cells as a model system, revealed that *P. aeruginosa* binding activates the PI3K signaling pathway that is necessary for its entry from the apical surface of polarized epithelial cells (Kierbel et al. 2005). Moreover *P. aeruginosa* subverts the PI3K/PtdIns(3,4,5)P3/Akt pathway to transform a patch of the apical surface into one with basolateral characteristics and to gain entry from the apical surface (Kierbel et al. 2007). Together these studies indicate that *P. Aeruginosa* subverted PtdIns (3,4,5)P3, a determinant of the basolateral membrane (Gassama-Diagne et al. 2006), to transform apical into basolateral membrane, creating a local microenvironment that facilitates its colonization and entry into the mucosal barrier. We surmise that in wounded or disrupted epithelium, the cells at the edge are poorly polarized and *P. aeruginosa* could efficiently recruit PI3K and generate PtdIns(3,4,5)P3 at the apical surface. On the other hand, as the mucosal barrier becomes increasingly differentiated and polarized, the scenario is not as efficient and correlates with the decreased susceptibility of the intact epithelium to *P. aeruginosa*-mediated invasion.

Hepatitis C virus (HCV) is a positive-stranded RNA virus belonging to the family *Flaviviridae*, which is a major public health problem and the leading cause of liver cirrhosis and cancer. The main target of HCV are hepatocytes that are highly polarized cells, with their plasma membranes being separated by tight junctions into apical (canalicular) and basolateral (sinusoidal) domains (Wang and Boyer 2004; Treyer and Müsch 2013). For its entry HCV exploits tight junction proteins such as claudin1 (Evans et al. 2007) and IFITM1 (Wilkins et al. 2013). HCV replication is intimately associated with specially arranged host intracellular membrane structures that establish a viral replication complex (Egger et al. 2002; Brass et al. 2009). An increasing body of literature reports that PIPs and their metabolizing enzymes are host factors essential for HCV replication (Berger et al. 2009; Borawski et al. 2009; Trotard et al. 2009; Berger et al. 2011; Reiss et al. 2011; Bianco et al. 2012; Hsu et al. 2010). In particular PI 4-kinases type III as PtdIns4P and PtdIns(4,5)P2 are required for HCV replication. In a recent publication, Awad et al. (2013) examined HCV core protein (HCP) localization and cell morphogenesis in MDCK cells grown in 3D culture. Even though MDCK cells don't present the complex polarized phenotype of hepatocytes, the system allows studies of HCP influence on cell polarization. MDCK cells expressing HCP formed multilumen cysts and further analysis indicated that in those multilumens cysts, expression of SHIP2 and PI(3,4)P2 were reduced at the basal membrane that contacted the ECM. In contrast a high level of expression of HCP was observed at the ECM contact. Interestingly, overexpression of SHIP2 cDNA in HCP-expressing cells restored single-lumen formation and polarity. Together these data indicate that HCV core is able to subvert SHIP2 to disrupt cell polarity in polarized cells (Awad et al. 2013).

Considered collectively these reports raise challenging questions that warrant further investigation. In pathological conditions, does HCV exploit SHIP2 signaling for its intracellular replication? Does HCV interact with PIs at the basal membrane in wounded or poorly polarized epithelium to enter the cell and invade? To date relatively few studies have been dedicated to unraveling the link between HCV and epithelial apicobasal polarity; nonetheless previous studies strongly suggest that appropriate polarization restricts HCV entry in epithelial cells (Mee et al. 2008, 2009; Snooks et al. 2008; Liu et al. 2009; Benedicto et al. 2011; Harris et al. 2013). Future studies on the subversion of epithelial polarity by pathogens through hijacking PI metabolism and signaling await careful investigation.

10.6 Conclusion

Extensive research on PIs over several decades has led to major advances in our understanding of their central role in cell signaling. While their importance in epithelial cell morphogenesis is becoming more established, we have fundamental gaps in our knowledge around their function. In particular, the mechanistic details that direct integration of the spatiotemporal distribution of these PIs to the cell polarization process are unclear. Important avenues for future investigation encompass developing a deeper understanding of the relationship of PIs with the cell cortex to spatiotemporally control cytoskeleton organization and in honing our understanding of their role in membrane trafficking, paying attention to their functional cooperation with Rab GTPases. Indeed there is a striking physical and functional relation between Rab effectors and PIs (Jean and Kiger 2012) that requires in-depth investigation. Moreover the contribution of PIs to membrane localization of the polarity regulators such as the Par and Scribble complexes also forms a part of this fundamental gap in knowledge.

There is an intimate relationship between loss of polarity in epithelial cells and the development of cancer, and most cancers are derived from epithelial cells (Royer and Lu 2011). Due to the essential role of PIs in regulation of epithelial cell signaling and polarity, there is growing interest in inhibiting PI3K signaling pathway in the area of drug development (Martini et al. 2013). Indeed PI3-kinase inhibitors are in clinical trials targeting many diseases such as cancer, autoimmunity, allergy, and metabolic disorders (Vanhaesebroeck et al. 2010; Banham-Hall et al. 2012). As we've outlined, PIs are coming to the fore as critical factors for numerous pathogens to infect epithelia. Clearly, there are a multitude of reasons to believe that PI research is one of the most promising avenues of scientific endeavor in terms of delivering important advances relevant to human disease.

Acknowledgments This work was supported by grants from the Association pour la Recherche sur le Cancer (ARC/SUBV/CKLQ6) to A.G.D. We acknowledge the support of the National Council for Scientific Research Lebanon for supporting A.A. A.M.S. is the recipient of a National Breast Cancer Foundation Early Career Researcher Fellowship.

References

Adler CE, Fetter RD, Bargmann CI (2006) UNC-6/Netrin induces neuronal asymmetry and defines the site of axon formation. Nat Neurosci 9(4):511–518

Altan-Bonnet N, Balla T (2012) Phosphatidylinositol 4-kinases: hostages harnessed to build panviral replication platforms. Trends Biochem Sci 37(7):293–302

Alvisi G, Madan V, Bartenschlager R (2011) Hepatitis C virus and host cell lipids: an intimate connection. RNA Biol 8(2):258–269

Awad A, Sar S, Barré R, Cariven C, Marin M, Salles JP et al (2013) SHIP2 regulates epithelial cell polarity through its lipid product, which binds to Dlg1, a pathway subverted by hepatitis C virus core protein. Mol Biol Cell 24(14):2171–2185

Backers K, Blero D, Paternotte N, Zhang J, Erneux C (2003) The termination of PI3K signalling by SHIP1 and SHIP2 inositol 5-phosphatases. Adv Enzyme Regul 43:15–28

Balla T (2013) Phosphoinositides: tiny lipids with giant impact on cell regulation. Physiol Rev 93 (3):1019–1137

Balla T, Várnai P (2009) Visualization of cellular phosphoinositide pools with GFP-fused protein-domains. Curr Protoc Cell Biol. Chapter 24:Unit 24.4

Banham-Hall E, Clatworthy MR, Okkenhaug K (2012) The therapeutic potential for PI3K inhibitors in autoimmune rheumatic diseases. Open Rheumatol J 6:245–258

Behnia R, Munro S (2005) Organelle identity and the signposts for membrane traffic. Nature 438 (7068):597–604

Benedicto I, Molina-Jiménez F, Moreno-Otero R, López-Cabrera M, Majano PL (2011) Interplay among cellular polarization, lipoprotein metabolism and hepatitis C virus entry. World J Gastroenterol 17(22):2683–2690

Berger KL, Cooper JD, Heaton NS, Yoon R, Oakland TE, Jordan TX et al (2009) Roles for endocytic trafficking and phosphatidylinositol 4-kinase III alpha in hepatitis C virus replication. Proc Natl Acad Sci U S A 106(18):7577–7582

Berger KL, Kelly SM, Jordan TX, Tartell MA, Randall G (2011) Hepatitis C virus stimulates the phosphatidylinositol 4-kinase III alpha-dependent phosphatidylinositol 4-phosphate production that is essential for its replication. J Virol 85(17):8870–8883

Berridge MJ, Irvine RF (1989) Inositol phosphates and cell signalling. Nature 341(6239):197–205

Bianco A, Reghellin V, Donnici L, Fenu S, Alvarez R, Baruffa C et al (2012) Metabolism of phosphatidylinositol 4-kinase IIIα-dependent PI4P Is subverted by HCV and is targeted by a 4-anilino quinazoline with antiviral activity. PLoS Pathog 8(3):e1002576

Bilder D, Li M, Perrimon N (2000) Cooperative regulation of cell polarity and growth by Drosophila tumor suppressors. Science 289(5476):113–116

Bishé B, Syed G, Siddiqui A (2012) Phosphoinositides in the hepatitis C virus life cycle. Viruses 4 (10):2340–2358

Bomsel M, Alfsen A (2003) Entry of viruses through the epithelial barrier: pathogenic trickery. Nat Rev Mol Cell Biol 4(1):57–68

Borawski J, Troke P, Puyang X, Gibaja V, Zhao S, Mickanin C et al (2009) Class III phosphatidylinositol 4-kinase alpha and beta are novel host factor regulators of hepatitis C virus replication. J Virol 83(19):10058–10074

Brass V, Gosert R, Moradpour D (2009) Investigation of the hepatitis C virus replication complex. Methods Mol Biol 510:195–209

Bruns JR, Ellis MA, Jeromin A, Weisz OA (2002) Multiple roles for phosphatidylinositol 4-kinase in biosynthetic transport in polarized Madin-Darby canine kidney cells. J Biol Chem 277 (3):2012–2018

Bryant DM, Mostov KE (2008) From cells to organs: building polarized tissue. Nat Rev Mol Cell Biol 9(11):887–901

Bryant DM, Datta A, Rodríguez-Fraticelli AE, Peränen J, Martín-Belmonte F, Mostov KE (2010) A molecular network for de novo generation of the apical surface and lumen. Nat Cell Biol 12 (11):1035–1045. doi:10.1038/ncb2106, PMID: 20890297

Burridge K, Wennerberg K (2004) Rho and Rac take center stage. Cell 116(2):167–179

Cai H, Devreotes PN (2011) Moving in the right direction: how eukaryotic cells migrate along chemical gradients. Semin Cell Dev Biol 22(8):834–841. doi:10.1016/j.semcdb.2011.07.020, PMID: 21821139

Cain RJ, Ridley AJ (2009) Phosphoinositide 3-kinases in cell migration. Biol Cell 101(1):13–29

Cantley LC (2002) The phosphoinositide 3-kinase pathway. Science 296(5573):1655–1657

Cerruti B, Puliafito A, Shewan AM, Yu W, Combes AN, Little MH et al (2013) Polarity, cell division, and out-of-equilibrium dynamics control the growth of epithelial structures. J Cell Biol 203(2):359–372

Cho W, Stahelin RV (2005) Membrane-protein interactions in cell signaling and membrane trafficking. Annu Rev Biophys Biomol Struct 34:119–151

Chukkapalli V, Hogue IB, Boyko V, Hu W-S, Ono A (2008) Interaction between the human immunodeficiency virus type 1 Gag matrix domain and phosphatidylinositol-(4,5)-bisphosphate is essential for efficient gag membrane binding. J Virol 82(5):2405–2417

Comer FI, Parent CA (2007) Phosphoinositides specify polarity during epithelial organ development. Cell 128(2):239–240

Cossart P, Sansonetti PJ (2004) Bacterial invasion: the paradigms of enteroinvasive pathogens. Science 304(5668):242–248

Cotter L, Ozçelik M, Jacob C, Pereira JA, Locher V, Baumann R et al (2010) Dlg1-PTEN interaction regulates myelin thickness to prevent damaging peripheral nerve overmyelination. Science 328(5984):1415–1418

Datta A, Bryant DM, Mostov KE (2011) Molecular regulation of lumen morphogenesis. Curr Biol 21(3):R126–R136. doi:10.1016/j.cub.2010.12.003, PMID: 21300279

De Graaf P, Zwart WT, van Dijken RAJ, Deneka M, Schulz TKF, Geijsen N et al (2004) Phosphatidylinositol 4-kinasebeta is critical for functional association of rab11 with the Golgi complex. Mol Biol Cell 15(4):2038–2047

De Matteis MA, Godi A (2004) PI-loting membrane traffic. Nat Cell Biol 6(6):487–492

Di Paolo G, De Camilli P (2006) Phosphoinositides in cell regulation and membrane dynamics. Nature 443(7112):651–657

Dixon MJ, Gray A, Schenning M, Agacan M, Tempel W, Tong Y et al (2012) IQGAP proteins reveal an atypical phosphoinositide (aPI) binding domain with a pseudo C2 domain fold. J Biol Chem 287(27):22483–22496

Dyson JM, O'Malley CJ, Becanovic J, Munday AD, Berndt MC, Coghill ID et al (2001) The SH2-containing inositol polyphosphate 5-phosphatase, SHIP-2, binds filamin and regulates submembraneous actin. J Cell Biol 155(6):1065–1079

Egger D, Wölk B, Gosert R, Bianchi L, Blum HE, Moradpour D et al (2002) Expression of hepatitis C virus proteins induces distinct membrane alterations including a candidate viral replication complex. J Virol 76(12):5974–5984, PMID: 12021330

Elong Edimo W, Schurmans S, Roger PP, Erneux C (2014) SHIP2 signaling in normal and pathological situations: its impact on cell proliferation. Adv Biol Regul 54:142–51

Elsum I, Yates L, Humbert PO, Richardson HE (2012) The Scribble-Dlg-Lgl polarity module in development and cancer: from flies to man. Essays Biochem 53:141–168

Engel JN (ed) (2002) Molecular pathogenesis of acute *Pseudomonas aeruginosa* infections. Kluwer, New York

Erneux C, Edimo WE, Deneubourg L, Pirson I (2011) SHIP2 multiple functions: a balance between a negative control of PtdIns(3,4,5)P3 level, a positive control of PtdIns(3,4)P2 production, and intrinsic docking properties. J Cell Biochem 112(9):2203–2209

Evans MJ, von Hahn T, Tscherne DM, Syder AJ, Panis M, Wölk B et al (2007) Claudin-1 is a hepatitis C virus co-receptor required for a late step in entry. Nature 446(7137):801–805

Flores AI, Narayanan SP, Morse EN, Shick HE, Yin X, Kidd G et al (2008) Constitutively active Akt induces enhanced myelination in the CNS. J Neurosci 28(28):7174–7183

Fukata M, Nakagawa M, Kaibuchi K (2003) Roles of Rho-family GTPases in cell polarisation and directional migration. Curr Opin Cell Biol 15(5):590–597

Gassama-Diagne A, Payrastre B (2009) Phosphoinositide signaling pathways: promising role as builders of epithelial cell polarity. Int Rev Cell Mol Biol 273:313–343

Gassama-Diagne A, Yu W, ter Beest M, Martin-Belmonte F, Kierbel A, Engel J et al (2006) Phosphatidylinositol-3,4,5-trisphosphate regulates the formation of the basolateral plasma membrane in epithelial cells. Nat Cell Biol 8(9):963–970

Geiser TK, Kazmierczak BI, Garrity-Ryan LK, Matthay MA, Engel JN (2001) Pseudomonas aeruginosa ExoT inhibits in vitro lung epithelial wound repair. Cell Microbiol 3(4):223–236

Geron E, Schejter ED, Shilo B-Z (2013) Directing exocrine secretory vesicles to the apical membrane by actin cables generated by the formin mDia1. Proc Natl Acad Sci U S A 110 (26):10652–10657

Gibson MC, Perrimon N (2003) Apicobasal polarization: epithelial form and function. Curr Opin Cell Biol 15(6):747–752

Goebbels S, Oltrogge JH, Kemper R, Heilmann I, Bormuth I, Wolfer S et al (2010) Elevated phosphatidylinositol 3,4,5-trisphosphate in glia triggers cell-autonomous membrane wrapping and myelination. J Neurosci 30(26):8953–8964

Goldstein B, Macara IG (2007) The PAR proteins: fundamental players in animal cell polarization. Dev Cell 13(5):609–622

Guillot C, Lecuit T (2013) Mechanics of epithelial tissue homeostasis and morphogenesis. Science 340(6137):1185–1189

Hales CM, Vaerman J-P, Goldenring JR (2002) Rab11 family interacting protein 2 associates with Myosin Vb and regulates plasma membrane recycling. J Biol Chem 277(52):50415–50421

Hammond GRV, Fischer MJ, Anderson KE, Holdich J, Koteci A, Balla T et al (2012) PI4P and PI (4,5)P2 are essential but independent lipid determinants of membrane identity. Science 337 (6095):727–730

Harlan JE, Hajduk PJ, Yoon HS, Fesik SW (1994) Pleckstrin homology domains bind to phosphatidylinositol-4,5-bisphosphate. Nature 371(6493):168–170

Harris HJ, Clerte C, Farquhar MJ, Goodall M, Hu K, Rassam P et al (2013) Hepatoma polarization limits CD81 and hepatitis C virus dynamics. Cell Microbiol 15(3):430–445

Heasman SJ, Ridley AJ (2008) Mammalian Rho GTPases: new insights into their functions from in vivo studies. Nat Rev Mol Cell Biol 9(9):690–701

Hicks AM, DeLong CJ, Thomas MJ, Samuel M, Cui Z (2006) Unique molecular signatures of glycerophospholipid species in different rat tissues analyzed by tandem mass spectrometry. Biochim Biophys Acta 1761(9):1022–1029

Houk AR, Jilkine A, Mejean CO, Boltyanskiy R, Dufresne ER, Angenent SB et al (2012) Membrane tension maintains cell polarity by confining signals to the leading edge during neutrophil migration. Cell 148(1–2):175–188

Hsu NY, Ilnytska O, Belov G, Santiana M, Chen YH, Takvorian PM, Pau C, van der Schaar H, Kaushik-Basu N, Balla T, Cameron CE, Ehrenfeld E, van Kuppeveld FJ, Altan-Bonnet N (2010) Viral reorganization of the secretory pathway generates distinct organelles for RNA replication. Cell 141(5):799–811. doi:10.1016/j.cell.2010.03.050, PMID: 20510927

Inlora J, Chukkapalli V, Derse D, Ono A (2011) Gag localization and virus-like particle release mediated by the matrix domain of human T-lymphotropic virus type 1 Gag are less dependent on phosphatidylinositol-(4,5)-bisphosphate than those mediated by the matrix domain of HIV-1 Gag. J Virol 85(8):3802–3810

Jacobson C, Mostov K (2007) Catch the KIF5B train to the apical surface. Dev Cell 13(4):457–458

Jean S, Kiger AA (2012) Coordination between RAB GTPase and phosphoinositide regulation and functions. Nat Rev Mol Cell Biol 13(7):463–470

Jeanes A, Smutny M, Leerberg JM, Yap AS (2009) Phosphatidylinositol 3′-kinase signalling supports cell height in established epithelial monolayers. J Mol Histol 40(5–6):395–405

Kato K, Yazawa T, Taki K, Mori K, Wang S, Nishioka T et al (2012) The inositol 5-phosphatase SHIP2 is an effector of RhoA and is involved in cell polarity and migration. Mol Biol Cell 23 (13):2593–2604

Kazmierczak BI, Mostov K, Engel JN (2001) Interaction of bacterial pathogens with polarized epithelium. Annu Rev Microbiol 55:407–435

Kierbel A, Gassama-Diagne A, Mostov K, Engel JN (2005) The phosphoinositol-3-kinase-protein kinase B/Akt pathway is critical for Pseudomonas aeruginosa strain PAK internalization. Mol Biol Cell 16(5):2577–2585

Kierbel A, Gassama-Diagne A, Rocha C, Radoshevich L, Olson J, Mostov K et al (2007) Pseudomonas aeruginosa exploits a PIP3-dependent pathway to transform apical into basolateral membrane. J Cell Biol 177(1):21–27

Kölsch V, Charest PG, Firtel RA (2008) The regulation of cell motility and chemotaxis by phospholipid signaling. J Cell Sci 121(Pt 5):551–559

Kotelevets L, van Hengel J, Bruyneel E, Mareel M, van Roy F, Chastre E (2005) Implication of the MAGI-1b/PTEN signalosome in stabilization of adherens junctions and suppression of invasiveness. FASEB J 19(1):115–117

Krahn MP, Wodarz A (2012) Phosphoinositide lipids and cell polarity: linking the plasma membrane to the cytocortex. Essays Biochem 53:15–27

Krahn MP, Klopfenstein DR, Fischer N, Wodarz A (2010) Membrane targeting of Bazooka/PAR-3 is mediated by direct binding to phosphoinositide lipids. Curr Biol 20(7):636–642

Krauss M, Haucke V (2007) Phosphoinositides: regulators of membrane traffic and protein function. FEBS Lett 581(11):2105–2111
Kutateladze TG (2010) Translation of the phosphoinositide code by PI effectors. Nat Chem Biol 6 (7):507–513
Lapierre LA, Kumar R, Hales CM, Navarre J, Bhartur SG, Burnette JO et al (2001) Myosin vb is associated with plasma membrane recycling systems. Mol Biol Cell 12(6):1843–1857
Laprise P, Viel A, Rivard N (2004) Human homolog of disc-large is required for adherens junction assembly and differentiation of human intestinal epithelial cells. J Biol Chem 279 (11):10157–10166
Lemmon MA (2008) Membrane recognition by phospholipid-binding domains. Nat Rev Mol Cell Biol 9(2):99–111
Lippincott-Schwartz J, Manley S (2009) Putting super-resolution fluorescence microscopy to work. Nat Methods 6(1):21–23
Liu J, Zuo X, Yue P, Guo W (2007) Phosphatidylinositol 4,5-bisphosphate mediates the targeting of the exocyst to the plasma membrane for exocytosis in mammalian cells. Mol Biol Cell 18 (11):4483–4492
Liu S, Yang W, Shen L, Turner JR, Coyne CB, Wang T (2009) Tight junction proteins claudin-1 and occludin control hepatitis C virus entry and are downregulated during infection to prevent superinfection. J Virol 83(4):2011–2014
Macklin WB (2010) The myelin brake: when enough is enough. Sci Signal 3(140):e32
Marée AFM, Grieneisen VA, Edelstein-Keshet L (2012) How cells integrate complex stimuli: the effect of feedback from phosphoinositides and cell shape on cell polarization and motility. PLoS Comput Biol 8(3):e1002402
Martin TF (1998) Phosphoinositide lipids as signaling molecules: common themes for signal transduction, cytoskeletal regulation, and membrane trafficking. Annu Rev Cell Dev Biol 14:231–264
Martin-Belmonte F, Gassama A, Datta A, Yu W, Rescher U, Gerke V et al (2007) PTEN-mediated apical segregation of phosphoinositides controls epithelial morphogenesis through Cdc42. Cell 128(2):383–397
Martini M, Ciraolo E, Gulluni F, Hirsch E (2013) Targeting PI3K in cancer: any good news? Front Oncol 3:108
Mason D, Mallo GV, Terebiznik MR, Payrastre B, Finlay BB, Brumell JH et al (2007) Alteration of epithelial structure and function associated with PtdIns(4,5)P2 degradation by a bacterial phosphatase. J Gen Physiol 129(4):267–283
Massarwa R, Schejter ED, Shilo B-Z (2009) Apical secretion in epithelial tubes of the Drosophila embryo is directed by the Formin-family protein Diaphanous. Dev Cell 16(6):877–888
Mee CJ, Grove J, Harris HJ, Hu K, Balfe P, McKeating JA (2008) Effect of cell polarization on hepatitis C virus entry. J Virol 82(1):461–470
Mee CJ, Harris HJ, Farquhar MJ, Wilson G, Reynolds G, Davis C et al (2009) Polarization restricts hepatitis C virus entry into HepG2 hepatoma cells. J Virol 83(12):6211–6221
Miyoshi J, Takai Y (2008) Structural and functional associations of apical junctions with cytoskeleton. Biochim Biophys Acta 1778(3):670–691
Muthuswamy SK, Xue B (2012) Cell polarity as a regulator of cancer cell behavior plasticity. Annu Rev Cell Dev Biol 28:599–625
Narayanan SP, Flores AI, Wang F, Macklin WB (2009) Akt signals through the mammalian target of rapamycin pathway to regulate CNS myelination. J Neurosci 29(21):6860–6870
Nelson WJ (2003) Adaptation of core mechanisms to generate cell polarity. Nature 422 (6933):766–774
Nelson CM (2013) Forces in epithelial origami. Dev Cell 26(6):554–556
Nelson CM, Bissell MJ (2006) Of extracellular matrix, scaffolds, and signaling: tissue architecture regulates development, homeostasis, and cancer. Annu Rev Cell Dev Biol 22:287–309
Nishio M, Watanabe K, Sasaki J, Taya C, Takasuga S, Iizuka R, Balla T, Yamazaki M, Watanabe H, Itoh R, Kuroda S, Horie Y, Förster I, Mak TW, Yonekawa H, Penninger JM, Kanaho Y,

Suzuki A, Sasaki T (2007) Control of cell polarity and motility by the PtdIns(3,4,5)P3 phosphatase SHIP1. Nat Cell Biol 9(1):36–44, PMID: 17173042

Noren NK, Niessen CM, Gumbiner BM, Burridge K (2001) Cadherin engagement regulates Rho family GTPases. J Biol Chem 276(36):33305–33308

Noseda R, Belin S, Piguet F, Vaccari I, Scarlino S, Brambilla P et al (2013) DDIT4/REDD1/RTP801 is a novel negative regulator of Schwann cell myelination. J Neurosci 33 (38):15295–15305

O'Brien LE, Jou TS, Pollack AL, Zhang Q, Hansen SH, Yurchenco P et al (2001) Rac1 orientates epithelial apical polarity through effects on basolateral laminin assembly. Nat Cell Biol 3 (9):831–838, PMID: 12094219

O'Brien LE, Zegers MMP, Mostov KE (2002) Opinion: Building epithelial architecture: insights from three-dimensional culture models. Nat Rev Mol Cell Biol 3(7):531–537

Ozato-Sakurai N, Fujita A, Fujimoto T (2011) The distribution of phosphatidylinositol 4,5-bisphosphate in acinar cells of rat pancreas revealed with the freeze-fracture replica labeling method. PLoS One 6(8):e23567

Parise LV, Lee J, Juliano RL (2000) New aspects of integrin signaling in cancer. Semin Cancer Biol 10(6):407–414

Payrastre B (2004) Phosphoinositides: lipid kinases and phosphatases. Methods Mol Biol 273:201–212

Pece S, Chiariello M, Murga C, Gutkind JS (1999) Activation of the protein kinase Akt/PKB by the formation of E-cadherin-mediated cell-cell junctions. Evidence for the association of phosphatidylinositol 3-kinase with the E-cadherin adhesion complex. J Biol Chem 274 (27):19347–19351

Pesesse X, Dewaste V, De Smedt F, Laffargue M, Giuriato S, Moreau C et al (2001) The Src homology 2 domain containing inositol 5-phosphatase SHIP2 is recruited to the epidermal growth factor (EGF) receptor and dephosphorylates phosphatidylinositol 3,4,5-trisphosphate in EGF-stimulated COS-7 cells. J Biol Chem 276(30):28348–28355

Pilot F, Philippe J-M, Lemmers C, Lecuit T (2006) Spatial control of actin organization at adherens junctions by a synaptotagmin-like protein Btsz. Nature 442(7102):580–584

Pinal N, Goberdhan DCI, Collinson L, Fujita Y, Cox IM, Wilson C et al (2006) Regulated and polarized PtdIns(3,4,5)P3 accumulation is essential for apical membrane morphogenesis in photoreceptor epithelial cells. Curr Biol 16(2):140–149

Pizarro-Cerdá J, Cossart P (2004) Subversion of phosphoinositide metabolism by intracellular bacterial pathogens. Nat Cell Biol 6(11):1026–1033

Prasad N, Topping RS, Decker SJ (2001) SH2-containing inositol 5′-phosphatase SHIP2 associates with the p130(Cas) adapter protein and regulates cellular adhesion and spreading. Mol Cell Biol 21(4):1416–1428

Reiss S, Rebhan I, Backes P, Romero-Brey I, Erfle H, Matula P et al (2011) Recruitment and activation of a lipid kinase by hepatitis C virus NS5A is essential for integrity of the membranous replication compartment. Cell Host Microbe 9(1):32–45

Reiter JF, Mostov K (2006) Vesicle transport, cilium formation, and membrane specialization: the origins of a sensory organelle. Proc Natl Acad Sci U S A 103(49):18383–18384

Rodríguez-Fraticelli AE, Martín-Belmonte F (2013) Mechanical control of epithelial lumen formation. Small GTPases 4(2):136–140

Roland JT, Kenworthy AK, Peranen J, Caplan S, Goldenring JR (2007) Myosin Vb interacts with Rab8a on a tubular network containing EHD1 and EHD3. Mol Biol Cell 18(8):2828–2837

Roland JT, Bryant DM, Datta A, Itzen A, Mostov KE, Goldenring JR (2011) Rab GTPase-Myo5B complexes control membrane recycling and epithelial polarization. Proc Natl Acad Sci U S A 108(7):2789–2794

Rousso T, Shewan AM, Mostov KE, Schejter ED, Shilo B-Z (2013) Apical targeting of the formin Diaphanous in Drosophila tubular epithelia. eLife 2:e00666

Royer C, Lu X (2011) Epithelial cell polarity: a major gatekeeper against cancer? Cell Death Differ 18(9):1470–1477

Salyers A, Whitt D (eds) (2002) Bacterial pathogenesis: a molecular approach. ASM press, Washington, DC

Santiago-Tirado FH, Legesse-Miller A, Schott D, Bretscher A (2011) PI4P and Rab inputs collaborate in myosin-V-dependent transport of secretory compartments in yeast. Dev Cell 20(1):47–59

Sasaki AT, Firtel RA (2006) Regulation of chemotaxis by the orchestrated activation of Ras, PI3K, and TOR. Eur J Cell Biol 85(9–10):873–895

Sasaki T, Takasuga S, Sasaki J, Kofuji S, Eguchi S, Yamazaki M et al (2009) Mammalian phosphoinositide kinases and phosphatases. Prog Lipid Res 48(6):307–343

Scheid MP, Huber M, Damen JE, Hughes M, Kang V, Neilsen P et al (2002) Phosphatidylinositol (3,4,5)P3 is essential but not sufficient for protein kinase B (PKB) activation; phosphatidylinositol (3,4)P2 is required for PKB phosphorylation at Ser-473: studies using cells from SH2-containing inositol-5-phosphatase knockout mice. J Biol Chem 277(11):9027–9035

Shewan A, Eastburn DJ, Mostov K (2011) Phosphoinositides in cell architecture. Cold Spring Harb Perspect Biol 3(8):a004796

Srinivasan S, Wang F, Glavas S, Ott A, Hofmann F, Aktories K, Kalman D, Bourne HR (2003) Rac and Cdc42 play distinct roles in regulating PI(3,4,5)P3 and polarity during neutrophil chemotaxis. J Cell Biol 160(3):375–385, PMID: 12551955

Snooks MJ, Bhat P, Mackenzie J, Counihan NA, Vaughan N, Anderson DA (2008) Vectorial entry and release of hepatitis A virus in polarized human hepatocytes. J Virol 82(17):8733–8742

Stenzel N, Fetzer CP, Heumann R, Erdmann KS (2009) PDZ-domain-directed basolateral targeting of the peripheral membrane protein FRMPD2 in epithelial cells. J Cell Sci 122 (Pt 18):3374–3384

Subauste MC, Nalbant P, Adamson ED, Hahn KM (2005) Vinculin controls PTEN protein level by maintaining the interaction of the adherens junction protein beta-catenin with the scaffolding protein MAGI-2. J Biol Chem 280(7):5676–5681

Szalinski CM, Guerriero CJ, Ruiz WG, Docter BE, Rbaibi Y, Pastor-Soler NM et al (2013) PIP5KIβ selectively modulates apical endocytosis in polarized renal epithelial cells. PLoS One 8(1):e53790

Takahama S, Hirose T, Ohno S (2008) aPKC restricts the basolateral determinant PtdIns(3,4,5)P3 to the basal region. Biochem Biophys Res Commun 368(2):249–255

Treyer A, Müsch A (2013) Hepatocyte polarity. Compr Physiol 3(1):243–287

Trotard M, Lepère-Douard C, Régeard M, Piquet-Pellorce C, Lavillette D, Cosset F-L et al (2009) Kinases required in hepatitis C virus entry and replication highlighted by small interference RNA screening. FASEB J 23(11):3780–3789

Tyler WA, Gangoli N, Gokina P, Kim HA, Covey M, Levison SW et al (2009) Activation of the mammalian target of rapamycin (mTOR) is essential for oligodendrocyte differentiation. J Neurosci 29(19):6367–6378

Van Zeijl L, Ponsioen B, Giepmans BNG, Ariaens A, Postma FR, Várnai P et al (2007) Regulation of connexin43 gap junctional communication by phosphatidylinositol 4,5-bisphosphate. J Cell Biol 177(5):881–891

Vanhaesebroeck B, Leevers SJ, Ahmadi K, Timms J, Katso R, Driscoll PC et al (2001) Synthesis and function of 3-phosphorylated inositol lipids. Annu Rev Biochem 70:535–602

Vanhaesebroeck B, Vogt PK, Rommel C (2010) PI3K: from the bench to the clinic and back. Curr Top Microbiol Immunol 347:1–19

Vanhaesebroeck B, Stephens L, Hawkins P (2012) PI3K signalling: the path to discovery and understanding. Nat Rev Mol Cell Biol 13(3):195–203

Várnai P, Lin X, Lee SB, Tuymetova G, Bondeva T, Spät A et al (2002) Inositol lipid binding and membrane localization of isolated pleckstrin homology (PH) domains. Studies on the PH domains of phospholipase C delta 1 and p130. J Biol Chem 277(30):27412–27422

Vázquez-Ulloa E, Lizano M, Avilés-Salas A, Alfaro-Moreno E, Contreras-Paredes A (2011) Abnormal distribution of hDlg and PTEN in premalignant lesions and invasive cervical cancer. Gynecol Oncol 122(3):663–668

Von Stein W, Ramrath A, Grimm A, Müller-Borg M, Wodarz A (2005) Direct association of Bazooka/PAR-3 with the lipid phosphatase PTEN reveals a link between the PAR/aPKC complex and phosphoinositide signaling. Development 132(7):1675–1686
Wang L, Boyer JL (2004) The maintenance and generation of membrane polarity in hepatocytes. Hepatology 39(4):892–899
Wang F, Herzmark P, Weiner OD, Srinivasan S, Servant G, Bourne HR (2002) Lipid products of PI(3)Ks maintain persistent cell polarity and directed motility in neutrophils. Nat Cell Biol 4 (7):513–518
Watton SJ, Downward J (1999) Akt/PKB localisation and 3′ phosphoinositide generation at sites of epithelial cell-matrix and cell-cell interaction. Curr Biol 9(8):433–436
Weber SS, Ragaz C, Hilbi H (2009) Pathogen trafficking pathways and host phosphoinositide metabolism. Mol Microbiol 71(6):1341–1352. doi:10.1111/j.1365-2958.2009.06608.x, PMID: 19208094
Weiger MC, Parent CA (2012) Phosphoinositides in chemotaxis. Subcell Biochem 59:217–254
Weiner OD, Neilsen PO, Prestwich GD, Kirschner MW, Cantley LC, Bourne HR (2002) A PtdInsP(3)- and Rho GTPase-mediated positive feedback loop regulates neutrophil polarity. Nat Cell Biol 4(7):509–513
Wilkins C, Woodward J, Lau DT-Y, Barnes A, Joyce M, McFarlane N et al (2013) IFITM1 is a tight junction protein that inhibits hepatitis C virus entry. Hepatology 57(2):461–469
Woodfield RJ, Hodgkin MN, Akhtar N, Morse MA, Fuller KJ, Saqib K et al (2001) The p85 subunit of phosphoinositide 3-kinase is associated with beta-catenin in the cadherin-based adhesion complex. Biochem J 360(Pt 2):335–344
Wu H, Feng W, Chen J, Chan L-N, Huang S, Zhang M (2007) PDZ domains of Par-3 as potential phosphoinositide signaling integrators. Mol Cell 28(5):886–898
Wullschleger S, Wasserman DH, Gray A, Sakamoto K, Alessi DR (2011) Role of TAPP1 and TAPP2 adaptor binding to PtdIns(3,4)P2 in regulating insulin sensitivity defined by knock-in analysis. Biochem J 434(2):265–274
Wymann MP, Marone R (2005) Phosphoinositide 3-kinase in disease: timing, location, and scaffolding. Curr Opin Cell Biol 17(2):141–149
Wymann MP, Schneiter R (2008) Lipid signalling in disease. Nat Rev Mol Cell Biol 9(2):162–176. doi:10.1038/nrm2335, PMID: 18216772
Xiong X, Xu Q, Huang Y, Singh RD, Anderson R, Leof E et al (2012) An association between type Iγ PI4P 5-kinase and Exo70 directs E-cadherin clustering and epithelial polarization. Mol Biol Cell 23(1):87–98
Yan Y, Denef N, Tang C, Schüpbach T (2011) Drosophila PI4KIIIalpha is required in follicle cells for oocyte polarization and Hippo signaling. Development 138(9):1697–1703
Yang HW, Shin M-G, Lee S, Kim J-R, Park WS, Cho K-H et al (2012) Cooperative activation of PI3K by Ras and Rho family small GTPases. Mol Cell 47(2):281–290
Yap AS, Kovacs EM (2003) Direct cadherin-activated cell signaling: a view from the plasma membrane. J Cell Biol 160(1):11–16
Yu W, Shewan AM, Brakeman P, Eastburn DJ, Datta A, Bryant DM et al (2008) Involvement of RhoA, ROCK I and myosin II in inverted orientation of epithelial polarity. EMBO Rep 9 (9):923–929
Zhang J, Liu Z, Rasschaert J, Blero D, Deneubourg L, Schurmans S et al (2007) SHIP2 controls PtdIns(3,4,5)P(3) levels and PKB activity in response to oxidative stress. Cell Signal 19 (10):2194–2200

Part III
Cell Polarity and Cell–Cell Interactions: The Role of Cell Adhesion

Chapter 11
Immunological Synapse Formation: Cell Polarity During T Cell–APC Interaction

Xin Liu and Morgan Huse

Abstract Intercellular communication is essential for coordinating a successful immune response. The immunological synapse provides a platform for such communication by coupling activated lymphocytes specifically to their target cells. The formation of an IS in T cells is triggered by stimulation of the T cell antigen receptor and depends on the rapid and polarized remodeling of the cytoskeleton. This dramatic change in cell polarity is required to achieve optimal T cell activation and function. In this chapter, we will discuss the structure of the IS, how it forms, and how it contributes to T cell function during immune responses.

Keywords Actin • Cell polarity • Centrosome • Immunological synapse • T cell

11.1 Synapse Formation in T Cells

Lymphocytes function by specific communication with small numbers of target cells in densely packed environments filled with largely irrelevant bystanders. To maintain specificity in this context, lymphocytes form a specialized cell–cell junction with their targets known as an immunological synapse (IS) (Dustin et al. 2010). IS formation restricts the scope of certain effector responses, thereby limiting inappropriate inflammation, cytotoxicity, and signaling in the surrounding tissue. The IS enables cytotoxic T lymphocytes (CTLs) and natural killer (NK) cells to kill infected or transformed target cells selectively (Stinchcombe and Griffiths

X. Liu (✉)
Immunology Program, Memorial Sloan-Kettering Cancer Center, 1275 York Avenue, Box 514, New York, NY 10065, USA

Immunology and Microbial Pathogenesis Graduate Program, Weill Cornell Graduate School of Medical Sciences, New York, NY 10065, USA
e-mail: xinliu1128@gmail.com

M. Huse
Immunology Program, Memorial Sloan-Kettering Cancer Center, 1275 York Avenue, Box 514, New York, NY 10065, USA
e-mail: husem@mskcc.org

K. Ebnet (ed.), *Cell Polarity 1*, DOI 10.1007/978-3-319-14463-4_11

2007). It has also been implicated in antigen gathering by B cells (Harwood and Batista 2011), the directional secretion of cytokines by T cells (Huse et al. 2006; Kupfer et al. 1991; Poo et al. 1988), and the specific delivery of T cell help to B cells (Boisvert et al. 2004).

This review will focus on the T cell IS, which is the most extensively studied lymphocyte synapse with respect to its ontogeny and function. Mature T cells are roughly divided into two subgroups based on their expression of the coreceptors CD4 and CD8. $CD4^+$ T cells, which are also called "helper T cells," coordinate immune responses by communicating with other leukocytes, typically via cell–cell contact or the secretion of specific cytokines. $CD8^+$ T cells, for their part, differentiate into CTLs capable of killing target cells presenting their cognate antigen.

In T cells, IS formation is initiated by T cell receptor (TCR) binding of cognate peptide major histocompatibility complexes (pMHC) on the surface of the antigen-presenting cell (APC). This is in many respects the signature moment in a T cell's life. In addition to driving IS assembly, antigen recognition induces the robust proliferation, differentiation, and cytokine secretion that are typically associated with lymphocyte activation. Within seconds of TCR engagement, the T cell begins to spread over the surface of the APC, forming a radially symmetric contact that evolves into a mature IS in a matter of minutes. The IS has long been characterized by its stereotyped concentric domains, referred to by some as supramolecular activation clusters (SMACs) (Dustin et al. 2010) (Fig. 11.1). At the very center of the interface is the central SMAC (cSMAC), which contains a large cluster of TCR. Costimulatory receptors such as CD2 and CD28, which boost lymphocyte activation in certain contexts, also accumulate in this region. Surrounding the cSMAC is the peripheral SMAC (pSMAC), which is dominated by the $\alpha_L\beta_2$ integrin LFA-1 and essentially serves as an adhesion ring. Surrounding the pSMAC at the very edge

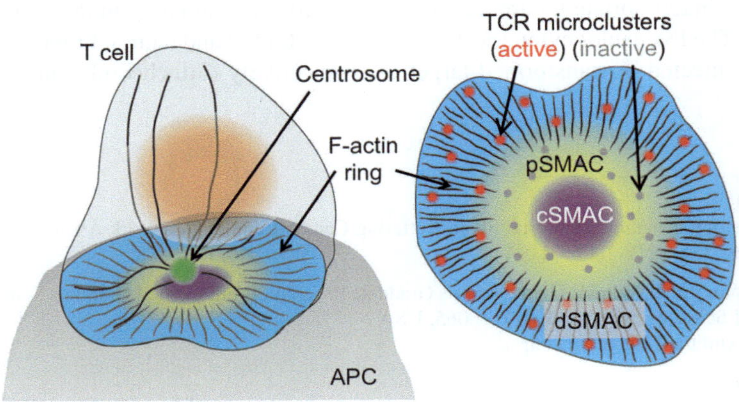

Fig. 11.1 Schematic diagram of the T cell IS. A side view is shown to the *left* and an en face view (from the perspective of the APC) to the *right*. Microtubules are drawn as *thick black lines* and F-actin is *thin gray lines*. The nucleus is *orange*. Other important components are labeled

of the synapse is the distal SMAC (dSMAC), which is enriched in filamentous actin (F-actin) and regulatory molecules such as the phosphatase CD45.

This reorganization of the T cell–APC interface is accompanied and enabled by the polarized remodeling of the T cell cytoskeleton (Gomez and Billadeau 2008; Huse et al. 2013) (Fig. 11.1). IS growth over the surface of the APC is driven by robust actin polymerization beneath the plasma membrane. Subsequently, this cortical F-actin resolves into the dSMAC, which maintains synaptic size and promotes the compartmentalization of cell surface proteins into their appropriate domains. Concomitant with F-actin ring formation, the centrosome (also called the microtubule-organizing center, or MTOC) reorients to a position just beneath the cSMAC, taking along with it the Golgi apparatus, endosomal compartment, and other vesicular organelles. This apposes the secretory machinery of the T cell with the plasma membrane at the center of the IS, which is relatively free of cortical F-actin and therefore available for vesicular fusion. In this manner, centrosome reorientation, coupled to synaptic F-actin ring formation, facilitates the release of cytokines and cytolytic factors toward the APC.

Although the polarized configuration of the centrosome, F-actin ring, and SMACs can persist for tens of minutes, even hours, the mature synapse is in actuality a highly dynamic structure requiring sustained antigenic input for its maintenance. Recognition of pMHC induces the formation of distinct TCR "microclusters" in the plasma membrane, which then traffic centripetally toward the center of the IS where they coalesce into the cSMAC (Yokosuka and Saito 2010) (Fig. 11.1). Studies from a number of labs indicate that TCR microclusters actively signal in the periphery but become downregulated in the cSMAC (Mossman et al. 2005; Vardhana et al. 2010; Varma et al. 2006; Yokosuka et al. 2005). Hence, new peripheral microclusters must be generated continuously in order to sustain the T cell activation that drives synaptic architecture. In turn, proper synaptic architecture shapes T cell activation and effector function. In the sections that follow, we will first discuss what is known about the molecular mechanisms coupling TCR activation to cytoskeletal reorganization at the IS. Then, we will discuss the functional consequences of these remodeling events.

11.2 Centrosome Reorientation

Centrosome polarization has been documented in a number of adherent cell types, including neurons, fibroblasts, and astrocytes, where it contributes to the directionality of cell migration (Li and Gundersen 2008). Studies in fibroblasts have demonstrated that this process results from the rearrangement of organelles, with minimal displacement of the centrosome itself (Gomes et al. 2005; Luxton et al. 2010). The reorientation response in T cells differs from these systems in two important respects. First, it features net translocation of the centrosome to the plasma membrane. Second, the process is much more dynamic; both the acquisition and the dissolution of the polarized state occur in minutes, as opposed to hours.

This reflects the transient nature of even the most stable lymphocyte cell–cell interactions. Hence, although one might expect that certain aspects of adherent cell polarity are conserved in T cells, it is reasonable to predict that there are other, more unique features that have been tailored to the specific needs of the T cell response. This expectation has, for the most part, been largely borne out by research.

11.2.1 Coupling TCR Activation to Polarizing Signals

It has been known for some time that TCR signaling drives centrosome reorientation to the IS. The centrosome is particularly responsive to the TCR and will faithfully polarize toward the APC even if integrins and costimulatory receptors are engaged more robustly at a different interface (Sedwick et al. 1999). TCR engagement rapidly induces the phosphorylation of immunotyrosine-based activation motifs (ITAMs) in its associated CD3 chains by the Src kinase Lck. These phosphorylated ITAMs then recruit the Syk family kinase Zap70, which in turn phosphorylates the membrane-associated scaffolding molecule LAT (linker for the activation of T cells). LAT then binds to another scaffolding molecule, Slp76, and the resulting LAT-Slp76 complex serves as a platform for recruitment of important downstream signaling molecules. These receptor-proximal components all colocalize with the TCR in signaling microclusters. Many of them, including Lck, Zap70, LAT, and Slp76, are also required for centrosome reorientation (Kuhne et al. 2003; Lowin-Kropf et al. 1998), indicative of the importance of TCR signaling for this process.

One of the key enzymes recruited to the LAT-Slp76 complex is phospholipase-Cγ (PLCγ), which hydrolyzes phosphatidylinositol 4,5-bisphosphate (PIP_2) in the plasma membrane to yield two second messengers, inositol trisphosphate (IP_3) and diacylglycerol (DAG) (Huse 2009). IP_3 diffuses into the cytoplasm, where it triggers calcium (Ca^{2+}) flux through the IP_3 receptor-STIM-Orai1 pathway. DAG, for its part, accumulates in the plasma membrane, forming a gradient that is centered at the IS (Spitaler et al. 2006). Pharmacological inhibition of PLCγ abrogates centrosome reorientation (Quann et al. 2009), strongly implicating one or both of these second messengers in the polarization response. Indeed, recent studies indicate that the synaptic DAG gradient plays an essential role (Quann et al. 2009) (Fig. 11.2). High-resolution imaging experiments have revealed that DAG accumulation at the IS consistently precedes centrosome reorientation by 10–15 s. Perturbations that disrupt the ability of the T cell to maintain or respond to synaptic DAG accumulation block polarization responses. Furthermore, localized photoactivation of caged DAG in the plasma membrane can induce transient centrosomal recruitment. The role of Ca^{2+} signaling, however, is more controversial, with certain studies suggesting it is necessary and others that it is dispensable (Kuhne et al. 2003; Quann et al. 2009; Yi et al. 2013; Maul-Pavicic et al. 2011). Interpretation of these results is complicated by the fact that Ca^{2+} flux is required

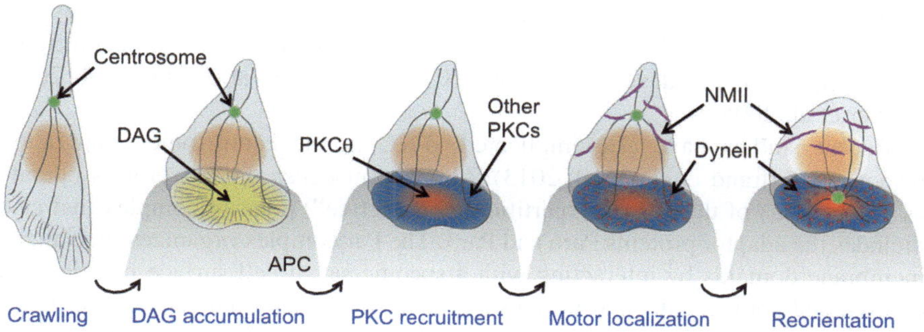

Fig. 11.2 Schematic diagram of the centrosome reorientation response. Microtubules are drawn as *thick black lines* and F-actin is *thin gray lines*. The nucleus is *orange*. Other important components are labeled

for strong adhesion to the APC (Huse et al. 2013). Hence, the centrosome reorientation phenotype observed in the absence of Ca^{2+} could be secondary to a defect in productive contact formation. Alternatively, it is possible that Ca^{2+} is not necessary for initial centrosome polarization, but is required for maintenance of the polarized state.

DAG transduces signals by recruiting proteins that contain DAG-binding C1 domains. Prominent among this class of molecules is the protein kinase C (PKC) family of serine–threonine kinases, which transduce plasma membrane-derived signals into a myriad of downstream responses (Newton 2010). PKCs are divided into three subfamilies based on their regulatory properties. Conventional PKCs (cPKCs—PKCα, PKCβ, and PKCγ) require both Ca^{2+} and DAG for their activation, novel PKCs (nPKCs—PKCδ, PKCε, PKCη, PKCθ) require DAG but not Ca^{2+}, and atypical PKCs (aPKCs, PKCζ, PKCι) require neither DAG nor Ca^{2+} and are instead regulated by protein–protein interactions. Consistent with the importance of synaptic DAG for T cell polarity, it was recently found that three nPKC isoforms, PKCε, PKCη, and PKCθ, are involved in centrosome reorientation (Quann et al. 2011). All three of these enzymes accumulate at the IS prior to the arrival of the centrosome, and they do so in a DAG-dependent manner (Fig. 11.2). PKCε and PKCη are recruited first (~15 s before the centrosome), and they occupy the entirety of the synaptic membrane. By contrast, PKCθ arrives ~5 s later and occupies a more restricted zone that includes only the cSMAC and the pSMAC. The distinct recruitment behavior displayed by these nPKCs is mirrored by differences in their functional properties. Thus, whereas siRNA-mediated suppression of PKCθ alone impairs centrosome polarization, simultaneous knockdown of both PKCε and PKCη is required to inhibit the response. Furthermore, T cells lacking both PKCε and PKCη also exhibit defective PKCθ recruitment to the IS. Taken together, these data indicate that PKCε and PKCη operate redundantly with each other and do so in a manner upstream of PKCθ. Although the molecular basis for this differential behavior is not clear, it is consistent with the fact that PKCε and PKCη are more closely related to each other than to PKCθ or any other member of the PKC family (Huse et al. 2013). Redundancy between PKCε and PKCη may also explain why mice lacking either isoform alone display weak or undetectable phenotypes in T cell development and TCR-induced activation (Fu et al. 2011; Gruber et al. 2005).

Recent studies have also implicated aPKC in the centrosome polarization response (Bertrand et al. 2010, 2013). In adherent cells, aPKC isoforms operate as components of the Par (for "partitioning defective") polarity complex that also includes the adaptor proteins Par6 and Par3. The Par complex organizes specialized membrane domains by interacting with a specific set of cell surface proteins and cytoskeletal components, and it plays a crucial role in the establishment of polarized structures such as neuronal axons and the apical surface of epithelial cells (Etienne-Manneville and Hall 2003; Li and Gundersen 2008). In T cells, both phosphorylated PKCζ and Par3 have been observed at the IS (Bertrand et al. 2010; Ludford-Menting et al. 2005), suggestive of a role in cytoskeletal polarization. Consistent with this notion, centrosome reorientation is hindered by pretreatment with a small molecule PKCζ inhibitor and also by expression of a dominant-negative PKCζ mutant (Bertrand et al. 2010, 2013). Furthermore, a dominant-negative version of Par1b, a kinase that shapes the localization of the Par complex, also disrupts centrosome polarization responses (Lin et al. 2009). Taken together, these data strongly suggest that aPKC signaling plays an important role in T cell polarity. Precisely how it is coupled to upstream signals emanating from the TCR is not known. In other systems, the Par complex is activated by the GTPase Cdc42 (Etienne-Manneville and Hall 2003). Whether Cdc42 contributes to centrosome polarization in T cells remains controversial, however (Chemin et al. 2012; Gomez et al. 2007; Stowers et al. 1995). Alternatively, it is possible that the Par complex might be regulated by lipid-based signaling. In that regard, it is interesting to note that phosphatidic acid (PA), a lipid second messenger that is generated from DAG by diacylglycerol kinases, is an allosteric activator of PKCζ and has also been shown to bind directly to Bazooka, the drosophila ortholog of Par3 (Limatola et al. 1994; Yu and Harris 2012). These results imply that aPKC, like the nPKCs, might also operate downstream of DAG in this system.

There is now good evidence that the recruitment and polarized activation of nPKCs and aPKCs is required for centrosome reorientation in T cells. Precisely how the activity of these enzymes contributes to the polarization process is still largely unclear and awaits the identification of relevant PKC substrates capable of transducing signals to the cytoskeletal machinery.

11.2.2 Molecular Motors That Move the Centrosome

Microtubules radiate from the centrosome with a uniform minus end to plus end configuration. Hence, a minus end-directed microtubule motor anchored at the IS could in principle drag the centrosome toward it, much like a fisherman reeling in his catch. The elegance of this model has motivated a number of labs to investigate cytoplasmic dynein, the sole minus end-directed microtubule motor available in the cytoplasm of most eukaryotic cells. Dynein is a large, multisubunit protein that carries out a diverse set of cellular functions, which range from organelle trafficking to positioning of the mitotic spindle (Kardon and Vale 2009). It is composed of two heavy chains, which contain the motor domains, and a number of intermediate and light chains, which stabilize the assembly and mediate interactions with cargo. TCR stimulation induces robust synaptic recruitment of dynein that precedes the arrival of the centrosome by ~5 s (Combs et al. 2006; Martin-Cofreces et al. 2008; Quann et al. 2009) (Fig. 11.2). Furthermore, suppression of dynein protein or dynein activity inhibits centrosome reorientation in primary T cells and T cell lines (Liu et al. 2013; Martin-Cofreces et al. 2008; Yi et al. 2013), indicative of an important role in the response. It has generally been thought that dynein moves the centrosome by walking along the microtubules that radiate from it. Recent data, however, suggests an alternative model in which dynein at the center of the IS captures microtubules and then induces their depolymerization from the plus end (Yi et al. 2013). This end-on shrinkage effectively drags the centrosome toward the IS.

Although there is general agreement that synaptic dynein accumulation is a prerequisite for centrosome reorientation, the molecular mechanisms mediating this polarized accumulation are not entirely clear. In Jurkat T cells, dynein recruitment to the IS requires ADAP (Combs et al. 2006), a scaffolding protein that associates with the TCR signaling complex via Slp76. ADAP knockdown impairs centrosome reorientation in this system, consistent with the notion that ADAP binding is crucial for dynein-mediated polarization. ADAP is not required for centrosome reorientation in primary mouse T cells (Combs et al. 2006), however, casting doubt on the generalizability of this mechanism. DAG and nPKC signaling also appears to play an important role. Indeed, TCR-induced dynein accumulation can be blocked both by disruption of the synaptic DAG gradient and by shRNA-mediated suppression of nPKCs (Liu et al. 2013). That being said, global activation or inhibition of PKCs has no effect on the distribution of cortical dynein in T cells (Liu et al. 2013). Taken together, these data suggest that the mechanism linking nPKCs and dynein is likely to be more complex than the simple linear activation of the latter by the former. It is tempting to speculate that the dynein recruitment system may be designed to read out the gradient of PKC activity rather than its absolute level.

Interestingly, a considerable amount of residual centrosome reorientation can be observed in the absence of dynein (Hashimoto-Tane et al. 2011; Liu et al. 2013; Martin-Cofreces et al. 2008), implying that other molecular motors also participate in the process. Consistent with this interpretation, it was recently shown that nonmuscle myosin II (NMII), an actin-based motor, collaborates with dynein to mediate centrosome reorientation (Liu et al. 2013). Simultaneous suppression of both dynein and NMII activity induces a much more pronounced polarization defect than that observed in cells lacking either motor alone. Although these functional data are indicative of a partially redundant relationship, imaging experiments suggest that dynein and NMII drive centrosome reorientation in different ways (Liu et al. 2013). Thus, whereas dynein accumulates at the site of TCR stimulation, NMII forms transient clusters in the membrane behind the centrosome (Fig. 11.2). These cortical NMII clusters are not induced by TCR signaling. Rather, they appear to form constitutively and are locally repressed at the site of TCR stimulation. This introduces asymmetry into the overall distribution of NMII, which is then coupled to centrosome movement. Cortical NMII dynamics appear to be regulated directly by nPKCs (Liu et al. 2013). TCR-induced depletion of NMII is blocked in cells lacking PKCε, PKCη, and PKCθ and in cells treated with a broad specificity PKC inhibitor. Furthermore, global PKC activation with phorbol ester drives NMII clusters off the membrane, whereas inhibition of PKC activity intensifies clustering. PKCs are known to regulate NMII activity and localization via phosphorylation of three amino acids at the N-terminus of the myosin regulatory light chain (MyoRLC) (Bengur et al. 1987; Ikebe et al. 1987; Ikebe and Reardon 1990; Komatsu and Ikebe 2007). Mutation of two out of three of these residues inhibits TCR-induced NMII remodeling (Liu et al. 2013), strongly suggesting that MyoRLC is a direct substrate of nPKCs in this system. Collectively, these results are consistent with a model whereby dynein "pulls" the centrosome from the front while NMII "pushes" it from behind (Fig. 11.2).

Precisely how NMII clusters move the centrosome is not entirely clear. It is known, however, that the NMII-dependent component of the response requires microtubules (Liu et al. 2013). This suggests that contractile forces induced by NMII in the cortex are propagated along microtubules to the centrosome. In that regard, it is interesting that several diaphanous formin proteins, including FMNL1, DIA1, and INF2, have been implicated in centrosome reorientation (Andres-Delgado et al. 2012; Gomez et al. 2007). Although they are best known as inducers unbranched F-actin growth, formins are also required for the formation of long-lived microtubules containing detyrosinated tubulin monomers (Chesarone et al. 2010). It has been proposed that these detyrosinated microtubules promote centrosome reorientation by "pushing" on the back of the cell (Andres-Delgado et al. 2013). Whether NMII-based contraction serves as the backstop for this force transduction remains to be seen and is an interesting topic for future research. It is worth noting that other regulators of microtubule stability and growth dynamics, including stathmin (Filbert et al. 2012), casein kinase 1δ (Zyss et al. 2011), and histone deacetylase 6 (Serrador et al. 2004), have also been implicated in the polarization response. Hence, centrosome dynamics in T cells depends not only on the distribution of molecular motors but also on the local disposition of microtubules. These overlapping and compensatory layers of control may appear somewhat excessive for a process as seemingly straightforward as centrosome reorientation. The T cell cytoplasm is a complex and crowded environment, however, and the machinery that moves the centrosome must be robust enough to overcome physical impediments that render one or more of its components inoperative.

11.3 F-Actin Ring Formation

In contrast to the centrosome, which in most cells is a unique entity, actin serves as a building block for myriad distinct cellular structures. This complicates experimental analysis, as care must be taken to focus on the relevant pool of F-actin in both space and time. It is becoming increasingly clear that, although there are a large number of potential F-actin regulators that play important roles in T cell activation, only a subset of these proteins are directly involved in the establishment and maintenance of the synaptic F-actin ring.

11.3.1 Regulation of Actin Polymerization by Rho Family GTPases

The Rho family GTPases Rac, Rho, and Cdc42 are widely recognized as "master regulators" of the cortical actin cytoskeleton (Jaffe and Hall 2005). Like all small GTPases, Rho family members cycle between an inactive, GDP-bound form and an active, GTP-bound form capable of recruiting downstream effector molecules. Rac activity is typically associated with the growth of sheetlike lamellipodial structures, while Rho generally promotes myosin-based contractility and the formation of stress fibers. Cdc42, for its part, drives fingerlike projections such as filopodia and has also been linked to polarity induction in multiple adherent cell types. Over the past decade, the roles played by each of these GTPases during T cell activation have been assessed, and consequently a coherent picture of how they influence synaptic F-actin dynamics is beginning to emerge.

IS growth and F-actin ring formation appear to be primarily dependent on Rac. This is not particularly surprising, given that the mature IS resembles a radially symmetric lamellipodium (Dustin 2007). TCR stimulation induces rapid and sustained Rac activation that occurs concurrently with IS formation (Ku et al. 2001; Le Floc'h et al. 2013; Sanui et al. 2003). In addition, shRNA-mediated suppression of either Rac1 or Rac2, the two isoforms expressed in T cells, reduces cell spreading and impairs F-actin ring formation (Le Floc'h et al. 2013). Simultaneous knockdown of both isoforms induces more pronounced defects in IS growth and organization, indicating that the two proteins function additively in this context. Expression of a constitutively active Rac1 mutant (G12V) has the opposite effect, robustly enhancing synaptic actin polymerization and IS growth (Le Floc'h et al. 2013).

Rac most likely mediates these effects through WAVE2, a member of the nucleation promoting factor (NPF) family of actin regulators (Derivery and Gautreau 2010) (Fig. 11.3). WAVE2 functions as part of a multiprotein assembly that also contains Sra-1, Nap1, Abi-1/2, and HSPC300. Activated Rac binds directly to Sra-1, thereby recruiting and organizing WAVE2-dependent F-actin growth. In Jurkat T cells, siRNA-mediated suppression of WAVE2 leads to a

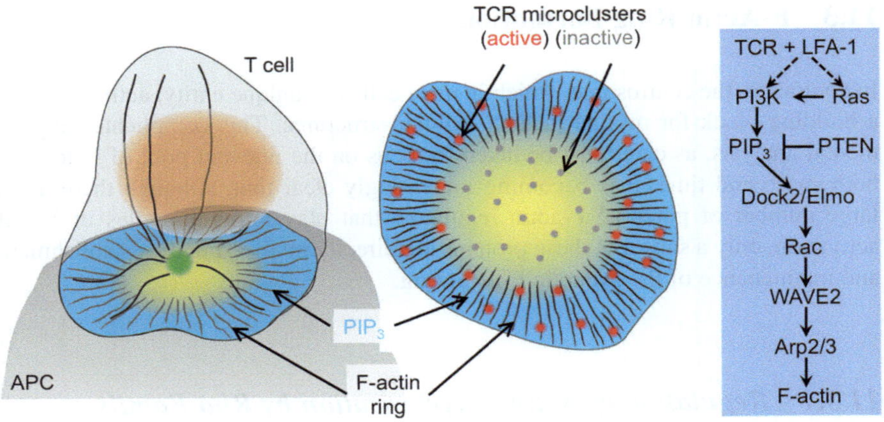

Fig. 11.3 Schematic diagram of synaptic F-actin ring formation. A side view is shown to the *left* and an en face view (from the perspective of the APC) in the center. Microtubules are drawn as *thick black lines* and F-actin is *thin gray lines*. The nucleus is *orange*. Other important components are labeled. A pathway diagram linking TCR and LFA-1 stimulation to F-actin polymerization is shown to the *right*

dramatic reduction in TCR-induced actin polymerization and cell spreading (Nolz et al. 2006; Zipfel et al. 2006), indicative of a crucial role for the protein in establishing IS architecture. Consistent with this interpretation, components of the WAVE2 complex localize to the leading edge of the synaptic lamellipodium during IS growth and also accumulate within the F-actin ring as the IS matures (Nolz et al. 2006; Zipfel et al. 2006; Le Floc'h et al. 2013). Most NPF proteins function by recruiting and activating the Arp2/3 complex, which induces actin polymerization from the sides of existing filaments, thereby generating branched arrays. In Jurkat cells, suppression of Arp2 or Arp3 dramatically reduces TCR-induced F-actin accumulation (Gomez et al. 2007), indicative of a key role for the complex in IS architecture. The Arp2/3 loss-of-function phenotype, however, is less pronounced than that of WAVE2; residual, filopodial structures persist at the T cell–APC interface, likely resulting from formin activity. Furthermore, WAVE2-deficient T cells exhibit additional signaling phenotypes that have not been reported in cells lacking Arp2 or Arp3 (Gomez et al. 2007; Nolz et al. 2006, 2008; Zipfel et al. 2006). Collectively, these results suggest that WAVE2 has both Arp2/3-dependent and Arp2/3-independent functions at the IS.

The role of Cdc42 in the establishment and maintenance of synaptic F-actin is, in contrast to Rac, largely unclear. Like Rac, Cdc42 is robustly activated by TCR signaling (Cannon et al. 2001). Loss-of-function experiments, however, have found that Cdc42 is not strictly required for T cell activation and synaptic F-actin accumulation (Chemin et al. 2012; Gomez et al. 2007). In T cells, Cdc42 is thought to signal through WASp, another prominent member of the NPF family (Derivery and Gautreau 2010). Mutations in WASp lead to Wiskott-Aldrich syndrome (WAS), an X-linked recessive disease in humans characterized by thrombocytopenia and immune deficiency (Ochs 1998). The majority of cellular WASp is associated with WIP (for WASp interacting protein), an adaptor protein that regulates WASp activity (Martinez-Quiles et al. 2001; Sasahara et al. 2002). Consistent with an important role in immune function, WASp-deficient T cells display defects in differentiation and effector responses (Dupre et al. 2002; Gallego et al. 1997; Molina et al. 1993; Snapper et al. 1998; Trifari et al. 2006; Zhang et al. 1999). Synaptic F-actin accumulation, however, is largely intact (Cannon and Burkhardt 2004), indicating that WASp is not required for the establishment of IS architecture. It does, however, appear to be necessary for IS stabilization and maintenance (Sims et al. 2007). WASp$^{-/-}$ T cells form unstable synapses on stimulatory lipid bilayers that quickly "break symmetry," enabling migration to other locations. T cells lacking PKCθ have the opposite phenotype, forming hyperstable, relatively immobile synapses. Importantly, PKCθ can modulate WASp activity via phosphorylation of WIP (Sasahara et al. 2002), providing a possible mechanistic basis for the reciprocal effects of the two proteins on IS stability. Although the functional significance of these observations remains to be seen, it is interesting to note that WASp$^{-/-}$ and PKCθ$^{-/-}$ T cells have reciprocal developmental phenotypes: loss of WASp inhibits helper T cell differentiation into the T_H1 lineage, while loss of PKCθ inhibits T_H2 development (Marsland et al. 2004; Trifari et al. 2006). This raises the intriguing possibility that IS stability may be coupled to T cell differentiation.

WASp$^{-/-}$ T cells also display a specific defect in IL-2 secretion that is seemingly independent of stable IS formation (Cannon and Burkhardt 2004; Zhang et al. 1999). Interestingly, it was recently found that loss of Cdc42 affects T cell secretion of IFNγ (Chemin et al. 2012). Whether these phenotypes are related is not known, and it is unclear if they have anything to do with actin dynamics. Finally, WASp has been implicated in ligand-induced internalization of the TCR (Zhang et al. 1999). Although the basis for this role is poorly understood, it is consistent with the fact that WASp colocalizes with the TCR in plasma membrane microclusters, in contrast to WAVE2, which accumulates in the synaptic F-actin ring.

Little is known about how Rho influences IS structure, if at all. RhoA, the predominant isoform expressed in T cells, is required for thymocyte development and appears to play an important role in the regulation of integrin-mediated adhesion during T cell migration (Corre et al. 2001; Galandrini et al. 1997; Henning et al. 1997; Laudanna et al. 1996). Loss-of-function experiments using a cell permeable form of a dominant-negative RhoA mutant, however, suggest that the protein is not required for F-actin accumulation at the IS (Tskvitaria-Fuller et al. 2006). Additional studies, preferably using T cells that lack endogenous RhoA, are required in order to draw a more definitive conclusion. Interestingly, a distantly related isoform, RhoG, has been implicated in TCR internalization and trafficking (Martinez-Martin et al. 2011). Whether it operates in the same downregulation pathway as Cdc42 and WASp is unclear, however. Taken together, the extant data indicate that Rac signaling is primarily responsible for synaptic growth and F-actin ring formation, with Cdc42 and Rho influencing other features, such as IS stability and the surface distribution of TCR.

11.3.2 Signaling Pathways That Shape GTPase Activity

As with centrosome reorientation, TCR signaling plays a critical role in driving F-actin dynamics at the IS. Indeed, a number of receptor-proximal kinases and adaptors, including Lck, Zap70, LAT, and Slp76, are absolutely required for synaptic F-actin accumulation and IS growth (Bubeck Wardenburg et al. 1998; Bunnell et al. 2001; Morgan et al. 2001; Zeng et al. 2003). Determining how these proteins influence F-actin regulators has been a subject of intense interest over the last decade. Much of the effort in this area has focused on guanine nucleotide exchange factors (GEFs) that might couple Rho GTPases to upstream signals emanating from the TCR. The Dbl-family GEF Vav is probably the most extensively studied protein in this class. It is a large, multisubunit adaptor that contains a number of protein–protein interaction modules in addition to its GEF domain, which can activate both Rac and Cdc42 (Tybulewicz 2005). Vav directly associates with the LAT-Slp76 complex and is an integral component of TCR microclusters (Huse 2009; Sylvain et al. 2011). As such, it is ideally positioned to transduce signals from receptor-proximal machinery to the actin cytoskeleton. Indeed, loss of Vav-1, the predominant isoform in T cells, inhibits synaptic F-actin enrichment and the formation of strong T cell–APC conjugates (Ardouin et al. 2003; Krawczyk et al. 2002; Miletic et al. 2009). However, Vav-1$^{-/-}$ T cells exhibit a wide range of additional phenotypes (Fischer et al. 1998; Penninger et al. 1999; Tarakhovsky et al. 1995; Zhang et al. 1995; Costello et al. 1999; Holsinger et al. 1998; Reynolds et al. 2002, 2004), including defects in LAT-Slp76 complex assembly and PLC-γ activation. This reduces downstream signaling through the Ca^{2+}, MAP kinase, and NF-κB pathways, leading to impaired proliferation and cytokine secretion. Vav-1$^{-/-}$ mice also display a partial block in thymocyte development. While these data leave little doubt that Vav is critically important for T cell activation, it is difficult to distinguish a role for Vav GEF activity in the direct activation of Rac or Cdc42 from a role as an essential scaffold within the LAT-Slp76 complex. Recently, this issue was addressed by reconstituting Vav-deficient primary T cells with a mutant of Vav-1 lacking GEF activity (Miletic et al. 2009). Surprisingly, this protein completely rescued T cell development and function. Although this result remains controversial, it has clearly cast doubt on the notion that Vav GEF activity is required for the establishment and maintenance of IS structure.

T cells also express Dock2, a Rac-specific GEF from the atypical CDM (CED-5, Dock180, myoblast city) family (Cote and Vuori 2007). Members of this family contain a GEF module called the DHR-2 (for Dock homology region 2) domain in addition to a conserved DHR-1 domain that binds to phosphatidylinositol trisphosphate (PIP_3). Certain CDM proteins, including Dock2, also associate constitutively with the adaptor Elmo, which stabilizes their expression and promotes their activity. T cells lacking Dock2 form miniaturized synapses that mediate suboptimal adhesion to the APC (Le Floc'h et al. 2013), indicative of a crucial role for the protein in synaptic F-actin dynamics. This structural phenotype is associated with dramatic reductions in both basal- and TCR-induced Rac activation (Le Floc'h et al. 2013; Sanui et al. 2003), as one would expect for a Rac-specific GEF. Importantly, other TCR-driven signaling events, including AKT phosphorylation, Ca^{2+} flux, and centrosome polarization, are unaffected by loss of Dock2, indicating that the disruption in IS architecture is not secondary to a defect in upstream TCR signaling. Consistent with this interpretation, the Dock2-Elmo complex does not localize in TCR microclusters, but rather in the periphery of the IS, overlying the F-actin ring (Le Floc'h et al. 2013) (Fig. 11.3). Hence, Dock2 specifically coordinates the actin cytoskeleton downstream of TCR proximal signaling events.

Recent studies have indicated that the synaptic accumulation pattern of Dock2 is dictated by PIP_3, which engages Dock2 via its DHR-1 domain (Le Floc'h et al. 2013). PIP_3 forms a striking annular gradient in the mature IS that superimposes with F-actin (Fig. 11.3). Blocking PIP_3 production with PI3K inhibitors or shRNA against key PI3K isoforms prevents Dock2/Elmo recruitment to the IS, reduces IS size, and abrogates TCR-induced Rac activation. Conversely, suppression of PTEN, a lipid phosphatase that antagonizes PI3K signaling in multiple contexts, enhances Rac activation and IS growth. Interestingly, inhibition or suppression of PI3K also disrupts F-actin ring formation, a phenotype that is not observed after depletion of Dock2. This suggests that other PIP_3-dependent regulators collaborate with the Dock2–Elmo complex to shape F-actin at the IS. In that regard, it is interesting to note that PI3K activity has also been linked to the activation of cofilin, an actin regulator that is required for IS formation and downstream effector responses (Samstag et al. 2013; Wabnitz et al. 2006).

Although there is some debate on the subject (Alcazar et al. 2007), most studies indicate that TCR-induced PIP_3 production is mediated by class IA PI3K isoforms (Garcon et al. 2008; Le Floc'h et al. 2013; Sauer et al. 2008). These heterodimeric enzymes contain one of three catalytic subunits (p110α, p110β, or p110δ) paired with a regulatory subunit in the p85/p55 family (Fruman and Bismuth 2009). p85/p55 proteins bear two SH2 domains that enable recognition of sequences containing phosphotyrosine (pTyr), and they generally couple PI3K activity to tyrosine kinase signaling pathways. Imaging experiments have demonstrated that class IA PI3K isoforms are recruited to TCR microclusters (Le Floc'h et al. 2013), which contain a number of pTyr containing proteins that can interact directly with p85 (Shim et al. 2004, 2011; Zhang et al. 1998). This process likely stimulates PI3K activity, as well, as it is known that p85/p55 engagement can allosterically activate the p110 subunit (Carpenter et al. 1993; Holt et al. 1994). PI3K is also regulated by the GTPase Ras, which in its GTP-loaded form can bind to a specific domain in the p110 subunit (Jimenez et al. 2002; Rodriguez-Viciana et al. 1994, 1996). Ras is strongly activated by TCR signaling and plays an important role in driving proliferative responses and cell survival (Genot and Cantrell 2000). Recent work has demonstrated that Ras also functions as a cytoskeletal organizer. It is required for

TCR-induced activation of PI3K, upregulation of Rac and cofilin, IS growth, and F-actin ring formation (Le Floc'h et al. 2013; Wabnitz et al. 2006).

Taken together, these results are consistent with a model whereby TCR stimulation induces recruitment of class IA PI3K isoforms into microclusters, where they encounter activated Ras and generate PIP_3. This PIP_3 then recruits Dock2-Elmo to the periphery of the IS in order to drive Rac-dependent actin polymerization, thereby establishing the synaptic F-actin ring (Fig. 11.3). The incorporation of class IA PI3K into TCR microclusters may, in part, explain why PIP_3 accumulation is annular; microclusters are more active in the periphery of the IS (Vardhana et al. 2010; Varma et al. 2006; Yokosuka et al. 2005) and presumably generate more PIP_3 in this zone. There is evidence that the same pathway also controls synaptic F-actin in natural killer cells (Sakai et al. 2013), indicating that this may be a general mechanism for shaping cell–cell interactions in lymphocytes. PI3K-Dock2 signaling plays a well-established role in leukocyte migration by establishing a protrusive leading edge lamellipodium (Fukui et al. 2001; Nishikimi et al. 2009; Nombela-Arrieta et al. 2004). The implication of this pathway in IS formation establishes a molecular link between these functionally disparate structures.

On stimulatory bilayers, robust F-actin ring formation requires co-engagement of the integrin LFA-1 (Le Floc'h et al. 2013). Although the pathway(s) linking integrins to synaptic F-actin are poorly understood, it has been shown in Jurkat cells that simultaneous stimulation of the TCR and LFA-1 is required for robust Ras activation at the plasma membrane (Mor et al. 2007). Hence, LFA-1 signaling, by activating Ras, could promote PI3K activity at the membrane, leading to activation of Rac in the IS periphery. Although this model is appealing, it is complicated by evidence indicating that Rac can also function upstream of LFA-1. TCR signaling drives the conversion of integrins from their bent, low affinity states to their extended, high affinity conformations, thereby enabling strong adhesion to the APC (Kinashi 2005). This "inside-out" activation process depends on the GTPase Rap1, a master regulator of integrin function. Studies have shown that WAVE2 induces the affinity maturation of LFA-1 by recruiting the Rap1 exchange factor C3G (Nolz et al. 2008). Rac-WAVE2 signaling also contributes to integrin function through the F-actin ring itself. Retrograde flow within this ring promotes LFA-1 clustering leading to stronger IS formation (Yi et al. 2012). Indeed, suppression of Arp2/3 components affects integrin dependent adhesion and conjugate formation to a similar extent as loss of WAVE2 (Gomez et al. 2007). Collectively, Ras-mediated signaling from LFA-1 to PI3K-Dock2 together with WAVE2-mediated signaling back to LFA-1 represents a putative positive feedback loop that couples integrin engagement to Rac-dependent F-actin growth. One can envision this loop playing an important role in locally accentuating adhesion and actin polymerization within the IS periphery.

It is noteworthy that both centrosome reorientation and F-actin ring formation, the signature cytoskeletal remodeling events associated with the IS, are guided by synaptic lipid gradients: DAG for the centrosome and PIP_3 for F-actin. This remarkable mechanistic parallel highlights the value of lipid second messenger signaling as a tool for patterning the cortical cytoskeleton. Lipid gradients can be formed quickly and are highly responsive to change (Swaney et al. 2010). As such, they are ideal for propagating signals within a cellular "neighborhood" in order to generate transient, polarized structures.

11.4 Purpose of Cytoskeletal Polarity at the T Cell IS

It was initially proposed that the large-scale clustering of TCR and LFA-1 observed in the cSMAC and pSMAC, respectively, was required to transduce strong activating signals (Grakoui et al. 1999; Monks et al. 1998). Subsequent studies, however, demonstrated conclusively that the synapse is a product of TCR signaling rather than its cause (Bunnell et al. 2002; Lee et al. 2002). That being the case, what purpose does it serve? The cellular reorganization that accompanies IS formation polarizes the T cell toward the APC and redistributes key organelles. This contributes to T cell activation in at least three important ways. First, it regulates the spatial distribution of receptor complexes so as to achieve optimal signal duration. Second, it enables the directional secretion of soluble cytokines and cytolytic molecules toward the APC, which preserves the specificity of effector responses. Finally, it plays a key role in T cell differentiation by establishing a polarized framework for asymmetric cell division. Below, we will discuss each of these features in turn.

11.4.1 The IS Regulates TCR Signaling

Concomitant with centrosome reorientation and F-actin ring formation, microclusters containing activated TCR, LAT, Slp76, and a number of other receptor-proximal signaling molecules begin to traffic toward the center of the IS (Bunnell et al. 2002; Campi et al. 2005; Yokosuka et al. 2005). It is generally thought that signaling from these microclusters is downregulated as they coalesce into the cSMAC. Zap70 and Slp76 appear to dissociate prior to cSMAC formation, and markers for active signal transduction, such as pTyr, are either absent or substantially reduced in centralized clusters relative to their peripheral counterparts (Campi et al. 2005; Yokosuka et al. 2005). Consistent with the idea that centralization reduces signaling, loss of CD2AP, a sorting protein required for cSMAC formation, actually enhances T cell activation (Lee et al. 2003). Furthermore, physically restraining centripetal microcluster motility appears to boost TCR signaling (Mossman et al. 2005). Studies suggest that the cSMAC is a focal point for the internalization and/or degradation of TCR signaling complexes. Markers for the endosomal compartment and multivesicular bodies have been localized to this central domain (Varma et al. 2006). In addition, both ubiquitin and TSG101, a member of the ESCRT (endosomal sorting complex required for transport) family of proteins, are required for cSMAC formation (Vardhana et al. 2010). The microcluster centralization and downregulation process imposes a defined lifetime on each TCR signaling module. Hence, continuous formation of new peripheral microclusters is likely to be required for sustained TCR signaling and maintenance of the IS.

Microcluster dynamics depend almost entirely on the cytoskeleton. At the periphery, centripetal motion is driven by retrograde flow of F-actin (Nguyen et al. 2008; Varma et al. 2006; Yi et al. 2012). Indeed, the application of agents that depolymerize F-actin, such as latrunculin A, arrests microcluster motility within seconds. As microclusters approach the cSMAC, however, F-actin becomes somewhat scarce. Recent work suggests that, in this zone, microclusters associate with dynein and traffic along microtubules radiating outward from centrosome (Hashimoto-Tane et al. 2011). Other studies suggest that dynein is not required and that motility near the central domain results from NMII-dependent assembly of peripheral F-actin into contractile arcs (Ilani et al. 2009; Yi et al. 2012). Still other studies argue that NMII plays no role in either TCR signaling or microcluster motility (Babich et al. 2012; Jacobelli et al. 2004). Additional work will obviously be required to resolve these issues. It is worth noting, however, that certain features of TCR microcluster centralization, including retrograde F-actin flow and dynein-dependent trafficking, have also been implicated in the gathering and internalization of receptor-bound antigen at the B cell IS (Harwood and Batista 2011; Schnyder et al. 2011). Hence, centripetal motility and endocytosis are likely to be conserved features of activating cell–cell contacts in lymphocytes.

Cytoskeletal polarity at the IS also appears to play an important role in the replenishment of depleted signaling components. It has been known for some time that the TCR and Lck are recycled back to the IS in intracellular vesicles (Das et al. 2004; Ehrlich et al. 2002). The centrosome plays an important role in this process by acting as a focal point for vesicular structures and thereby targeting them to the synaptic membrane. Recent studies suggest that this process is decidedly more complex than initially imagined. Indeed, whereas the TCR requires synaptotagmin-7 and Ca^{2+} for its trafficking, Lck is transported in a Ca^{2+}-independent manner by vesicles containing the MAL protein (Soares et al. 2013). TCR recycling also involves the intraflagellar transport (IFT) proteins IFT20, IFT57, and IFT88 (Finetti et al. 2009). These factors are best known for their roles moving molecules into and out of the primary cilium, a centrosome-based sensory structure that is present in most eukaryotic cells but, interestingly, absent in lymphocytes (Christensen et al. 2008). This result establishes an intriguing link between the primary cilium and the IS and suggests that the IS may represent an alternative cilium of sorts (Griffiths et al. 2010).

Other cell surface receptors also form microclusters at the IS, which share some, but not all, of the dynamic properties exhibited by the TCR and its associated proteins. Both LFA-1 and the costimulatory receptor CD28 form ligand-induced clusters that traffic centripetally toward the center of the contact. Neither structure incorporates into the cSMAC, however. LFA-1 clusters arrest in the pSMAC to form a dense adhesion ring (Yi et al. 2012), while CD28 accumulates in a distinct annulus between the pSMAC and the cSMAC (Yokosuka et al. 2008). The segregation of CD28 and LFA-1 from the TCR may allow these receptors to escape the downregulation that occurs at the cSMAC and thereby signal in a more sustained manner. Indeed, the profound dephosphorylation of signaling proteins observed for centralized TCR microclusters is not observed in the CD28 annulus (Yokosuka et al. 2008). The mechanistic basis for differential synaptic localization is in general not well understood and remains an interesting topic for future investigation.

11.4.2 The IS Facilitates Polarized Secretion

The combination of centrosome reorientation and F-actin ring formation positions vesicular organelles in close apposition to the relatively F-actin free plasma membrane at the center of the IS. This is generally thought to enable directional secretion toward the APC (Huse et al. 2013; Stinchcombe and Griffiths 2007). The molecular machinery that mediates the release of cytokines and cytolytic factors from T cells has been described in detail in recent publications (de Saint Basile et al. 2010; Duitman et al. 2011; Stow and Murray 2013) and will not be discussed at length in this chapter. We will focus instead on the purpose of directional secretion and the role of synaptic architecture in guiding this process.

Directional release of cytokines into the synaptic space was first suggested by imaging studies demonstrating that compartments containing nascent IL-2 and IL-4 accumulate just beneath the synaptic membrane (Kupfer et al. 1991, 1994) (Fig. 11.4). These studies were consistent with experiments showing that T cells, when attached to porous membranes coated with stimulatory antibodies, tend to secrete cytokines through the membranes (Poo et al. 1988). It was later shown that T cells use at least two directionally distinct pathways for cytokine secretion (Huse et al. 2006). One pathway targets cytokines like IL-2 and IFNγ to the IS, while another releases TNF-α and certain chemokines in an unpolarized, multidirectional manner. This discovery suggested a mechanism by which T cells could shape the scope of intercellular communication. Thus, molecules mediating "private" communication between the T cell and the APC would be secreted synaptically. Meanwhile "public" factors designed to affect cells distal to the T cell–APC conjugate (e.g., inflammatory molecules or chemokines) would be released multidirectionally. While this model is intellectually appealing, recent studies have cast doubt on the notion that synaptic secretion actually constrains cytokine diffusion. In ex vivo cultures of $CD8^+$ CTLs and astrocytes, CTL-derived IFNγ was observed to activate APCs at a great distance from the IS (Sanderson et al. 2012). Similar observations were reported for $CD4^+$ T cells using an in vivo model of *Leishmania major* infection (Muller et al. 2012). It is possible that directional secretion into the IS does limit the scope of cytokine-mediated communication in certain contexts, where cell–cell adhesion is particularly tight or the amount of secreted cytokine is very small. At the moment, however, the exact purpose of this process is quite mysterious.

Cytolytic perforins and granzymes, for their part, are stored in specialized secretory lysosomes called lytic granules (de Saint Basile et al. 2010). TCR

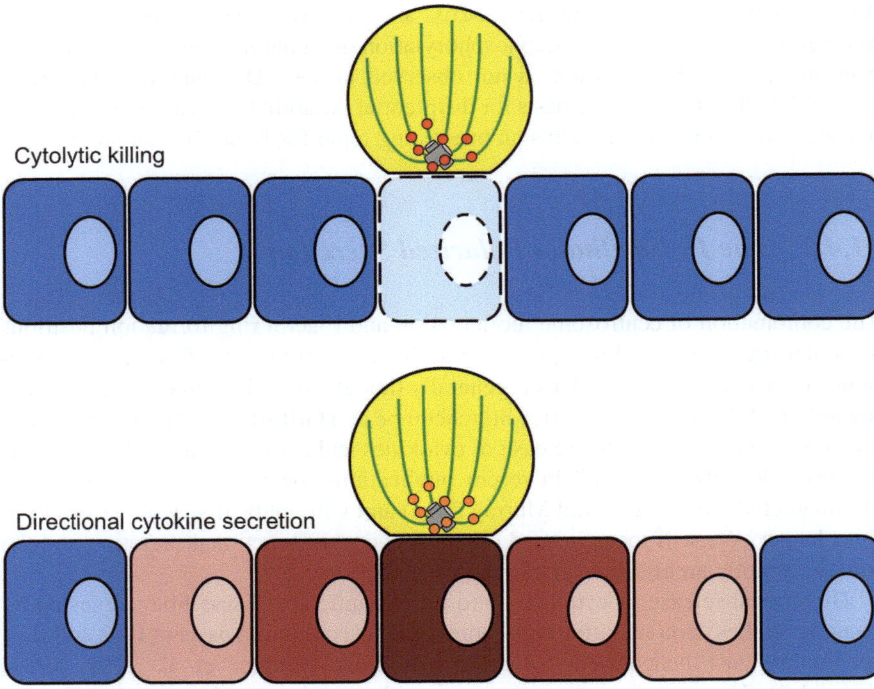

Fig. 11.4 Schematic diagrams of synaptic secretion. T cells are depicted in *yellow*, with the centrosome as two orthogonal *gray cylinders* and microtubules as *green lines*. The APC is depicted as a rectangular cell embedded in a line of bystanders. Above, directional secretion of lytic granules (*red*) ensures the specific killing of the APC. Below, directional secretion of cytokine (*orange* compartments) induces the selective activation of the APC and cells close to it (gradient of *brown* colors)

stimulation induces the Ca^{2+}-dependent trafficking of these granules toward the centrosome, which delivers them to the IS (Beal et al. 2009). Once there, they fuse with the synaptic membrane, releasing their contents into a "synaptic cleft" just adjacent to the cSMAC (Stinchcombe et al. 2001) (Fig. 11.4). Secreted perforins form membrane pores in the APC that induce, through a series of steps, the delivery of granzymes into the cytoplasm, where they trigger apoptosis by cleaving a number of substrates (Chowdhury and Lieberman 2008; Thiery et al. 2011). Unlike cytokines, the effects of this targeted secretory event do remain localized. Only the APC that is actively engaged by the CTL undergoes apoptosis. This specificity probably reflects a predisposition of perforins to bind to the first membrane they encounter (Pipkin and Lieberman 2007) rather than the ability of the IS to physically restrict their diffusion. Recent studies have shown that CTLs and NK cells are capable of inducing APC death in the absence of centrosome polarization (Bertrand et al. 2013; Chauveau et al. 2010). Taken together, these results suggest that synaptic lytic granule exocytosis functions to ensure the safety of innocent bystander cells rather than to focus the strength of cytolytic secretion. That being said, there are indications that the IS, by virtue of its polarized cytoskeletal architecture, can also enhance the power of each "lytic hit" (Le Floc'h et al. 2013). CTLs lacking Dock2, which form small, weakly adhesive synapses, display a marked reduction in killing potential. Conversely, suppression of PTEN, which enhances IS size, dramatically increases both the speed and the efficiency of cytotoxic responses. Importantly, loss of Dock2 or PTEN has no effect on TCR-induced degranulation, indicating that the observed killing phenotypes are not caused by changes in levels of secretion. Rather, these results suggest that synaptic architecture, specifically the size and adhesive strength of the synaptic F-actin ring, plays an important role in modulating cytotoxic efficiency. The mechanistic basis for this relationship remains to be explored. One particularly intriguing possibility is that the IS contributes an important physical component to the killing process, either holding or deforming the APC in a manner that optimizes the cytotoxic effects of degranulation. Investigating hypotheses like this will require methods that enable the quantitative, high-resolution analysis of CTL–APC interactions.

11.4.3 Polarity Promotes T Cell Differentiation

TCR activation induces robust T cell proliferation that occurs over a period of days. Imaging studies have indicated that this proliferative burst begins while the T cell is attached to its APC, implying that the IS might influence the cell divisions that occur. Indeed, work from multiple labs suggests that these mitotic events are oriented orthogonally to the IS and that they lead to asymmetric inheritance of several key intracellular and cell surface proteins (Chang et al. 2007, 2011; Oliaro et al. 2010) (Fig. 11.5). Conspicuous among these are the components of the Par/PKCζ and Scribble/Dlg complexes, which are known to control asymmetric division in other cell types. In dividing cells, Scribble/Dlg appears to localize to the membrane facing the APC, whereas Par/PKCζ accumulates in the rear. Interestingly, this accumulation pattern, which is observed after about 10 h of TCR stimulation, is essentially the reverse of what is seen in early (~30 min) synapses, which contain Par3 and phosphorylated PKCζ but exclude Scribble (Ludford-Menting et al. 2005). The molecular mechanisms that drive this remarkable inversion are not known. Regardless, these polarity complexes do appear to be important for asymmetric T cell division. Pharmacological inhibition or siRNA-mediated suppression of PKCζ, for example, abrogates the differential accumulation of polarity determinants in dividing T cells (Chang et al. 2011; Oliaro et al. 2010).

Asymmetric cell division contributes to the differentiation of multiple cell types by ensuring the unequal inheritance of key signaling proteins and transcriptional regulators (Betschinger and Knoblich 2004). It appears to play a similar role in naïve T cells, which differentiate during their antigen-induced proliferative burst to become either effector or memory cells. Effector cells are short-lived (1–3 weeks)

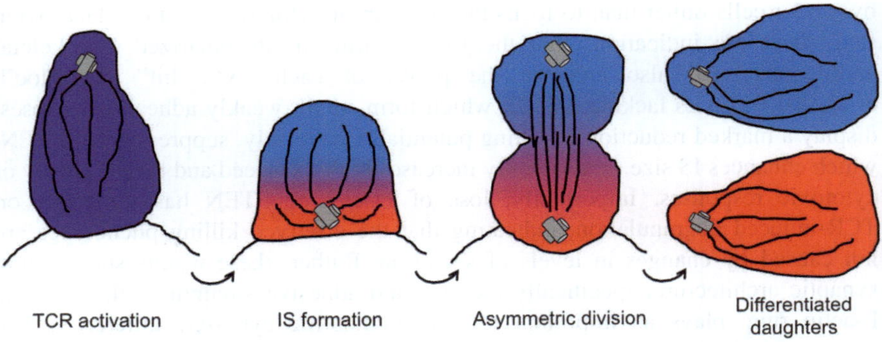

Fig. 11.5 Schematic diagram of asymmetric T cell division. The dividing T cell is shown above and the boundary of the APC below. The centrosome is represented as two orthogonal *gray cylinders* and microtubules as *thick black lines*. Asymmetric partitioning of cellular contents is depicted as a separation of the T cell into *red* and *blue halves*

and mediate rapid responses to pathogens, whereas memory cells persist for years and confer protection against future infections. Studies have shown that the acquisition of a robust memory pool is tightly correlated with asymmetric cell division (Chang et al. 2007; King et al. 2012). The daughter cell oriented more proximally to the IS appears to adopt an effector cell fate, whereas the distal daughter cell exhibits the properties of a memory cell precursor. Recent results suggest that this distinction results, at least in part, from asymmetric partitioning of the lineage specifying transcription factor Tbet, which accumulates preferentially in the proximal daughter (Chang et al. 2011). This hypothesis remains to be rigorously tested, however. Indeed, there is much about the mechanisms and physiological functions of T cell asymmetric cell division that we still do not understand. It will no doubt be the focus of intense research in the coming years.

11.5 Concluding Remarks

Although the field has made considerable progress, there is still a great deal that we do not understand about the molecular mechanisms of T cell polarity. It remains largely unclear, for example, how synaptic PKC activity is translated into centrosome reorientation and also how cortical F-actin is cleared from the center of the IS. Even less is known about how these processes contribute to T cell function in physiological settings. In the coming years, it is our hope that a deeper mechanistic understanding of the signaling pathways guiding microtubule and F-actin remodeling at the IS will identify molecular handles that can be used for selective perturbation of these processes in vivo. This should enable scientists to address the real purpose(s) of synaptic T cell polarity during immune responses.

References

Alcazar I, Marques M, Kumar A, Hirsch E, Wymann M, Carrera AC, Barber DF (2007) Phosphoinositide 3-kinase gamma participates in T cell receptor-induced T cell activation. J Exp Med 204(12):2977–2987. doi:10.1084/jem.20070366, jem.20070366 [pii]

Andres-Delgado L, Anton OM, Alonso MA (2013) Centrosome polarization in T cells: a task for formins. Front Immunol 4:191. doi:10.3389/fimmu.2013.00191

Andres-Delgado L, Anton OM, Bartolini F, Ruiz-Saenz A, Correas I, Gundersen GG, Alonso MA (2012) INF2 promotes the formation of detyrosinated microtubules necessary for centrosome reorientation in T cells. J Cell Biol 198(6):1025–1037. doi:10.1083/jcb.201202137

Ardouin L, Bracke M, Mathiot A, Pagakis SN, Norton T, Hogg N, Tybulewicz VL (2003) Vav1 transduces TCR signals required for LFA-1 function and cell polarization at the immunological synapse. Eur J Immunol 33(3):790–797

Babich A, Li S, O'Connor RS, Milone MC, Freedman BD, Burkhardt JK (2012) F-actin polymerization and retrograde flow drive sustained PLCgamma1 signaling during T cell activation. J Cell Biol 197(6):775–787. doi:10.1083/jcb.201201018, jcb.201201018 [pii]

Beal AM, Anikeeva N, Varma R, Cameron TO, Vasiliver-Shamis G, Norris PJ, Dustin ML, Sykulev Y (2009) Kinetics of early T cell receptor signaling regulate the pathway of lytic granule delivery to the secretory domain. Immunity 31(4):632–642. doi:10.1016/j.immuni.2009.09.004, S1074-7613(09)00416-6 [pii]

Bengur AR, Robinson EA, Appella E, Sellers JR (1987) Sequence of the sites phosphorylated by protein kinase C in the smooth muscle myosin light chain. J Biol Chem 262(16):7613–7617

Bertrand F, Esquerre M, Petit AE, Rodrigues M, Duchez S, Delon J, Valitutti S (2010) Activation of the ancestral polarity regulator protein kinase C zeta at the immunological synapse drives polarization of Th cell secretory machinery toward APCs. J Immunol 185(5):2887–2894. doi:10.4049/jimmunol.1000739, jimmunol.1000739 [pii]

Bertrand F, Muller S, Roh KH, Laurent C, Dupre L, Valitutti S (2013) An initial and rapid step of lytic granule secretion precedes microtubule organizing center polarization at the cytotoxic T lymphocyte/target cell synapse. Proc Natl Acad Sci U S A 110(15):6073–6078. doi:10.1073/pnas.1218640110

Betschinger J, Knoblich JA (2004) Dare to be different: asymmetric cell division in Drosophila C. elegans and vertebrates. Curr Biol 14(16):R674–R685. doi:10.1016/j.cub.2004.08.017

Boisvert J, Edmondson S, Krummel MF (2004) Immunological synapse formation licenses CD40-CD40L accumulations at T-APC contact sites. J Immunol 173(6):3647–3652

Bubeck Wardenburg J, Pappu R, Bu JY, Mayer B, Chernoff J, Straus D, Chan AC (1998) Regulation of PAK activation and the T cell cytoskeleton by the linker protein SLP-76. Immunity 9(5):607–616

Bunnell SC, Hong DI, Kardon JR, Yamazaki T, McGlade CJ, Barr VA, Samelson LE (2002) T cell receptor ligation induces the formation of dynamically regulated signaling assemblies. J Cell Biol 158(7):1263–1275

Bunnell SC, Kapoor V, Trible RP, Zhang W, Samelson LE (2001) Dynamic actin polymerization drives T cell receptor-induced spreading: a role for the signal transduction adaptor LAT. Immunity 14(3):315–329

Campi G, Varma R, Dustin ML (2005) Actin and agonist MHC-peptide complex-dependent T cell receptor microclusters as scaffolds for signaling. J Exp Med 202(8):1031–1036

Cannon JL, Burkhardt JK (2004) Differential roles for Wiskott-Aldrich syndrome protein in immune synapse formation and IL-2 production. J Immunol 173(3):1658–1662

Cannon JL, Labno CM, Bosco G, Seth A, McGavin MH, Siminovitch KA, Rosen MK, Burkhardt JK (2001) Wasp recruitment to the T cell:APC contact site occurs independently of Cdc42 activation. Immunity 15(2):249–259

Carpenter CL, Auger KR, Chanudhuri M, Yoakim M, Schaffhausen B, Shoelson S, Cantley LC (1993) Phosphoinositide 3-kinase is activated by phosphopeptides that bind to the SH2 domains of the 85-kDa subunit. J Biol Chem 268(13):9478–9483

Chang JT, Ciocca ML, Kinjyo I, Palanivel VR, McClurkin CE, Dejong CS, Mooney EC, Kim JS, Steinel NC, Oliaro J, Yin CC, Florea BI, Overkleeft HS, Berg LJ, Russell SM, Koretzky GA, Jordan MS, Reiner SL (2011) Asymmetric proteasome segregation as a mechanism for unequal partitioning of the transcription factor T-bet during T lymphocyte division. Immunity 34 (4):492–504. doi:10.1016/j.immuni.2011.03.017

Chang JT, Palanivel VR, Kinjyo I, Schambach F, Intlekofer AM, Banerjee A, Longworth SA, Vinup KE, Mrass P, Oliaro J, Killeen N, Orange JS, Russell SM, Weninger W, Reiner SL (2007) Asymmetric T lymphocyte division in the initiation of adaptive immune responses. Science 315(5819):1687–1691

Chauveau A, Aucher A, Eissmann P, Vivier E, Davis DM (2010) Membrane nanotubes facilitate long-distance interactions between natural killer cells and target cells. Proc Natl Acad Sci U S A 107(12):5545–5550. doi:10.1073/pnas.0910074107

Chemin K, Bohineust A, Dogniaux S, Tourret M, Guegan S, Miro F, Hivroz C (2012) Cytokine secretion by CD4+ T cells at the immunological synapse requires Cdc42-dependent local actin remodeling but not microtubule organizing center polarity. J Immunol 189(5):2159–2168. doi:10.4049/jimmunol.1200156

Chesarone MA, DuPage AG, Goode BL (2010) Unleashing formins to remodel the actin and microtubule cytoskeletons. Nat Rev Mol Cell Biol 11(1):62–74. doi:10.1038/nrm2816, nrm2816 [pii]

Chowdhury D, Lieberman J (2008) Death by a thousand cuts: granzyme pathways of programmed cell death. Annu Rev Immunol 26:389–420. doi:10.1146/annurev.immunol.26.021607.090404

Christensen ST, Pedersen SF, Satir P, Veland IR, Schneider L (2008) The primary cilium coordinates signaling pathways in cell cycle control and migration during development and tissue repair. Curr Top Dev Biol 85:261–301. doi:10.1016/S0070-2153(08)00810-7, S0070-2153(08)00810-7 [pii]

Combs J, Kim SJ, Tan S, Ligon LA, Holzbaur EL, Kuhn J, Poenie M (2006) Recruitment of dynein to the Jurkat immunological synapse. Proc Natl Acad Sci U S A 103(40):14883–14888

Corre I, Gomez M, Vielkind S, Cantrell DA (2001) Analysis of thymocyte development reveals that the GTPase RhoA is a positive regulator of T cell receptor responses in vivo. J Exp Med 194(7):903–914

Costello PS, Walters AE, Mee PJ, Turner M, Reynolds LF, Prisco A, Sarner N, Zamoyska R, Tybulewicz VL (1999) The Rho-family GTP exchange factor Vav is a critical transducer of T cell receptor signals to the calcium, ERK, and NF-kappaB pathways. Proc Natl Acad Sci U S A 96(6):3035–3040

Cote JF, Vuori K (2007) GEF what? Dock180 and related proteins help Rac to polarize cells in new ways. Trends Cell Biol 17(8):383–393. doi:10.1016/j.tcb.2007.05.001, S0962-8924(07) 00157-2 [pii]

Das V, Nal B, Dujeancourt A, Thoulouze MI, Galli T, Roux P, Dautry-Varsat A, Alcover A (2004) Activation-induced polarized recycling targets T cell antigen receptors to the immunological synapse; involvement of SNARE complexes. Immunity 20(5):577–588

de Saint Basile G, Menasche G, Fischer A (2010) Molecular mechanisms of biogenesis and exocytosis of cytotoxic granules. Nat Rev 10(8):568–579. doi:10.1038/nri2803

Derivery E, Gautreau A (2010) Generation of branched actin networks: assembly and regulation of the N-WASP and WAVE molecular machines. Bioessays 32(2):119–131. doi:10.1002/bies. 200900123

Duitman EH, Orinska Z, Bulfone-Paus S (2011) Mechanisms of cytokine secretion: a portfolio of distinct pathways allows flexibility in cytokine activity. Eur J Cell Biol 90(6–7):476–483. doi:10.1016/j.ejcb.2011.01.010

Dupre L, Aiuti A, Trifari S, Martino S, Saracco P, Bordignon C, Roncarolo MG (2002) Wiskott-Aldrich syndrome protein regulates lipid raft dynamics during immunological synapse formation. Immunity 17(2):157–166

Dustin ML (2007) Cell adhesion molecules and actin cytoskeleton at immune synapses and kinapses. Curr Opin Cell Biol 19(5):529–533. doi:10.1016/j.ceb.2007.08.003, S0955-0674(07)00120-2 [pii]

Dustin ML, Chakraborty AK, Shaw AS (2010) Understanding the structure and function of the immunological synapse. Cold Spring Harb Perspect Biol 2(10):a002311. doi:10.1101/cshperspect.a002311, cshperspect.a002311 [pii]

Ehrlich LI, Ebert PJ, Krummel MF, Weiss A, Davis MM (2002) Dynamics of p56lck translocation to the T cell immunological synapse following agonist and antagonist stimulation. Immunity 17(6):809–822

Etienne-Manneville S, Hall A (2003) Cell polarity: Par6, aPKC and cytoskeletal crosstalk. Curr Opin Cell Biol 15(1):67–72

Filbert EL, Le Borgne M, Lin J, Heuser JE, Shaw AS (2012) Stathmin regulates microtubule dynamics and microtubule organizing center polarization in activated T cells. J Immunol 188(11):5421–5427. doi:10.4049/jimmunol.1200242

Finetti F, Paccani SR, Riparbelli MG, Giacomello E, Perinetti G, Pazour GJ, Rosenbaum JL, Baldari CT (2009) Intraflagellar transport is required for polarized recycling of the TCR/CD3 complex to the immune synapse. Nat Cell Biol 11(11):1332–1339. doi:10.1038/ncb1977

Fischer KD, Kong YY, Nishina H, Tedford K, Marengere LE, Kozieradzki I, Sasaki T, Starr M, Chan G, Gardener S, Nghiem MP, Bouchard D, Barbacid M, Bernstein A, Penninger JM (1998) Vav is a regulator of cytoskeletal reorganization mediated by the T-cell receptor. Curr Biol 8(10):554–562

Fruman DA, Bismuth G (2009) Fine tuning the immune response with PI3K. Immunol Rev 228(1):253–272. doi:10.1111/j.1600-065X.2008.00750.x, IMR750 [pii]

Fu G, Hu J, Niederberger-Magnenat N, Rybakin V, Casas J, Yachi PP, Feldstein S, Ma B, Hoerter JA, Ampudia J, Rigaud S, Lambolez F, Gavin AL, Sauer K, Cheroutre H, Gascoigne NR (2011) Protein kinase C eta is required for T cell activation and homeostatic proliferation. Sci Signal 4(202):84

Fukui Y, Hashimoto O, Sanui T, Oono T, Koga H, Abe M, Inayoshi A, Noda M, Oike M, Shirai T, Sasazuki T (2001) Haematopoietic cell-specific CDM family protein DOCK2 is essential for lymphocyte migration. Nature 412(6849):826–831. doi:10.1038/35090591, 35090591 [pii]

Galandrini R, Henning SW, Cantrell DA (1997) Different functions of the GTPase Rho in prothymocytes and late pre-T cells. Immunity 7(1):163–174

Gallego MD, Santamaria M, Pena J, Molina IJ (1997) Defective actin reorganization and polymerization of Wiskott-Aldrich T cells in response to CD3-mediated stimulation. Blood 90(8):3089–3097

Garcon F, Patton DT, Emery JL, Hirsch E, Rottapel R, Sasaki T, Okkenhaug K (2008) CD28 provides T-cell costimulation and enhances PI3K activity at the immune synapse independently of its capacity to interact with the p85/p110 heterodimer. Blood 111(3):1464–1471. doi:10.1182/blood-2007-08-108050, blood-2007-08-108050 [pii]

Genot E, Cantrell DA (2000) Ras regulation and function in lymphocytes. Curr Opin Immunol 12(3):289–294, S0952-7915(00)00089-3 [pii]

Gomes ER, Jani S, Gundersen GG (2005) Nuclear movement regulated by Cdc42, MRCK, myosin, and actin flow establishes MTOC polarization in migrating cells. Cell 121(3):451–463

Gomez TS, Billadeau DD (2008) T cell activation and the cytoskeleton: you can't have one without the other. Adv Immunol 97:1–64

Gomez TS, Kumar K, Medeiros RB, Shimizu Y, Leibson PJ, Billadeau DD (2007) Formins regulate the actin-related protein 2/3 complex-independent polarization of the centrosome to the immunological synapse. Immunity 26(2):177–190

Grakoui A, Bromley SK, Sumen C, Davis MM, Shaw AS, Allen PM, Dustin ML (1999) The immunological synapse: a molecular machine controlling T cell activation. Science 285(5425):221–227

Griffiths GM, Tsun A, Stinchcombe JC (2010) The immunological synapse: a focal point for endocytosis and exocytosis. J Cell Biol 189(3):399–406. doi:10.1083/jcb.201002027

Gruber T, Thuille N, Hermann-Kleiter N, Leitges M, Baier G (2005) Protein kinase Cepsilon is dispensable for TCR/CD3-signaling. Mol Immunol 42(3):305–310

Harwood NE, Batista FD (2011) The cytoskeleton coordinates the early events of B-cell activation. Cold Spring Harb Perspect Biol 3(2), 10.1101/cshperspect.a002360

Hashimoto-Tane A, Yokosuka T, Sakata-Sogawa K, Sakuma M, Ishihara C, Tokunaga M, Saito T (2011) Dynein-driven transport of T cell receptor microclusters regulates immune synapse formation and T cell activation. Immunity 34(6):919–931. doi:10.1016/j.immuni.2011.05.012, S1074-7613(11)00228-7 [pii]

Henning SW, Galandrini R, Hall A, Cantrell DA (1997) The GTPase Rho has a critical regulatory role in thymus development. EMBO J 16(9):2397–2407. doi:10.1093/emboj/16.9.2397

Holsinger LJ, Graef IA, Swat W, Chi T, Bautista DM, Davidson L, Lewis RS, Alt FW, Crabtree GR (1998) Defects in actin-cap formation in Vav-deficient mice implicate an actin requirement for lymphocyte signal transduction. Curr Biol 8(10):563–572

Holt KH, Olson L, Moye-Rowley WS, Pessin JE (1994) Phosphatidylinositol 3-kinase activation is mediated by high-affinity interactions between distinct domains within the p110 and p85 subunits. Mol Cell Biol 14(1):42–49

Huse M (2009) The T-cell-receptor signaling network. J Cell Sci 122(Pt 9):1269–1273

Huse M, Le Floc'h A, Liu X (2013) From lipid second messengers to molecular motors: microtubule-organizing center reorientation in T cells. Immunol Rev 256(1):95–106. doi:10.1111/imr.12116

Huse M, Lillemeier BF, Kuhns MS, Chen DS, Davis MM (2006) T cells use two directionally distinct pathways for cytokine secretion. Nat Immunol 7(3):247–255

Ikebe M, Hartshorne DJ, Elzinga M (1987) Phosphorylation of the 20,000-dalton light chain of smooth muscle myosin by the calcium-activated, phospholipid-dependent protein kinase. Phosphorylation sites and effects of phosphorylation. J Biol Chem 262(20):9569–9573

Ikebe M, Reardon S (1990) Phosphorylation of bovine platelet myosin by protein kinase C. Biochemistry 29(11):2713–2720

Ilani T, Vasiliver-Shamis G, Vardhana S, Bretscher A, Dustin ML (2009) T cell antigen receptor signaling and immunological synapse stability require myosin IIA. Nat Immunol 10 (5):531–539. doi:10.1038/ni.1723, ni.1723 [pii]

Jacobelli J, Chmura SA, Buxton DB, Davis MM, Krummel MF (2004) A single class II myosin modulates T cell motility and stopping, but not synapse formation. Nat Immunol 5(5):531–538. doi:10.1038/ni1065, ni1065 [pii]

Jaffe AB, Hall A (2005) Rho GTPases: biochemistry and biology. Annu Rev Cell Dev Biol 21:247–269

Jimenez C, Hernandez C, Pimentel B, Carrera AC (2002) The p85 regulatory subunit controls sequential activation of phosphoinositide 3-kinase by Tyr kinases and Ras. J Biol Chem 277 (44):41556–41562. doi:10.1074/jbc.M205893200, M205893200 [pii]

Kardon JR, Vale RD (2009) Regulators of the cytoplasmic dynein motor. Nat Rev Mol Cell Biol 10(12):854–865. doi:10.1038/nrm2804, nrm2804 [pii]

Kinashi T (2005) Intracellular signalling controlling integrin activation in lymphocytes. Nat Rev 5 (7):546–559. doi:10.1038/nri1646, nri1646 [pii]

King CG, Koehli S, Hausmann B, Schmaler M, Zehn D, Palmer E (2012) T cell affinity regulates asymmetric division, effector cell differentiation, and tissue pathology. Immunity 37 (4):709–720. doi:10.1016/j.immuni.2012.06.021

Komatsu S, Ikebe M (2007) The phosphorylation of myosin II at the Ser1 and Ser2 is critical for normal platelet-derived growth factor induced reorganization of myosin filaments. Mol Biol Cell 18(12):5081–5090. doi:10.1091/mbc.E06-12-1076, E06-12-1076 [pii]

Krawczyk C, Oliveira-dos-Santos A, Sasaki T, Griffiths E, Ohashi PS, Snapper S, Alt F, Penninger JM (2002) Vav1 controls integrin clustering and MHC/peptide-specific cell adhesion to antigen-presenting cells. Immunity 16(3):331–343

Ku GM, Yablonski D, Manser E, Lim L, Weiss A (2001) A PAK1-PIX-PKL complex is activated by the T-cell receptor independent of Nck, Slp-76 and LAT. EMBO J 20(3):457–465. doi:10.1093/emboj/20.3.457

Kuhne MR, Lin J, Yablonski D, Mollenauer MN, Ehrlich LI, Huppa J, Davis MM, Weiss A (2003) Linker for activation of T cells, zeta-associated protein-70, and Src homology 2 domain-containing leukocyte protein-76 are required for TCR-induced microtubule-organizing center polarization. J Immunol 171(2):860–866
Kupfer A, Mosmann TR, Kupfer H (1991) Polarized expression of cytokines in cell conjugates of helper T cells and splenic B cells. Proc Natl Acad Sci U S A 88(3):775–779
Kupfer H, Monks CR, Kupfer A (1994) Small splenic B cells that bind to antigen-specific T helper (Th) cells and face the site of cytokine production in the Th cells selectively proliferate: immunofluorescence microscopic studies of Th-B antigen-presenting cell interactions. J Exp Med 179(5):1507–1515
Laudanna C, Campbell JJ, Butcher EC (1996) Role of Rho in chemoattractant-activated leukocyte adhesion through integrins. Science 271(5251):981–983
Le Floc'h A, Tanaka Y, Bantilan NS, Voisinne G, Altan-Bonnet G, Fukui Y, Huse M (2013) Annular PIP3 accumulation controls actin architecture and modulates cytotoxicity at the immunological synapse. J Exp Med 210(12):2721–2737. doi:10.1084/jem.20131324
Lee KH, Dinner AR, Tu C, Campi G, Raychaudhuri S, Varma R, Sims TN, Burack WR, Wu H, Wang J, Kanagawa O, Markiewicz M, Allen PM, Dustin ML, Chakraborty AK, Shaw AS (2003) The immunological synapse balances T cell receptor signaling and degradation. Science 302(5648):1218–1222
Lee KH, Holdorf AD, Dustin ML, Chan AC, Allen PM, Shaw AS (2002) T cell receptor signaling precedes immunological synapse formation. Science 295(5559):1539–1542
Li R, Gundersen GG (2008) Beyond polymer polarity: how the cytoskeleton builds a polarized cell. Nat Rev Mol Cell Biol 9(11):860–873
Limatola C, Schaap D, Moolenaar WH, van Blitterswijk WJ (1994) Phosphatidic acid activation of protein kinase C-zeta overexpressed in COS cells: comparison with other protein kinase C isotypes and other acidic lipids. Biochem J 304(Pt 3):1001–1008
Lin J, Hou KK, Piwnica-Worms H, Shaw AS (2009) The polarity protein Par1b/EMK/MARK2 regulates T cell receptor-induced microtubule-organizing center polarization. J Immunol 183 (2):1215–1221. doi:10.4049/jimmunol.0803887, jimmunol.0803887 [pii]
Liu X, Kapoor TM, Chen JK, Huse M (2013) Diacylglycerol promotes centrosome polarization in T cells via reciprocal localization of dynein and myosin II. Proc Natl Acad Sci U S A 110 (29):11976–11981. doi:10.1073/pnas.1306180110
Lowin-Kropf B, Shapiro VS, Weiss A (1998) Cytoskeletal polarization of T cells is regulated by an immunoreceptor tyrosine-based activation motif-dependent mechanism. J Cell Biol 140 (4):861–871
Ludford-Menting MJ, Oliaro J, Sacirbegovic F, Cheah ET, Pedersen N, Thomas SJ, Pasam A, Iazzolino R, Dow LE, Waterhouse NJ, Murphy A, Ellis S, Smyth MJ, Kershaw MH, Darcy PK, Humbert PO, Russell SM (2005) A network of PDZ-containing proteins regulates T cell polarity and morphology during migration and immunological synapse formation. Immunity 22(6):737–748
Luxton GW, Gomes ER, Folker ES, Vintinner E, Gundersen GG (2010) Linear arrays of nuclear envelope proteins harness retrograde actin flow for nuclear movement. Science 329 (5994):956–959. doi:10.1126/science.1189072
Marsland BJ, Soos TJ, Spath G, Littman DR, Kopf M (2004) Protein kinase C theta is critical for the development of in vivo T helper (Th)2 cell but not Th1 cell responses. J Exp Med 200 (2):181–189
Martin-Cofreces NB, Robles-Valero J, Cabrero JR, Mittelbrunn M, Gordon-Alonso M, Sung CH, Alarcon B, Vazquez J, Sanchez-Madrid F (2008) MTOC translocation modulates IS formation and controls sustained T cell signaling. J Cell Biol 182(5):951–962
Martinez-Martin N, Fernandez-Arenas E, Cemerski S, Delgado P, Turner M, Heuser J, Irvine DJ, Huang B, Bustelo XR, Shaw A, Alarcon B (2011) T cell receptor internalization from the immunological synapse is mediated by TC21 and RhoG GTPase-dependent phagocytosis. Immunity 35(2):208–222. doi:10.1016/j.immuni.2011.06.003, S1074-7613(11)00233-0 [pii]

Martinez-Quiles N, Rohatgi R, Anton IM, Medina M, Saville SP, Miki H, Yamaguchi H, Takenawa T, Hartwig JH, Geha RS, Ramesh N (2001) WIP regulates N-WASP-mediated actin polymerization and filopodium formation. Nat Cell Biol 3(5):484–491. doi:10.1038/35074551

Maul-Pavicic A, Chiang SC, Rensing-Ehl A, Jessen B, Fauriat C, Wood SM, Sjoqvist S, Hufnagel M, Schulze I, Bass T, Schamel WW, Fuchs S, Pircher H, McCarl CA, Mikoshiba K, Schwarz K, Feske S, Bryceson YT, Ehl S (2011) ORAI1-mediated calcium influx is required for human cytotoxic lymphocyte degranulation and target cell lysis. Proc Natl Acad Sci U S A 108(8):3324–3329. doi:10.1073/pnas.1013285108

Miletic AV, Graham DB, Sakata-Sogawa K, Hiroshima M, Hamann MJ, Cemerski S, Kloeppel T, Billadeau DD, Kanagawa O, Tokunaga M, Swat W (2009) Vav links the T cell antigen receptor to the actin cytoskeleton and T cell activation independently of intrinsic Guanine nucleotide exchange activity. PloS One 4(8):e6599. doi:10.1371/journal.pone.0006599

Molina IJ, Sancho J, Terhorst C, Rosen FS, Remold-O'Donnell E (1993) T cells of patients with the Wiskott-Aldrich syndrome have a restricted defect in proliferative responses. J Immunol 151(8):4383–4390

Monks CR, Freiberg BA, Kupfer H, Sciaky N, Kupfer A (1998) Three-dimensional segregation of supramolecular activation clusters in T cells. Nature 395(6697):82–86

Mor A, Campi G, Du G, Zheng Y, Foster DA, Dustin ML, Philips MR (2007) The lymphocyte function-associated antigen-1 receptor costimulates plasma membrane Ras via phospholipase D2. Nat Cell Biol 9(6):713–719

Morgan MM, Labno CM, Van Seventer GA, Denny MF, Straus DB, Burkhardt JK (2001) Superantigen-induced T cell:B cell conjugation is mediated by LFA-1 and requires signaling through Lck, but not ZAP-70. J Immunol 167(10):5708–5718

Mossman KD, Campi G, Groves JT, Dustin ML (2005) Altered TCR signaling from geometrically repatterned immunological synapses. Science 310(5751):1191–1193

Muller AJ, Filipe-Santos O, Eberl G, Aebischer T, Spath GF, Bousso P (2012) CD4+ T cells rely on a cytokine gradient to control intracellular pathogens beyond sites of antigen presentation. Immunity 37, 10.1016/j.immuni.2012.05.015

Newton AC (2010) Protein kinase C: poised to signal. Am J Physiol Endocrinol Metab 298(3): E395–E402. doi:10.1152/ajpendo.00477.2009, ajpendo.00477.2009 [pii]

Nguyen K, Sylvain NR, Bunnell SC (2008) T cell costimulation via the integrin VLA-4 inhibits the actin-dependent centralization of signaling microclusters containing the adaptor SLP-76. Immunity 28(6):810–821. doi:10.1016/j.immuni.2008.04.019, S1074-7613(08)00237-9 [pii]

Nishikimi A, Fukuhara H, Su W, Hongu T, Takasuga S, Mihara H, Cao Q, Sanematsu F, Kanai M, Hasegawa H, Tanaka Y, Shibasaki M, Kanaho Y, Sasaki T, Frohman MA, Fukui Y (2009) Sequential regulation of DOCK2 dynamics by two phospholipids during neutrophil chemotaxis. Science 324(5925):384–387. doi:10.1126/science.1170179, 1170179 [pii]

Nolz JC, Gomez TS, Zhu P, Li S, Medeiros RB, Shimizu Y, Burkhardt JK, Freedman BD, Billadeau DD (2006) The WAVE2 complex regulates actin cytoskeletal reorganization and CRAC-mediated calcium entry during T cell activation. Curr Biol 16(1):24–34

Nolz JC, Nacusi LP, Segovis CM, Medeiros RB, Mitchell JS, Shimizu Y, Billadeau DD (2008) The WAVE2 complex regulates T cell receptor signaling to integrins via Abl- and CrkL-C3G-mediated activation of Rap1. J Cell Biol 182(6):1231–1244

Nombela-Arrieta C, Lacalle RA, Montoya MC, Kunisaki Y, Megias D, Marques M, Carrera AC, Manes S, Fukui Y, Martinez AC, Stein JV (2004) Differential requirements for DOCK2 and phosphoinositide-3-kinase gamma during T and B lymphocyte homing. Immunity 21 (3):429–441. doi:10.1016/j.immuni.2004.07.012, S1074761304001992 [pii]

Ochs HD (1998) The Wiskott-Aldrich syndrome. Semin Hematol 35(4):332–345

Oliaro J, Van Ham V, Sacirbegovic F, Pasam A, Bomzon Z, Pham K, Ludford-Menting MJ, Waterhouse NJ, Bots M, Hawkins ED, Watt SV, Cluse LA, Clarke CJ, Izon DJ, Chang JT, Thompson N, Gu M, Johnstone RW, Smyth MJ, Humbert PO, Reiner SL, Russell SM (2010) Asymmetric cell division of T cells upon antigen presentation uses multiple conserved mechanisms. J Immunol 185(1):367–375. doi:10.4049/jimmunol.0903627, jimmunol.0903627 [pii]

Penninger JM, Fischer KD, Sasaki T, Kozieradzki I, Le J, Tedford K, Bachmaier K, Ohashi PS, Bachmann MF (1999) The oncogene product Vav is a crucial regulator of primary cytotoxic T cell responses but has no apparent role in CD28-mediated co-stimulation. Eur J Immunol 29 (5):1709–1718. doi:10.1002/(SICI)1521-4141(199905)29:05<1709::AID-IMMU1709>3.0.CO;2-O

Pipkin ME, Lieberman J (2007) Delivering the kiss of death: progress on understanding how perforin works. Curr Opin Immunol 19(3):301–308. doi:10.1016/j.coi.2007.04.011, S0952-7915(07)00063-5 [pii]

Poo WJ, Conrad L, Janeway CA Jr (1988) Receptor-directed focusing of lymphokine release by helper T cells. Nature 332(6162):378–380

Quann EJ, Liu X, Altan-Bonnet G, Huse M (2011) A cascade of protein kinase C isozymes promotes cytoskeletal polarization in T cells. Nat Immunol 12(7):647–654. doi:10.1038/ni.2033, ni.2033 [pii]

Quann EJ, Merino E, Furuta T, Huse M (2009) Localized diacylglycerol drives the polarization of the microtubule-organizing center in T cells. Nat Immunol 10(6):627–635

Reynolds LF, de Bettignies C, Norton T, Beeser A, Chernoff J, Tybulewicz VL (2004) Vav1 transduces T cell receptor signals to the activation of the Ras/ERK pathway via LAT, Sos, and RasGRP1. J Biol Chem 279(18):18239–18246. doi:10.1074/jbc.M400257200

Reynolds LF, Smyth LA, Norton T, Freshney N, Downward J, Kioussis D, Tybulewicz VL (2002) Vav1 transduces T cell receptor signals to the activation of phospholipase C-gamma1 via phosphoinositide 3-kinase-dependent and -independent pathways. J Exp Med 195 (9):1103–1114

Rodriguez-Viciana P, Warne PH, Dhand R, Vanhaesebroeck B, Gout I, Fry MJ, Waterfield MD, Downward J (1994) Phosphatidylinositol-3-OH kinase as a direct target of Ras. Nature 370 (6490):527–532

Rodriguez-Viciana P, Warne PH, Vanhaesebroeck B, Waterfield MD, Downward J (1996) Activation of phosphoinositide 3-kinase by interaction with Ras and by point mutation. EMBO J 15 (10):2442–2451

Sakai Y, Tanaka Y, Yanagihara T, Watanabe M, Duan X, Terasawa M, Nishikimi A, Sanematsu F, Fukui Y (2013) The Rac activator DOCK2 regulates natural killer cell-mediated cytotoxicity in mice through the lytic synapse formation. Blood 122(3):386–393. doi:10.1182/blood-2012-12-475897

Samstag Y, John I, Wabnitz GH (2013) Cofilin: a redox sensitive mediator of actin dynamics during T-cell activation and migration. Immunol Rev 256(1):30–47. doi:10.1111/imr.12115

Sanderson NS, Puntel M, Kroeger KM, Bondale NS, Swerdlow M, Iranmanesh N, Yagita H, Ibrahim A, Castro MG, Lowenstein PR (2012) Cytotoxic immunological synapses do not restrict the action of interferon-gamma to antigenic target cells. Proc Natl Acad Sci U S A 109 (20):7835–7840. doi:10.1073/pnas.1116058109, 1116058109 [pii]

Sanui T, Inayoshi A, Noda M, Iwata E, Oike M, Sasazuki T, Fukui Y (2003) DOCK2 is essential for antigen-induced translocation of TCR and lipid rafts, but not PKC-theta and LFA-1, in T cells. Immunity 19(1):119–129, S1074761303001699 [pii]

Sasahara Y, Rachid R, Byrne MJ, de la Fuente MA, Abraham RT, Ramesh N, Geha RS (2002) Mechanism of recruitment of WASP to the immunological synapse and of its activation following TCR ligation. Mol Cell 10(6):1269–1281

Sauer S, Bruno L, Hertweck A, Finlay D, Leleu M, Spivakov M, Knight ZA, Cobb BS, Cantrell D, O'Connor E, Shokat KM, Fisher AG, Merkenschlager M (2008) T cell receptor signaling controls Foxp3 expression via PI3K, Akt, and mTOR. Proc Natl Acad Sci U S A 105 (22):7797–7802. doi:10.1073/pnas.0800928105, 0800928105 [pii]

Schnyder T, Castello A, Feest C, Harwood NE, Oellerich T, Urlaub H, Engelke M, Wienands J, Bruckbauer A, Batista FD (2011) B cell receptor-mediated antigen gathering requires ubiquitin ligase Cbl and adaptors Grb2 and Dok-3 to recruit dynein to the signaling microcluster. Immunity 34(6):905–918. doi:10.1016/j.immuni.2011.06.001

Sedwick CE, Morgan MM, Jusino L, Cannon JL, Miller J, Burkhardt JK (1999) TCR, LFA-1, and CD28 play unique and complementary roles in signaling T cell cytoskeletal reorganization. J Immunol 162(3):1367–1375

Serrador JM, Cabrero JR, Sancho D, Mittelbrunn M, Urzainqui A, Sanchez-Madrid F (2004) HDAC6 deacetylase activity links the tubulin cytoskeleton with immune synapse organization. Immunity 20(4):417–428

Shim EK, Jung SH, Lee JR (2011) Role of two adaptor molecules SLP-76 and LAT in the PI3K signaling pathway in activated T cells. J Immunol 186(5):2926–2935. doi:10.4049/jimmunol.1001785, jimmunol.1001785 [pii]

Shim EK, Moon CS, Lee GY, Ha YJ, Chae SK, Lee JR (2004) Association of the Src homology 2 domain-containing leukocyte phosphoprotein of 76 kD (SLP-76) with the p85 subunit of phosphoinositide 3-kinase. FEBS Lett 575(1–3):35–40. doi:10.1016/j.febslet.2004.07.090, S0014579304010324 [pii]

Sims TN, Soos TJ, Xenias HS, Dubin-Thaler B, Hofman JM, Waite JC, Cameron TO, Thomas VK, Varma R, Wiggins CH, Sheetz MP, Littman DR, Dustin ML (2007) Opposing effects of PKCtheta and WASp on symmetry breaking and relocation of the immunological synapse. Cell 129(4):773–785

Snapper SB, Rosen FS, Mizoguchi E, Cohen P, Khan W, Liu CH, Hagemann TL, Kwan SP, Ferrini R, Davidson L, Bhan AK, Alt FW (1998) Wiskott-Aldrich syndrome protein-deficient mice reveal a role for WASP in T but not B cell activation. Immunity 9(1):81–91

Soares H, Henriques R, Sachse M, Ventimiglia L, Alonso MA, Zimmer C, Thoulouze MI, Alcover A (2013) Regulated vesicle fusion generates signaling nanoterritories that control T cell activation at the immunological synapse. J Exp Med 210(11):2415–2433. doi:10.1084/jem.20130150

Spitaler M, Emslie E, Wood CD, Cantrell D (2006) Diacylglycerol and protein kinase D localization during T lymphocyte activation. Immunity 24(5):535–546

Stinchcombe JC, Bossi G, Booth S, Griffiths GM (2001) The immunological synapse of CTL contains a secretory domain and membrane bridges. Immunity 15(5):751–761

Stinchcombe JC, Griffiths GM (2007) Secretory mechanisms in cell-mediated cytotoxicity. Annu Rev Cell Dev Biol 23:495–517

Stow JL, Murray RZ (2013) Intracellular trafficking and secretion of inflammatory cytokines. Cytokine Growth Factor Rev 24(3):227–239. doi:10.1016/j.cytogfr.2013.04.001

Stowers L, Yelon D, Berg LJ, Chant J (1995) Regulation of the polarization of T cells toward antigen-presenting cells by Ras-related GTPase CDC42. Proc Natl Acad Sci U S A 92(11):5027–5031

Swaney KF, Huang CH, Devreotes PN (2010) Eukaryotic chemotaxis: a network of signaling pathways controls motility, directional sensing, and polarity. Annu Rev Biophys 39:265–289. doi:10.1146/annurev.biophys.093008.131228

Sylvain NR, Nguyen K, Bunnell SC (2011) Vav1-mediated scaffolding interactions stabilize SLP-76 microclusters and contribute to antigen-dependent T cell responses. Sci Signal 4(163):ra14, doi:10.1126/scisignal.2001178

Tarakhovsky A, Turner M, Schaal S, Mee PJ, Duddy LP, Rajewsky K, Tybulewicz VL (1995) Defective antigen receptor-mediated proliferation of B and T cells in the absence of Vav. Nature 374(6521):467–470. doi:10.1038/374467a0

Thiery J, Keefe D, Boulant S, Boucrot E, Walch M, Martinvalet D, Goping IS, Bleackley RC, Kirchhausen T, Lieberman J (2011) Perforin pores in the endosomal membrane trigger the release of endocytosed granzyme B into the cytosol of target cells. Nat Immunol 12(8):770–777. doi:10.1038/ni.2050

Trifari S, Sitia G, Aiuti A, Scaramuzza S, Marangoni F, Guidotti LG, Martino S, Saracco P, Notarangelo LD, Roncarolo MG, Dupre L (2006) Defective Th1 cytokine gene transcription in

CD4+ and CD8+ T cells from Wiskott-Aldrich syndrome patients. J Immunol 177 (10):7451–7461

Tskvitaria-Fuller I, Seth A, Mistry N, Gu H, Rosen MK, Wulfing C (2006) Specific patterns of Cdc42 activity are related to distinct elements of T cell polarization. J Immunol 177 (3):1708–1720

Tybulewicz VL (2005) Vav-family proteins in T-cell signalling. Curr Opin Immunol 17 (3):267–274

Vardhana S, Choudhuri K, Varma R, Dustin ML (2010) Essential role of ubiquitin and TSG101 protein in formation and function of the central supramolecular activation cluster. Immunity 32 (4):531–540. doi:10.1016/j.immuni.2010.04.005, S1074-7613(10)00130-5 [pii]

Varma R, Campi G, Yokosuka T, Saito T, Dustin ML (2006) T cell receptor-proximal signals are sustained in peripheral microclusters and terminated in the central supramolecular activation cluster. Immunity 25(1):117–127

Wabnitz GH, Nebl G, Klemke M, Schroder AJ, Samstag Y (2006) Phosphatidylinositol 3-kinase functions as a Ras effector in the signaling cascade that regulates dephosphorylation of the actin-remodeling protein cofilin after costimulation of untransformed human T lymphocytes. J Immunol 176(3):1668–1674, 176/3/1668 [pii]

Yi J, Wu X, Chung AH, Chen JK, Kapoor TM, Hammer JA (2013) Centrosome repositioning in T cells is biphasic and driven by microtubule end-on capture-shrinkage. J Cell Biol 202 (5):779–792. doi:10.1083/jcb.201301004

Yi J, Wu XS, Crites T, Hammer JA 3rd (2012) Actin retrograde flow and actomyosin II arc contraction drive receptor cluster dynamics at the immunological synapse in Jurkat T cells. Mol Biol Cell 23(5):834–852. doi:10.1091/mbc.E11-08-0731, mbc.E11-08-0731 [pii]

Yokosuka T, Kobayashi W, Sakata-Sogawa K, Takamatsu M, Hashimoto-Tane A, Dustin ML, Tokunaga M, Saito T (2008) Spatiotemporal regulation of T cell costimulation by TCR-CD28 microclusters and protein kinase C theta translocation. Immunity 29(4):589–601

Yokosuka T, Saito T (2010) The immunological synapse, TCR microclusters, and T cell activation. Curr Top Microbiol Immunol 340:81–107

Yokosuka T, Sakata-Sogawa K, Kobayashi W, Hiroshima M, Hashimoto-Tane A, Tokunaga M, Dustin ML, Saito T (2005) Newly generated T cell receptor microclusters initiate and sustain T cell activation by recruitment of Zap70 and SLP-76. Nat Immunol 6(12):1253–1262

Yu CG, Harris TJ (2012) Interactions between the PDZ domains of Bazooka (Par-3) and phosphatidic acid: in vitro characterization and role in epithelial development. Mol Biol Cell 23 (18):3743–3753. doi:10.1091/mbc.E12-03-0196

Zeng R, Cannon JL, Abraham RT, Way M, Billadeau DD, Bubeck-Wardenberg J, Burkhardt JK (2003) SLP-76 coordinates Nck-dependent Wiskott-Aldrich syndrome protein recruitment with Vav-1/Cdc42-dependent Wiskott-Aldrich syndrome protein activation at the T cell-APC contact site. J Immunol 171(3):1360–1368

Zhang J, Shehabeldin A, da Cruz LA, Butler J, Somani AK, McGavin M, Kozieradzki I, dos Santos AO, Nagy A, Grinstein S, Penninger JM, Siminovitch KA (1999) Antigen receptor-induced activation and cytoskeletal rearrangement are impaired in Wiskott-Aldrich syndrome protein-deficient lymphocytes. J Exp Med 190(9):1329–1342

Zhang R, Alt FW, Davidson L, Orkin SH, Swat W (1995) Defective signalling through the T- and B-cell antigen receptors in lymphoid cells lacking the vav proto-oncogene. Nature 374 (6521):470–473. doi:10.1038/374470a0

Zhang W, Sloan-Lancaster J, Kitchen J, Trible RP, Samelson LE (1998) LAT: the ZAP-70 tyrosine kinase substrate that links T cell receptor to cellular activation. Cell 92(1):83–92

Zipfel PA, Bunnell SC, Witherow DS, Gu JJ, Chislock EM, Ring C, Pendergast AM (2006) Role for the Abi/wave protein complex in T cell receptor-mediated proliferation and cytoskeletal remodeling. Curr Biol 16(1):35–46. doi:10.1016/j.cub.2005.12.024

Zyss D, Ebrahimi H, Gergely F (2011) Casein kinase I delta controls centrosome positioning during T cell activation. J Cell Biol 195(5):781–797. doi:10.1083/jcb.201106025

Chapter 12
Homotypic Cell–Cell Interactions and Apicobasal Polarity in Epithelial Cells and Endothelial Cells

Benjamin Franz Brinkmann, Hüseyin Tuncay, and Klaus Ebnet

Abstract Epithelial and endothelial cells are embedded in sheets of cells and have a pronounced apicobasal polarity. The development of apicobasal polarity is regulated by a conserved set of cell polarity proteins which regulate different aspects of cellular polarization in various contexts. In epithelial and endothelial cells, apicobasal polarity is initiated when cells start to form cell–cell contacts and long before the cells have developed mature cell–cell junctions and have acquired a fully polarized phenotype. Cell–cell adhesion has turned out to be intimately involved in the regulation of polarization. In this chapter, we will highlight the molecular mechanisms through which cell–cell adhesion molecules act in concert with cell polarity proteins to regulate various aspects of cellular polarization in vertebrate epithelial and endothelial cells.

Keywords Apicobasal polarity • Junctional adhesion molecule • Lumen formation • Par proteins • Spindle orientation • Tight junctions

12.1 Introduction

Epithelial and endothelial cells are polarized cells with a distinctive asymmetry in the plasma membrane. The asymmetry is characterized by a selective localization of lipids and membrane proteins in the apical versus the basolateral membrane (Simons and Fuller 1985), with the apical domain being defined as the membrane domain that faces the free space and the basolateral membrane being defined as the membrane that is in contact with either neighboring cells or with the extracellular matrix (ECM) (Nelson 2003) (Fig. 12.1). This asymmetry in the plasma membrane, also called apicobasal membrane polarity, is the basis for functional differences of the membrane domains which in turn are the prerequisites for the functioning of the

B.F. Brinkmann • H. Tuncay • K. Ebnet (✉)
Institute-associated Research Group "Cell adhesion and cell polarity," Institute of Medical Biochemistry, ZMBE, University of Münster, Von-Esmarch-Str. 56, D-48149 Münster, Germany
e-mail: ebnetk@uni-muenster.de

K. Ebnet (ed.), *Cell Polarity 1*, DOI 10.1007/978-3-319-14463-4_12

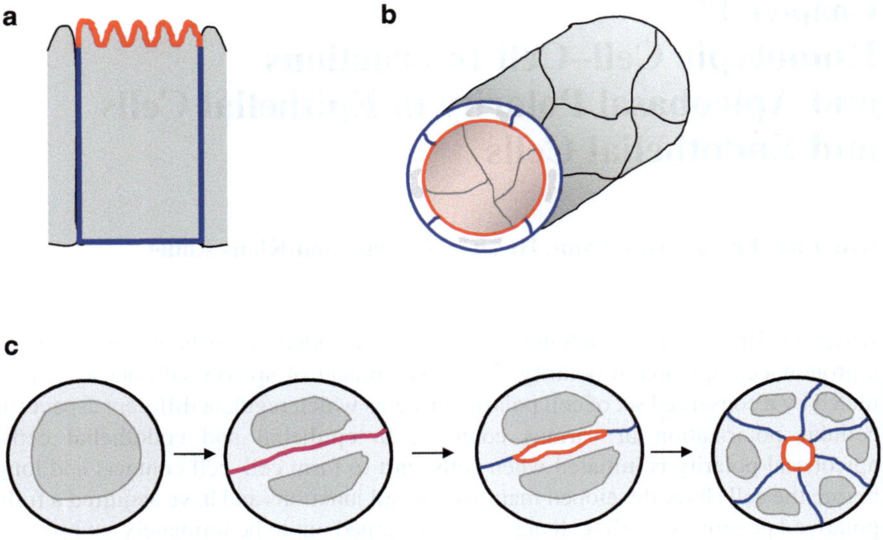

Fig. 12.1 Apicobasal membrane polarity in polarized epithelial cells. (**a**, **b**) The apical membrane domain (*red*) is free of contact (**a**) and faces the lumen of lumen-containing organs (**b**); the basolateral membrane domain (*blue*) faces other cells and the extracellular matrix. (**c**) Lumen formation is initiated at sites of cell–cell contact. Polarization begins with the first cell division which generates a first site of cell–cell contact (*magenta*). Apical membrane markers co-localize with cell adhesion complexes (*blue*) at these contact sites. Lumenogenesis proceeds by separation of the opposed membranes and segregation of cell adhesion complexes from apical membranes forming apical (*red*) and basolateral (*blue*) membrane domains

organ, such as uptake and vectorial transport of nutrients, formation of a selective barrier for hydrophilic substances, or the regulated passage of cells during inflammation.

The formation of membrane asymmetry is initiated when epithelial or endothelial cells form cell–cell contacts. When polarizing epithelial cells are kept as single cells in suspension, apical or basolateral marker proteins are evenly distributed along the entire cell surface. Only when cells form cell–cell contacts these markers segregate to the apical and basolateral membrane domains (Rodriguez-Boulan et al. 1983; Wang et al. 1990). A new cell–cell contact area can also be generated when single cells grown in a three-dimensional environment divide to generate two daughter cells (Fig. 12.1). Cell–cell adhesion thus generates a cell surface area that is in contact with other cells which allows the cell to distinguish this area from the free cell surface area as a first step during the development of cell surface polarity. Once established, these sites of cell–cell adhesion serve as targeting patches for the delivery of proteins selectively localized at the bounded surfaces. Signals derived from sites of cell–cell adhesion but also from cell–matrix adhesion regulate the maturation of intercellular junctions when cells are grown in two dimensions or when cells are grown in three dimensions. This chapter addresses the molecular mechanisms through which adhesion molecules localized at epithelial and endothelial cell contacts contribute to the formation of stable cell–cell contacts and the development of apicobasal polarity.

12.2 Homotypic Cell–Cell Contacts of Epithelial and Endothelial Cells

Epithelial and endothelial cells form sheets in which individual cells are connected to their neighbors through homotypic cell–cell contacts. On the basis of their appearance at the ultrastructural level, their molecular composition, and their biological function, several substructures can be distinguished within the cell contacts, the tight junctions (TJ), adherens junctions (AJ), and desmosomes. The TJs restrict the diffusion of molecules though intercellular junctions and are responsible for the barrier function of epithelial or endothelial sheets (Schneeberger and Lynch 2004). In addition, TJs prevent the intermixing of apical and basolateral membrane domains, which is important to maintain surface polarity (van Meer and Simons 1986). AJs serve to connect the intercellular junctions to the actin cytoskeleton (Gumbiner 1996) and play important role during tissue morphogenesis (Gumbiner 2005). Finally, desmosomes are anchoring sites for intermediate filaments which provide resistance to shearing forces to epithelial monolayers (Green and Simpson 2007). In all three structures, the physical contact between adjacent cells is mediated by cell–cell adhesion molecules, of which some build up the barrier, some have a prominent role in mediating physical adhesion, and others have primarily a regulatory role as signaling molecules. All adhesion molecules that are localized at specific structures are directly associated with peripheral membrane proteins through their cytoplasmic domains (Fig. 12.2). These interactions serve several purposes: First, the recruitment of cytoplasmic proteins to the specific site of cell–cell contact. Second, the formation of a cytoplasmic plaque structure to bring various integral membrane proteins in close proximity to support their interactions. Third, the generation of a physical link between cell surface and the actin cytoskeleton. The adhesion molecules which have a prominent role in the development of apicobasal membrane polarity will be discussed in more detail in this chapter.

12.3 Adhesion Molecules at Tight Junctions and Their Associated Cytoplasmic Proteins

The major family of integral membrane proteins at the TJs consists of proteins with four transmembrane domains and includes occludin, claudins, and tricellulin. Claudins form a family of 26 members in humans and undergo homophilic and heterophilic interactions in *cis* and in *trans*. These *cis*- and *trans*-interactions result

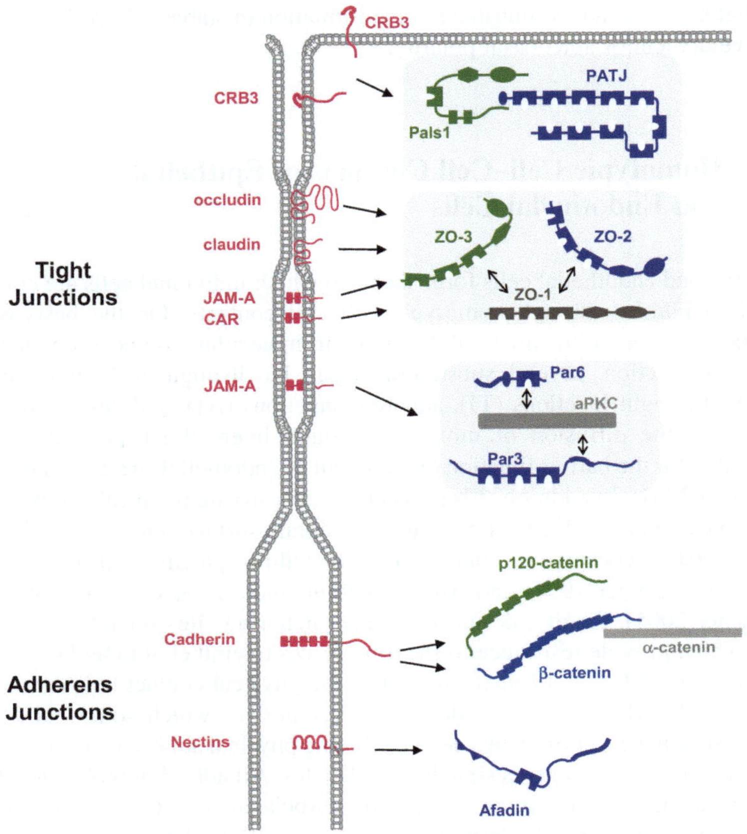

Fig. 12.2 Integral membrane proteins at tight junctions and adherens junctions. A number of integral membrane proteins localize at tight junctions and adherens junctions of epithelial and endothelial cells. The integral membrane proteins interact with cytoplasmic scaffolding proteins as indicated by *arrows*. The major protein complexes at tight junctions are indicated by *gray boxes*. In many cases, the cytoplasmic scaffolding proteins link the integral membrane proteins to the actin cytoskeleton. Note that the protein–protein interactions depicted are dynamic and that the composition of the major protein complexes at tight junctions as well as their interaction with integral membrane proteins is subject to change depending on the state of junction maturation and during junction remodeling

in the formation of the TJ strands, i.e., regions where the two outer membrane leaflets of adjacent cells are in direct physical contact (Tsukita et al. 2001). The TJ strands form a barrier for ions and small hydrophilic substances (Van Itallie and Anderson 2006; Tsukita et al. 2008). Claudins as well as occludin associate with members of the membrane-associated guanylate kinase (MAGUK) family of scaffolding proteins including zonula occludens (ZO)-1, ZO-2, and ZO-3 (Itoh et al. 1999) (Fig. 12.2), and this interaction is most likely necessary for the assembly of claudins into TJ strands. In the absence of these three MAGUK proteins, epithelial cells fail to form TJ strands and to develop a functional epithelial barrier (Umeda et al. 2006). The interaction between claudins and ZO proteins thus illustrates one example that scaffolding proteins regulate the functional activity of the integral membrane proteins.

A second family of adhesion molecules at TJs belongs to the CD2 subfamily of the immunoglobulin (Ig) superfamily, the junctional adhesion molecule (JAM) family (Ebnet et al. 2004). The JAM family consists of eight members (JAM-A, JAM-B, JAM-C, JAM4, JAM-L, ESAM, CAR, CLMP). It should be noted that not all JAMs are expressed by epithelial cells, and, among those which are, not all localize at TJs. In addition, some (e.g., JAM-A) are enriched but not exclusively localized at TJs. As opposed to claudins, JAMs are not components of the TJ strands but localize to areas at TJs where the lipid bilayers of the adjacent cells are not in a direct physical contact (Tsukita et al. 2001) (Fig. 12.2). The three JAM family members JAM-A, JAM-B, and JAM-C directly interact with the polarity protein PAR-3 (Ebnet et al. 2001, 2003) (Fig. 12.2). PAR-3 is a member of the highly conserved PAR-3-aPKC-PAR-6 polarity protein complex which is involved in the regulation of cell polarity in different contexts such as apicobasal polarity in epithelial cells, neuronal polarity in neuronal cells, or directed cell migration in epithelial cells and astrocytes (Macara 2004; Suzuki and Ohno 2006). The two PAR proteins in this complex, PAR-3 and PAR-6, regulate the localization and the activity, respectively, of aPKC. In the absence of Rho family GTPases, aPKC is inactive. The binding of GTP-bound Cdc42 or Rac1 to PAR-6 results in a conformational change of PAR-6 which releases aPKC from restraints and allows it to adopt the active conformation (Yamanaka et al. 2001). As we will see below, the direct interaction of PAR-3 with JAM-A most likely serves to recruit the PAR-3-aPKC-PAR-6 complex to nascent cell junctions. This interaction represents an example that adhesion molecules can act as positional cue for the localization of the associated cytoplasmic protein.

Finally, a third integral membrane protein at epithelial TJs is the vertebrate homologue of *Drosophila* Crumbs, called CRB3 (Tepass et al. 1990; Makarova et al. 2003). Although the role of the extracellular domain of Crumbs has been unclear for a long time, recent evidence indicates that *Drosophila* Crumbs undergoes homophilic interaction in *trans* (Thompson et al. 2013). As opposed to *Drosophila* Crumbs, vertebrate CRB3 has only a very short extracellular domain consisting of 33 amino acids (AA), and it is unclear if this short extracellular domain mediates homophilic interactions. CRB3 assembles another cell polarity protein complex consisting of CRB3 and the two scaffolding proteins Pals1 and PATJ (Roh et al. 2002) (Fig. 12.2). CRB3 contains a PDZ domain-binding motif at its C-terminus through which it directly interacts with the PDZ domain of Pals1, which in turn interacts with the multi-PDZ domain protein PATJ (Roh et al. 2002). In addition to the interaction with Pals1, CRB3 also interacts with PAR-6 (Lemmers et al. 2004). It should be noted that CRB3 does not exclusively localize to TJs, but is also localized at the apical membrane (Lemmers et al. 2004).

12.4 Adhesion Molecules at Adherens Junctions and Their Associated Cytoplasmic Proteins

AJs contain two families of adhesion molecules, cadherins and nectins (Pokutta and Weis 2007; Takai et al. 2008). Classical cadherins contain five characteristic ~110 amino acid domains in the extracellular part of the protein, and E-cadherin is a prototypic member of classical cadherins that is expressed in epithelial cells. In the presence of Ca^{2+}, the cadherin ectodomains interact in *trans* to mediate *trans*-homophilic interactions and to promote mechanical cell–cell adhesion (Brasch et al. 2012). In addition, the extracellular domains interact in *cis* resulting in clusters of oligomers which explain the parallel alignment of the plasma membranes from apposed cells in AJs (Harrison et al. 2011). The cytoplasmic domain of E-cadherin associates with two members of the armadillo family of proteins, p120ctn and β-catenin, through the juxtamembrane and the C-terminal domain of the cytoplasmic domain, respectively (Fig. 12.2). The interaction with β-catenin serves to link the AJs to the actin cytoskeleton, probably through the actin-binding proteins α-catenin and EPLIN, whereas the association with p120ctn regulates the stability and turnover rate of E-cadherin at cell–cell contacts (Pokutta and Weis 2007; Abe and Takeichi 2008). The coupling of sites of cell adhesion to the underlying actin cytoskeleton by β-catenin is required to help in the organization of cell structure as well as during morphogenesis to transmit forces within the epithelial tissue. The second family of adhesion molecules that is localized at AJs comprises the Ig-SF members nectins (Takai et al. 2008) (Fig. 12.2). Nectins contain three Ig-like domains, form *cis*-dimers through the second Ig domain, and undergo *trans*-homophilic and heterophilic interactions through the first, membrane-distal Ig domain (Aricescu and Jones 2007). The large number of possible homophilic and heterophilic interactions between nectins and nectin-like molecules probably serves to regulate cell–cell interactions in different tissues and between different cells types (Rikitake et al. 2012). Similar to JAMs, nectins contain a PDZ domain-binding motif at the C-terminus through which they can interact with PDZ domain-containing proteins. The major nectin-binding protein is afadin, a scaffolding protein with a C-terminal F-actin-binding region (Mandai et al. 1997; Takahashi et al. 1999). One function of afadin is to connect nectins to the underlying actin cytoskeleton.

12.5 The Process of Cell–Cell Contact Formation

The formation of new epithelial cell–cell contacts is a process that occurs during embryonic development. The majority of epithelial cells are generated by elongation and branching of the primary epithelial-like sheets derived from the ectoderm and the endoderm. However, a number of epithelial sheets originate from mesenchymal cells in the mesoderm including the endothelium, the tubular cells of the kidney, the epithelial cells in male and female reproductive tracts, or the trophectoderm during implantation of the embryo (Ekblom 1989). Mesenchymal cells are not directly attached to each other, are highly motile, and display no apicobasal polarity. In contrast, epithelial cells are embedded in a cellular sheet through cell–cell contacts, display a prominent apicobasal membrane polarity, and are sessile. The transition from a mesenchymal cell into an epithelial cell—also called mesenchymal to epithelial transition (MET)—is accompanied by the formation of new cell–cell contacts, altered interaction with the extracellular environment, and a dramatic change in cell morphology. The reverse process, epithelial to mesenchymal transition (EMT), also occurs during embryonic development and regulates mesoderm formation, neural crest formation, or the formation of the cardiac valves or neural derivatives (Thiery and Sleeman 2006; Yang and Weinberg 2008). Importantly, both MET and EMT contribute to the development and progression of carcinoma, whereby EMT allows cells to escape from the primary site and adopt a migratory phenotype and MET allows these cells to form cellular aggregates and metastases at remote sites (Yang and Weinberg 2008).

The process of MET cannot be mimicked in vitro as it is a dynamic process and regulated by a plethora of signals present in the embryonic tissue. However, several aspects of MET can be addressed experimentally using epithelial cell lines. A frequently used cell line is the kidney-derived Madin-Darby canine kidney cell line (MDCK). MDCK cells share many aspects of polarized epithelial cells, such as stable cell–cell contacts, formation of a monolayer with barrier-forming TJs, apicobasal membrane polarity, and the development of cysts when grown in a three-dimensional matrix (Simons and Fuller 1985; O'Brien et al. 2002). The new formation of cell–cell contacts can be induced by seeding cells at a low density or alternatively by mechanically introducing an artificial wound into the monolayer and to allow the cells to close the gap (Adams et al. 1996; Yonemura et al. 1995). When cells are fixed and stained at different time points after seeding or wounding, the process of junction formation can be analyzed by the appearance of cell adhesion molecules and their associated cytoplasmic proteins at the nascent contacts. This approach has led to the model of a stepwise formation and maturation of cell–cell junctions (Fig. 12.3), according to which opposing cells first form numerous transient contacts which are subsequently stabilized to form adhesive microdomains or "puncta" (McNeill et al. 1993; Adams et al. 1996). Puncta contain several proteins which in mature junctions are localized at AJs, which has led to the term "primordial, spotlike adherens junctions" (pAJs) (Yonemura et al. 1995). After some time, the puncta-like contacts spread along the opposed membranes by a zippering activity, resulting in a linear appearance of cell–cell junctions. Finally, when cells have assembled stable, linear cell–cell junctions, the junctions mature in vertical direction, and proteins start to segregate from each other, resulting in the formation of specific substructures like TJs, AJs, and desmosomes.

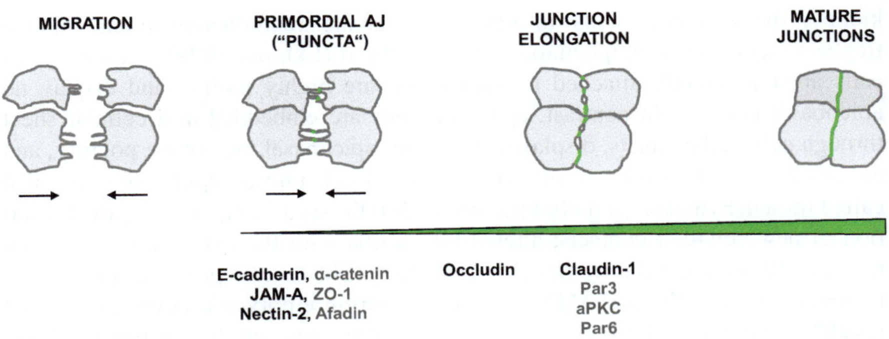

Fig. 12.3 Stages of cell–cell contact formation. Migrating epithelial cells form dynamic protrusions to sample the environment for other epithelial cell surfaces. Upon encountering protrusions of cells in close proximity, cell adhesion molecules like E-cadherin, JAM-A, or nectin-2 stabilize these contacting membrane areas by homophilic and perhaps heterophilic interactions, which leads to the formation of primordial, spotlike AJs or "puncta." Stabilization of cell–cell contacts is further enhanced by the association of the adhesion molecules with actin filaments, which is mediated by cytoplasmic proteins like α-catenin, ZO-1, or afadin. Subsequently, the spotlike contact sites gradually fuse to more linear patterns of cell junctions, and concomitantly new molecules are recruited into the maturing intercellular junctions. Occludin is recruited after pAJs have formed but before claudins and before members of the Par polarity protein complex. The recruitment and activation of the Par complex is a key step for the development of pAJs into mature junctions. The green bar reflects the increase in the contacting membrane area during junction maturation

12.6 The Role of Cell Adhesion Molecules During Cell–Cell Contact Formation and Development of Apicobasal Polarity

Scratch-wounding assays or sparse cell seeding combined with immunofluorescence microscopy has revealed a stepwise appearance of adhesion molecules and peripheral membrane proteins at sites of cell–cell contact during contact formation (Fig. 12.3). Among the first proteins which are detectable at the pAJs are the adhesion molecules E-cadherin, JAM-A, and nectin-2 but also the peripheral membrane proteins α-catenin, β-catenin, afadin, and ZO-1 (Adams et al. 1996; Yonemura et al. 1995; Asakura et al. 1999; Ando-Akatsuka et al. 1999; Ebnet et al. 2001). All peripheral membrane proteins are binding partners for one or more adhesion molecules. For example, E-cadherin interacts with α-catenin and β-catenin (Pokutta and Weis 2007), JAM-A interacts with afadin and ZO-1 (Ebnet et al. 2000), and nectin-2 interacts with afadin (Takahashi et al. 1999). As individual pAJs are associated with a bundle of actin filaments (Adams et al. 1996), the interaction with the cytoplasmic proteins that contain actin-binding sites serves to strengthen and stabilize initially weak cell–cell contacts. It is unclear if in each molecular pair the adhesion molecule recruits the cytoplasmic partner or if vice versa the cytoplasmic protein recruits the adhesion molecule. Recent studies showed that the ectodomain of E-cadherin is sufficient to localize at pAJs and to stimulate pAJs formation, suggesting that cytoplasmic interactions are not required for its stable localization at pAJs (Ozaki et al. 2010; Hong et al. 2010). On the other hand, some evidence exists that afadin recruits the adhesion molecules JAM-A and nectin-2 to intercellular junctions, at least in heterologous expression systems (Ebnet et al. 2000; Takahashi et al. 1999). So it might depend on the adhesive strength of an adhesion molecule if its stable localization at pAJs requires cytoplasmic interaction. Following this initial localization at pAJs, the next integral membrane protein that is recruited is occludin (Ando-Akatsuka et al. 1999) (Fig. 12.3). Occludin has most likely a signaling activity during the formation of TJs (Yu et al. 2005). Its early localization at pAJs shortly after the onset of maturation therefore suggests that it also might have a signaling role during maturation of early cell junctions. After the localization of occludin at maturing cell–cell junctions, the next adhesion molecule that appears is claudin-1 (Suzuki et al. 2002). Claudins might be recruited through their association with ZO-1 or the related ZO-2 and/or ZO-3, since ZO proteins are also required to assemble claudins into TJ strands (Umeda et al. 2006). Interestingly, the PAR polarity protein complex is recruited considerably later than the molecules to which the PAR complex can bind, i.e., JAM-A and nectin-2 (Suzuki et al. 2002). PAR-3 appears simultaneously with claudin-1, and probably with a further delay, PAR-6 and aPKC appear at cell–cell junctions which have adopted now a clear linear arrangement (Suzuki et al. 2002).

The PAR complex most likely plays a central role for the maturation of pAJs into functional intercellular junctions of polarized epithelial cells. Expression of a dominant-negative mutant of aPKC does not affect the formation of pAJs, but blocks their maturation into linear junctions (Suzuki et al. 2001, 2002). Therefore, the correct localization and local activation of the PAR polarity complex at pAJs is critical for the development of functional intercellular junctions. Based on the current literature, the following model can be proposed which illustrates the functions of cell adhesion molecules in the process of cell–cell contact formation and the development of apicobasal polarity. In contact-naïve migrating cells, the cells sample the environment by forming long protrusions (Yonemura et al. 1995). These protrusions are transient and will be retracted unless adhesion molecules expressed at their tips or along their sides undergo homophilic interaction with protrusions from other cells. The association of the adhesion molecules with actin filaments through cytoplasmic proteins like α-catenin, ZO-1, or afadin probably contributes further to their stabilization. Once stable pAJs have been formed, the adhesion molecules now have additional functions. One of these functions is to recruit the PAR complex to the stabilized pAJs, and this function is most likely mediated by JAM-A (Ebnet et al. 2001). Since PAR-3 and aPKC appear with slightly different kinetics, it is assumed that PAR-3 is recruited first by JAM-A to generate a platform for the recruitment of PAR-6-aPKC and thus for the assembly of a PAR-3-aPKC-PAR-6 complex at pAJs (Suzuki et al. 2002). After its assembly at pAJs, the PAR complex is activated by the binding of Cdc42 or Rac1 to PAR-6 (Yamanaka et al. 2001; Joberty et al. 2000; Lin et al. 2000). How these two Rho family GTPases are activated is incompletely understood. However, it is very likely that adhesion molecules present at pAJs trigger their activation. E-cadherin can activate Rac1 (Nakagawa et al. 2001; Kovacs et al. 2002; Ehrlich et al. 2002; Yamada and Nelson 2007). Intriguingly, active Rac1 is concentrated at the earliest sites of cell–cell contacts, and its activity diminishes at more mature cell–cell contact regions (Ehrlich et al. 2002; Yamada and Nelson 2007; Kitt and Nelson 2011), suggesting that Rac1 activity is particularly important during the first steps of junctional maturation. However, Cdc42 cannot be excluded as alternative activator of the PAR complex at pAJs. Cdc42 can principally be activated by nectins (Fukuhara et al. 2004; Fukuyama et al. 2005), and Cdc42 is required for junction formation (Yamanaka et al. 2006; Du et al. 2009; Elbediwy et al. 2012), but a localization at pAJs has not been demonstrated yet.

The next step in the process is the regulation of junctional maturation by activated aPKC. The identification of aPKC substrates is key to the understanding of this process. Among the substrates identified so far is the Ser/Thr kinase PAR-1 (Hurov et al. 2004; Suzuki et al. 2004). Phosphorylation by aPKC results in a protein 14-3-3-dependent removal of PAR-1 from the membrane (Hurov et al. 2004; Suzuki et al. 2004) and thus in a PAR-1-free membrane compartment. At the same time, the phosphorylation of PAR-3 by PAR-1 results in inactivation of the PAR-3-aPKC-PAR-6 complex (Benton and St Johnston 2003). This reciprocal phosphorylation between PAR-1 and the PAR-3-aPKC-PAR-6 complex provides the basis for a biochemical mechanism to generate membrane asymmetry. A second aPKC substrate is PAR-3 (Nagai-Tamai et al. 2002). aPKC phosphorylates PAR-3 at Ser827, and this phosphorylation reduces the stability of the ternary PAR-3-aPKC-PAR-6 complex, suggesting that a Par-6-aPKC unit operates independently of PAR-3 once aPKC is activated. Interestingly, in *Drosophila* epithelial cells, the phosphorylation of PAR-3/Bazooka by aPKC is necessary to establish a boundary between apical domain and lateral domain (Morais-de-Sa et al. 2010). Although not demonstrated yet for vertebrate epithelial cells, it is likely that aPKC-mediated phosphorylation of PAR-3 plays a key role in the establishment of apicobasal membrane polarity in epithelial cells. Another aPKC substrate is the protein Numb (Smith et al. 2007). Phosphorylation of Numb by aPKC prevents the diffusion of Numb into the apical membrane domain (Smith et al. 2007). Whether Numb plays a regulatory role for junctional maturation is unclear. The fourth substrate of aPKC during junction formation is JAM-A (Iden et al. 2012). JAM-A is phosphorylated by aPKC at Ser285 in cells, and this phosphorylation regulates the timely maturation of cell–cell junctions. Interestingly, JAM-A present at pAJs is not phosphorylated at Ser285, and phosphorylation at Ser285 only occurs after the recruitment of aPKC to slightly more mature junctions. How Ser285-phosphorylated JAM-A regulates junctional maturation is unclear, but it is likely that phosphorylated JAM-A recruits cytoplasmic proteins that might promote the maturation through unknown mechanisms. JAM-A phosphorylation not only regulates junction maturation but also epithelial barrier function, indicating that JAM-A phosphorylation by aPKC regulates the formation of functional TJs as well (Iden et al. 2012).

Interestingly, the development of cell surface polarity starts very early and almost simultaneously with the formation of E-cadherin-based cell–cell contacts. Studies with aquaporin-3 (AQP-3) and AQP-5, which are markers for the basolateral and the apical membrane domain, respectively, have shown that AQP-3 but not AQP-5 co-distributes with E-cadherin at early cell–cell contact sites (Nejsum and Nelson 2007). Components of the cellular machinery which regulates vesicle trafficking from the post-Golgi network to specific sites at the membrane, such as microtubules and the exocyst and SNARE complexes, localize to nascent cell–cell contacts similarly early (Nejsum and Nelson 2007). These observations thus indicate that E-cadherin sites of cell adhesion provide an extrinsic signal to generate a targeting patch for various components of polarized protein transport, such as microtubules, the exocyst, and SNARE complexes, for the selective targeting of basolateral proteins. They also indicate that cell surface polarization is initiated already at the level of early cell–cell contact formation, long before cells have developed fully mature intercellular junctions.

In summary, adhesion molecules play several important roles during formation and maturation of cell–cell contacts. The first important function is the formation of the first sites of cell–cell contacts through homophilic interactions and their stabilization by association with actin fibers through adapter proteins. E-cadherin, JAM-A, and nectin-2 interact with actin-binding proteins and could mediate this function. The second function is the recruitment of key regulators of cell–cell contact maturation to pAJs, and this function regulates the localization of the regulatory machinery at the right sites. JAM-A as recruiter of the PAR-aPKC complex is one example. The third important function is the activation of the regulatory machinery at the correct subcellular localization. E-cadherin is the major candidate molecule to activate aPKC through its ability to activate Rac1. Finally, as a fourth function, adhesion molecules might also influence the levels of their associated proteins at the transcriptional, translational, or posttranslational level. Recent evidence indicates that CRB3 regulates the levels of the two CRB3 complex members Pals1 and PATJ but not of other proteins at TJs like occludin or ZO-1 (Whiteman et al. 2013).

After the establishment of stable cell–cell junctions and the formation of functional TJs, the role of cell adhesion molecules for apicobasal polarity is less well understood. One role might still be to localize cytoplasmic proteins to sites of cell–cell adhesion. However, it seems that in fully polarized cells, the complexity of protein interactions is much higher than during the process of junction formation (Hurd et al. 2003b; Wells et al. 2006), which might allow many proteins to be localized to TJs even in the absence of their direct interaction with an integral membrane protein. A still unresolved issue is the nature of the intramembrane diffusion barrier which in vertebrate epithelial cells is localized at the TJs. Cells lacking the ZO proteins ZO-1, ZO-2, and ZO-3 have no TJ strands; the TJ-associated cell adhesion molecules claudins, occludin, and JAM-A are not present at cell contacts; and the barrier function for the paracellular diffusion of small hydrophilic molecules is completely lost (Umeda et al. 2006). Surprisingly, the intramembrane diffusion barrier is unaffected by the loss of TJ strands (Umeda et al. 2006). This indicates that apicobasal membrane polarity does not depend on the presence of TJ strands. It is therefore unclear how intramembrane diffusion is regulated. The diffusion could have a physical nature as suggested for the diffusion barrier at the axonal hillock of neurons (Nakada et al. 2003) or could be purely biochemical in nature as described for the mutual exclusion of PAR proteins (Hao et al. 2006; Benton and St Johnston 2003; Morais-de-Sa et al. 2010; Hurd et al. 2003a; Hurov et al. 2004; Suzuki et al. 2004).

12.7 Loss of Cell–Cell Adhesion and Cell Polarity During EMT

As pointed out before, EMT is a process during which cells convert from a static epithelial to a migratory mesenchymal phenotype, and when inappropriately activated, EMT can contribute to cancer metastasis (Yang and Weinberg 2008). Hallmarks of EMT are a loss of cell–cell adhesion and a loss of apicobasal polarity. Stimuli which trigger EMT frequently target cell adhesion molecules as well as cell polarity proteins.

EMT can be triggered by a variety of extracellular signals, including soluble growth factors like members of the transforming growth factor (TGF)β family, of the fibroblast growth factor (FGF) family, of the epidermal growth factor (EGF) family, or of the Wnt family, which cooperate with signals derived from the ECM-like hyaluronic acid or collagen (Thiery and Sleeman 2006). The activation of intracellular signaling pathways by these factors results in the activation of transcription factors of the Snail and ZEB families which in turn can repress the expression of proteins that regulate cell–cell adhesion. The primary target of these transcription factors is E-cadherin (Cano et al. 2000; Comijn et al. 2001; Peinado et al. 2007), and the loss of E-cadherin together with the new expression of mesenchymal markers like the intermediate filament protein vimentin is a hallmark of EMT (Thiery and Sleeman 2006). Reduced levels of E-cadherin will not only affect cell cohesion but also trigger new gene expression as a result of enhanced β-catenin-dependent nuclear signaling through LEF/TCF4 transcription factors. However, E-cadherin is not the only adhesion molecule that is targeted for repression during EMT. The transcription factor ZEB1 represses the genes encoding occludin, tricellulin, claudin-7, CRB3, and JAM-A (Aigner et al. 2007). Among these, CRB3 has been found to be negatively regulated by Snail at the level of protein stability as well, probably through a mechanism that involves differential glycosylation of CRB3 (Harder et al. 2012). It should be noted that the repression of adhesion molecule gene expression during EMT may not only be regulated by direct repression of target genes but also indirectly by upregulation of certain microRNA (miRNA) species. The two miRNAs miR-9 and miR-661 which target E-cadherin and nectin-1, respectively, were found to be upregulated during EMT (Ma et al. 2010; Vetter et al. 2010). Together these observations indicate that a number of cell–cell adhesion molecules are subject to downregulation during EMT, suggesting an important role of these molecules for the maintenance of the epithelial phenotype. This is further supported by the recent observations that the adhesion molecules claudin-3 and claudin-4 prevent mesenchymal characteristics by positively regulating the levels of E-cadherin and negatively regulating the expression of the EMT transcription factor Twist (Lin et al. 2013).

In most cases, the mechanisms through which cell adhesion molecules regulate EMT are incompletely understood. However, in the case of occludin, a mechanism has been delineated that underlines the intimate relation between cell adhesion molecules and cell polarity proteins and their role during EMT. TGFβ induces EMT by signaling through the two TGFβ receptors TβRI and TβRII (Huber et al. 2005). One of these two receptors, TβRI, is directly associated with occludin, thereby recruiting TβRI to TJs (Barrios-Rodiles et al. 2005). TGFβ induces a heterodimerization of TβRI and TβRII, resulting in the formation of an active TβR complex. Interestingly, TβRI is also directly associated with PAR-6, which allows a direct phosphorylation of PAR-6 at Ser345 by the TβR complex (Ozdamar et al. 2005). Ser345 phosphorylation of PAR-6 results in the recruitment of the ubiquitin ligase Smurf1 which triggers the localized ubiquitination and subsequent degradation of RhoA (Ozdamar et al. 2005). Since TJ integrity depends on RhoA activity (Terry et al. 2011), the local degradation of RhoA results in the dissolution of TJs as a step contributing to EMT. The association of occludin with TβRI highlights one principal mechanism through which cell adhesion molecules regulate the formation or dissolution of epithelial cell–cell contacts, which is the recruitment of signaling molecules to specific subcellular sites.

It should be noted that many of the findings regarding the molecular mechanisms underlying EMT have been obtained from cultured cells that were stimulated in vitro with specific factors. As mentioned above, epithelial cells in vivo are exposed to a much wider variety of stimuli, and the effect of a given EMT inducer depends on the context. Nevertheless, the demonstrated role of cell adhesion molecules and cell polarity proteins during tumor development (Huber et al. 2005; McSherry et al. 2009; Zhang et al. 2013); see also the book chapters 6, 7, and 8 of Vol. 2 of this book) strongly suggests that adhesion molecules cooperate with cell polarity proteins during EMT.

12.8 Adhesion Molecules, Endothelial Polarity, and Vascular Lumen Formation

Although endothelial and epithelial junctions share common features, there are differences in junctional organization. Endothelial cells have AJs and TJs, but lack desmosomes. In addition, while in polarized epithelial cells the TJs are at the most apical sites of cell–cell contacts and the AJ are localized below, the two types of structures are frequently intermingled in endothelial cells (Bazzoni and

Dejana 2004). However, similar to epithelial cells which line the lumen in lumenized organs such as in the kidney tubules or in the intestine, endothelial cells lining the lumen of the blood and lymphatics vessels display apicobasal membrane polarity, and the border between these two domains is most likely at the most apicolateral region of the cell–cell contact.

As expected from the similarities between epithelial and endothelial cells, it is not surprising that adhesion molecules were found to be involved in endothelial lumen formation, which most likely reflects a role of these proteins in endothelial polarization. Among the adhesion molecules identified are the three JAMs JAM-A, JAM-B, and JAM-C. Interestingly, downregulation or inactivation of JAM-B or JAM-C leads to impaired lumen formation in three-dimensional collagen matrices, whereas the inactivation of JAM-A has the opposite effect. It has been suggested that JAM-B and JAM-C are part of a lumen-regulating signaling complex consisting of Cdc42, Par3, Par6B, MT1-MMP, and α2β1-integrin (Sacharidou et al. 2010). Since endothelial polarity and lumen formation is the topic of a specific chapter in this book, the reader is referred to chapter 9 of Vol. 1 of this book.

One interesting observation shall be discussed in more detail as it relates to one specific function of adhesion molecules during the initiation of lumen formation. Although demonstrated for vascular endothelial cells, the underlying molecular mechanism might be applicable to epithelial lumen formation as well. The major cell–cell adhesion molecule at endothelial cell–cell contacts is VE-cadherin (Giannotta et al. 2013). Inactivation of the VE-cadherin gene in mice or gene silencing in zebrafish lead to vascular defects and disturbed lumen formation (Carmeliet et al. 1999; Montero-Balaguer et al. 2009). Recent studies imply that VE-cadherin has a crucial role in establishing endothelial membrane polarity during lumen formation (Fig. 12.4). During the development of the mouse aorta, VE-cadherin is localized at cell–cell contacts in a cord of vascular cells and marks the sites of cell–cell contact even in the absence of a lumen (Strilic et al. 2009). VE-cadherin is required for the localization of the CD34-sialomucins CD34 and podocalyxin at the cell–cell contacts, whose extracellular domains are decorated with oligosaccharides that contain a large number of negatively charged sialic acids (Nielsen and McNagny 2008). Once recruited and inserted in the membrane, the negatively charged extracellular domains of the sialomucins result in electrostatic surface repulsion leading to membrane separation and formation of an intercellular space (Strilic et al. 2010). These repulsive forces seem to overcome the adhesive forces mediated by *trans*-homophilic VE-cadherin interaction since VE-cadherin relocalizes to the lateral regions of cell–cell contacts. The subsequent recruitment of moesin and its phosphorylation by PKC links this membrane domain to the F-actin cytoskeleton, which allows VEGF-A-induced shape changes through nmMyosin II. As a result of these actomyosin-induced changes in the cell shapes, the apical membrane domains separate further away from each other and the lumen size increases (Zeeb et al. 2010). Consistent with these observations in the murine aorta, a study in zebrafish shows that VE-cadherin and moesin are crucial for polarization and lumen formation in intersomitic vessels and the dorsal aorta during zebrafish development (Wang et al. 2010). In further support of VE-cadherin as

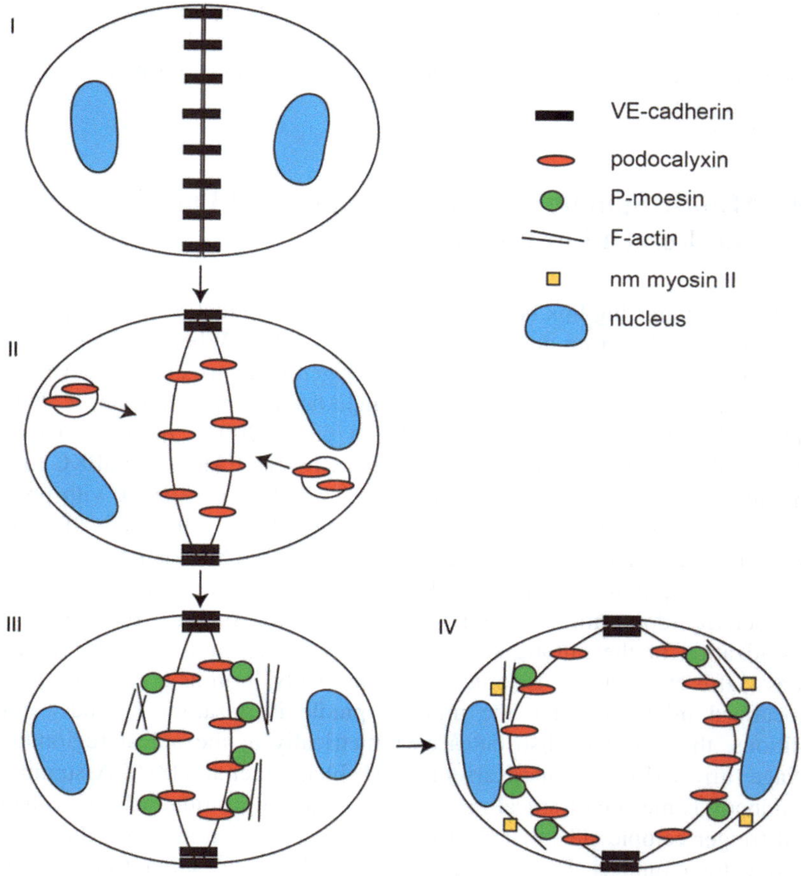

Fig. 12.4 Mechanism of lumen formation in the developing mouse aorta. Lumen formation requires stable intercellular junctions which are mainly mediated by VE-cadherin. The sialomucin podocalyxin that is initially stored in vesicles is recruited by VE-cadherin to sites of cell–cell adhesion. Repulsive forces generated by the negatively charged extracellular domains of podocalyxin molecules from opposing cells induces the separation of the two membranes. At the same time, VE-cadherin is relocalized from the center of the junctional region to the periphery. The phosphorylation of the actin-binding protein moesin by PKC generates binding sites for F-actin leading to the enrichment of F-actin at the newly formed apical membrane domain. Finally, angiogenic growth factors such as VEGF induce the phosphorylation of the myosin regulatory light chain (MLC) and the recruitment of nonmuscle myosin II to the apical F-actin cytoskeleton, and the resulting actomyosin contractile activity results in endothelial shape changes required for lumen formation. This model is based on Strilic et al. (2009)

critical regulator of vascular lumen formation, downregulation of VE-cadherin in cultured endothelial cells disturbs endothelial polarity and impairs lumen formation in three-dimensional collagen gels (Hayashi et al. 2013). This activity of VE-cadherin depends at least partially on its ability to interact with the cell polarity protein Par3,

Par6, and Pals1 (Iden et al. 2006) (Brinkmann B.F. et al. unpublished observations). VE-cadherin thus represents another example how adhesion molecules cooperate with cell polarity proteins to regulate apicobasal polarity and lumen formation.

12.9 Mitotic Spindle Orientation, Epithelial Polarity, and Lumen Formation

Recently, an interesting link between mitotic spindle orientation and the process of lumen formation has been uncovered (Jaffe et al. 2008). When epithelial cells are cultured in a three-dimensional matrix of collagen or matrigel, they develop into spheroid-like structures consisting of a single-layered epithelial sheet that surrounds a fluid-filled single lumen (O'Brien et al. 2002). In individual cells, PAR-3 localizes to the basolateral membrane, whereas PAR-6 and aPKC localize to the apical membrane (Durgan et al. 2011), which is consistent with previous studies showing that PAR-6-aPKC separate from PAR-3 after activation of the complex by Cdc42 and/or Rac1 (Yamanaka et al. 2001). Surprisingly, knockdown of Cdc42 as ubiquitous regulator of cell polarity and putative upstream regulator of aPKC activity does not alter apicobasal membrane polarity but results in a mislocalization of the mitotic spindle and in the formation of multiple ectopic lumens (Jaffe et al. 2008). Ectopic lumens probably form as a consequence of a mislocalized midbody when the mitotic spindle is misaligned. Under normal conditions, the midbody distributes asymmetrically at the apical region of the cleavage furrow (Reinsch and Karsenti 1994; Fleming et al. 2007). Assuming that the midbody is the site of apical membrane patches deposition, its mislocalization would trigger ectopic sites of apical membrane deposition and thus the development of ectopic lumens. These observations therefore suggest that Cdc42 regulates the stable interaction of aster microtubules with the lateral cortex to align the spindle axis parallel to the plane of the cell sheet.

The planar orientation of the spindle is mediated through the stable interaction of the microtubule plus ends with the lateral cortex. This interaction is mediated through the dynein–dynactin motor protein complex, a microtubule minus end-directed motor (Dujardin and Vallee 2002). Dynein–dynactin can interact with cortical dynein-interacting proteins which are enriched at the midcortex, and at the same time it can interact with the microtubule plus ends. The motor activity results in tension toward the centrosome and in torque on the mitotic apparatus that finally aligns the spindle apparatus in the plane of the sheet (Laan et al. 2012). Two pathways have been described as to how the plus ends of the aster microtubules stably interact with the lateral cell cortex during mitosis to align the spindle axis in the plane of the epithelial sheet (Fig. 12.5). One pathway depends on a ternary protein complex consisting of NuMA, LGN, and Gαi (Kotak and Gonczy 2013). Dynein interacts with NuMA (Kotak et al. 2012), which interacts with LGN (Du and Macara 2004). LGN can simultaneously interact with NuMA and with

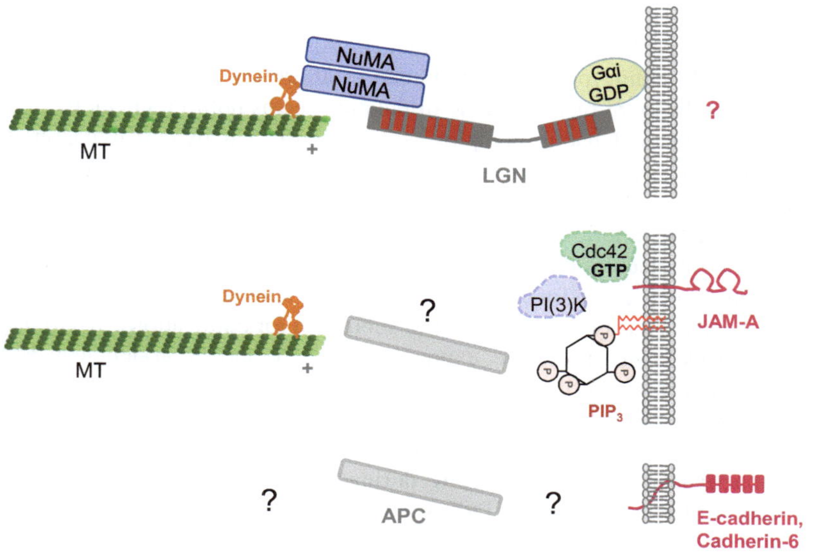

Fig. 12.5 Cell adhesion molecules and mitotic spindle orientation. Two mechanisms have been described which involve the immobilization of the MT-associated dynein–dynactin motor protein complex at the lateral cortex during mitosis. In the NuMA-LGN-Gαi pathway (*top*), MT-associated dynein is connected to the cortex through a ternary complex consisting of NuMA, LGN, and Gαi, in which Gαi is directly anchored to the lipid bilayer through myristoylation. In the Cdc42-PI(3)K-PtdIns(3,4,5)P3 pathway (*middle*), dynein is recruited to PtdIns(3,4,5)P3 through an as yet unidentified linker molecule. Extrinsic signals that regulate planar spindle orientation include JAM-A, which operates through the Cdc42-PI(3)K-PtdIns (3,4,5)P3 pathway, and E-cadherin and Cadherin-6, which cooperatively regulate spindle orientation through a pathway which involves the adenomatous polyposis coli (APC) protein but not the dynein–dynactin complex

myristoylated, membrane-bound Gαi-GDP (Du et al. 2001; Du and Macara 2004; Zheng et al. 2010) which is inserted in the membrane through myristoylation and thus anchors the entire complex at the membrane. The second pathway is less well understood. This pathway depends on the localization of the phosphoinositide PtdIns(3,4,5)P3 as well as on the reorganization of the cortical actin cytoskeleton (Toyoshima et al. 2007; Toyoshima and Nishida 2007; Mitsushima et al. 2009) at the lateral midcortex. Cdc42 turned out as an upstream regulator of this pathway, as Cdc42 is required for the generation of PtdIns(3,4,5)P3 through activation of PI(3) K and is required to reorganize the actin cytoskeleton through p21-activated kinase 2 (PAK2) and the Rac1 GEF βPIX (Mitsushima et al. 2009). How PtdIns(3,4,5)P3 interacts with dynein/dynactin is still unknown.

Although cell–cell contacts have been proposed to regulate mitotic spindle orientation in cell sheets almost two decades ago (Goldstein 1995), the nature of the cell adhesion molecules that provide cortical cues for planar spindle orientation has remained unclear. More recently, cadherins have been identified to regulate planar spindle orientation in MDCK cells (den Elzen et al. 2009). E-cadherin and cadherin-6 act in a redundant manner to regulate the localization of adenomatous polyposis coli (APC) at the cell–cell contacts. APC can interact with the microtubule plus end-binding protein EB1 (Green et al. 2005). However, the cadherin-dependent pathway of planar spindle orientation seems not to regulate the interaction of the dynein–dynactin complex with the cortex as dynein–dynactin is not mislocalized in cells expressing a dominant-negative mutant of E-cadherin (den Elzen et al. 2009). Although the mechanisms on how cadherins regulate the planar orientation of the mitotic spindle are still unclear, these findings support the role of cell adhesion molecules during planar spindle orientation. We have recently identified an unexpected role for JAM-A during planar spindle orientation in polarized epithelial cells (Tuncay et al., manuscript submitted for publication). We found that JAM-A knockdown or ectopic expression of a dimerization-deficient JAM-A mutant results in a multilumenal phenotype and in spindle misorientation when MDCK cells are grown in a three-dimensional collagen gel, which is similar to what has been observed upon Cdc42 knockdown (Jaffe et al. 2008). We also found that JAM-A activates Cdc42 and PI(3)K during mitosis and regulates the cortical localization of dynactin and of PtdIns(3,4,5)P3 at the cortex. These observations thus indicate that the Cdc42-dependent pathway of planar spindle orientation is regulated by JAM-A, and they identify JAM-A as the first adhesion molecule that regulates the stable interaction of the dynein–dynactin complex with the cortex during mitosis. It is to be expected that additional cell adhesion molecules will be identified which regulate signaling events to govern the correct immobilization of the aster microtubules at the lateral cortex.

12.10 Conclusion

Epithelial cell polarization is a highly dynamic process and involves a complex series of signaling events that occur at the cell–cell contact sites. It starts with the formation of early, primordial cell–cell contacts and is finished when cells have developed fully mature intercellular junction with substructures such as TJs, AJs, and desmosomes. Among these structures, the TJs are particularly important since they present the boundary between apical and basolateral membrane domains and harbor—even though molecularly not understood—the diffusion barrier for lipids and intramembrane proteins. Adhesion molecules are involved in the development of epithelial cell polarity at various steps of the process and through different mechanisms. One important function of cell adhesion molecules is their ability to interact in a trans-homophilic and trans-heterophilic manner which in combination with their intracellular association with the actin cytoskeleton generates a mechanically resistant sheet of cells. A second important function is their ability to interact intracellularly with cell polarity proteins which regulates the correct subcellular localization of the polarity regulators. This function as “positional cue” for cell polarity proteins is probably among their most important functions for epithelial cell polarization. A third function is their ability to stimulate signaling events, which is due to their property to interact with scaffolding proteins for signaling molecules such as small GTPases and their regulators. In combination, the latter two functions regulate the precise spatial activation of cell polarity regulators. These different functions place cell adhesion molecules at a central position in the process of epithelial and endothelial cell polarization.

According to this widely accepted model on the role of cell adhesion during epithelial and endothelial polarization, cell adhesion initiates and is thus a prerequisite for polarization. However, it should be noted that not all aspects of epithelial cell polarity depend on the cell–cell adhesion and that some features of the polarized state can develop in contact-naïve cells. For example, absorptive epithelial cells develop a dense array of microvilli at the apical membrane domain, the brush border. The activation of the Ser/Thr kinase LKB1, originally identified as the polarity regulator PAR-4 in *C. elegans* (Kemphues et al. 1988), in E-cadherin mutant epithelial cells results in the rapid formation of the brush border which normally develops at the apical membrane only after long-term culture at confluency, and this occurs in single cells in the absence of cell–cell contacts (Baas et al. 2004). In addition, activation of LKB1 also induces the relocalization of junctional proteins around the periphery of the brush border as well as the sorting of apical and basolateral markers to the apical membrane (defined by the position of the brush border) and the basolateral membrane (defined as the membrane domain excluded from the brush border), respectively. These findings indicate that some aspects of epithelial cell polarity can develop in a cell-autonomous manner without a need for cell–cell adhesion. A second aspect on the role of cell–cell adhesion during cellular polarization that should be noted is the interrelation between cell adhesion molecules and polarity proteins. According to the standard model, cell adhesion drives polarization, but there are instances where polarity proteins seem to act upstream of cell adhesion molecules. For example, the two cell polarity proteins Pals1 and Scribble regulate the trafficking and stable localization of E-cadherin at intercellular junctions (Wang et al. 2007; Lohia et al. 2012). In addition, the polarity protein PATJ regulates endocytosis of CRB3, probably to fine-tune the levels of CRB3 at apical cell contacts, and in addition PATJ restricts the localization of occludin and ZO-3 to TJs (Michel et al. 2005). Together, these observations indicate that the activity of cell polarity proteins may not only be secondary to cell–cell contact formation, and they suggest that it might depend on the state of polarization if cell adhesion precedes polarization or if polarity regulates adhesion. These examples also illustrate that the process of cellular polarization is a highly dynamic process that depends on feedback mechanisms. The most likely scenario is that cell–cell adhesion is of particular importance during early steps of the polarization process. Once the initial cell–cell contacts have been formed and polarization has been initiated, polarity proteins can regulate cell adhesion as a positive feedback to reinforce the polarization process. Thus, cell adhesion molecules and cell polarity proteins cooperate intimately to govern the complex process of epithelial and endothelial cell polarity.

Acknowledgements This work was supported by grants from the Deutsche Forschungsgemeinschaft (EB 160/4-1 and 160/4-2) and from the Medical Faculty of the University of Münster (IZKF Eb2/028/09 and Eb2/020/14).

References

Abe K, Takeichi M (2008) EPLIN mediates linkage of the cadherin catenin complex to F-actin and stabilizes the circumferential actin belt. Proc Natl Acad Sci U S A 105(1):13–19

Adams CL, Nelson WJ, Smith SJ (1996) Quantitative analysis of cadherin-catenin-actin reorganization during development of cell-cell adhesion. J Cell Biol 135(6 Pt 2):1899–1911

Aigner K, Dampier B, Descovich L, Mikula M, Sultan A, Schreiber M, Mikulits W, Brabletz T, Strand D, Obrist P, Sommergruber W, Schweifer N, Wernitznig A, Beug H, Foisner R, Eger A (2007) The transcription factor ZEB1 (deltaEF1) promotes tumour cell dedifferentiation by repressing master regulators of epithelial polarity. Oncogene 26(49):6979–6988. doi:10.1038/sj.onc.1210508

Ando-Akatsuka Y, Yonemura S, Itoh M, Furuse M, Tsukita S (1999) Differential behavior of E-cadherin and occludin in their colocalization with ZO-1 during the establishment of epithelial cell polarity. J Cell Physiol 179(2):115–125

Aricescu AR, Jones EY (2007) Immunoglobulin superfamily cell adhesion molecules: zippers and signals. Curr Opin Cell Biol 19(5):543–550. doi:10.1016/j.ceb.2007.09.010

Asakura T, Nakanishi H, Sakisaka T, Takahashi K, Mandai K, Nishimura M, Sasaki T, Takai Y (1999) Similar and differential behaviour between the nectin-afadin-ponsin and cadherin-catenin systems during the formation and disruption of the polarized junctional alignment in epithelial cells. Genes Cells 4(10):573–581

Baas AF, Kuipers J, van der Wel NN, Batlle E, Koerten HK, Peters PJ, Clevers HC (2004) Complete polarization of single intestinal epithelial cells upon activation of LKB1 by STRAD. Cell 116(3):457–466

Barrios-Rodiles M, Brown KR, Ozdamar B, Bose R, Liu Z, Donovan RS, Shinjo F, Liu Y, Dembowy J, Taylor IW, Luga V, Przulj N, Robinson M, Suzuki H, Hayashizaki Y, Jurisica I, Wrana JL (2005) High-throughput mapping of a dynamic signaling network in mammalian cells. Science 307(5715):1621–1625

Bazzoni G, Dejana E (2004) Endothelial cell-to-cell junctions: molecular organization and role in vascular homeostasis. Physiol Rev 84(3):869–901. doi:10.1152/physrev.00035.2003

Benton R, St Johnston D (2003) Drosophila PAR-1 and 14-3-3 inhibit Bazooka/PAR-3 to establish complementary cortical domains in polarized cells. Cell 115(6):691–704

Brasch J, Harrison OJ, Honig B, Shapiro L (2012) Thinking outside the cell: how cadherins drive adhesion. Trends Cell Biol 22(6):299–310. doi:10.1016/j.tcb.2012.03.004

Cano A, Perez-Moreno MA, Rodrigo I, Locascio A, Blanco MJ, del Barrio MG, Portillo F, Nieto MA (2000) The transcription factor snail controls epithelial-mesenchymal transitions by repressing E-cadherin expression. Nat Cell Biol 2(2):76–83

Carmeliet P, Lampugnani MG, Moons L, Breviario F, Compernolle V, Bono F, Balconi G, Spagnuolo R, Oostuyse B, Dewerchin M, Zanetti A, Angellilo A, Mattot V, Nuyens D, Lutgens E, Clotman F, de Ruiter MC, Gittenberger-de Groot A, Poelmann R, Lupu F, Herbert JM, Collen D, Dejana E (1999) Targeted deficiency or cytosolic truncation of the VE-cadherin gene in mice impairs VEGF-mediated endothelial survival and angiogenesis. Cell 98(2):147–157

Comijn J, Berx G, Vermassen P, Verschueren K, van Grunsven L, Bruyneel E, Mareel M, Huylebroeck D, van Roy F (2001) The two-handed E box binding zinc finger protein SIP1 downregulates E-cadherin and induces invasion. Mol Cell 7(6):1267–1278 den Elzen N, Buttery CV, Maddugoda MP, Ren G, Yap AS (2009) Cadherin adhesion receptors orient the mitotic spindle during symmetric cell division in mammalian epithelia. Mol Biol Cell 20(16):3740–3750

Du D, Pedersen E, Wang Z, Karlsson R, Chen Z, Wu X, Brakebusch C (2009) Cdc42 is crucial for the maturation of primordial cell junctions in keratinocytes independent of Rac1. Exp Cell Res 315(8):1480–1489

Du Q, Macara IG (2004) Mammalian Pins is a conformational switch that links NuMA to heterotrimeric G proteins. Cell 119(4):503–516

Du Q, Stukenberg PT, Macara IG (2001) A mammalian partner of inscuteable binds NuMA and regulates mitotic spindle organization. Nat Cell Biol 3(12):1069–1075. doi:10.1038/ncb1201-1069

Dujardin DL, Vallee RB (2002) Dynein at the cortex. Curr Opin Cell Biol 14(1):44–49

Durgan J, Kaji N, Jin D, Hall A (2011) Par6B and atypical PKC (aPKC) regulate mitotic spindle orientation during epithelial morphogenesis. J Biol Chem 286(14):12461–12471

Ebnet K, Aurrand-Lions M, Kuhn A, Kiefer F, Butz S, Zander K, Meyer Zu Brickwedde MK, Suzuki A, Imhof BA, Vestweber D (2003) The junctional adhesion molecule (JAM) family members JAM-2 and JAM-3 associate with the cell polarity protein PAR-3: a possible role for JAMs in endothelial cell polarity. J Cell Sci 116:3879–3891

Ebnet K, Schulz CU, Meyer Zu Brickwedde MK, Pendl GG, Vestweber D (2000) Junctional adhesion molecule interacts with the PDZ domain-containing proteins AF-6 and ZO-1. J Biol Chem 275(36):27979–27988

Ebnet K, Suzuki A, Horikoshi Y, Hirose T, Meyer Zu Brickwedde MK, Ohno S, Vestweber D (2001) The cell polarity protein ASIP/PAR-3 directly associates with junctional adhesion molecule (JAM). EMBO J 20(14):3738–3748

Ebnet K, Suzuki A, Ohno S, Vestweber D (2004) Junctional adhesion molecules (JAMs): more molecules with dual functions? J Cell Sci 117(1):19–29

Ehrlich JS, Hansen MD, Nelson WJ (2002) Spatio-temporal regulation of Rac1 localization and lamellipodia dynamics during epithelial cell-cell adhesion. Dev Cell 3(2):259–270

Ekblom P (1989) Developmentally regulated conversion of mesenchyme to epithelium. FASEB J 3(10):2141–2150

Elbediwy A, Zihni C, Terry SJ, Clark P, Matter K, Balda MS (2012) Epithelial junction formation requires confinement of Cdc42 activity by a novel SH3BP1 complex. J Cell Biol 198(4):677–693

Fleming ES, Zajac M, Moschenross DM, Montrose DC, Rosenberg DW, Cowan AE, Tirnauer JS (2007) Planar spindle orientation and asymmetric cytokinesis in the mouse small intestine. J Histochem Cytochem 55(11):1173–1180. doi:10.1369/jhc.7A7234.2007

Fukuhara T, Shimizu K, Kawakatsu T, Fukuyama T, Minami Y, Honda T, Hoshino T, Yamada T, Ogita H, Okada M, Takai Y (2004) Activation of Cdc42 by trans interactions of the cell adhesion molecules nectins through c-Src and Cdc42-GEF FRG. J Cell Biol 166(3):393–405

Fukuyama T, Ogita H, Kawakatsu T, Fukuhara T, Yamada T, Sato T, Shimizu K, Nakamura T, Matsuda M, Takai Y (2005) Involvement of the c-Src-Crk-C3G-Rap1 signaling in the nectin-induced activation of Cdc42 and formation of adherens junctions. J Biol Chem 280(1):815–825

Giannotta M, Trani M, Dejana E (2013) VE-cadherin and endothelial adherens junctions: active guardians of vascular integrity. Dev Cell 26(5):441–454. doi:10.1016/j.devcel.2013.08.020

Goldstein B (1995) Cell contacts orient some cell division axes in the Caenorhabditis elegans embryo. J Cell Biol 129(4):1071–1080

Green KJ, Simpson CL (2007) Desmosomes: new perspectives on a classic. J Invest Dermatol 127 (11):2499–2515. doi:10.1038/sj.jid.5701015

Green RA, Wollman R, Kaplan KB (2005) APC and EB1 function together in mitosis to regulate spindle dynamics and chromosome alignment. Mol Biol Cell 16(10):4609–4622. doi:10.1091/mbc.E05-03-0259

Gumbiner BM (1996) Cell adhesion: the molecular basis of tissue architecture and morphogenesis. Cell 84(3):345–357

Gumbiner BM (2005) Regulation of cadherin-mediated adhesion in morphogenesis. Nat Rev Mol Cell Biol 6(8):622–634

Hao Y, Boyd L, Seydoux G (2006) Stabilization of cell polarity by the C. elegans RING protein PAR-2. Dev Cell 10(2):199–208

Harder JL, Whiteman EL, Pieczynski JN, Liu CJ, Margolis B (2012) Snail destabilizes cell surface Crumbs3a. Traffic 13(8):1170–1185. doi:10.1111/j.1600-0854.2012.01376.x

Harrison OJ, Jin X, Hong S, Bahna F, Ahlsen G, Brasch J, Wu Y, Vendome J, Felsovalyi K, Hampton CM, Troyanovsky RB, Ben-Shaul A, Frank J, Troyanovsky SM, Shapiro L, Honig B (2011) The extracellular architecture of adherens junctions revealed by crystal structures of type I cadherins. Structure 19(2):244–256. doi:10.1016/j.str.2010.11.016

Hayashi M, Majumdar A, Li X, Adler J, Sun Z, Vertuani S, Hellberg C, Mellberg S, Koch S, Dimberg A, Koh GY, Dejana E, Belting HG, Affolter M, Thurston G, Holmgren L, Vestweber D, Claesson-Welsh L (2013) VE-PTP regulates VEGFR2 activity in stalk cells to establish endothelial cell polarity and lumen formation. Nat Commun 4:1672. doi:10.1038/ncomms2683

Hong S, Troyanovsky RB, Troyanovsky SM (2010) Spontaneous assembly and active disassembly balance adherens junction homeostasis. Proc Natl Acad Sci U S A 107(8):3528–3533. doi:10.1073/pnas.0911027107

Huber MA, Kraut N, Beug H (2005) Molecular requirements for epithelial-mesenchymal transition during tumor progression. Curr Opin Cell Biol 17(5):548–558

Hurd TW, Fan S, Liu CJ, Kweon HK, Hakansson K, Margolis B (2003a) Phosphorylation-dependent binding of 14-3-3 to the polarity protein Par3 regulates cell polarity in mammalian epithelia. Curr Biol 13(23):2082–2090

Hurd TW, Gao L, Roh MH, Macara IG, Margolis B (2003b) Direct interaction of two polarity complexes implicated in epithelial tight junction assembly. Nat Cell Biol 5(2):137–142

Hurov JB, Watkins JL, Piwnica-Worms H (2004) Atypical PKC phosphorylates PAR-1 kinases to regulate localization and activity. Curr Biol 14(8):736–741

Iden S, Misselwitz S, Peddibhotla SS, Tuncay H, Rehder D, Gerke V, Robenek H, Suzuki A, Ebnet K (2012) aPKC phosphorylates JAM-A at Ser285 to promote cell contact maturation and tight junction formation. J Cell Biol 196(5):623–639. doi:10.1083/jcb.201104143

Iden S, Rehder D, August B, Suzuki A, Wolburg-Buchholz K, Wolburg H, Ohno S, Behrens J, Vestweber D, Ebnet K (2006) A distinct PAR complex associates physically with VE-cadherin in vertebrate endothelial cells. EMBO Rep 7(12):1239–1246

Itoh M, Furuse M, Morita K, Kubota K, Saitou M, Tsukita S (1999) Direct binding of three tight junction-associated MAGUKs, ZO-1, ZO-2, and ZO-3, with the COOH termini of claudins. J Cell Biol 147(6):1351–1363

Jaffe AB, Kaji N, Durgan J, Hall A (2008) Cdc42 controls spindle orientation to position the apical surface during epithelial morphogenesis. J Cell Biol 183(4):625–633

Joberty G, Petersen C, Gao L, Macara IG (2000) The cell-polarity protein Par6 links Par3 and atypical protein kinase C to Cdc42. Nat Cell Biol 2(8):531–539

Kemphues KJ, Priess JR, Morton DG, Cheng NS (1988) Identification of genes required for cytoplasmic localization in early C. elegans embryos. Cell 52(3):311–320

Kitt KN, Nelson WJ (2011) Rapid suppression of activated Rac1 by cadherins and nectins during de novo cell-cell adhesion. PLoS One 6(3):e17841

Kotak S, Busso C, Gonczy P (2012) Cortical dynein is critical for proper spindle positioning in human cells. J Cell Biol 199(1):97–110. doi:10.1083/jcb.201203166

Kotak S, Gonczy P (2013) Mechanisms of spindle positioning: cortical force generators in the limelight. Curr Opin Cell Biol 25(6):741–748. doi:10.1016/j.ceb.2013.07.008

Kovacs EM, Ali RG, McCormack AJ, Yap AS (2002) E-cadherin homophilic ligation directly signals through Rac and phosphatidylinositol 3-kinase to regulate adhesive contacts. J Biol Chem 277(8):6708–6718

Laan L, Pavin N, Husson J, Romet-Lemonne G, van Duijn M, Lopez MP, Vale RD, Julicher F, Reck-Peterson SL, Dogterom M (2012) Cortical dynein controls microtubule dynamics to generate pulling forces that position microtubule asters. Cell 148(3):502–514

Lemmers C, Michel D, Lane-Guermonprez L, Delgrossi MH, Medina E, Arsanto JP, Le Bivic A (2004) CRB3 binds directly to Par6 and regulates the morphogenesis of the tight junctions in mammalian epithelial cells. Mol Biol Cell 15(3):1324–1333

Lin D, Edwards AS, Fawcett JP, Mbamalu G, Scott JD, Pawson T (2000) A mammalian PAR-3-PAR-6 complex implicated in Cdc42/Rac1 and aPKC signalling and cell polarity. Nat Cell Biol 2(8):540–547

Lin X, Shang X, Manorek G, Howell SB (2013) Regulation of the epithelial-mesenchymal transition by claudin-3 and claudin-4. PLoS One 8(6):e67496. doi:10.1371/journal.pone.0067496

Lohia M, Qin Y, Macara IG (2012) The Scribble polarity protein stabilizes E-cadherin/p120-catenin binding and blocks retrieval of E-cadherin to the Golgi. PLoS One 7(11):e51130. doi:10.1371/journal.pone.0051130

Ma L, Young J, Prabhala H, Pan E, Mestdagh P, Muth D, Teruya-Feldstein J, Reinhardt F, Onder TT, Valastyan S, Westermann F, Speleman F, Vandesompele J, Weinberg RA (2010) miR-9, a MYC/MYCN-activated microRNA, regulates E-cadherin and cancer metastasis. Nat Cell Biol 12(3):247–256. doi:10.1038/ncb2024

Macara IG (2004) Parsing the polarity code. Nat Rev Mol Cell Biol 5(3):220–231

Makarova O, Roh MH, Liu CJ, Laurinec S, Margolis B (2003) Mammalian Crumbs3 is a small transmembrane protein linked to protein associated with Lin-7 (Pals1). Gene 302(1–2):21–29

Mandai K, Nakanishi H, Satoh A, Obaishi H, Wada M, Nishioka H, Itoh M, Mizoguchi A, Aoki T, Fujimoto T, Matsuda Y, Tsukita S, Takai Y (1997) Afadin: a novel actin filament-binding protein with one PDZ domain localized at cadherin-based cell-to-cell adherens junction. J Cell Biol 139(2):517–528, Published erratum appears in J Cell Biol 1997 Nov 17;139(4):1060

McNeill H, Ryan TA, Smith SJ, Nelson WJ (1993) Spatial and temporal dissection of immediate and early events following cadherin-mediated epithelial cell adhesion. J Cell Biol 120 (5):1217–1226

McSherry EA, McGee SF, Jirstrom K, Doyle EM, Brennan DJ, Landberg G, Dervan PA, Hopkins AM, Gallagher WM (2009) JAM-A expression positively correlates with poor prognosis in breast cancer patients. Int J Cancer 125(6):1343–1351. doi:10.1002/ijc.24498

Michel D, Arsanto JP, Massey-Harroche D, Beclin C, Wijnholds J, Le Bivic A (2005) PATJ connects and stabilizes apical and lateral components of tight junctions in human intestinal cells. J Cell Sci 118(Pt 17):4049–4057

Mitsushima M, Toyoshima F, Nishida E (2009) Dual role of Cdc42 in spindle orientation control of adherent cells. Mol Cell Biol 29(10):2816–2827

Montero-Balaguer M, Swirsding K, Orsenigo F, Cotelli F, Mione M, Dejana E (2009) Stable vascular connections and remodeling require full expression of VE-cadherin in zebrafish embryos. PLoS One 4(6):e5772

Morais-de-Sa E, Mirouse V, St Johnston D (2010) aPKC phosphorylation of Bazooka defines the apical/lateral border in Drosophila epithelial cells. Cell 141(3):509–523

Nagai-Tamai Y, Mizuno K, Hirose T, Suzuki A, Ohno S (2002) Regulated protein-protein interaction between aPKC and PAR-3 plays an essential role in the polarization of epithelial cells. Genes Cells 7(11):1161–1171

Nakada C, Ritchie K, Oba Y, Nakamura M, Hotta Y, Iino R, Kasai RS, Yamaguchi K, Fujiwara T, Kusumi A (2003) Accumulation of anchored proteins forms membrane diffusion barriers during neuronal polarization. Nat Cell Biol 5(7):626–632

Nakagawa M, Fukata M, Yamaga M, Itoh N, Kaibuchi K (2001) Recruitment and activation of Rac1 by the formation of E-cadherin-mediated cell-cell adhesion sites. J Cell Sci 114 (Pt 10):1829–1838

Nejsum LN, Nelson WJ (2007) A molecular mechanism directly linking E-cadherin adhesion to initiation of epithelial cell surface polarity. J Cell Biol 178(2):323–335

Nelson WJ (2003) Epithelial cell polarity from the outside looking in. News Physiol Sci 18:143–146
Nielsen JS, McNagny KM (2008) Novel functions of the CD34 family. J Cell Sci 121(Pt 22):3683–3692. doi:10.1242/jcs.037507
O'Brien LE, Zegers MM, Mostov KE (2002) Opinion: building epithelial architecture: insights from three-dimensional culture models. Nat Rev Mol Cell Biol 3(7):531–537
Ozaki C, Obata S, Yamanaka H, Tominaga S, Suzuki ST (2010) The extracellular domains of E- and N-cadherin determine the scattered punctate localization in epithelial cells and the cytoplasmic domains modulate the localization. J Biochem 147(3):415–425. doi:10.1093/jb/mvp192
Ozdamar B, Bose R, Barrios-Rodiles M, Wang HR, Zhang Y, Wrana JL (2005) Regulation of the polarity protein Par6 by TGFbeta receptors controls epithelial cell plasticity. Science 307 (5715):1603–1609
Peinado H, Olmeda D, Cano A (2007) Snail, Zeb and bHLH factors in tumour progression: an alliance against the epithelial phenotype? Nat Rev Cancer 7(6):415–428
Pokutta S, Weis WI (2007) Structure and mechanism of cadherins and catenins in cell-cell contacts. Annu Rev Cell Dev Biol 23:237–261
Reinsch S, Karsenti E (1994) Orientation of spindle axis and distribution of plasma membrane proteins during cell division in polarized MDCKII cells. J Cell Biol 126(6):1509–1526
Rikitake Y, Mandai K, Takai Y (2012) The role of nectins in different types of cell-cell adhesion. J Cell Sci 125(Pt 16):3713–3722
Rodriguez-Boulan E, Paskiet KT, Sabatini DD (1983) Assembly of enveloped viruses in Madin-Darby canine kidney cells: polarized budding from single attached cells and from clusters of cells in suspension. J Cell Biol 96(3):866–874
Roh MH, Makarova O, Liu CJ, Shin K, Lee S, Laurinec S, Goyal M, Wiggins R, Margolis B (2002) The Maguk protein, Pals1, functions as an adapter, linking mammalian homologues of Crumbs and Discs Lost. J Cell Biol 157(1):161–172
Sacharidou A, Koh W, Stratman AN, Mayo AM, Fisher KE, Davis GE (2010) Endothelial lumen signaling complexes control 3D matrix-specific tubulogenesis through interdependent Cdc42- and MT1-MMP-mediated events. Blood 115(25):5259–5269
Schneeberger EE, Lynch RD (2004) The tight junction: a multifunctional complex. Am J Physiol Cell Physiol 286(6):C1213–C1228. doi:10.1152/ajpcell.00558.2003
Simons K, Fuller SD (1985) Cell surface polarity in epithelia. Annu Rev Cell Biol 1:243–288. doi:10.1146/annurev.cb.01.110185.001331
Smith CA, Lau KM, Rahmani Z, Dho SE, Brothers G, She YM, Berry DM, Bonneil E, Thibault P, Schweisguth F, Le Borgne R, McGlade CJ (2007) aPKC-mediated phosphorylation regulates asymmetric membrane localization of the cell fate determinant Numb. EMBO J 26(2):468–480
Strilic B, Eglinger J, Krieg M, Zeeb M, Axnick J, Babal P, Muller DJ, Lammert E (2010) Electrostatic cell-surface repulsion initiates lumen formation in developing blood vessels. Curr Biol 20(22):2003–2009. doi:10.1016/j.cub.2010.09.061
Strilic B, Kucera T, Eglinger J, Hughes MR, McNagny KM, Tsukita S, Dejana E, Ferrara N, Lammert E (2009) The molecular basis of vascular lumen formation in the developing mouse aorta. Dev Cell 17(4):505–515. doi:10.1016/j.devcel.2009.08.011
Suzuki A, Hirata M, Kamimura K, Maniwa R, Yamanaka T, Mizuno K, Kishikawa M, Hirose H, Amano Y, Izumi N, Miwa Y, Ohno S (2004) aPKC acts upstream of PAR-1b in both the establishment and maintenance of mammalian epithelial polarity. Curr Biol 14(16):1425–1435
Suzuki A, Ishiyama C, Hashiba K, Shimizu M, Ebnet K, Ohno S (2002) aPKC kinase activity is required for the asymmetric differentiation of the premature junctional complex during epithelial cell polarization. J Cell Sci 115(Pt 18):3565–3573
Suzuki A, Ohno S (2006) The PAR-aPKC system: lessons in polarity. J Cell Sci 119(Pt 6):979–987
Suzuki A, Yamanaka T, Hirose T, Manabe N, Mizuno K, Shimizu M, Akimoto K, Izumi Y, Ohnishi T, Ohno S (2001) Atypical protein kinase C is involved in the evolutionary conserved

PAR protein complex and plays a critical role in establishing epithelia-specific junctional structures. J Cell Biol 152(6):1183–1196

Takahashi K, Nakanishi H, Miyahara M, Mandai K, Satoh K, Satoh A, Nishioka H, Aoki J, Nomoto A, Mizoguchi A, Takai Y (1999) Nectin/PRR: an immunoglobulin-like cell adhesion molecule recruited to cadherin-based adherens junctions through interaction with Afadin, a PDZ domain-containing protein. J Cell Biol 145(3):539–549

Takai Y, Miyoshi J, Ikeda W, Ogita H (2008) Nectins and nectin-like molecules: roles in contact inhibition of cell movement and proliferation. Nat Rev Mol Cell Biol 9(8):603–615

Tepass U, Theres C, Knust E (1990) Crumbs encodes an EGF-like protein expressed on apical membranes of Drosophila epithelial cells and required for organization of epithelia. Cell 61 (5):787–799

Terry SJ, Zihni C, Elbediwy A, Vitiello E, Leefa Chong San IV, Balda MS, Matter K (2011) Spatially restricted activation of RhoA signalling at epithelial junctions by p114RhoGEF drives junction formation and morphogenesis. Nat Cell Biol 13(2):159–166

Thiery JP, Sleeman JP (2006) Complex networks orchestrate epithelial-mesenchymal transitions. Nat Rev Mol Cell Biol 7(2):131–142

Thompson BJ, Pichaud F, Roper K (2013) Sticking together the Crumbs – an unexpected function for an old friend. Nat Rev Mol Cell Biol 14(5):307–314. doi:10.1038/nrm3568

Toyoshima F, Matsumura S, Morimoto H, Mitsushima M, Nishida E (2007) PtdIns(3,4,5)P3 regulates spindle orientation in adherent cells. Dev Cell 13(6):796–811

Toyoshima F, Nishida E (2007) Integrin-mediated adhesion orients the spindle parallel to the substratum in an EB1- and myosin X-dependent manner. EMBO J 26(6):1487–1498

Tsukita S, Furuse M, Itoh M (2001) Multifunctional strands in tight junctions. Nat Rev Mol Cell Biol 2(4):285–293

Tsukita S, Yamazaki Y, Katsuno T, Tamura A, Tsukita S (2008) Tight junction-based epithelial microenvironment and cell proliferation. Oncogene 27(55):6930–6938

Umeda K, Ikenouchi J, Katahira-Tayama S, Furuse K, Sasaki H, Nakayama M, Matsui T, Tsukita S, Furuse M (2006) ZO-1 and ZO-2 independently determine where claudins are polymerized in tight-junction strand formation. Cell 126(4):741–754

Van Itallie CM, Anderson JM (2006) Claudins and epithelial paracellular transport. Annu Rev Physiol 68:403–429

van Meer G, Simons K (1986) The function of tight junctions in maintaining differences in lipid composition between the apical and the basolateral cell surface domains of MDCK cells. EMBO J 5(7):1455–1464

Vetter G, Saumet A, Moes M, Vallar L, Le Bechec A, Laurini C, Sabbah M, Arar K, Theillet C, Lecellier CH, Friederich E (2010) miR-661 expression in SNAI1-induced epithelial to mesenchymal transition contributes to breast cancer cell invasion by targeting Nectin-1 and StarD10 messengers. Oncogene 29(31):4436–4448. doi:10.1038/onc.2010.181

Wang AZ, Ojakian GK, Nelson WJ (1990) Steps in the morphogenesis of a polarized epithelium. I. Uncoupling the roles of cell-cell and cell-substratum contact in establishing plasma membrane polarity in multicellular epithelial (MDCK) cysts. J Cell Sci 95(Pt 1):137–151

Wang Q, Chen XW, Margolis B (2007) PALS1 regulates E-cadherin trafficking in mammalian epithelial cells. Mol Biol Cell 18(3):874–885

Wang Y, Kaiser MS, Larson JD, Nasevicius A, Clark KJ, Wadman SA, Roberg-Perez SE, Ekker SC, Hackett PB, McGrail M, Essner JJ (2010) Moesin1 and Ve-cadherin are required in endothelial cells during in vivo tubulogenesis. Development 137(18):3119–3128. doi:10.1242/dev.048785

Wells CD, Fawcett JP, Traweger A, Yamanaka Y, Goudreault M, Elder K, Kulkarni S, Gish G, Virag C, Lim C, Colwill K, Starostine A, Metalnikov P, Pawson T (2006) A Rich1/Amot complex regulates the Cdc42 GTPase and apical-polarity proteins in epithelial cells. Cell 125 (3):535–548

Whiteman EL, Fan S, Harder JL, Walton KD, Liu CJ, Soofi A, Fogg VC, Hershenson MB, Dressler GR, Deutsch GH, Gumucio DL, Margolis B (2013) Crumbs3 is essential for proper epithelial development and viability. Mol Cell Biol. doi:10.1128/MCB. 00999-13

Yamada S, Nelson WJ (2007) Localized zones of Rho and Rac activities drive initiation and expansion of epithelial cell-cell adhesion. J Cell Biol 178(3):517–527

Yamanaka T, Horikoshi Y, Izumi N, Suzuki A, Mizuno K, Ohno S (2006) Lgl mediates apical domain disassembly by suppressing the PAR-3-aPKC-PAR-6 complex to orient apical membrane polarity. J Cell Sci 119(Pt 10):2107–2118

Yamanaka T, Horikoshi Y, Suzuki A, Sugiyama Y, Kitamura K, Maniwa R, Nagai Y, Yamashita A, Hirose T, Ishikawa H, Ohno S (2001) Par-6 regulates aPKC activity in a novel way and mediates cell-cell contact-induced formation of epithelial junctional complex. Genes Cells 6:721–731

Yang J, Weinberg RA (2008) Epithelial-mesenchymal transition: at the crossroads of development and tumor metastasis. Dev Cell 14(6):818–829. doi:10.1016/j.devcel.2008.05.009

Yonemura S, Itoh M, Nagafuchi A, Tsukita S (1995) Cell-to-cell adherens junction formation and actin filament organization: similarities and differences between non-polarized fibroblasts and polarized epithelial cells. J Cell Sci 108(Pt 1):127–142

Yu AS, McCarthy KM, Francis SA, McCormack JM, Lai J, Rogers RA, Lynch RD, Schneeberger EE (2005) Knockdown of occludin expression leads to diverse phenotypic alterations in epithelial cells. Am J Physiol Cell Physiol 288(6):C1231–C1241

Zeeb M, Strilic B, Lammert E (2010) Resolving cell-cell junctions: lumen formation in blood vessels. Curr Opin Cell Biol 22(5):626–632. doi:10.1016/j.ceb.2010.07.003

Zhang M, Luo W, Huang B, Liu Z, Sun L, Zhang Q, Qiu X, Xu K, Wang E (2013) Overexpression of JAM-A in non-small cell lung cancer correlates with tumor progression. PLoS One 8(11): e79173. doi:10.1371/journal.pone.0079173

Zheng Z, Zhu H, Wan Q, Liu J, Xiao Z, Siderovski DP, Du Q (2010) LGN regulates mitotic spindle orientation during epithelial morphogenesis. J Cell Biol 189(2):275–288

Chapter 13
Cell–Cell Interactions, Cell Polarity, and the Blood–Testis Barrier

Elizabeth I. Tang, Dolores D. Mruk, Will M. Lee, and C. Yan Cheng

Abstract The study of polarity and its role in cell–cell interactions, such as spermatid polarity and adhesion in the seminiferous epithelium, and blood–testis barrier (BTB) in the testis during the epithelial cycle of spermatogenesis in recent years has yielded some unexpected and interesting observations. Similar to other polarized tissues, Sertoli and germ cells in the seminiferous epithelium express many of the component proteins of the Par-, Scribble- and Crumb-based polarity protein complexes. These polarity proteins are working in concert with non-receptor protein kinases, adhesion proteins, and cytoskeletons to confer spermatid and Sertoli cell polarity, and these proteins are also involved in germ cell transport in the epithelium during the epithelial cycle. In this review, we summarize the latest findings in the field. Based on the available data in the literature, it is increasingly clear that polarity proteins are crucial in (1) conferring spermatid and Sertoli cell polarity, (2) regulating spermatid adhesion and transport, and (3) regulating BTB dynamics in the testis during the epithelial cycle. We also highlight specific areas of research that deserve attention in future years. This information should be helpful to investigators in other blood–tissue and epithelial barriers in the field.

Keywords Blood–testis barrier • Cell polarity • Cell–cell interactions • Ectoplasmic specialization • F-actin • Germ cell • Microtubule • Sertoli cell • Spermatogenesis • Testis

---

This work was supported by grants from the National Institutes of Health (NICHD, U54 HD029990, Project 5 to C.Y.C.; R01 HD056034 to C.Y.C.); NSFC/RGC Joint Research Scheme (N_HKU 717/12 to W.M.L.), Hong Kong Research Grants Council General Research Fund (771513 to W.M.L.), and University of Hong Kong CRCG Seed Funding (to W.M.L.).

E.I. Tang • D.D. Mruk • C.Y. Cheng (✉)
The Mary M. Wohlford Laboratory for Male Contraceptive Research, Center for Biomedical Research, Population Council, 1230 York Avenue, New York, NY 10065, USA
e-mail: Y-Cheng@popcbr.rockefeller.edu; ccheng@rockefeller.edu

W.M. Lee
School of Biological Sciences, University of Hong Kong, Hong Kong, China

K. Ebnet (ed.), *Cell Polarity 1*, DOI 10.1007/978-3-319-14463-4_13

13.1 Introduction

13.1.1 Concept of Cell–Cell Interactions, Cell Polarity, and the BTB in the Testis

The two major functions of the mammalian testis are to produce testosterone, a hormone synthesized by the interstitial Leydig cells to maintain sexual characteristics of the male, and spermatozoa in the seminiferous tubules for fertilization to safeguard the continuation of a species. One man produces >300 million spermatozoa, as compared to ~10 and 50 million in a male mouse and rat, respectively, via spermatogenesis each day following puberty until death (Auharek et al. 2011; Johnson et al. 1980; Mauss and Hackstedt 1972). This thus illustrates there are tightly regulated and robust cellular events that take place in the testis to maintain this tempo of sperm output. Spermatogenesis is a complex cellular process, and it can be divided into three discrete events: (1) mitosis, for self-renewal of spermatogonial stem cells (SSC) and spermatogonia; (2) meiosis, to ensure that the amount of genetic material carried in each spermatid is half that of spermatocytes; and (3) spermiogenesis, for the development of spermatids into functional spermatozoa, such that during fertilization, the fusion of a single sperm and an ovum reconstitute the genetic material similar to other somatic cells. These events all take place in the seminiferous epithelium of the seminiferous tubule—the functional unit in the testis that produces sperm—via the epithelial cycle of spermatogenesis (Fig. 13.1). These events are supported by testosterone produced by Leydig cells in the interstitium and also estrogen produced by Sertoli and germ cells (Sharpe 1994; O'Donnell et al. 2001; Carreau and Hess 2010; Carreau et al. 2010; Verhoeven et al. 2010; Wang et al. 2009; McLachlan et al. 2002; Mruk and Cheng 2004b). Recent studies have shown that peritubular myoid cells in the tunica propria also play an important role in supporting spermatogenesis (Welsh et al. 2009; Qian et al. 2013). It is conceivable that there are extensive cell–cell interactions in the testis, in particular at the Sertoli–Sertoli and Sertoli–germ cell interface (Fig. 13.2) during the epithelial cycle of spermatogenesis (Fig. 13.1).

As noted in Fig. 13.2 illustrating the cross section of a typical seminiferous tubule, such as in the rat testis, during the 14 stages of epithelial cycle from I through XIV (Fig. 13.1), elongating/elongated spermatids are highly polarized cells in which their heads are all pointing toward the basement membrane (BM) with their tails toward the tubule lumen (Fig. 13.2), such that the maximal number of developing spermatids can be packed in the tubule to allow the simultaneous development of millions of germ cells. It is noted that spermatid polarity is affected during toxicant- or drug-induced disruption in spermatogenesis such as following treatments with cadmium or adjudin [1-(2,4-dichlorobenzyl)-1H-indazole-3-carbohydrazide, a potential male contraceptive under development in our laboratory (Cheng et al. 2005, 2011a)] in which exposure of rats to these toxicants induces misorientation of elongating/elongated spermatids in the epithelium (Cheng et al. 2011b; Wong et al. 2008b). This loss of spermatid polarity appears to be

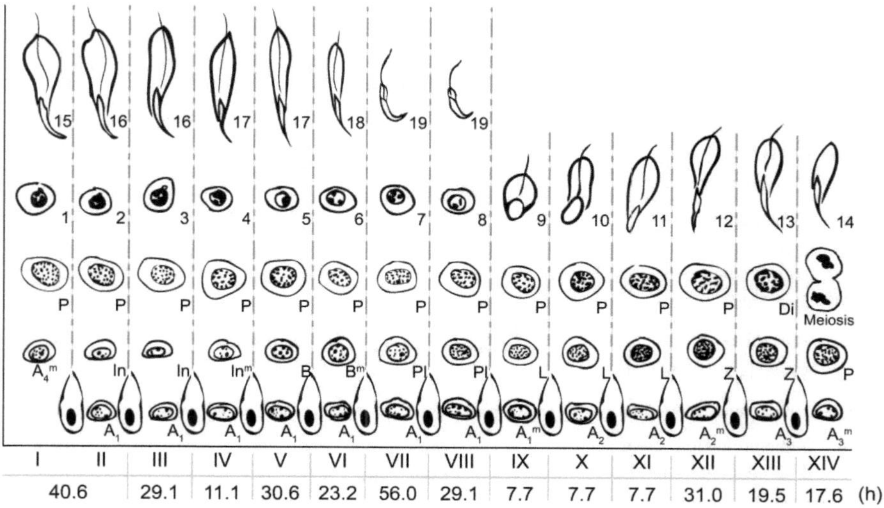

Fig. 13.1 Stages of the seminiferous epithelial cycle of spermatogenesis. The epithelial cycle of spermatogenesis in the rat is divided into 14 (I–XIV) stages and lasts 12.8 days, and the duration, in hour (h), of each cycle from I through XIV is shown. It is crucial to note that each stage does not contain one type of germ cell; rather each stage is unique and contains germ cells at different stages in their maturation and development. For instance, in the seminiferous epithelium of a stage VIII tubule, only type A1 spermatogonia, preleptotene spermatocytes, pachytene spermatocytes, and step 8 and 19 spermatids are found. However, a type A1 spermatogonium (at the *bottom* of the figure on the *left panel*) will go through the epithelial cycle ~4.5 times until it finally matures into step 19 spermatid, taking a total of ~58 days to complete. Spermatogonia, type A (A) containing A_1, A_2, A_3 and A_4, type A_1 undergo mitosis (A_1^m), type A2 undergo mitosis (A_2^m), type A3 undergo mitosis (A_3^m), type A4 undergo mitosis (A_4^m), intermediate (In), type In undergo mitosis (In^m), and type B (B), type B undergo mitosis (B^m). Spermatocytes, preleptotene (Pl), leptotene (L), zygotene (Z), pachytene (P), and diplotene (Di). Spermatids, round spermatids (steps 1–8), and elongate spermatids (steps 9–19). This figure was prepared based on the earlier reports and reviews (Dym and Clermont 1970; de Kretser and Kerr 1988; Mruk et al. 2008). Sertoli cell nucleus near the base of the seminiferous epithelium in each stage of the cycle is also shown

related to the subsequent loss of spermatid adhesion in the epithelium, causing their eventual departure from the testis prematurely (Wong and Cheng 2009; Wong et al. 2008a), thereby leading to infertility (Siu et al. 2009; Mok et al. 2011).

Besides spermatids, Sertoli cells in the epithelium are also highly polarized cells in which their nuclei are located restrictively near the basement membrane (Fig. 13.1). The ultrastructures that constitute the blood–testis barrier (BTB), such as the actin-based tight junction (TJ), basal ectoplasmic specialization (basal ES), and gap junction, as well as the intermediate-based desmosome, are also restricted near the basement membrane which also physically divides the seminiferous epithelium into the basal and the adluminal (apical) compartments (Fig. 13.2). On the other hand, the testis-specific actin-rich anchoring device only found at the Sertoli–spermatid (steps 8–19) interface is restricted to the adluminal compartment (Fig. 13.2). Collectively, these findings illustrate that the Sertoli cell is a highly polarized cell type in the epithelium. Furthermore, Sertoli cells are cyclic in nature during the epithelial cycle. For instance, the protein secretory and phagocytic

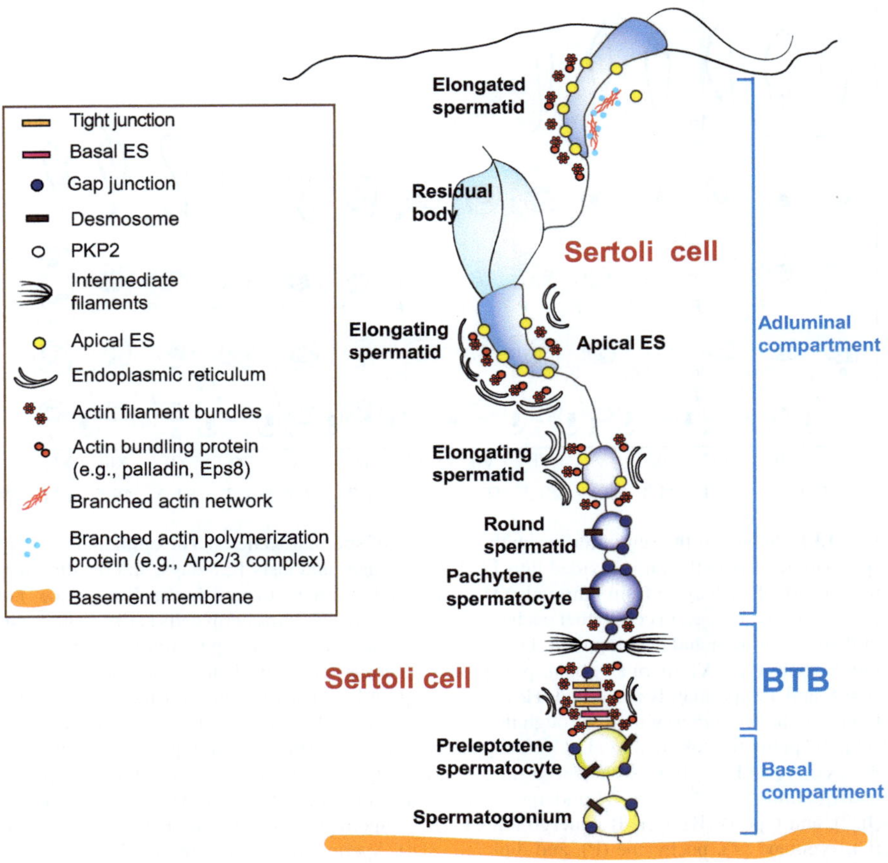

Fig. 13.2 A schematic drawing of the cross section of a seminiferous tubule, illustrating different cell junctions at the Sertoli–Sertoli and Sertoli–germ cell interface in the rat testis. The BTB, which is constituted by coexisting actin-based TJ (tight junction), basal ES (ectoplasmic specialization) and gap junction, as well as intermediate filament-based desmosome, physically divides the seminiferous epithelium into the basal and the adluminal compartment. The basal compartment is in close contact with the basement membrane, a modified form of extracellular matrix in the mammalian testis. Apical ES is restricted to the Sertoli–spermatid interface of step 8–19 spermatids, whereas gap junction and desmosome are at the Sertoli–spermatid interface of step 1–7 spermatids. Preleptotene spermatocytes transformed from type B spermatogonia at stage VII of the cycle will be transported across the BTB at late VII–VIII of the cycle (see Fig. 13.1). Thus, the actin-based cytoskeleton undergoes extensive restructuring, and this event is tightly coordinated with the tubulin-based cytoskeleton which serves as the track for the transport of germ cells across the BTB. This mechanism is also used to transport developing spermatids across the adluminal compartment during spermiogenesis

activities of Sertoli cells are stage-dependent. It is also noted that at each stage of the epithelial cycle, Sertoli cells are catered to specific cellular events that are unique to that specific stage. For example, in the rat testis, meiosis takes place at stage XIV of the cycle (Fig. 13.1), whereas the release of sperm at spermiation and the transit of preleptotene spermatocytes at the BTB take place at stage VIII of the cycle (Figs. 13.1 and 13.3). Thus, at stage VIII, Sertoli cells are working in concert with elongated spermatid to prepare for apical ES degeneration at the adluminal compartment, and Sertoli cells per se are also undergoing restructuring in which a "new" BTB behind and the "old" BTB above the preleptotene spermatocytes (note: preleptotene spermatocytes transformed from type B spermatocytes are the only germ cell type that are transported across the BTB to enter the adluminal compartment for further development) in transit are being assembled and disassembled, respectively. Furthermore, these spermatocytes are connected in "clones" via intercellular bridges (Weber and Russell 1987; Fawcett 1961), making their transport across the BTB a highly coordinated and regulated cellular event. These findings thus illustrate Sertoli cells are highly cyclic in function during the epithelial cycle in which the Sertoli cell performs different functions according to the cycle of the epithelium (Fig. 13.1). As such, in order for spermatogenesis to proceed flawlessly during the epithelial cycle, Sertoli cells must be highly polarized, and perhaps compartmentalized, ultrastructurally and functionally, so that different functions can be performed at discrete domains of a Sertoli cell throughout spermatogenesis. This is particularly important since it is known that a single Sertoli cell is in close contact with ~30–50 germ cells at different stages of their development (Wong and Russell 1983; Weber et al. 1983) to support these germ cells functionally, structurally, and nutritionally, as each type of germ cell has distinct functional needs that are different from each other.

13.2 The Seminiferous Epithelial Cycle of Spermatogenesis and Ectoplasmic Specialization

13.2.1 The Epithelial Cycle of Spermatogenesis

The seminiferous epithelial cycle refers to the specific pattern of cellular association between Sertoli and germ cells and the cellular events pertinent to spermatogenesis that take place within the epithelium (de Kretser and Kerr 1988; Hess and de Franca 2008; Mruk et al. 2008). The cycle of events is divided into 14 stages (I–XIV) in the rat (Fig. 13.1) versus 12 (I–XII) and 6 (I–VI) stages in the mouse and human testis, respectively (Hess and de Franca 2008; Amann 2008; Mruk et al. 2008; Parvinen and Vanha-Perttula 1972). Each stage has its own defining characteristics; for instance, in the rat testis, preleptotene spermatocytes are found only in stages VII–VIII of the cycle, spermiation is restricted to stage VIII (lasts ~29.1 h), and meiosis is limited to stage XIV (lasts ~17.6 h) (Fig. 13.2). The duration of a complete epithelial cycle (stages I–XIV) is about 12.8 days in the rat testis. However, a type A1 spermatogonium takes ~58 days to develop into a step 19 (mature) spermatid. Thus, a type A1 spermatogonium goes through the

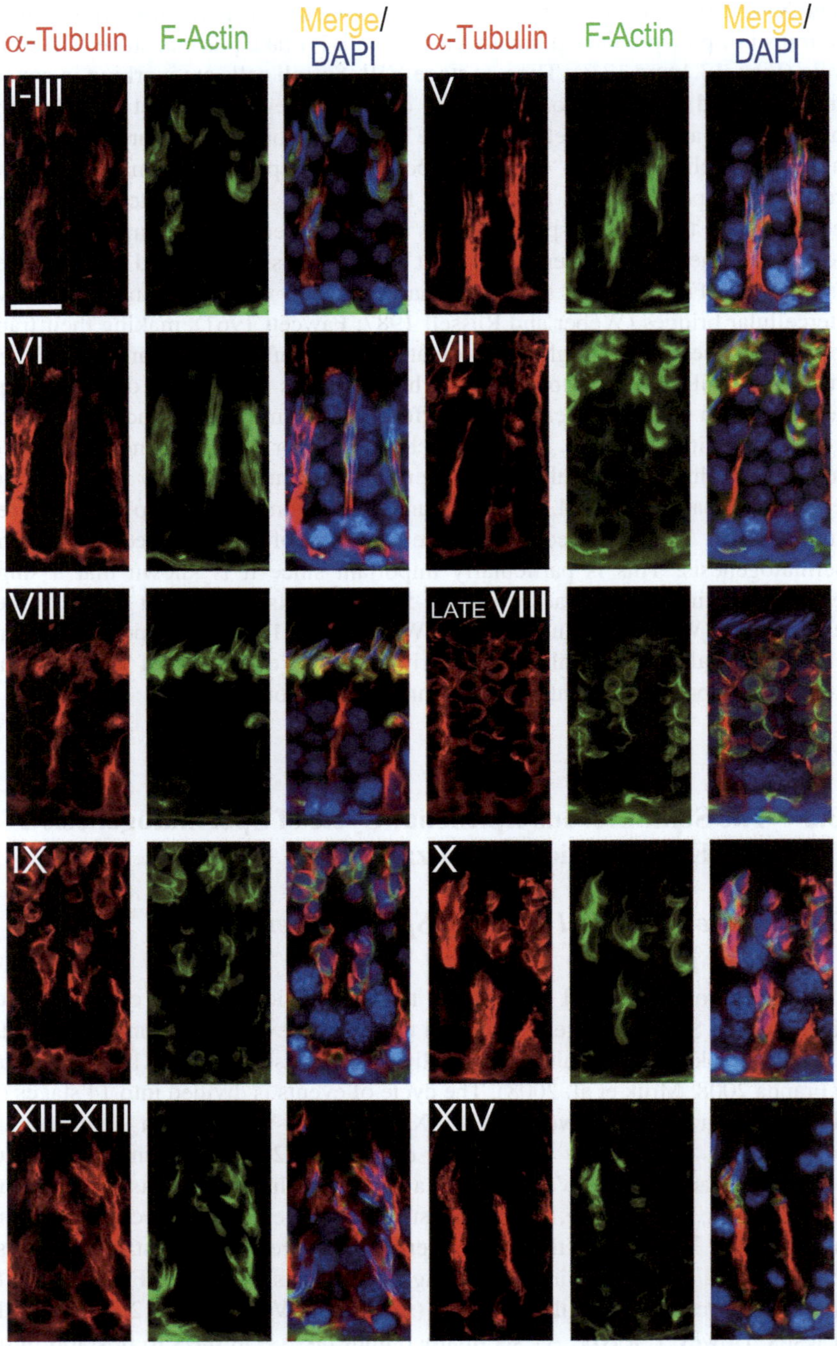

Fig. 13.3 Stage specific expression of actin and α-tubulin. Dual-labeled immunofluorescence analysis was performed in which α-tubulin (*red*) was found to co-localize with and F-actin (*green*) throughout stages I–XIV of the seminiferous epithelial cycle in the rat. Stage I–XIV of the epithelial cycle can also be found in Fig. 13.1. Scale bar, 30 μm

epithelial cycle a total of 4.5 times (Fig. 13.1). In short, Sertoli-germ cell associations differ with each stage, as each stage contains specific types of germ cells progressing through the epithelial cycle (Fig. 13.1). If a discrete section of a seminiferous tubule at stage VIII is visualized under a transillumination microscope, one will witness the progression to stages IX through XIV and then I through VIII again in 12.8 days (Fig. 13.1) (Parvinen 1982; Mruk et al. 2008; Lie et al. 2011).

13.2.2 Ectoplasmic Specialization

When the cross section of a seminiferous tubule, such as in the rat testis, is examined under electron microscope, a testis-specific actin-rich anchoring junction called ectoplasmic specialization (ES) is readily noted which is known to confer cell polarity, cell adhesion, and cellular structural support (Russell and Peterson 1985; Vogl et al. 2008; Cheng and Mruk 2002; Mruk and Cheng 2004a). In the testis, ES is either found in the basal compartment at the Sertoli cell–cell interface at the BTB called basal ES or in the adluminal (apical) compartment at the Sertoli–spermatid interface called apical ES (Fig. 13.2). ES is typified by the presence of actin filament bundles that lie perpendicular to the apposing plasma membranes of either Sertoli–Sertoli (basal ES) or Sertoli–spermatid (apical ES), which are sandwiched in between cisternae of the endoplasmic reticulum and the apposing plasma membranes (Fig. 13.2). Ultrastructurally, apical and basal ES are indistinguishable except that since actin filament bundles are only found in Sertoli cells and not in germ cells, a single array of actin filament bundles is found at the apical ES, whereas a double array of actin microfilaments is found at the basal ES. Once apical ES appears at the interface of Sertoli cells and step 8 spermatids in stage VIII tubules, it becomes the only anchoring device from steps 8 to 19 spermatids, replacing desmosome and gap junction at the Sertoli–spermatid (1–7) interface, and persists until its degeneration at spermiation. Unlike apical ES, basal ES at the Sertoli cell–cell interface never exists alone; rather it coexists with either TJ or gap junction, which together with desmosome constitutes the BTB, one of the tightest blood–tissue barriers in the mammalian body (Franca et al. 2012; Cheng and Mruk 2012; Pelletier 2011; Wong and Cheng 2005) (Fig. 13.2). In fact, when the force that is required to disrupt adhesion induced by apical ES versus desmosome is compared and quantified, ES is at least twice as strong as desmosome to confer steps 8–19 spermatid adhesion to the Sertoli cell (Wolski et al. 2005). This observation is unusual since desmosome, the predominant anchoring junction in skin, is considered to be the strongest adhesive junction in the mammalian body (Green and Simpson 2007; Thomason et al. 2010; Jamora and Fuchs 2002). This unusual strength of the ES apparently is contributed by the extensive network of actin filament bundles (Russell et al. 1988) (Fig. 13.2). It is of interest to note that while the appearance of apical ES at the Sertoli–spermatid (step 8) interface at stage VIII of the cycle replaces desmosome and gap junction between Sertoli cells and step 1–7 spermatids, apical ES is a hybrid atypical adherens junction (AJ) (Wong et al. 2008a; Yan et al. 2007) since it is constituted by proteins that are found in AJ (e.g., nectins, afadins, N-cadherin, β-catenin), TJ (e.g., JAM-C, CAR), gap junction (e.g., connexin 43), and focal adhesion complex (or focal contact) (e.g., α6β1-integrin, laminin-α3,β3,γ3, FAK, c-Src, c-Yes) (Cheng and Mruk 2010). Regardless of the ultrastructural similarity between the apical and basal ES, proteins that are found at the basal ES/BTB are quite different from those at the apical ES (Cheng and Mruk 2010).

13.2.3 The Uniqueness of the Seminiferous Epithelium Versus Other Epithelia

As noted in Fig. 13.1, the seminiferous epithelium is unique compared to other epithelia due to the seminiferous epithelial cycle of spermatogenesis. This is a unique cellular process in the mammalian body due to the cyclic events of the epithelial cycle such that millions of sperm can be produced from the tubules via spermatogenesis. Furthermore, as germ cells change shape, orientation, and size during maturation, they undergo extensive "adhesion" and "de-adhesion," which involves major restructuring of junctions at the cell–cell interface. Additionally, germ cells are transported progressively from the basal to the adluminal compartment across the epithelium throughout the epithelial cycle so that step 19 spermatids can be lined up at the adluminal edge of the apical compartment to prepare for their release once they are transformed into spermatozoa at late stage VIII of the cycle (Figs. 13.1 and 13.2). Thus, it is envisioned that actin filament bundles at the ES must undergo cyclic reorganization from their "bundled" to "de-bundled" configuration and vice versa to facilitate the transport of preleptotene spermatocytes and spermatids across the BTB and the epithelium. It is of interest to note that germ cells per se are not motile cells, since they lack cellular structures, such as lamellipodia and filopodia found in fibroblasts, macrophages, and keratinocytes, and they rely solely on the Sertoli cell for their transport across the BTB and the epithelium. Thus, the transport of spermatids across the epithelium during spermiogenesis requires the presence of tubulin-based cytoskeleton/microtubules which serve as the track for cargo (e.g., spermatids) to be transported across the epithelium. In short, the actin- and tubulin-based cytoskeletons are working in concert via yet-to-be defined mechanism(s) to facilitate germ cell transport across the epithelium.

13.3 Polarity Proteins on Cell Polarity and Cell Adhesion During Spermatogenesis

13.3.1 *Why Is Polarity Important During Spermatogenesis?*

Spermatids are highly polarized cells in which the heads of spermatids point toward the basement membrane, while the tails point toward the tubule lumen. Spermatid polarity is crucial to maximize the production of sperm during spermatogenesis since the arrangement of polarized spermatids in the epithelium as noted in Fig. 13.1 allows the maximal number of spermatids that can be packed and developed simultaneously in the epithelium in the tubule. Similarly, polarized Sertoli cells can coordinate cellular events across the epithelium more efficiently so that signals can be sent across the Sertoli cell orderly and Sertoli cell can also communicate with its neighboring cells effectively during the epithelial cycle. In the testis, cell polarity is conferred by the Par (partitioning defective)-based protein complex (e.g., Par6, Par3, Cdc42, aPKC, Pals1, and PatJ), the Scribble-based complex (e.g., Scribble, Dlg, Lgl), and the Crumbs-based complex (e.g., CRB-3, Pals1, Patj) (Wong and Cheng 2009; Wong et al. 2008a). Each of these polarity protein complexes recruits its own binding partners, thus conferring cellular asymmetry; this is because a multiprotein complex can be effectively created for each protein complex and also because there is mutually exclusive distribution between Par-/CRB- and Scribble-based protein complexes across an epithelial cell, such as the Sertoli cell (Iden and Collard 2008; Assemat et al. 2008; Head et al. 2013; Goldstein and Macara 2007). During spermiogenesis, most of the cytosol is eliminated from the developing spermatids and transported to the residual body to be scavenged and cleaned up by the Sertoli cell (Fig. 13.2). Thus, there is scant cytosol remaining in the more mature spermatids, such as step 8–19 spermatids, particularly in the head region where apical ES is present both to anchor spermatids to the Sertoli cell and to confer spermatid polarity (*note*: acrosome that is found at the spermatid head represents a giant proteasome containing acrosin, a serine protease with trypsin-like specificity, and is to be used by the sperm to penetrate the zona pellucida at fertilization (Honda et al. 2002)). The mechanism(s) through which polarity proteins expressed by germ cells (Wong et al. 2008b; Su et al. 2012b) involved in conferring or regulating spermatid polarity is still not known.

13.3.2 *Role of Par-Based Polarity Proteins on Cell Adhesion and Polarity in the Testis*

Studies in the testis have shown that a knockdown of either Par3 or Par6 specifically by RNAi without detectable off-target effects impedes Sertoli cell TJ barrier. In these studies, proteins at the Sertoli cell–cell interface, such as JAM-A and α-catenin, became mis-localized, as these proteins no longer tightly localized to the Sertoli cell BTB, but relocated to the cell cytosol, thereby destabilizing the BTB function via a loss of Sertoli cell adhesion (Wong et al. 2008b). Furthermore, it appears that Par3 and Par6 regulate the localization of different BTB proteins differentially since the knockdown of Par3, but not Par6, induces mis-localization of ZO-1, whereas the knockdown of Par6, but not Par3, induces mis-localization of N-cadherin selectively (Wong et al. 2008b). These changes in protein localization appear to be the result of changes in the kinetics of endocytic vesicle-mediated protein trafficking, since the knockdown of Par5 (also known as 14-3-3) is found to accelerate the endocytosis of JAM-A and N-cadherin, thereby destabilizing the Sertoli cell BTB (Wong et al. 2009), illustrating the role of polarity proteins in cell adhesion function. Par6 was also found to be crucial to confer spermatid polarity in the rat testis. Adjudin induces spermatid loss; however, prior to spermatid loss, there is a loss of spermatid polarity as evidenced by the presence of misoriented spermatids in rats treated with the drug. Treatment of rats with adjudin was associated with considerable loss of Par6 surrounding the spermatid heads, which were pointed in all directions in the epithelium (Wong et al. 2008b). More important, this loss of spermatid polarity occurs before a disruption of spermatid adhesion onto the Sertoli cell in the epithelium is detected, seeming to suggest that spermatid polarity and adhesion are two intimately related events regulated by polarity proteins, such as Par3 and Par6, during spermatogenesis in the testis.

13.3.3 Role of Scribble-Based Polarity on Cell Polarity and Adhesion Is Mediated by Changes in the Actin-Based Cytoskeleton in the Testis

Scribble, Lgl (lethal giant larvae), and Dlg (discs large) are found to be expressed by both Sertoli and germ cells in the rat testis (Su et al. 2012b). Scribble is localized most notably at the Sertoli cell–cell interface when Sertoli cells establish a functional TJ-permeability barrier in vitro but it is also found in the cell cytosol (Su et al. 2012b). Scribble is also localized predominantly to the BTB in the seminiferous epithelium in vivo in virtually all stages of the epithelial cycle in the rat testis (Su et al. 2012b), illustrating polarity protein Scribble is involved in Sertoli cell polarity and adhesion, and it is involved in BTB dynamics during the epithelial cycle. While the knockdown of Scribble or Dlg1 alone fails to modulate the Sertoli cell TJ-barrier function, the simultaneous knockdown of Scribble and its two integral component proteins Dlg1 and Lgl2 by RNAi using specific siRNA duplexes with no detectable off-target effects is shown to promote the Sertoli cell TJ-permeability barrier, making it “tighter” (Su et al. 2012b), illustrating its role in inducing BTB restructuring during the epithelial cycle. This promoting effect of Scribble on the Sertoli cell BTB function is supported by studies using immunofluorescence microscopy since a considerable increase in occludin and β-catenin localization to the Sertoli cell–cell interface is found in the Sertoli cell epithelium following the simultaneous knockdown of Scribble, Dlg1 and Lgl2 (Su et al. 2012b). More important, when Scribble/Dlg1/Lgl2 is knocked down in the testis in vivo, an increase in occludin localization to the BTB is detected in stage VIII tubules. This change is accompanied by a loss of spermatid polarity in the adluminal compartment in which spermatids no longer orientate properly with their heads pointing toward the basement membrane; instead spermatid alignment is disarrayed, concomitant with a downregulation of laminin-γ3 (Su et al. 2012b), an apical ES protein limited to the spermatid which forms a bona fide adhesion complex with α6β1-integrin in the Sertoli cell (Yan and Cheng 2006; Koch et al. 1999). Interestingly, this loss of spermatid polarity, downregulation of laminin-γ3 at the apical ES, and the increase in occludin at the BTB are associated with changes in the organization of F-actin at these sites; when F-actin is visualized by rhodamine phalloidin, more F-actin is found at the BTB whereas reduced levels of F-actin are detected at the apical ES (Su et al. 2012b). Collectively, these findings thus illustrate that the Scribble-based polarity protein complex supports spermatid polarity and adhesion, while it also promotes BTB dynamics possibly via restructuring, illustrating its antagonistic effects on the apical and basal ES in the testis during the epithelial cycle.

13.4 Actin- and Microtubule-Based Cytoskeletons and Their Role in Cell–Cell Interactions, Cell Polarity, and BTB Function

13.4.1 Cross Talk Between Actin and Tubulin Cytoskeletons

Across different cell types, both the actin filament and microtubule networks play critical roles in a variety of processes such as cell division, cell polarization, transport, and migration. Although actin- and tubulin-based cytoskeletons are usually portrayed as functionally independent networks, recent research has begun to unravel the cooperative interaction between them via some routes of communication or cross talk (Gavin 1997; Goode et al. 2000). The Sertoli cell is a highly polarized cell, as earlier described, and also is very dynamic via cyclic changes functionally, adapting to the evolving shape of germ cells as they progress through different stages of the epithelial cycle. This is made possible due to the dynamicity of both actin and microtubule (MT) cytoskeletons, both of which exemplify stage specificity within the seminiferous epithelial cycle. For example, at late stage VIII in the rat, both actin and tubulin (e.g., α-tubulin) levels diminish significantly at the apical ES at the onset of the spermiation (Fig. 13.3). Spermiation is a highly regulated temporal event; if not tightly regulated, it can result in premature or delayed release of spermatids (O'Donnell et al. 2011; Mruk and Cheng 2004b). As shown in Fig. 13.3, actin and α-tubulin co-localize in the

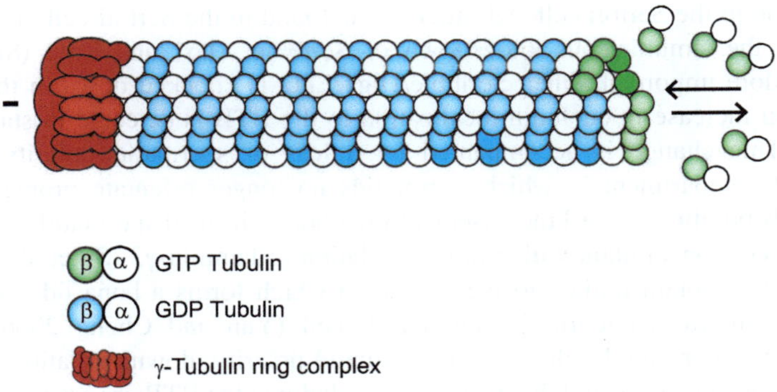

Fig. 13.4 Microtubule structure in mammalian cells including the Sertoli cell in the testis. Microtubules (MTs) are assembled from αβ-tubulin dimers. β-tubulin designates the MT plus (+) end, and α-tubulin the minus (−) end. Polymerization of a MT occurs through the interaction of the α-subunit of an incoming dimer with the β-subunit of a preexisting dimer on a MT protofilament. GTP is bound to both α and β-subunits; however, only the β-tubulin GTP exhibits GTPase activity. GTP hydrolysis occurs as the MT is assembled, leaving most of the MT comprised of GDP-tubulin and the growing plus end as the only region where GTP is still bound to β-tubulin, known as the GTP-cap. γ-tubulin forms a protein complex called the γ-tubulin ring complex and is responsible for nucleation and stabilization of MTs

majority of the 14 stages of the epithelial cycle in the rat testis (*note:* α- and β-tubulin are the structural components of tubulin-based cytoskeleton; see Fig. 13.4). Though this is a morphological observation, it lays the foundation for expanding this information on how actin and MT cytoskeletons work in concert to regulate spermatogenesis.

There are a number of MT regulatory proteins that not only associate with MTs but also with the actin cytoskeleton. For example, CLIP-170 (cytoplasmic linker protein of 170 kDa) is one MT regulatory protein that associates with myosin VI, which is an ATP-dependent actin motor protein, thus linking the actin and MT networks together. CLIP-170 is also known to be involved in the dynein and dynactin pathway (Akhmanova et al. 2005). In general, motor proteins are either plus- or minus-end-directed along MTs. Some formins, such as mDia1, are actin-nucleating proteins that also can stabilize MTs. Both CLIP-170 and mDia1 are present in testis and thus may be the subject of future studies on cross talk between the actin and tubulin networks in regulating key events of spermatogenesis. Table 13.1 summarizes results of findings that illustrate the likely function of microtubule regulatory proteins in microtubule dynamics and their role in male fertility via the use of genetic (such as gene knockout KO or N-ethyl-N-nitrosourea-induced mutation) or knockdown (KD) models. However, the molecular mechanism(s) by which these proteins regulate spermatogenesis remain largely unknown.

Table 13.1 KO/KD studies of microtubule regulatory proteins

Protein	Function(s)	KO/KD/ Mutation	Phenotype	Reference(s)
CLASP1 & 2	MT stability: promote MT rescue, suppress MT catastrophe	KD	Decreased axon and dendrite length in neurons; defects in mitotic spindle and delayed metaphase	(Drabek et al. 2012; Mimori-Kiyosue et al. 2006)
CLIP-170	Regulation of MT dynamics, cell polarity, MT-dependent organelle transport	KO	Male mice were viable, but subfertile with abnormal spermatid head shape	(Akhmanova et al. 2005)
EB1	Regulation of MT dynamics and interaction with cellular elements	KD	Disruption in PKC signaling pathways, cell migration, spindle pole movement	(Schober et al. 2012; Bruning-Richardson et al. 2011)
KATNAL1	MT severing	ENU-induced mutation	Premature release of immature sperm and male infertility	(Smith et al. 2012)
KATNB1	MT severing	ENU-induced mutation	Mice were sterile; decrease in sperm production; abnormal spermatid head shape	(O'Donnell et al. 2012)
MCAK	Regulation of MT dynamics and mitosis, depolymerizes MTs	KD	Defects in chromosome congression and segregation	(Stout et al. 2006; Ganem et al. 2005; Kline-Smith et al. 2004)
mDia1	Actin polymerization and MT stabilization	KD	Decreased MT stabilization	(Bartolini et al. 2012)
Tau	MT assembly and stabilization; predominant in neurons	KO	Mice develop normally, delay in neuronal maturation, some behavioral effects reported	(Dawson et al. 2001; Harada et al. 1994; Denk and Wade-Martins 2009)
Stathmin	Regulation of MT dynamics and mitosis	KO	Mice displayed normal phenotype, but developed axonopathy of CNS and PNS as they aged	(Schubart et al. 1996; Liedtke et al. 2002)
		KD	Delayed mitotic entry in HeLa cells	(Silva and Cassimeris 2013)

(continued)

Table 13.1 (continued)

Protein	Function(s)	KO/KD/ Mutation	Phenotype	Reference(s)
XMAP215	Regulation of MT dynamics, MT polymerase	KD	Reduced axon growth	(Lowery et al. 2013)

CLASP CLIP-170 associated protein, *CLIP-170* cytoskeletal linker protein 170, *EB1* end-binding protein 1, *KATNAL1* katanin p60 subunit A-like1, *KATNB1* p80 regulatory subunit of katanin, *MCAK* mitotic centromere-associated kinesin, *mDia1* mammalian protein diaphanous homolog 1, *XMAP215 Xenopus* microtubule-associated protein 215; KO, knock-out; KD, knockdown; ENU, N-ethyl-N-nitrosourea.

13.4.2 Functional Role of Actin- and Tubulin-Based Cytoskeletons in the Testis

The current understanding of spermatid transport via MTs and the involvement of actin-based cytoskeleon in the seminiferous epithelium remains elusive. Microtubules are abundant in Sertoli cells, and research thus far has implicated their role by serving as the tracks for translocation of spermatids throughout the epithelial cycle. As germ cells mature, they adopt an elongate shape; concomitant with elongation, spermatids are enveloped by the Sertoli cell, and these cells, unlike macrophages, neutrophils, and fibroblasts, are not motile cells. Instead, their transport across the epithelium during the epithelial cycle relies completely on Sertoli cells, and as such, it is logical to speculate that the "cargoes" (i.e., germ cells) require the presence of a "track" (i.e., polarized microtubules) for their transport (Redenbach and Vogl 1991). Indeed, microtubules, which are intrinsically polar, are arranged with their plus and minus ends directed basally and apically, respectively, within the Sertoli cell. Studies have revealed the presence of microtubule motor proteins in the testis, such as dynein and kinesin, which are MT minus and plus-end-directed motor proteins (Hall et al. 1992), supporting the idea that microtubules are in part responsible for the organized movement of spermatids across the seminiferous epithelium. Dynein is classified as a minus-end-directed motor protein, but when it forms a complex with dynactin, an adaptor, it can also be targeted to the plus end (Kardon and Vale 2009). During the epithelial cycle, germ cells are transported progressively up from the basal to the apical region of the seminiferous epithelium; however, during stage V in the rat, developing spermatids actually return to the basal region and are found deep inside the Sertoli cell crypts. This suggests the importance of bidirectional transport along the MTs during spermatogenesis. It has been proposed that the cytoplasmic side of the endoplasmic reticulum (ER) of the apical ES comes into contact with the microtubules, MTs thus confer ER the ability to move along the tracks while the actin filaments of the apical ES anchor the spermatid (Hall et al. 1992; Russell 1977).

Microtubules are polar cylindrical structures made from protofilaments of α-β-tubulin heterodimers. Protofilaments, which are assembled by the head-to-tail addition of αβ-tubulin subunits, arrange laterally to form a hollow tube (Fig. 13.3). MTs possess a property called dynamic instability, which describes both the polymerization and depolymerization that occurs at the plus end of MTs (Wade 2007; Mitchison and Kirschner 1984). As listed in Table 13.1, there are a number of proteins that regulate microtubules. For example, MT plus-end tracking proteins (+TIPs) such as CLIP-170 and EB1 can stabilize MTs. It is conceivable that these proteins are helping to stabilize MTs as spermatids are transported directionally across the epithelium. Stabilizing agents ensure a "steady ride" as the spermatids migrate ultimately toward the apex of the epithelium. Severing proteins, another type of MT regulatory protein, like katanin, which has been studied in the testis (Smith et al. 2012; O'Donnell et al. 2012), may play an important role in regulating transport. Tubulin expression is stage specific, as immunohistological staining show (Fig. 13.3), for instance, in some stages of the epithelial cycle, the length of the MTs that appears as "stalks" in the seminiferous epithelium in late VIII-IX stages are shorter than in others such as at stages V-VII (Wenz and Hess 1998). This observation may be attributed to severing proteins which as the name suggests, sever or cut the MT, to promote generation of new MTs. Transport which occurs across the seminiferous epithelium is a continuous process; thus it is probable that generation of new MTs coincides with spermatid movement.

In epithelia, apicobasal polarity requires specific targeting of proteins to both the apical and basal regions of the cell type. In the testis, protein trafficking in the seminiferous epithelium is crucial for regulation of discrete cellular events of spermatogenesis, such as mitosis, meiosis, spermiogenesis, and spermiation. Most of the current research on protein trafficking and cell–cell communication in the testis has only begun to elucidate the intimate relationship between the actin cytoskeleton and regulation of these cellular events (Su et al. 2013). Although most studies focus on actin dynamics and protein trafficking, there is much yet to be uncovered. In addition, the microtubule cytoskeleton is also at play in regulating the events of spermatogenesis, but how it does so is still the subject of future research.

Cell junction protein recruitment and endocytosis are two types of processes, not only found in the seminiferous epithelium but across all epithelia. Polarity proteins such as Par3/Par6, Scribble, Lgl, 14-3-3, and Cdc42 are involved in regulating these processes in the testis (Wong et al. 2008b; Su et al. 2012b). For example, Par3/Par6, which are established polarity proteins, has been shown to confer spermatid adhesion at the apical ES (Wong et al. 2008a). Changes in cell adhesion regulated by Par proteins may likely coordinate protein endocytosis, which is in part regulated by the actin network (Wong et al. 2009). Microtubules also play a role in endocytosis, via transport of endosomes and lysosomes (Matteoni and Kreis 1987), but the mechanism in the testis has yet to be defined. As previously mentioned, CLIP-170 is one MT regulatory protein that is involved in MT dynamics. This protein was first discovered for its role in linking endocytic vesicles to the MTs (Pierre et al. 1992); in addition to its role as a + TIP, it is a likely player in endocytic trafficking in the testis.

13.5 Polarity Proteins and Cytoskeletons in the Apical ES–BTB–Basement Membrane (BM) Functional Axis

13.5.1 The Apical ES–BTB–BM Functional Axis

The concept of a local functional axis that coordinates and regulates cellular events taking place across the seminiferous epithelium during the epithelial cycle was first reported in 2008 (Yan et al. 2008). It was noted that overexpression of fragments of laminin chains at the apical ES or purified recombinant proteins perturbed the Sertoli cell TJ-permeability barrier function by downregulating expression of proteins at the BTB, such as occludin and N-cadherin, but also β1-integrin at the hemidesmosome (HD) at the Sertoli-BM interface (Yan et al. 2008). These findings thus illustrate that matrix metalloproteinase-2 (MMP-2) which is highly expressed at the apical ES at stage VIII of the epithelial cycle (Siu and Cheng 2004) is capable of inducing cleavage of laminin chains at the apical ES during its degeneration to prepare for spermiation to release biologically active fragments. These autocrine-like fragments in turn induce BTB restructuring to facilitate the transit of preleptotene spermatocytes across the BTB. These fragments also perturb HD function, which creates a positive regulatory loop to further potentiate BTB restructuring. This possibility was confirmed by silencing β1-integrin at the HD in Sertoli cells, which indeed was found to perturb the Sertoli cell TJ-permeability function (Yan et al. 2008). Taken collectively, these data illustrate the presence of a functional axis that links cellular events that occur at the apical ES in the adluminal compartment at spermiation with BTB restructuring near the BM in the seminiferous epithelium, and also HD in the BM, at stage VIII of the epithelial cycle. Subsequently studies have shown that MMP-9 is also capable of inducing cleavage of collagen chains, mostly collagen α3(IV), in the BM to release the NC1 (non-collagenous 1) domain peptide, which was shown to perturb the Sertoli cell TJ barrier function (Wong and Cheng 2013). While the purified NC1 recombinant protein was shown not to downregulate the expression of BTB-associated proteins such as CAR-ZO-1 and N-cadherin-β-catenin, it effectively induced mis-localization of these proteins at the Sertoli cell BTB, so that they no longer localized predominantly at the Sertoli cell–cell interface; instead, they were relocalized to the cell cytosol, thereby destabilizing the Sertoli cell TJ-barrier function (Wong and Cheng 2013). Studies using the phthalate-toxicant model have confirmed the presence of this local functional axis in which a disruption of the apical ES by phthalate induces BTB restructuring and can compromise its integrity (Yao et al. 2009, 2010). Collectively, these data demonstrate unequivocally that there is a functional autocrine-based regulatory axis that coordinates cellular events, such as spermiation and BTB restructuring, which take place simultaneously at the opposite ends of the epithelium at stage VIII of the epithelial cycle . In brief, apical ES degeneration as well as HD/BM restructuring contribute to BTB restructuring, which is further induced by re-organization of collagen network in the BM at the tunica propria by generating NC1 domain-containing peptide. A recent study has identified the biologically active domain of the laminin-γ3 chain that induces BTB restructuring, and synthetic peptide based on this functional domain designated F5-peptide is known to perturb BTB integrity in vivo reversibly and it also induces germ cell loss from the epithelium, illustrating its potential to serve as an endogenously produced male contraceptive peptide (Su et al. 2012a). These findings using F5-peptide based on the biologically active fragment of laminin-γ3 chain illustrate that the apical ES-BTB-BM axis can be a target of male contraceptive development. In fact, studies have shown that this functional axis is a target of environmental toxicants, such as phthalates, BPA, cadmium, and PFOS (perfluorooctane sulfonate) (Wan et al. 2013b; Mazaud-Guittot 2011). A disruption of the critical regulatory components in this axis following exposure of men to these toxicants is likely the cause of reduced semen quality and sperm count as recently reported (Rolland et al. 2013; Toft et al. 2012).

13.5.2 Role of Focal Adhesion Kinase (FAK), Polarity Proteins, and Cytoskeletons at the Apical ES–BTB–BM Functional Axis

While the apical ES–BTB–BM functional is crucial to coordinate cellular events that take place in the seminiferous epithelium during the epithelial cycle, the molecules that are involved in the regulation and the underlying molecular mechanism(s) remain unknown. Recent studies have shown that non-receptor protein tyrosine kinases, such as FAK (Lie et al. 2012) and c-Yes (Xiao et al. 2013); polarity proteins, such as Par6 (Xiao et al. 2013); and actin-based cytoskeleton (Su et al. 2012a) are critical players in this functional axis. For instance, studies using different mutants of FAK have shown that p-FAK-Tyr^{407} and p-FAK-Tyr^{397} are crucial to the integrity of the BTB and apical ES, respectively (Lie et al. 2012). In fact, these two phosphorylated forms of FAK are shown to display antagonistic effects on the BTB integrity in which p-FAK-Tyr^{407} promotes whereas p-FAK-Tyr^{397} disrupts the Sertoli cell TJ-permeability barrier function (Lie et al. 2012). However, p-FAK-Tyr^{397} plays a dominant role in maintaining spermatid adhesion via its effects on the adhesive function of the apical ES (Wan et al. 2013a). Thus, p-FAK-Tyr^{407} and p-FAK-Tyr^{397} likely serve as the molecular "switches" by turning "on" and "off" cell adhesion function at the Sertoli cell BTB and also Sertoli–spermatid interface along the apical ES–BTB–BM axis. It has been reported that overexpression of p-FAK-Tyr^{407} that promotes BTB integrity can block the F5-peptide-induced Sertoli cell TJ-permeability barrier disruption (Su et al. 2012a). Also, the F5-peptide-mediated BTB disruption and spermatid loss in vivo is accompanied by a mis-localization of p-FAK-Tyr^{407} in which this activated form of FAK is no longer restricted tightly to the BTB and the apical ES (Su et al. 2012a). This, in turn, perturbs the organization of F-actin at the apical ES and the BTB (Su et al. 2012a), such that actin filament bundles fail to be properly reorganized at both sites in response to the epithelial cycle of spermatogenesis, likely the result of a disruption in actin polymerization. Thus, a failure in F-actin organization can no longer support adhesive function at the BTB and the apical ES, leading to unwanted BTB and apical ES restructuring or degeneration. This postulate is supported by studies in vitro since the promoting effects of p-FAK-Tyr^{407} on BTB integrity is mediated, at least in part, via the Arp2/3-N-WASP protein complex that alters the kinetics of branched actin polymerization (Lie et al. 2012; Cheng et al. 2013). In this context, it is of interest to note that since the F5-peptide administered to the testis can be rapidly metabolized and cleared, its disruptive effects on spermatogenesis are reversible and germ cells gradually re-populate the epithelium (Su et al. 2012a).

13.5.3 The Role of c-Yes, p-FAK-Tyr^{407}, Par6, and F-actin on BTB and Apical ES Function at the Apical ES–BTB Axis

Studies have shown that FAK is the putative substrate of Src family kinases (SFK) such as c-Src and c-Yes in most epithelia including the seminiferous epithelium (Zhao and Guan 2010; Boutros et al. 2008; Xiao et al. 2012). Also, multiple proteins at the BTB and apical ES are binding partners of SFK and/or FAK (Xiao et al. 2012; Li et al. 2013; Cheng and Mruk 2012). In fact, the dual FAK/Src complex is one of the primary targets of chemotherapy (Bolos et al. 2010) and inflammatory and autoimmune diseases (Lowell 2011). Thus, it is not surprising that c-Yes is recently shown to be a crucial player in the apical ES–BTB axis (Xiao et al. 2013). For instance, a knockdown of c-Yes by RNAi was shown to perturb the Sertoli cell TJ-barrier function both in vitro and in vivo, mediated via a disorganization of F-actin at the BTB, in which actin microfilaments no longer tightly restricted to the BTB near the BM (Xiao et al. 2013). These findings are consistent with an earlier report by using SU6656, a selective inhibitor of c-Yes, to probe the role of c-Yes in modulating F-actin organization in Sertoli cells (Xiao et al. 2011). Interestingly, the knockdown of c-Yes in the testis that affects the BTB integrity also perturbs apical ES function, disrupting spermatid polarity and adhesion, which is mediated by a mis-localization of p-FAK-Tyr^{407} and also polarity protein Par6 (Xiao et al. 2013), illustrating there is a feedback loop between the apical ES and the BTB. These changes, namely, mis-localization of p-FAK-Tyr^{407} and Par6 at the apical ES, thus impede actin microfilaments at the apical ES, leading to mis-localization of adhesion protein nectin-3, causing defects in spermatid transport and spermiation, so that elongated spermatids are entrapped in the seminiferous epithelium even at the site close to the BM in stage IX tubules (Xiao et al. 2013). These data thus illustrate the intimate functional relationship between FAK/SKF (e.g., p-FAK-Tyr^{407}, c-Yes), polarity proteins (e.g., Par6), and cytoskeletons (e.g., actin microfilaments). Any changes on the cross talk between these proteins would impede cell adhesion function at the apical ES and/or the BTB, illustrating their pivotal role in the apical ES–BTB–BM functional axis.

13.6 Concluding Remarks and Future Perspectives

Findings discussed herein thus illustrate the intimate relationship between polarity proteins, cell–cell interactions at the Sertoli–Sertoli, and Sertoli–spermatid interface and cytoskeleton in the seminiferous epithelium during the epithelial cycle. It is also noted that non-receptor protein tyrosine kinases, in particular FAK and SFK (e.g., c-Yes and c-Src), are intimately involved in these events. At the time of this writing, no concrete data were found in the literature providing credible information regarding the mechanism(s) by which polarity proteins regulate cytoskeleton and vice versa in the testis. Except for in studies using different animal models, such as the adjudin model, it was shown that the loss of spermatid polarity due to a downregulation of Par6 was likely mediated by adjudin-induced truncation and defragmentation of actin filament bundles at the apical ES (Wong et al. 2008b). This thus destabilized the actin-based adhesion protein complexes at the apical ES, such as integrin–laminin, nectin–afadin, leading to premature loss of spermatids from the epithelium, analogous to "spermiation." Also, we have yet to integrate the concept regarding the role of tubulin-based cytoskeleton into the biology of spermatid transport using the apical ES and the biology of preleptotene spermatocyte transport at the BTB using the basal ES, and how actin- and tubulin-based cytoskeletons are working in concert to regulate germ cell transport. For instance, several actin regulatory proteins have been identified and studied in the testis; virtually no tubulin regulatory proteins have been investigated in the testis except for several microtubule motor proteins, such as dynein and kinesin. This is an area of research that deserves some attention in future years.

Disclosure Statement The authors have nothing to disclose.

References

Akhmanova A, Mausset-Bonnefont AL, van Cappellen W, Keijzer N, Hoogenraad CC, Stepanova T, Drabek K, van der Wees J, Mommaas M, Onderwater J, van der Meulen H, Tanenbaum ME, Medema RH, Hoogerbrugge J, Vreeburg J, Uringa EJ, Grootegoed JA, Grosveld F, Galjart N (2005) The microtubule plus-end-tracking protein CLIP-170 associates with the spermatid manchette and is essential for spermatogenesis. Genes Dev 19:2501–2515. doi:10.1101/gad.344505

Amann RP (2008) The cycle of the seminiferous epithelium in humans: a need to revisit? J Androl 29:469–487

Assemat E, Bazellieres E, Pallesi-Pocachard E, Le Bivic A, Massey-Harroche D (2008) Polarity complex proteins. Biochim Biophys Acta 1778:614–630

Auharek SA, Avelar GF, Lara NLM, Sharpe RM, Franca LR (2011) Sertoli cell numbers and spermatogenic efficiency are increased in inducible nitric oxide synthase (iNOS) mutant-mice. Int J Androl 34:e621–e629

Bartolini F, Ramalingam N, Gundersen GG (2012) Actin-capping protein promotes microtubule stability by antagonizing the actin activity of mDia1. Mol Biol Cell 23:4032–4040

Bolos V, Gasent JM, Lopez-Tarruella S, Grande E (2010) The dual kinase complex FAK-Src as a promising therapeutic target in cancer. Onco Targets Ther 3:83–97
Boutros T, Chevet E, Metrakos P (2008) Mitogen-activated protein (MAP) kinase/MAP kinase phosphatase regulation: roles in cell growth, death, and cancer. Pharmacol Rev 60:261–310
Bruning-Richardson A, Langford KJ, Ruane P, Lee T, Askham JM, Morrison EE (2011) EB1 is required for spindle symmetry in mammalian mitosis. PLoS One 6:e28884
Carreau S, Hess RA (2010) Oestrogens and spermatogenesis. Philos Trans R Soc Lond B Biol Sci 365:1517–1535
Carreau S, Wolczynski S, Galeraud-Denis I (2010) Aromatase, estrogens and human male reproduction. Phil Trans R Soc Lond B Biol Sci 365:1571–1579
Cheng CY, Lie PPY, Wong EWP, Mruk DD (2013) Focal adhesion kinase and actin regulatory/binding proteins that modulate F-actin organization at the tissue barrier. Lesson from the testis. Tissue Barriers 1:e24252
Cheng CY, Lie PPY, Wong EWP, Mruk DD, Silvestrini B (2011a) Adjudin disrupts spermatogenesis via the action of some unlikely partners: Eps8, Arp2/3 complex, drebrin E, PAR6 and 14-3-3. Spermatogenesis 1:291–297
Cheng CY, Mruk DD (2002) Cell junction dynamics in the testis: Sertoli-germ cell interactions and male contraceptive development. Physiol Rev 82:825–874
Cheng CY, Mruk DD (2010) A local autocrine axis in the testes that regulates spermatogenesis. Nat Rev Endocrinol 6:380–395
Cheng CY, Mruk DD (2012) The blood-testis barrier and its implication in male contraception. Pharmacol Rev 64:16–64
Cheng CY, Mruk DD, Silvestrini B, Bonanomi M, Wong CH, Siu MKY, Lee NPY, Mo MY (2005) AF-2364 [1-(2,4-dichlorobenzyl)-*1H*-indazole-3-carbohydrazide] is a potential male contraceptive: a review of recent data. Contraception 72:251–261
Cheng CY, Wong EWP, Lie PPY, Li MWM, Su L, Siu ER, Yan HHN, Mannu J, Mathur PP, Bonanomi M, Silvestrini B, Mruk DD (2011b) Environmental toxicants and male reproductive function. Spermatogenesis 1:2–13
Dawson HN, Ferreira A, Eyster MV, Ghoshal N, Binder LI, Vitek MP (2001) Inhibition of neuronal maturation in primary hippocampal neurons from tau deficient mice. J Cell Sci 114(Pt 6):1179–1187
de Kretser DM, Kerr JB (1988) The cytology of the testis. In: Knobil E, Neill JB, Ewing LL, Greenwald GS, Markert CL, Pfaff DW (eds) The physiology of reproduction, vol 1. Raven, New York, pp 837–932
Denk F, Wade-Martins R (2009) Knock-out and transgenic mouse models of tauopathies. Neurobiol Aging 30(1):1–13. doi:10.1016/j.neurobiolaging.2007.05.010
Drabek K, Gutierrez L, Vermeij M, Clapes T, Patel SR, Boisset JC, van Haren J, Pereira AL, Liu Z, Akinci U, Nikolic T, van Ijcken W, van den Hout M, Meinders M, Melo C, Sambade C, Drabek D, Hendriks RW, Philipsen S, Mommaas M, Grosveld F, Maiato H, Italiano JE Jr, Robin C, Galjart N (2012) The microtubule plus-end tracking protein CLASP2 is required for hematopoiesis and hematopoietic stem cell maintenance. Cell Rep 2:781–788
Dym M, Clermont Y (1970) Role of spermatogonia in the repair of the seminiferous epithelium following X-irradiation of the rat testis. Am J Anat 128:265–282
Fawcett DW (1961) Intercellular bridges. Exp Cell Res 8:174–187
Franca LR, Auharek SA, Hess RA, Dufour JM, Hinton BT (2012) Blood-tissue barriers: morphofunctional and immunological aspects of the blood-testis and blood-epididymal barriers. Adv Exp Med Biol 763:237–259
Ganem NJ, Upton K, Compton DA (2005) Efficient mitosis in human cells lacking poleward microtubule flux. Curr Biol 15:1827–1832
Gavin RH (1997) Microtubule-microfilament synergy in the cytoskeleton. Int Rev Cytol 173: 207–242
Goldstein B, Macara IG (2007) The PAR proteins: fundamental players in animal cell polarization. Dev Cell 13:609–622

Goode BL, Drubin DG, Barnes G (2000) Functional cooperation between the microtubule and actin cytoskeletons. Curr Opin Cell Biol 12:63–71

Green KJ, Simpson CL (2007) Desmosomes: new perspectives on a classic. J Invest Dermatol 127: 2499–2515

Hall ES, Eveleth J, Jiang C, Redenbach DM, Boekelheide K (1992) Distribution of the microtubule-dependent motors cytoplasmic dynein and kinesin in rat testis. Biol Reprod 46:817–828

Harada A, Oguchi K, Okabe S, Kuno J, Terada S, Ohshima T, Sato-Yoshitake R, Takei Y, Noda T, Hirokawa N (1994) Altered microtubule organization in small-calibre axons of mice lacking tau protein. Nature 369:488–491

Head BP, Patel HH, Insel PA (2013) Interaction of membrane/lipid rafts with the cytoskeleton. Impact on signaling and function: membrane/lipid rafts, mediators of cytoskeletal arrangement and cell signaling. Biochim Biophys Acta (in press), doi: 10.1016/jbbamem201307018

Hess RA, de Franca LR (2008) Spermatogenesis and cycle of the seminiferous epithelium. Adv Exp Med Biol 636:1–15

Honda A, Siruntawineti J, Baba T (2002) Role of acrosomal matrix proteases in sperm-zona pellucida interactions. Hum Reprod Update 8:405–412

Iden S, Collard JG (2008) Crosstalk between small GTPases and polarity proteins in cell polarization. Nat Rev Mol Cell Biol 9:846–859

Jamora C, Fuchs E (2002) Intercellular adhesion, signaling and the cytoskeletons. Nat Cell Biol 4: E101–E108

Johnson L, Petty CS, Neaves WB (1980) A comparative study of daily sperm production and testicular composition in humans and rats. Biol Reprod 22:1233–1243

Kardon JR, Vale RD (2009) Regulators of the cytoplasmic dynein motor. Nat Rev Mol Cell Biol 10:854–865

Kline-Smith SL, Khodjakov A, Hergert P, Walczak CE (2004) Depletion of centromeric MCAK leads to chromosome congression and segregation defects due to improper kinetochore attachments. Mol Biol Cell 15:1146–1159

Koch M, Olson P, Albus A, Jin W, Hunter D, Brunken W, Burgeson R, Champliaud M (1999) Characterization and expression of the laminin $\gamma3$ chain: a novel, non-basement membrane-associated, laminin chain. J Cell Biol 145:605–618

Li SYT, Mruk DD, Cheng CY (2013) Focal adhesion kinase is a regulator of F-actin dynamics. New insights from studies in the testis. Spermatogenesis 3:e25385

Lie PPY, Cheng CY, Mruk DD (2011) The biology of the desmosome-like junction: a versatile anchoring junction and signal transducer in the seminiferous epithelium. Int Rev Cell Mol Biol 286:223–269

Lie PPY, Mruk DD, Mok KW, Su L, Lee WM, Cheng CY (2012) Focal adhesion kinase-Tyr^{407} and -Tyr^{397} exhibit antagonistic effects on blood-testis barrier dynamics in the rat. Proc Natl Acad Sci U S A 109:12562–12567

Liedtke W, Leman EE, Fyffe RE, Raine CS, Schubart UK (2002) Stathmin-deficient mice develop an age-dependent axonopathy of the central and peripheral nervous systems. Am J Pathol 160:469–480

Lowell CA (2011) Src-family and Syk kinases in activating and inhibitory pathways in innate immune cells: signaling cross talk. Cold Spring Harb Perspect Biol 3:a002352

Lowery LA, Stout A, Faris AE, Ding L, Baird MA, Davidson MW, Danuser G, Van Vactor D (2013) Growth cone-specific functions of XMAP215 in restricting microtubule dynamics and promoting axonal outgrowth. Neural Dev 8:22

Matteoni R, Kreis TE (1987) Translocation and clustering of endosomes and lysosomes depends on microtubules. J Cell Biol 105:1253–1265

Mauss J, Hackstedt G (1972) The effect of unilateral orchiectomy and unilateral cryptorchidism on sperm output in the rat. J Reprod Fertil 30:289–292

Mazaud-Guittot S (2011) Dissecting the phthalate-induced Sertoli cell injury: the fragile balance of proteases and their inhibitors. Biol Reprod 85:1091–1093

McLachlan RI, O'Donnell L, Meachem SJ, Stanton PG, De Kretser DM, Pratis K, Robertson DM (2002) Identification of specific sites of hormonal regulation in spermatogenesis in rats, monkeys, and man. Recent Prog Horm Res 57:149–179

Mimori-Kiyosue Y, Grigoriev I, Sasaki H, Matsui C, Akhmanova A, Tsukita S, Vorobjev I (2006) Mammalian CLASPs are required for mitotic spindle organization and kinetochore alignment. Genes Cells 11:845–857

Mitchison T, Kirschner M (1984) Dynamic instability of microtubule growth. Nature 312:237–242

Mok KW, Mruk DD, Lie PPY, Lui WY, Cheng CY (2011) Adjudin, a potential male contraceptive, exerts its effects locally in the seminiferous epithelium of mammalian testes. Reproduction 141:571–580

Mruk DD, Cheng CY (2004a) Cell-cell interactions at the ectoplasmic specialization in the testis. Trends Endocrinol Metab 15:439–447

Mruk DD, Cheng CY (2004b) Sertoli-Sertoli and Sertoli-germ cell interactions and their significance in germ cell movement in the seminiferous epithelium during spermatogenesis. Endocr Rev 25:747–806

Mruk DD, Silvestrini B, Cheng CY (2008) Anchoring junctions as drug targets: role in contraceptive development. Pharmacol Rev 60:146–180

O'Donnell L, Nicholls PK, O'Bryan MK, McLachlan RI, Stanton PG (2011) Spermiation: the process of sperm release. Spermatogenesis 1:14–35

O'Donnell L, Rhodes D, Smith SJ, Merriner DJ, Clark BJ, Borg C, Whittle B, O'Connor AE, Smith LB, McNally FJ, de Kretser DM, Goodnow CC, Ormandy CJ, Jamsai D, O'Bryan MK (2012) An essential role for katanin p80 and microtubule severing in male gamete production. PLoS Genet 8:e1002698

O'Donnell L, Robertson KM, Jones ME, Simpson ER (2001) Estrogen and spermatogenesis. Endocr Rev 22:289–318

Parvinen M (1982) Regulation of the seminiferous epithelium. Endocr Rev 3:404–417

Parvinen M, Vanha-Perttula T (1972) Identification and enzyme quantification of the stages of the seminiferous epithelial wave in the rat. Anat Rec 174:435–450

Pelletier RM (2011) The blood-testis barrier: the junctional permeability, the proteins and the lipids. Prog Histochem Cytochem 46:49–127

Pierre P, Scheel J, Rickard JE, Kreis TE (1992) CLIP-170 links endocytic vesicles to microtubules. Cell 70:887–900

Qian Y, Liu S, Guan Y, Pan H, Guan X, Qiu Z, Li L, Gao N, Zhao Y, Li X, Lu Y, Liu M, Li D (2013) Lgr4-mediated Wnt/β-catenin signaling in pertibular myoid cells is essential for spermatogenesis. Development 140:1751–1761

Redenbach DM, Vogl AW (1991) Microtubule polarity in Sertoli cells: a model for microtubule-based spermatid transport. Eur J Cell Biol 54:277–290

Rolland M, Le Moal J, Wagner V, Royere D, De Mouzon J (2013) Decline in semen concentration and morphology in a sample of 26,609 men close to general population between 1989 and 2005 in France. Hum Reprod 28:462–470

Russell L (1977) Observations on rat Sertoli ectoplasmic ("junctional") specializations in their association with germ cells of the rat testis. Tissue Cell 9(3):475–498

Russell LD, Goh JC, Rashed RMA, Vogl AW (1988) The consequences of actin disruption at Sertoli ectoplasmic specialization sites facing spermatids after in vivo exposure of rat testis to cytochalasin D. Biol Reprod 39:105–118

Russell LD, Peterson RN (1985) Sertoli cell junctions: morphological and functional correlates. Int Rev Cytol 94:177–211

Schober JM, Kwon G, Jayne D, Cain JM (2012) The microtubule-associated protein EB1 maintains cell polarity through activation of protein kinase C. Biochem Biophys Res Commun 417: 67–72

Schubart UK, Yu J, Amat JA, Wang Z, Hoffmann MK, Edelmann W (1996) Normal development of mice lacking metablastin (P19), a phosphoprotein implicated in cell cycle regulation. J Biol Chem 271:14062–14066

Sharpe RM (1994) Regulation of spermatogenesis. In: Knobil E, Neill JD (eds) The physiology of reproduction. Raven Press, New York, pp 1363–1434

Silva VC, Cassimeris L (2013) Stathmin and microtubules regulate mitotic entry in HeLa cells by controlling activation of both Aurora kinase A and Plk1. Mol Biol Cell 24:3819–3831

Siu ER, Mruk DD, Porto CS, Cheng CY (2009) Cadmium-induced testicular injury. Toxicol Appl Pharmacol 238:240–249

Siu MKY, Cheng CY (2004) Interactions of proteases, protease inhibitors, and the β1 integrin/laminin γ3 protein complex in the regulation of ectoplasmic specialization dynamics in the rat testis. Biol Reprod 70:945–964

Smith LB, Milne L, Nelson N, Eddie S, Brown P, Atanassova N, O'Bryan MK, O'Donnell L, Rhodes D, Wells S, Napper D, Nolan P, Lalanne Z, Cheeseman M, Peters J (2012) KATNAL1 regulation of sertoli cell microtubule dynamics is essential for spermiogenesis and male fertility. PLoS Genet 8:e1002697

Stout JR, Rizk RS, Kline SL, Walczak CE (2006) Deciphering protein function during mitosis in PtK cells using RNAi. BMC Cell Biol 7:26

Su L, Mruk DD, Lie PPY, Silvestrini B, Cheng CY (2012a) A peptide derived from laminin-γ3 reversibly impairs spermatogenesis in rats. Nat Commun 3:1185

Su W, Mruk DD, Cheng CY (2013) Regulation of actin dynamics and protein trafficking during spermatogenesis-insights into a complex process. Crit Rev Biochem Mol Biol 48(2):153–172

Su WH, Wong EWP, Mruk DD, Cheng CY (2012b) The Scribble/Lgl/Dlg polarity protein complex is a regulator of blood-testis barrier dynamics and spermatid polarity during spermatogenesis. Endocrinology 153:6041–6053

Thomason HA, Scothern A, McHarg S, Garrod DR (2010) Desmosomes: adhesive strength and signalling in health and disease. Biochem J 429:419–433

Toft G, Jonsson BA, Lindh CH, Giwercman A, Spano M, Heederik D, Lenters V, Vermeulen R, Rylander L, Pedersen HS, Ludwicki JK, Zviezdai V, Bonde JP (2012) Exposure to perfluorinated compounds and human semen quality in Arctic and European populations. Hum Reprod 27:2532–2540

Verhoeven G, Willems A, Denolet E, Swinnen JV, De Gendt K (2010) Androgens and spermatogenesis: lessons from transgenic mouse models. Phil Trans R Soc Lond B Biol Sci 365: 1537–1556

Vogl AW, Vaid KS, Guttman JA (2008) The Sertoli cell cytoskeleton. Adv Exp Med Biol 636: 186–211

Wade RH (2007) Microtubules: an overview. Meth Mol Med 137:1–16

Wan HT, Mruk DD, Li SYT, Mok KW, Lee WM, Wong CKC, Cheng CY (2013a) p-FAK-Tyr^{397} regulates spermatid adhesion in the rat testis via its effects on F-actin organization at the ectoplasmic specialization. Am J Physiol Endocrinol Metab 305:E687–E699

Wan HT, Mruk DD, Wong CKC, Cheng CY (2013b) The apical ES-BTB-BM functional axis is an emerging target for toxicant-induced infertility. Trends Mol Med 19:396–405

Wang RS, Yeh S, Tzeng CR, Chang C (2009) Androgen receptor roles in spermatogenesis and fertility: lessons from testicular cell-specific androgen receptor knockout mice. Endocr Rev 30: 119–132

Weber JE, Russell LD (1987) A study of intercellular bridges during spermatogenesis in the rat. Am J Anat 180:1–24

Weber JE, Russell LD, Wong V, Peterson RN (1983) Three dimensional reconstruction of a rat stage V Sertoli cell: II. Morphometry of Sertoli-Sertoli and Sertoli-germ cell relationships. Am J Anat 167:163–179

Welsh M, Saunders PT, Atanassova N, Sharpe RM, Smith LB (2009) Androgen action via testicular peritubular myoid cells is essential for male fertility. FASEB J 23:4218–4230

Wenz JR, Hess RA (1998) Characterization of stage-specific tyrosinated alpha-tubulin immunoperoxidase staining patterns in Sertoli cells of rat seminiferous tubules by light microscopic image analysis. Tissue Cell 30:492–501

Wolski KM, Perrault C, Tran-Son-Tay R, Cameron DF (2005) Strength measurement of the Sertoli-spermatid junctional complex. J Androl 26:354–359

Wong CH, Cheng CY (2005) The blood-testis barrier: its biology, regulation and physiological role in spermatogenesis. Curr Topics Dev Biol 71:263–296

Wong EWP, Cheng CY (2009) Polarity proteins and cell-cell interactions in the testis. Int Rev Cell Mol Biol 278:309–353

Wong EWP, Cheng CY (2013) NC1 domain of collagen α3(IV) derived from the basement membrane regulates Sertoli cell blood-testis barrier dynamics. Spermatogenesis 3:e25465

Wong EWP, Mruk DD, Cheng CY (2008a) Biology and regulation of ectoplasmic specialization, an atypical adherens junction type, in the testis. Biochem Biophys Acta 1778:692–708

Wong EWP, Mruk DD, Lee WM, Cheng CY (2008b) Par3/Par6 polarity complex coordinates apical ectoplasmic specialization and blood-testis barrier restructuring during spermatogenesis. Proc Natl Acad Sci U S A 105:9657–9662

Wong EWP, Sun S, Li MWM, Lee WM, Cheng CY (2009) 14-3-3 protein regulates cell adhesion in the seminiferous epithelium of rat testes. Endocrinology 150:4713–4723

Wong V, Russell LD (1983) Three-dimensional reconstruction of a rat stage V Sertoli cell: I. Methods, basic configuration, and dimensions. Am J Anat 167:143–161

Xiao X, Mruk DD, Cheng CY (2013) c-Yes regulates cell adhesion at the apical ectoplasmic specialization-blood-testis barrier axis via its effects on protein recruitment and distribution. Am J Physiol Endocrinol Metab 304:E145–E159

Xiao X, Mruk DD, Cheng FL, Cheng CY (2012) c-Src and c-Yes are two unlikely partners of spermatogenesis and their roles in blood-testis barrier dynamics. Adv Exp Med Biol 763: 295–317

Xiao X, Mruk DD, Lee WM, Cheng CY (2011) c-Yes regulates cell adhesion at the blood-testis barrier and the apical ectoplasmic specialization in the seminiferous epithelium of rat testes. Int J Biochem Cell Biol 43:651–665

Yan HHN, Cheng CY (2006) Laminin α3 forms a complex with β3 and γ3 chains that serves as the ligand for α6β1-integrin at the apical ectoplasmic specialization in adult rat testes. J Biol Chem 281:17286–17303

Yan HHN, Mruk DD, Lee WM, Cheng CY (2007) Ectoplasmic specialization: a friend or a foe of spermatogenesis? BioEssays 29:36–48

Yan HHN, Mruk DD, Wong EWP, Lee WM, Cheng CY (2008) An autocrine axis in the testis that coordinates spermiation and blood-testis barrier restructuring during spermatogenesis. Proc Natl Acad Sci U S A 105:8950–8955

Yao PL, Lin YC, Richburg JH (2009) TNFα-mediated disruption of spermatogenesis in response to Sertoli cell injury in rodents is partially regulated by MMP2. Biol Reprod 80:581–589

Yao PL, Lin YC, Richburg JH (2010) Mono-(2-ethylhexyl) phthalate-induced disruption of junctional complexes in the seminiferous epithelium of the rodent testis is mediated by MMP2. Biol Reprod 82:516–527

Zhao X, Guan JL (2010) Focal adhesion kinase and its signaling pathways in cell migration and angiogenesis. Adv Drug Deliv Rev 63:610–615

Part IV
Protein Sorting and Distribution in Polarized Epithelial Cells

Chapter 14
Membrane Traffic and Apicobasal Polarity in *Drosophila* Epithelial Cells

Barry J. Thompson

Abstract Previous chapters have introduced the core determinants of apicobasal polarity in different cell types. This chapter will examine how these determinants become localised to particular membrane domains in *Drosophila*. In some cell types, such as neuroblasts, polarity does not rely on transmembrane proteins or membrane trafficking. In contrast, epithelial cell polarity involves several key transmembrane proteins, such as Crumbs and E-cadherin, and their distribution depends critically on the exocytic and endocytic machinery. In addition, certain types of epithelial cells secrete an apical extracellular matrix and a basement membrane matrix, processes that must involve polarised exocytic delivery of matrix components. Disruption of membrane trafficking can therefore lead to loss of epithelial polarity and tumour-like phenotypes in *Drosophila*.

Keywords Epithelial polarity • Drosophila • Crumbs • E-cadherin • Endocytosis • Exocytosis

14.1 Polarity Without Membrane Trafficking: *Drosophila* Neuroblasts

Any discussion of the role of transmembrane proteins in cell polarity must begin with the observation that some cell types, such as *Drosophila* neuroblasts, do not require transmembrane proteins or the endocytic machinery to polarise (Fig. 14.1). Neuroblasts divide asymmetrically in a stem cell pattern to produce one differentiating daughter cell and one neuroblast (Homem and Knoblich 2012; Wodarz and Gonzalez 2006). Polarity in neuroblasts only appears during mitosis, in order to segregate cell fate determinants basally, so that they are inherited by the differentiating daughter cell following cytokinesis (Wirtz-Peitz et al. 2008). These fate determinants include Miranda, Prospero, Brat and Numb (a regulator of Delta–Notch signalling) (Homem and Knoblich 2012). The apical domain of neuroblasts

B.J. Thompson (✉)
Epithelial Biology Laboratory, Cancer Research UK—London Research Institute, 44 Lincoln's Inn Fields, London WC2A 3LY, UK
e-mail: barry.thompson@cancer.org.uk

K. Ebnet (ed.), *Cell Polarity 1*, DOI 10.1007/978-3-319-14463-4_14

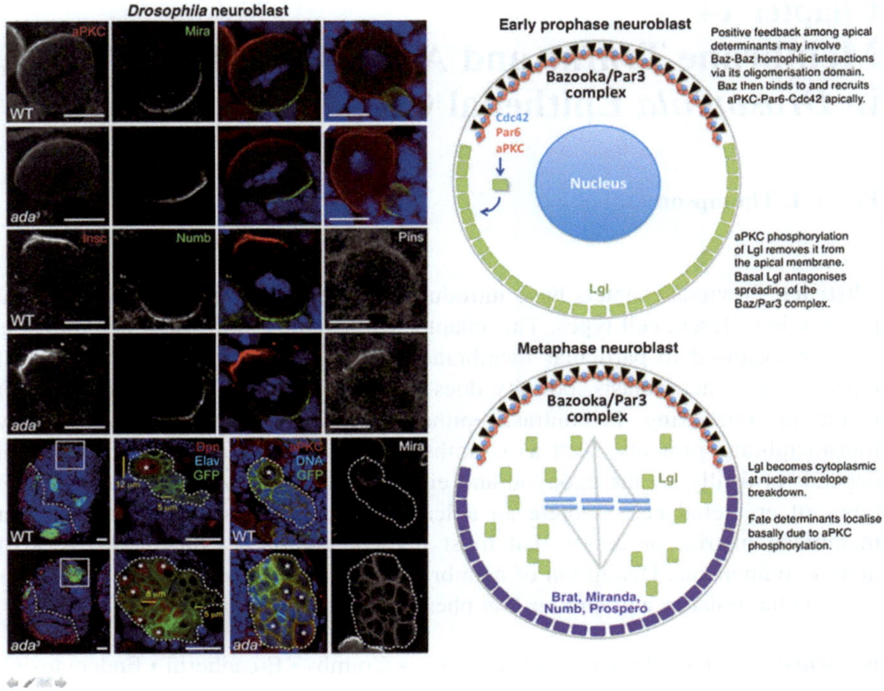

Fig. 14.1 Clathrin-mediated endocytosis is dispensable for polarity in neuroblasts

is determined by the Bazooka/Par3–aPKC–Par6–Cdc42 complex, which appears to self-recruit to the plasma membrane and acts in a mutually antagonistic manner with the transiently basal Lgl protein (Homem and Knoblich 2012).

None of the key determinants of neuroblast polarity are transmembrane proteins, suggesting that polarisation of neuroblasts is determined purely by association and dissociation of these proteins from the plasma membrane. In keeping with this notion, the endocytic machinery is dispensable for neuroblast polarity (Halbsgut et al. 2011) (Fig. 14.1). In the absence of clathrin-mediated endocytosis, determinants of neuroblast cell fate are normally polarised (Halbsgut et al. 2011) (Fig. 14.1). The same mutants do exhibit severe *numb*-like phenotypes later in the process of cellular differentiation, due to mis-regulation of Delta–Notch signalling in daughter cells (Fig. 14.1) (Berdnik et al. 2002). This observation demonstrates that cell polarity need not depend on membrane trafficking in all cell types.

Why neuroblasts do not make use of transmembrane proteins for their polarisation remains a mystery. One suggestion is that polarisation in neuroblasts only occurs during cell division, a time when cellular endosomes can undergo dramatic reorganisation to accommodate mitosis and cytokinesis. Another possibility is that the endosomal machinery itself may be asymmetrically distributed during neuroblast division to help bias Delta–Notch signalling following cytokinesis (Coumailleau et al. 2009). Whatever the case, neuroblasts prove that the Bazooka/Par3 complex is able to polarise without the help of the membrane trafficking machinery (Fig. 14.1).

14.2 Polarity via Membrane Trafficking: *Drosophila* Epithelial Cells

In *Drosophila* epithelia, membrane trafficking is essential to organise polarity. RNAi knockdown or mutation of the endocytic machinery causes a loss of epithelial polarity, with apical and basolateral determinants overlapping around the plasma membrane (Fig. 14.2) (Lu and Bilder 2005; Thompson et al. 2005; Vaccari and Bilder 2005; Menut et al. 2007; Fletcher et al. 2012). As a result of the loss of polarity, epithelial cells round up and pile atop one another, resembling a carcinoma. In both the *Drosophila* follicle cell epithelium and imaginal disc epithelia, the mutant tissue also overproliferates to form a tumour-like mass of cells (Fig. 14.2). In the case of imaginal disc tumours, this results in a delay in larval pupariation and developmental arrest as giant-sized pupae (Fig. 14.2). Notably,

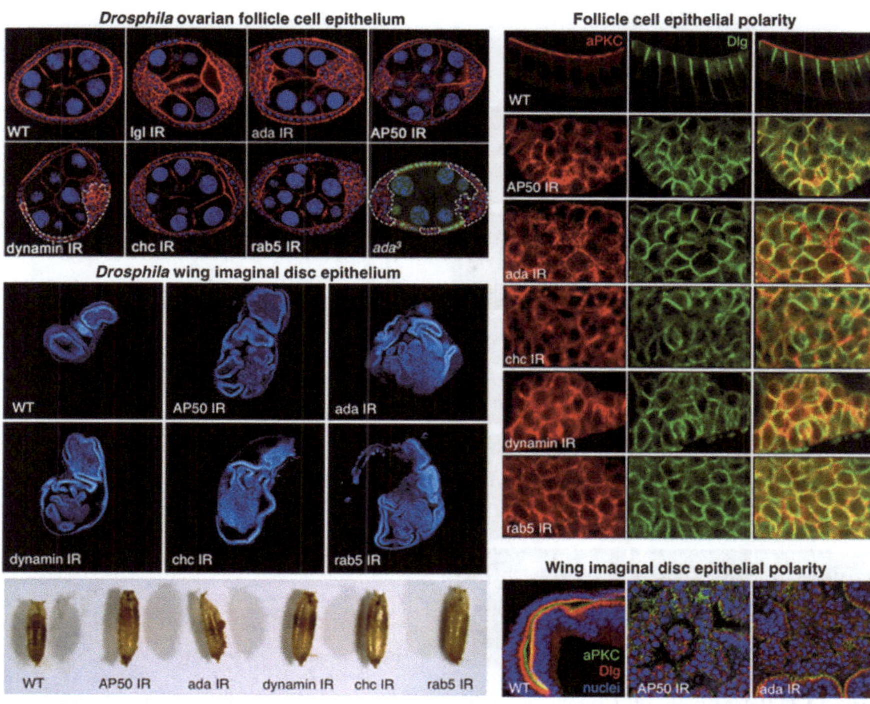

Fig. 14.2 Clathrin-mediated endocytosis is essential for epithelia polarity in *Drosophila* ovarian follicle and imaginal disc epithelia

small clones of endocytic mutants do not form tumours because they are rapidly eliminated by neighbouring wild-type cells in a protective process known as cell competition (Thompson et al. 2005; Vaccari and Bilder 2005; Menut et al. 2007).

Phenotypically, loss of the endocytic machinery closely resembles loss of the basolateral polarity determinants Lgl (lethal giant larvae), Dlg (discs large) or Scrib (scribble) (Bryant and Schmidt 1990; Woods and Bryant 1991; Woods et al. 1996; Bilder and Perrimon 2000; Bilder et al. 2000). These basolateral determinants normally act to antagonise the localisation of the apical determinants to the plasma membrane (Bilder and Perrimon 2000; Bilder et al. 2000; Tanentzapf and Tepass 2003). This suggests that the primary function of the endocytic pathway in epithelial polarity is to restrict the spreading of the apical determinants (Fig. 14.2).

Of the apical polarity determinants, only one molecule—Crumbs—is a transmembrane protein. Thus, trafficking of Crumbs is likely to be central in maintaining epithelial polarity (Fig. 14.3). Consistent with this idea, Crumbs accumulates around the entire plasma membrane when endocytosis is inhibited in *Drosophila* epithelia (Lu and Bilder 2005; Fletcher et al. 2012). Furthermore, overexpression of Crumbs is sufficient to cause ectopic spreading of the other apical determinants into the basolateral domain and thus a loss of polarity and tumour-like phenotype (Wodarz et al. 1995; Lu and Bilder 2005; Fletcher et al. 2012). Given the importance of Crumbs trafficking to epithelial polarity, a detailed examination of how this molecule is localised by the exocytic and endocytic machinery is warranted.

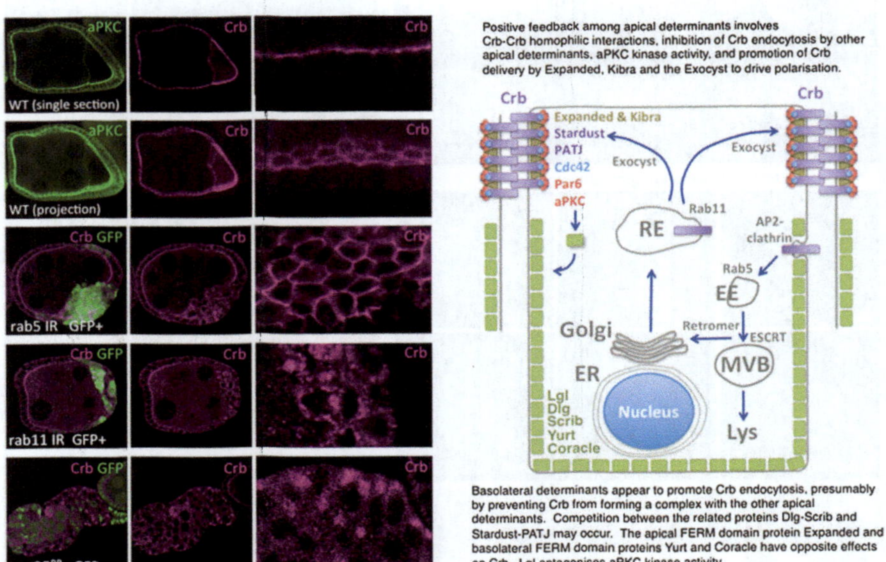

Fig. 14.3 Polarisation of Crumbs in epithelial cells occurs through regulated membrane trafficking

14.3 Trafficking and Localisation of Crumbs in *Drosophila* Epithelia

It is now clear that Crumbs localisation depends on continuous clathrin/AP2-mediated endocytosis and on the early endosome protein Rab5 (Lu and Bilder 2005; Fletcher et al. 2012; Menut et al. 2007). Thus, Crumbs must be removed from the plasma membrane to prevent its accumulation and ectopic spreading around the membrane. Once in early endosomes, Crumbs—like other transmembrane proteins—can be trafficked via multivesicular bodies (MVBs) to the lysosome for degradation. In mutants of the ESCRT machinery, which sorts proteins into the MVB for degradation, Crumbs accumulates with other transmembrane proteins on the outer limiting membrane of abnormal MVBs (Vaccari and Bilder 2005; Thompson et al. 2005). Thus, a significant proportion of endocytosed Crumbs must traffic through the MVB (Fig. 14.3).

An alternative route for endocytosed Crumbs molecules is to be recycled to the plasma membrane. An important mediator of Crumbs recycling is the retromer complex, which retrieves transmembrane proteins from the endocytic pathway and returns them to the Golgi for re-secretion to the plasma membrane (Pocha et al. 2011; Zhou et al. 2011). In the absence of retromer function, all endocytosed Crumbs is degraded in the lysosome, such that the total level of Crumbs remaining at the apical domain is reduced (Pocha et al. 2011; Zhou et al. 2011) (Fig. 14.3).

From the Golgi, both newly synthesised Crumbs and retromer-recycled Crumbs must travel to the plasma membrane via the secretory pathway. A critical way station on this journey is the Rab11 'recycling endosome', which is essential for Crumbs to reach the plasma membrane. In the absence of Rab11, Crumbs accumulates in abnormal and swollen endosomes rather than at the plasma membrane (Roeth et al. 2009; Fletcher et al. 2012). Trafficking of Crumbs from Rab11 endosomes to the plasma membrane appears to be promoted by the exocyst complex, as Crumbs levels are reduced in Exo84 mutant embryos (Blankenship et al. 2007) (Fig. 14.3). However, whether the exocyst is strictly essential for Crumbs to localise apically in all epithelia remains to be explored.

How Crumbs becomes polarised and maintains its apical localisation is an important unanswered question. Computer modelling suggests that polarity arises from the combination of a positive feedback loop among apical determinants together with mutual antagonism between apical and basolateral determinants (Fletcher et al. 2012). In the case of Crumbs, positive feedback could be achieved firstly by Crumbs–Crumbs homomeric interactions via its large extracellular domain, which would stabilise neighbouring Crumbs molecules at the plasma membrane (Fletcher et al. 2012; Letizia et al. 2013; Roper 2012; Hafezi et al. 2012; Thompson et al. 2013). Secondly, Crumbs recruits Stardust, PATJ and other apical determinants, including the key aPKC–Par6–Cdc42 complex, which is essential to confer apical membrane domain identity. Thus, aPKC–Par6–Cdc42 may help recruit the exocyst complex and potentially other factors that could promote polarised delivery of Crumbs and prevent endocytosis of Crumbs (Fletcher et al. 2012). Understanding the molecular nature of this positive feedback loop is critical to understanding Crumbs polarisation.

The mutual antagonism between apical and basolateral determinants is partly understood for the aPKC–Lgl interaction, in which aPKC phosphorylates Lgl to remove it from the apical domain, while Lgl inhibits aPKC kinase activity to prevent ectopic spreading of the apical domain (Betschinger et al. 2003; Atwood and Prehoda 2009). This aPKC–Lgl mutual antagonism is actively involved in polarising the Bazooka/Par3 complex in neuroblasts, yet may also be relevant for Crumbs polarisation in epithelia, since Crumbs localisation depends on aPKC kinase activity and since ectopic activation of aPKC kinase activity causes ectopic spreading of Crumbs (Fletcher et al. 2012).

How the other basolateral determinants, Dlg and Scrib, contribute to mutual antagonism with the apical determinants is not understood. However, one attractive hypothesis is that Dlg and Scrib may directly compete with Stardust and PATJ, respectively, as the proteins have very similar domain structures. In addition, the basolateral FERM domain proteins Yurt and Coracle contribute to restricting Crumbs to the apical domain, presumably by antagonising Crumbs localisation to the basolateral membrane (Laprise et al. 2006, 2009). In contrast, the apical FERM domain protein Expanded promotes Crumbs localisation to the apical membrane by operating in parallel with the Kibra protein (Fletcher et al. 2012). Crumbs has a FERM-binding domain in its intracellular tail, so these different FERM domain proteins may bind to Crumbs to either promote or inhibit its endocytosis from the plasma membrane. Further work is needed to understand the mechanism of action of these FERM domain proteins in Crumbs polarisation and how they are themselves localised to either the apical or basolateral domains.

Together, the above findings provide an emerging picture of how Crumbs can maintain its polarised subcellular location in epithelia (Fig. 14.3). How the polarisation of Crumbs is initiated is less clear, but its localisation appears to follow that of the Bazooka/Par3 complex, which appears to polarise first in early embryos and follicle cell epithelia and with which the Crumbs complex has an important and complex relationship.

14.4 The Crumbs and Bazooka/Par3 Complexes Are Interlinked Yet Redundant Systems

The Crumbs–Stardust–PATJ complex and Bazooka/Par3 complex are similar in that they both act to recruit aPKC–Par6–Cdc42 to the apical plasma membrane domain. This fact means that polarisation of one complex can recruit aPKC–Par6–Cdc42 to then assist the polarisation of the other complex to the same location. Importantly, removal of both Crumbs (or Stardust) and Bazooka/Par3 is necessary to completely abolish the localisation of aPKC–Par6–Cdc42 to the plasma membrane and apical domain identity (Tanentzapf and Tepass 2003; Fletcher

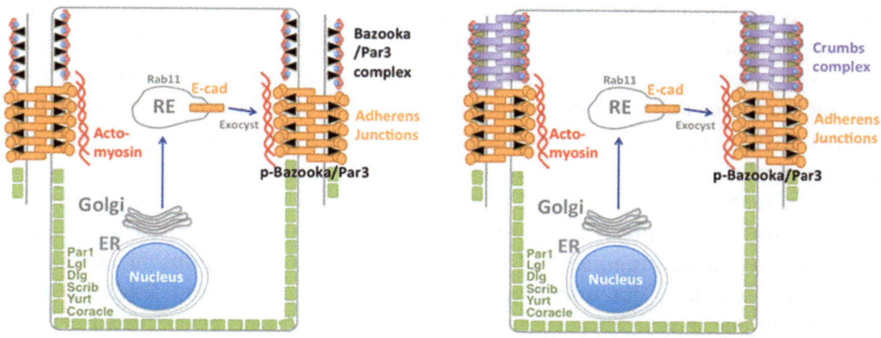

Fig. 14.4 Either the Bazooka/Par3 complex or the Crumbs complex is sufficient to maintain epithelial polarity and localise adherens junctions

et al. 2012). Thus, the two complexes act in a semi-redundant fashion, and Crumbs and Bazooka/Par3 can be dispensable for maintaining apicobasal polarity in many epithelia—unlike aPKC–Par6–Cdc42, which is always essential (Fig. 14.4).

The fact that only one of either the Crumbs complex or Bazooka/Par3 complex is normally required to maintain apicobasal polarity may explain a curious interaction between the two complexes. When Crumbs is expressed in cells, it tends to promote aPKC phosphorylation of Bazooka/Par3 on a site that disrupts its association with the rest of the apical determinants at the plasma membrane (Morais-de-Sa et al. 2010; Walther and Pichaud 2010). The result of this phosphorylation event is that Bazooka/Par3 tends to associate with adherens junctions, through its ability to bind directly to them (Morais-de-Sa et al. 2010; Walther and Pichaud 2010) (Fig. 14.4). In the absence of Crumbs, more Bazooka/Par3 complexes form at the apical domain, yet some Bazooka/Par3 is still phosphorylated by aPKC and localises to adherens junctions (Morais-de-Sa et al. 2010). Thus, the Bazooka/Par3 exists in two pools, one in a complex with other apical determinants at the apical domain and the other at adherens junctions.

The pool of Bazooka/Par3 at adherens junctions is able to perform distinct functions from the one that normally maintains apicobasal polarity in parallel with the Crumbs complex. For example, during embryonic gastrulation, morphogenetic movements involve planar polarisation of Bazooka/Par3 at adherens junctions (Zallen and Wieschaus 2004; Simoes Sde et al. 2010). A similar role for Bazooka/Par3 at junctions appears during retinal morphogenesis in *Drosophila* eye development (Walther and Pichaud 2010). It is no surprise, then, that these two points in development are those at which Crumbs begins to have an essential function in maintaining polarity (Campbell et al. 2009; Pellikka et al. 2002; Walther and Pichaud 2010).

The Crumbs complex can also adopt a planar polarised localisation at certain points in development. During the formation of epithelial tubes, such as the salivary gland placode or the tracheal pits of the *Drosophila* embryo, Crumbs expression is elevated within a patch of cells compared to the surrounding neighbours (Roper

2012; Letizia et al. 2011). This difference in Crumbs expression levels creates a planar polarisation of Crumbs in the cells at the boundary (Roper 2012). The planar polarisation of Crumbs then directs formation of an actin cable that generates contractile forces that drive invagination of the epithelium to initiate tube formation (Roper 2012). Bazooka/Par3 presumably maintains apicobasal polarity, while Crumbs performs these planar polarity functions.

The redundancy between the Crumbs complex and the Bazooka/Par3 complex also reveals that Crumbs has an additional function in regulating the Hippo signalling pathway, a key regulator of tissue growth. Loss of Crumbs in *Drosophila* imaginal discs does not affect apicobasal polarity, but instead leads to tissue overgrowth due to activation of the Hippo signal transducer Yorkie (YAP/TAZ in mammals) (Ling et al. 2010; Robinson et al. 2010; Chen et al. 2010). This signalling role for Crumbs is thought to be mediated by the apical FERM domain protein Expanded, which binds directly to the Crumbs intracellular domain (Ling et al. 2010). Why an apical polarity determinant like Crumbs must signal to the Hippo pathway remains a major mystery, but several interesting functions have been proposed.

The first hypothesis is that Crumbs–Expanded–Hippo signalling is primarily involved in signalling to the cytoskeleton rather than the nucleus. Hippo signalling involves activation of the upstream kinase Hippo (MST1/2 in humans) and the downstream kinase Warts (LATS1/2 in humans). Warts not only phosphorylates the nuclear effector Yorkie but also phosphorylates the actin cytoskeleton regulator Ena (VASP in humans) (Lucas et al. 2013). Thus, Crumbs–Expanded–Hippo signalling can help to polarise the actin cytoskeleton, a function that seems particularly important for cells that rely upon a dramatically polarised actin cytoskeleton, such as collectively migrating *Drosophila* border cells (Lucas et al. 2013). In these cells, the role of Warts phosphorylation of Yorkie appears to be primarily involved in providing a negative feedback loop to limit pathway activity and the speed of migration (Lucas et al. 2013). In addition, Hippo signalling can also regulate the actin cytoskeleton in the imaginal disc epithelia, even though the major function of this pathway is to regulate the tissue growth in imaginal discs (Fernandez et al. 2011).

The second hypothesis is that Crumbs–Expanded–Hippo signalling is involved in regulating cell competition in imaginal disc epithelia (Hafezi et al. 2012). Cell competition is a short-range signalling process whereby healthy cells kill their unhealthy neighbours. When populations of cells express different levels of Crumbs, cells with high Crumbs tend to act as 'losers' that are killed by neighbouring 'winner' cells with low Crumbs (Hafezi et al. 2012). Thus, Crumbs may be involved in sensing differences between winner and loser cells during cell competition.

The above discussion highlights the crucial role that Crumbs plays in apicobasal polarity and planar polarity in epithelia. Once epithelial cells have obtained their polarity, they use this information to polarise many other molecules that direct cell behaviour. One of the most important molecules acting downstream of the core epithelial polarity determinants is E-cadherin, which forms adherens junctions to organise the cytoskeleton, connects neighbouring cells and directs epithelial cell shape. How E-cadherin is trafficked in *Drosophila* epithelia will be considered in the next section.

14.5 Trafficking and Localisation of E-Cadherin in *Drosophila* Epithelia

E-cadherin is a transmembrane protein that forms homophilic adhesive interactions that connect neighbouring epithelial cells at adherens junctions (Meng and Takeichi 2009). The cytoplasmic tail of E-cadherin binds to the actin cytoskeleton via the beta-catenin and alpha-catenin proteins. A ring of filamentous actin and myosin-II forms around adherens junctions to provide tensile strength to the epithelium. The balance between tension and adhesion forces at adherens junctions is responsible for determining the polygonal shapes of epithelial cells at the apical surface. Subtle alterations in tension and adhesion by planar polarised cues underlie many cell shape changes and cell–cell rearrangements during development, although the nature of these upstream cues remains unclear in most cases (Vichas and Zallen 2011). Complete absence of E-cadherin, beta-catenin or alpha-catenin causes a failure to maintain epithelial integrity and leads to a rounding up of epithelial cells that then pile atop one another (Tepass et al. 1996; Peifer 1995; Sarpal et al. 2012).

The strong cell shape phenotype observed in the absence of adherens junctions suggests that one of the major functions of epithelial polarity determinants is to localise E-cadherin to the correct position within cells. It is striking that adherens junctions are generally most concentrated right at the boundary between the apical and basolateral membrane domains (Fig. 14.4). However, very little is known about how determinants of apicobasal polarity might control the trafficking of E-cadherin. One possibility is that polarity determinants might influence the formation of the apical actomyosin ring, which is then responsible for binding to and stabilising adherens junctions at the correct position (Gomez et al. 2011). But this possibility remains to be explored experimentally in *Drosophila*.

Much more attention has been paid to more immediate regulators of actomyosin contractility and their role in organising adherens junctions. In particular, the small GTPase Rho (also called Rho1 or RhoA) is known to be important for actomyosin contractility and junction maintenance in many contexts, including embryos, imaginal discs and the pupal retina (Magie et al. 1999, 2002; Bloor and Kiehart 2002; Fox et al. 2005; Yan et al. 2009; Warner and Longmore 2009). Rho localises to the adherens junctions and to the entire apical surface of epithelial cells. The two key effectors of Rho are the formin Diaphanous, which promotes actin polymerisation, and the Rho-kinase, which phosphorylates and activates myosin-II. However, the precise contribution of Rho-kinase and Diaphanous in the regulation of adherens junctions remains controversial. Evidence in retinal epithelia suggests that removal of both Rho-kinase and Diaphanous does not affect Rho's ability to maintain adherens junctions (Warner and Longmore 2009). Furthermore, planar polarised activation of Rho-kinase and Dia during convergent extension movements in the *Drosophila* embryo leads to endocytosis of E-cadherin and shrinkage of cell–cell junctions, rather than stabilisation of E-cadherin at the membrane (Levayer et al. 2011). Similarly, activation of Rho-kinase and Dia across the entire apical surface during ventral furrow invagination in the embryo also leads to shrinkage of adherens junctions and apical surface area (Mason et al. 2013). Thus, further work is necessary to understand the basic mechanism by which Rho acts to stabilise adherens junctions in most epithelial cells and why high levels of Rho activity seem to have a destabilising effect on adherens junctions.

As a transmembrane protein, E-cadherin polarisation must depend on the membrane trafficking machinery. Delivery of E-cadherin to the plasma membrane appears to require the exocyst complex, as E-cadherin accumulates in intracellular vesicles in embryos, follicle cells, imaginal discs and pupal notum epithelia lacking exocyst components (Blankenship et al. 2007; Langevin et al. 2005; Wan et al. 2013; Guichard et al. 2010). However, cells lacking exocyst function still appear to have normal epithelial shape, indicating that functional adherens junctions are still present in these cells (Blankenship et al. 2007; Langevin et al. 2005; Wan et al. 2013; Guichard et al. 2010). Delivery of E-cadherin also appears to depend to some extent on Rab11 recycling endosomes in the embryo (Roeth et al. 2009). However, a complication in interpreting data on the role of membrane trafficking in polarising adherens junctions in the embryo is that effects on E-cadherin may be a consequence of loss of Crumbs, which is essential for maintaining polarity during embryo morphogenesis (when the Baz complex is planar polarised) (Blankenship et al. 2007; Roeth et al. 2009; Campbell et al. 2009; Zallen and Wieschaus 2004). Thus, trafficking of E-cadherin may be best studied in tissues where Crumbs and Baz act redundantly, so that effects on Crumbs do not disturb the overall epithelial polarity. For example, in the follicle cell epithelium, the retromer complex is required to recycle Crumbs and thus maintain Crumbs levels at the apical domain, but appears to be dispensable for adherens junction formation as epithelial cell shape is unaffected (Zhou et al. 2011). Thus, further work is necessary to understand the extent to which different membrane trafficking pathways are required to localise E-cadherin to adherens junctions (Fig. 14.4).

14.6 The Multiple Roles of Bazooka/Par3 in Establishment and Maintenance of Adherens Junctions

Bazooka/Par3 has several distinct roles in regulating adherens junctions. The first of these is due to its primary function in recruiting aPKC–Par6–Cdc42—in parallel with Crumbs—to maintain apicobasal polarity. Without aPKC–Par6–Cdc42 to maintain apicobasal polarity, adherens junctions cannot be maintained in their usual location. However, since Bazooka/Par3 acts redundantly with Crumbs, it is actually dispensable for maintaining adherens junctions in tissues where Crumbs is expressed, such as imaginal disc epithelia or follicle cell epithelia. Crumbs is not expressed during the very earliest establishment of epithelial polarity in the blastoderm embryo, and here Bazooka/Par3 has an essential role (presumably with aPKC–Par6–Cdc42) in defining apicobasal polarity and directing the localisation of adherens junctions.

The second role of Bazooka/Par3 appears to be independent of aPKC–Par6–Cdc42. A pool of Bazooka/Par3 that is phosphorylated by aPKC localises to adherens junctions rather than forming a complex with aPKC–Par6–Cdc42 at the apical domain. Note also that Bazooka/Par3 does not localise to the basolateral membrane due to phosphorylation by the basolateral Par1 kinase (Benton and St Johnston 2003). The direct function of Bazooka/Par3 at the adherens junction is still unclear, but it is suggestive that planar polarisation of junctional Bazooka/Par3 correlates with planar polarised junctional rearrangements during embryonic convergent extension movements (Zallen and Wieschaus 2004; Simoes Sde et al. 2010). In addition, the localisation of Bazooka/Par3 correlates with junctional rearrangements during embryonic ventral furrow invagination, and, in this case, Bazooka/Par3 is required for the junctional rearrangements involved (Wang et al. 2012). In Par1 mutant embryos, spreading of Bazooka/Par3 basolaterally leads to a basolateral re-localisation of adherens junctions (Wang et al. 2012). How Bazooka/Par3 is able to promote localisation of E-cadherin is not known, but one possibility is that it induces clustering of E-cadherin through its oligomerisation domain. However, it is interesting that this function of Bazooka/Par3 tends to occur during periods of cellular morphogenetic change, when the actin cytoskeleton is undergoing dynamic alterations (and thus may not be sufficient to sustain the normal localisation of adherens junctions) and E-cadherin is also altering its localisation.

This second role of Bazooka/Par3 may explain an important element of the *cdc42* mutant phenotype. Unlike in *aPKC* or *Par6* null mutants, which disperse adherens junctions, cells mutant for *cdc42* retain a single small patch of E-cadherin at the surface for some time before this patch is finally internalised as one giant endosome (Fletcher et al. 2012; Harris and Tepass 2008). This E-cadherin patch appears to form in *cdc42* mutants for the following reasons. Firstly, Cdc42 is required to activate aPKC kinase activity and to maintain Crumbs at the plasma membrane (Fletcher et al. 2012). Secondly, Cdc42 is required for Bazooka/Par3 to form a normal apical domain (note that the polarisation of Bazooka/Par3 does not depend on aPKC kinase activity, so this function must require other effectors of Cdc42) (Fletcher et al. 2012; Kim et al. 2009). Thirdly, in the absence of Cdc42 and a normal apical domain, Bazooka/Par3 can still bind directly to both adherens junctions and to aPKC–Par6, so that the residual patch of E-cadherin also co-localises with aPKC–Par6 (note that because the aPKC kinase is inactive in *cdc42* mutants, Bazooka/Par3 is not phosphorylated by aPKC) (Harris and Tepass 2008; Fletcher et al. 2012). Fourthly, the Bazooka/Par3–aPKC–Par6 complex maintains a single patch of E-cadherin at the membrane for a sustained period despite the complete absence of normal apicobasal polarity.

These findings suggest that Bazooka/Par3 has only a limited ability to cluster E-cadherin on its own (as in *aPKC* or *Par6* mutants), but that the complex of Bazooka/Par3 with aPKC–Par6 complex has an enhanced ability (as in *cdc42* mutants) (Fletcher et al. 2012; Harris and Tepass 2008). When Cdc42 is also present to form the Bazooka/Par3–aPKC–Par6–Cdc42 complex, fully normal apicobasal polarity emerges and active aPKC kinase then segregates Bazooka/Par3 into two pools, an apical Bazooka/Par3–aPKC–Par6–Cdc42 complex and a junctional, phosphorylated Bazooka/Par3 (Morais-de-Sa et al. 2010). Further work is needed to understand how phosphorylation of Bazooka/Par3 gives rise to these two distinct pools.

Notably, the principles governing how E-cadherin becomes polarised in cells may be similar in many respects to those governing polarisation of Crumbs. Both Crumbs and E-cadherin are delivered apically by Rab11 and the exocyst machinery, and both form apical–lateral cell–cell junctions with neighbouring cells via homotypic interactions in *cis* and *trans*. The two domains do not overlap, suggesting that the Crumbs complex helps prevent apical spreading of adherens junctions, just as basolateral determinants must help prevent basal spreading of adherens junctions. This mechanism appears to involve aPKC phosphorylation of Bazooka/Par3 to exclude it from the Crumbs domain and Par1 phosphorylation of Bazooka/Par3 to exclude it from the basolateral domain (Morais-de-Sa et al. 2010; Benton and St Johnston 2003) but must also involve other mechanisms, as Bazooka/Par3 is not strictly essential to maintain adherens junctions once they are formed. It is possible that the actomyosin ring that forms around adherens junctions may also be able to sustain these junctions in the absence of other inputs from the apicobasal polarity determinants, but further work is necessary to explore this notion in vivo.

14.7 Elaborating Apicobasal Polarity: Polarised Trafficking of Other Transmembrane Proteins and Matrix Proteins in *Drosophila* Epithelia

The apicobasal polarisation of all other transmembrane proteins, endosomal trafficking pathways and the cytoskeleton depends on the core apicobasal polarity determinants. Many of the other polarised transmembrane proteins mediate the polarised functions of epithelial cells. In general, relatively little is understood of how these effectors become localised to their respective positions within epithelia, but the process must involve polarised membrane trafficking in some way.

At the basal surface of certain *Drosophila* epithelia is a basement membrane extracellular matrix (ECM) that is composed of collagen IV, perlecan, laminins and nidogens. In some tissues, such as imaginal disc epithelia, the collagen IV and perlecan are secreted by another tissue (the fat body) and then bind to the exposed basal surface of the imaginal disc epithelia (Pastor-Pareja and Xu 2011). Once formed, the basement membrane provides a mechanical constraint that helps to shape the tissue (Pastor-Pareja and Xu 2011). In other tissues, such as the follicle cell epithelium that surrounds the ovary, the collagen IV is secreted by the follicle cells themselves but also provides a mechanical constraint to help shape the tissue (Haigo and Bilder 2011; Lerner et al. 2013). Trafficking of collagen IV, perlecan and laminin to the basal surface requires Rab10, which is normally found in basal endosomes (Lerner et al. 2013). In the absence of Rab10, basement membrane proteins begin to be trafficked apically (Lerner et al. 2013). Another molecule that is critical to prevent apical trafficking of basement membrane proteins is Crag, a cytoplasmic DENN domain protein that could potentially regulate Rab10 in some way (Denef et al. 2008; Lerner et al. 2013). The basement membrane is attached to the base of epithelial cells via transmembrane proteins of the integrin family (which binds to collagen IV) and dystroglycan (which binds to perlecan) (Haigo and Bilder 2011; Schneider et al. 2006; Deng et al. 2003; Mirouse et al. 2009). The basal localisation of these transmembrane proteins appears to depend on the basal secretion or localisation of their extracellular matrix ligands, but might also depend on basally directed secretion route.

At the apical surface of externally facing *Drosophila* epithelia is a secreted cuticle exoskeleton composed of chitin polymers. Just beneath this layer of cuticle lies the apical extracellular matrix (aECM), which is composed of the zona pellucida (ZP) domain proteins Piopio, Papillote and Dumpy. Dumpy is a very large transmembrane protein that appears to link epithelial cells to the fibres of the aECM and the overlying cuticle (Wilkin et al. 2000; Jazwinska et al. 2003; Bokel et al. 2005). The cuticle provides a rigid exoskeleton that protects the animal from the external environment, while the aECM is important for connecting epithelial cells to the cuticle. Very little is known about how these apically secreted proteins are trafficked to the correct location within epithelial cells, but presumably this occurs through an apical endosome compartment of some sort.

On the lateral sides of epithelial cells, just beneath the adherens junctions, lie the septate junctions. Septate junctions are composed of claudins, four transmembrane proteins that provide a tight seal between neighbouring epithelial cells known as the 'paracellular permeability barrier' to prevent diffusion of solutes through the epithelial layer. This epithelial barrier function only arises in certain epithelial tissues, but is critical for the blood–brain barrier, for tracheal tube size control and for the function of the intestinal epithelium (Yanagihashi et al. 2012; Izumi et al. 2012; Banerjee et al. 2008; Stork et al. 2008; Baumann 2001). The major claudins in *Drosophila* are Kune-kune, Megatrachea and Sinuous, while other transmembrane proteins involved in septate junctions include Neuroglian, Neurexin, Contactin, Lachesin, Na/K ATPase, Gliotactin, Fasciclin III, Wunen, Mesh and Snakeskin as well as the membrane-associated proteins Varicose, Coracle, Scrib and Dlg (Woods et al. 1997; Izumi et al. 2012; Yanagihashi et al. 2012; Ile et al. 2012; Banerjee et al. 2011; Nelson et al. 2010; Laval et al. 2008; Bachmann et al. 2008; Moyer and Jacobs 2008; Wu et al. 2004, 2007; Strigini et al. 2006; Faivre-Sarrailh et al. 2004; Llimargas et al. 2004; Behr et al. 2003; Paul et al. 2003;

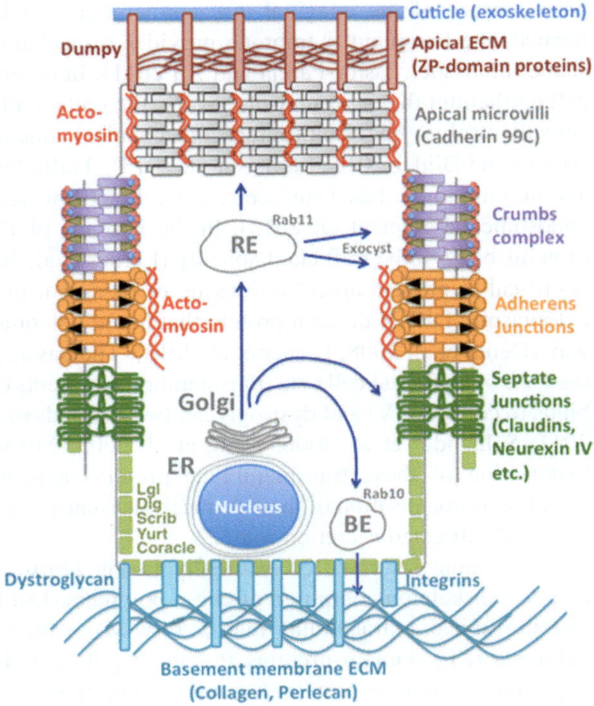

Fig. 14.5 Apicobasal polarity determinants can direct formation of apical microvilli, apical cuticle and ECM, septate junctions and basement membrane ECM in certain *Drosophila* cell types

Genova and Fehon 2003). A secreted protein, Boudin, is also involved (Hijazi et al. 2009). Importantly, these molecules can be expressed and localised differently in different tissue contexts, although much more remains to be learned about how and why septate junction components are used in each tissue type.

In summary, a wide range of transmembrane proteins and secreted proteins are polarised in various epithelial cells to enable epithelial cell morphogenesis and function (Fig. 14.5). Some proteins not discussed here are the transporter proteins that must polarise the transport of solutes across the epithelia in the intestine and certain other epithelia. Very little is known about the role of transporters in *Drosophila* physiology and metabolism. How such transporters might polarise is yet another outstanding question in *Drosophila* epithelial biology. Another set of proteins not discussed here are the planar polarised Frizzled–Flamingo–Strabismus planar polarity proteins and the Dachsous–Fat cadherin planar polarity proteins whose distributions are similar to adherens junctions except that they are additionally polarised in the plane of the epithelium. How this planar polarity comes about is also a major mystery that must involve regulated membrane trafficking in some way.

14.8 Conclusions

In summary, membrane trafficking has a profound role in apicobasal polarity in *Drosophila* epithelia. Firstly, the trafficking machinery polarises Crumbs, a key apical determinant. Part of the Crumbs polarisation process involves polarisation of the exocyst complex, which then delivers Crumbs to the apical domain in a positive feedback loop. The polarised exocyst complex then helps deliver adherens junctions to the boundary between the apical and basolateral determinants. Adherens junctions then organise the actin cytoskeleton and the overall shape of the epithelial cell and its interactions with neighbours. Whether the exocyst is the only mechanism localising Crumbs and adherens junctions in *Drosophila* remains unclear. Many other transmembrane proteins and secreted proteins are then localised apically, basally, laterally or in a planar polarised manner by a polarised membrane trafficking machinery whose nature is still very poorly understood. However, progress is being made here, and we can expect many insights into how different epithelial cells acquire their unique characteristics to emerge over the coming years.

Acknowledgements I thank Georgina Fletcher and Eliana Lucas for helping in preparing the figures. I also apologise to my colleagues working with *C. elegans*, or mammalian cells in culture, whose important work could not be cited here due to space limitations.

References

Atwood SX, Prehoda KE (2009) aPKC phosphorylates Miranda to polarize fate determinants during neuroblast asymmetric cell division. Curr Biol 19(9):723–729. doi:10.1016/j.cub.2009.03.056

Bachmann A, Draga M, Grawe F, Knust E (2008) On the role of the MAGUK proteins encoded by Drosophila varicose during embryonic and postembryonic development. BMC Dev Biol 8:55. doi:10.1186/1471-213X-8-55

Banerjee S, Bainton RJ, Mayer N, Beckstead R, Bhat MA (2008) Septate junctions are required for ommatidial integrity and blood-eye barrier function in Drosophila. Dev Biol 317(2):585–599. doi:10.1016/j.ydbio.2008.03.007

Banerjee S, Paik R, Mino RE, Blauth K, Fisher ES, Madden VJ, Fanning AS, Bhat MA (2011) A Laminin G-EGF-Laminin G module in Neurexin IV is essential for the apico-lateral localization of Contactin and organization of septate junctions. PLoS One 6(10):e25926. doi:10.1371/journal.pone.0025926

Baumann O (2001) Posterior midgut epithelial cells differ in their organization of the membrane skeleton from other Drosophila epithelia. Exp Cell Res 270(2):176–187. doi:10.1006/excr.2001.5343

Behr M, Riedel D, Schuh R (2003) The claudin-like megatrachea is essential in septate junctions for the epithelial barrier function in Drosophila. Dev Cell 5(4):611–620

Benton R, St Johnston D (2003) Drosophila PAR-1 and 14-3-3 inhibit Bazooka/PAR-3 to establish complementary cortical domains in polarized cells. Cell 115(6):691–704

Berdnik D, Torok T, Gonzalez-Gaitan M, Knoblich JA (2002) The endocytic protein alpha-Adaptin is required for numb-mediated asymmetric cell division in Drosophila. Dev Cell 3(2):221–231

Betschinger J, Mechtler K, Knoblich JA (2003) The Par complex directs asymmetric cell division by phosphorylating the cytoskeletal protein Lgl. Nature 422(6929):326–330. doi:10.1038/nature01486

Bilder D, Perrimon N (2000) Localization of apical epithelial determinants by the basolateral PDZ protein Scribble. Nature 403(6770):676–680. doi:10.1038/35001108

Bilder D, Li M, Perrimon N (2000) Cooperative regulation of cell polarity and growth by Drosophila tumor suppressors. Science 289(5476):113–116

Blankenship JT, Fuller MT, Zallen JA (2007) The Drosophila homolog of the Exo84 exocyst subunit promotes apical epithelial identity. J Cell Sci 120(Pt 17):3099–3110. doi:10.1242/jcs.004770

Bloor JW, Kiehart DP (2002) Drosophila RhoA regulates the cytoskeleton and cell-cell adhesion in the developing epidermis. Development 129(13):3173–3183

Bokel C, Prokop A, Brown NH (2005) Papillote and Piopio: Drosophila ZP-domain proteins required for cell adhesion to the apical extracellular matrix and microtubule organization. J Cell Sci 118(Pt 3):633–642. doi:10.1242/jcs.01619

Bryant PJ, Schmidt O (1990) The genetic control of cell proliferation in Drosophila imaginal discs. J Cell Sci Suppl 13:169–189

Campbell K, Knust E, Skaer H (2009) Crumbs stabilises epithelial polarity during tissue remodelling. J Cell Sci 122(Pt 15):2604–2612. doi:10.1242/jcs.047183

Chen CL, Gajewski KM, Hamaratoglu F, Bossuyt W, Sansores-Garcia L, Tao C, Halder G (2010) The apical-basal cell polarity determinant Crumbs regulates Hippo signaling in Drosophila. Proc Natl Acad Sci U S A 107(36):15810–15815. doi:10.1073/pnas.1004060107

Coumailleau F, Furthauer M, Knoblich JA, Gonzalez-Gaitan M (2009) Directional Delta and Notch trafficking in Sara endosomes during asymmetric cell division. Nature 458 (7241):1051–1055. doi:10.1038/nature07854

Denef N, Chen Y, Weeks SD, Barcelo G, Schupbach T (2008) Crag regulates epithelial architecture and polarized deposition of basement membrane proteins in Drosophila. Dev Cell 14 (3):354–364. doi:10.1016/j.devcel.2007.12.012

Deng WM, Schneider M, Frock R, Castillejo-Lopez C, Gaman EA, Baumgartner S, Ruohola-Baker H (2003) Dystroglycan is required for polarizing the epithelial cells and the oocyte in Drosophila. Development 130(1):173–184

Faivre-Sarrailh C, Banerjee S, Li J, Hortsch M, Laval M, Bhat MA (2004) Drosophila contactin, a homolog of vertebrate contactin, is required for septate junction organization and paracellular barrier function. Development 131(20):4931–4942. doi:10.1242/dev.01372

Fernandez BG, Gaspar P, Bras-Pereira C, Jezowska B, Rebelo SR, Janody F (2011) Actin-Capping Protein and the Hippo pathway regulate F-actin and tissue growth in Drosophila. Development 138(11):2337–2346. doi:10.1242/dev.063545

Fletcher GC, Lucas EP, Brain R, Tournier A, Thompson BJ (2012) Positive feedback and mutual antagonism combine to polarize Crumbs in the Drosophila follicle cell epithelium. Curr Biol 22(12):1116–1122. doi:10.1016/j.cub.2012.04.020

Fox DT, Homem CC, Myster SH, Wang F, Bain EE, Peifer M (2005) Rho1 regulates Drosophila adherens junctions independently of p120ctn. Development 132(21):4819–4831. doi:10.1242/dev.02056

Genova JL, Fehon RG (2003) Neuroglian, Gliotactin, and the Na^{+}/K^{+} ATPase are essential for septate junction function in Drosophila. J Cell Biol 161(5):979–989. doi:10.1083/jcb.200212054

Gomez GA, McLachlan RW, Yap AS (2011) Productive tension: force-sensing and homeostasis of cell-cell junctions. Trends Cell Biol 21(9):499–505. doi:10.1016/j.tcb.2011.05.006

Guichard A, McGillivray SM, Cruz-Moreno B, van Sorge NM, Nizet V, Bier E (2010) Anthrax toxins cooperatively inhibit endocytic recycling by the Rab11/Sec15 exocyst. Nature 467 (7317):854–858. doi:10.1038/nature09446

Hafezi Y, Bosch JA, Hariharan IK (2012) Differences in levels of the transmembrane protein Crumbs can influence cell survival at clonal boundaries. Dev Biol 368(2):358–369. doi:10.1016/j.ydbio.2012.06.001

Haigo SL, Bilder D (2011) Global tissue revolutions in a morphogenetic movement controlling elongation. Science 331(6020):1071–1074. doi:10.1126/science.1199424

Halbsgut N, Linnemannstons K, Zimmermann LI, Wodarz A (2011) Apical-basal polarity in Drosophila neuroblasts is independent of vesicular trafficking. Mol Biol Cell 22 (22):4373–4379. doi:10.1091/mbc.E11-03-0219

Harris KP, Tepass U (2008) Cdc42 and Par proteins stabilize dynamic adherens junctions in the Drosophila neuroectoderm through regulation of apical endocytosis. J Cell Biol 183 (6):1129–1143. doi:10.1083/jcb.200807020

Hijazi A, Masson W, Auge B, Waltzer L, Haenlin M, Roch F (2009) boudin is required for septate junction organisation in Drosophila and codes for a diffusible protein of the Ly6 superfamily. Development 136(13):2199–2209. doi:10.1242/dev.033845

Homem CC, Knoblich JA (2012) Drosophila neuroblasts: a model for stem cell biology. Development 139(23):4297–4310. doi:10.1242/dev.080515

Ile KE, Tripathy R, Goldfinger V, Renault AD (2012) Wunen, a Drosophila lipid phosphate phosphatase, is required for septate junction-mediated barrier function. Development 139 (14):2535–2546. doi:10.1242/dev.077289

Izumi Y, Yanagihashi Y, Furuse M (2012) A novel protein complex, Mesh-Ssk, is required for septate junction formation in the Drosophila midgut. J Cell Sci 125(Pt 20):4923–4933. doi:10.1242/jcs.112243

Jazwinska A, Ribeiro C, Affolter M (2003) Epithelial tube morphogenesis during Drosophila tracheal development requires Piopio, a luminal ZP protein. Nat Cell Biol 5(10):895–901. doi:10.1038/ncb1049

Kim S, Gailite I, Moussian B, Luschnig S, Goette M, Fricke K, Honemann-Capito M, Grubmuller H, Wodarz A (2009) Kinase-activity-independent functions of atypical protein kinase C in Drosophila. J Cell Sci 122(Pt 20):3759–3771. doi:10.1242/jcs.052514

Langevin J, Morgan MJ, Sibarita JB, Aresta S, Murthy M, Schwarz T, Camonis J, Bellaiche Y (2005) Drosophila exocyst components Sec5, Sec6, and Sec15 regulate DE-Cadherin trafficking from recycling endosomes to the plasma membrane. Dev Cell 9(3):365–376

Laprise P, Beronja S, Silva-Gagliardi NF, Pellikka M, Jensen AM, McGlade CJ, Tepass U (2006) The FERM protein Yurt is a negative regulatory component of the Crumbs complex that controls epithelial polarity and apical membrane size. Dev Cell 11(3):363–374. doi:10.1016/j.devcel.2006.06.001

Laprise P, Lau KM, Harris KP, Silva-Gagliardi NF, Paul SM, Beronja S, Beitel GJ, McGlade CJ, Tepass U (2009) Yurt, Coracle, Neurexin IV and the Na(+), K(+)-ATPase form a novel group of epithelial polarity proteins. Nature 459(7250):1141–1145. doi:10.1038/nature08067

Laval M, Bel C, Faivre-Sarrailh C (2008) The lateral mobility of cell adhesion molecules is highly restricted at septate junctions in Drosophila. BMC Cell Biol 9:38. doi:10.1186/1471-2121-9-38

Lerner DW, McCoy D, Isabella AJ, Mahowald AP, Gerlach GF, Chaudhry TA, Horne-Badovinac S (2013) A Rab10-dependent mechanism for polarized basement membrane secretion during organ morphogenesis. Dev Cell 24(2):159–168. doi:10.1016/j.devcel.2012.12.005

Letizia A, Sotillos S, Campuzano S, Llimargas M (2011) Regulated Crb accumulation controls apical constriction and invagination in Drosophila tracheal cells. J Cell Sci 124(Pt 2):240–251. doi:10.1242/jcs.073601

Letizia A, Ricardo S, Moussian B, Martin N, Llimargas M (2013) A functional role of the extracellular domain of Crumbs in cell architecture and apicobasal polarity. J Cell Sci 126 (Pt 10):2157–2163. doi:10.1242/jcs.122382

Levayer R, Pelissier-Monier A, Lecuit T (2011) Spatial regulation of Dia and Myosin-II by RhoGEF2 controls initiation of E-cadherin endocytosis during epithelial morphogenesis. Nat Cell Biol 13(5):529–540. doi:10.1038/ncb2224

Ling C, Zheng Y, Yin F, Yu J, Huang J, Hong Y, Wu S, Pan D (2010) The apical transmembrane protein Crumbs functions as a tumor suppressor that regulates Hippo signaling by binding to Expanded. Proc Natl Acad Sci U S A 107(23):10532–10537. doi:10.1073/pnas.1004279107

Llimargas M, Strigini M, Katidou M, Karagogeos D, Casanova J (2004) Lachesin is a component of a septate junction-based mechanism that controls tube size and epithelial integrity in the Drosophila tracheal system. Development 131(1):181–190. doi:10.1242/dev.00917

Lu H, Bilder D (2005) Endocytic control of epithelial polarity and proliferation in Drosophila. Nat Cell Biol 7(12):1232–1239. doi:10.1038/ncb1324

Lucas EP, Khanal I, Gaspar P, Fletcher GC, Polesello C, Tapon N, Thompson BJ (2013) The Hippo pathway polarizes the actin cytoskeleton during collective migration of Drosophila border cells. J Cell Biol 201(6):875–885. doi:10.1083/jcb.201210073

Magie CR, Meyer MR, Gorsuch MS, Parkhurst SM (1999) Mutations in the Rho1 small GTPase disrupt morphogenesis and segmentation during early Drosophila development. Development 126(23):5353–5364

Magie CR, Pinto-Santini D, Parkhurst SM (2002) Rho1 interacts with p120ctn and alpha-catenin, and regulates cadherin-based adherens junction components in Drosophila. Development 129 (16):3771–3782

Mason FM, Tworoger M, Martin AC (2013) Apical domain polarization localizes actin-myosin activity to drive ratchet-like apical constriction. Nat Cell Biol 15(8):926–936. doi:10.1038/ncb2796

Meng W, Takeichi M (2009) Adherens junction: molecular architecture and regulation. Cold Spring Harb Perspect Biol 1(6):a002899. doi:10.1101/cshperspect.a002899

Menut L, Vaccari T, Dionne H, Hill J, Wu G, Bilder D (2007) A mosaic genetic screen for Drosophila neoplastic tumor suppressor genes based on defective pupation. Genetics 177 (3):1667–1677. doi:10.1534/genetics.107.078360

Mirouse V, Christoforou CP, Fritsch C, St Johnston D, Ray RP (2009) Dystroglycan and perlecan provide a basal cue required for epithelial polarity during energetic stress. Dev Cell 16 (1):83–92. doi:10.1016/j.devcel.2008.11.006

Morais-de-Sa E, Mirouse V, St Johnston D (2010) aPKC phosphorylation of Bazooka defines the apical/lateral border in Drosophila epithelial cells. Cell 141(3):509–523. doi:10.1016/j.cell.2010.02.040

Moyer KE, Jacobs JR (2008) Varicose: a MAGUK required for the maturation and function of Drosophila septate junctions. BMC Dev Biol 8:99. doi:10.1186/1471-213X-8-99

Nelson KS, Furuse M, Beitel GJ (2010) The Drosophila Claudin Kune-kune is required for septate junction organization and tracheal tube size control. Genetics 185(3):831–839. doi:10.1534/genetics.110.114959

Pastor-Pareja JC, Xu T (2011) Shaping cells and organs in Drosophila by opposing roles of fat body-secreted Collagen IV and perlecan. Dev Cell 21(2):245–256. doi:10.1016/j.devcel.2011.06.026

Paul SM, Ternet M, Salvaterra PM, Beitel GJ (2003) The Na^{+}/K^{+} ATPase is required for septate junction function and epithelial tube-size control in the Drosophila tracheal system. Development 130(20):4963–4974. doi:10.1242/dev.00691

Peifer M (1995) Cell adhesion and signal transduction: the Armadillo connection. Trends Cell Biol 5(6):224–229

Pellikka M, Tanentzapf G, Pinto M, Smith C, McGlade CJ, Ready DF, Tepass U (2002) Crumbs, the Drosophila homologue of human CRB1/RP12, is essential for photoreceptor morphogenesis. Nature 416(6877):143–149. doi:10.1038/nature721

Pocha SM, Wassmer T, Niehage C, Hoflack B, Knust E (2011) Retromer controls epithelial cell polarity by trafficking the apical determinant Crumbs. Curr Biol 21(13):1111–1117. doi:10.1016/j.cub.2011.05.007

Robinson BS, Huang J, Hong Y, Moberg KH (2010) Crumbs regulates Salvador/Warts/Hippo signaling in Drosophila via the FERM-domain protein Expanded. Curr Biol 20(7):582–590. doi:10.1016/j.cub.2010.03.019

Roeth JF, Sawyer JK, Wilner DA, Peifer M (2009) Rab11 helps maintain apical crumbs and adherens junctions in the Drosophila embryonic ectoderm. PLoS One 4(10):e7634. doi:10.1371/journal.pone.0007634

Roper K (2012) Anisotropy of Crumbs and aPKC drives myosin cable assembly during tube formation. Dev Cell 23(5):939–953. doi:10.1016/j.devcel.2012.09.013

Sarpal R, Pellikka M, Patel RR, Hui FY, Godt D, Tepass U (2012) Mutational analysis supports a core role for Drosophila alpha-catenin in adherens junction function. J Cell Sci 125 (Pt 1):233–245. doi:10.1242/jcs.096644

Schneider M, Khalil AA, Poulton J, Castillejo-Lopez C, Egger-Adam D, Wodarz A, Deng WM, Baumgartner S (2006) Perlecan and Dystroglycan act at the basal side of the Drosophila follicular epithelium to maintain epithelial organization. Development 133(19):3805–3815. doi:10.1242/dev.02549

Simoes Sde M, Blankenship JT, Weitz O, Farrell DL, Tamada M, Fernandez-Gonzalez R, Zallen JA (2010) Rho-kinase directs Bazooka/Par-3 planar polarity during Drosophila axis elongation. Dev Cell 19(3):377–388. doi:10.1016/j.devcel.2010.08.011

Stork T, Engelen D, Krudewig A, Silies M, Bainton RJ, Klambt C (2008) Organization and function of the blood-brain barrier in Drosophila. J Neurosci 28(3):587–597. doi:10.1523/JNEUROSCI. 4367-07.2008

Strigini M, Cantera R, Morin X, Bastiani MJ, Bate M, Karagogeos D (2006) The IgLON protein Lachesin is required for the blood-brain barrier in Drosophila. Mol Cell Neurosci 32 (1–2):91–101. doi:10.1016/j.mcn.2006.03.001

Tanentzapf G, Tepass U (2003) Interactions between the crumbs, lethal giant larvae and bazooka pathways in epithelial polarization. Nat Cell Biol 5(1):46–52. doi:10.1038/ncb896

Tepass U, Gruszynski-DeFeo E, Haag TA, Omatyar L, Torok T, Hartenstein V (1996) shotgun encodes Drosophila E-cadherin and is preferentially required during cell rearrangement in the neurectoderm and other morphogenetically active epithelia. Genes Dev 10(6):672–685

Thompson BJ, Mathieu J, Sung HH, Loeser E, Rorth P, Cohen SM (2005) Tumor suppressor properties of the ESCRT-II complex component Vps25 in Drosophila. Dev Cell 9(5):711–720. doi:10.1016/j.devcel.2005.09.020

Thompson BJ, Pichaud F, Roper K (2013) Sticking together the Crumbs—an unexpected function for an old friend. Nat Rev Mol Cell Biol 14(5):307–314. doi:10.1038/nrm3568

Vaccari T, Bilder D (2005) The Drosophila tumor suppressor vps25 prevents nonautonomous overproliferation by regulating notch trafficking. Dev Cell 9(5):687–698. doi:10.1016/j.devcel.2005.09.019

Vichas A, Zallen JA (2011) Translating cell polarity into tissue elongation. Semin Cell Dev Biol 22(8):858–864. doi:10.1016/j.semcdb.2011.09.013

Walther RF, Pichaud F (2010) Crumbs/DaPKC-dependent apical exclusion of Bazooka promotes photoreceptor polarity remodeling. Curr Biol 20(12):1065–1074. doi:10.1016/j.cub.2010.04.049

Wan P, Wang D, Luo J, Chu D, Wang H, Zhang L, Chen J (2013) Guidance receptor promotes the asymmetric distribution of exocyst and recycling endosome during collective cell migration. Development 140(23):4797–4806. doi:10.1242/dev.094979

Wang YC, Khan Z, Kaschube M, Wieschaus EF (2012) Differential positioning of adherens junctions is associated with initiation of epithelial folding. Nature 484(7394):390–393. doi:10.1038/nature10938

Warner SJ, Longmore GD (2009) Distinct functions for Rho1 in maintaining adherens junctions and apical tension in remodeling epithelia. J Cell Biol 185(6):1111–1125. doi:10.1083/jcb.200901029

Wilkin MB, Becker MN, Mulvey D, Phan I, Chao A, Cooper K, Chung HJ, Campbell ID, Baron M, MacIntyre R (2000) Drosophila dumpy is a gigantic extracellular protein required to maintain tension at epidermal-cuticle attachment sites. Curr Biol 10(10):559–567

Wirtz-Peitz F, Nishimura T, Knoblich JA (2008) Linking cell cycle to asymmetric division: Aurora-A phosphorylates the Par complex to regulate Numb localization. Cell 135 (1):161–173. doi:10.1016/j.cell.2008.07.049

Wodarz A, Gonzalez C (2006) Connecting cancer to the asymmetric division of stem cells. Cell 124(6):1121–1123. doi:10.1016/j.cell.2006.03.004

Wodarz A, Hinz U, Engelbert M, Knust E (1995) Expression of crumbs confers apical character on plasma membrane domains of ectodermal epithelia of Drosophila. Cell 82(1):67–76

Woods DF, Bryant PJ (1991) The discs-large tumor suppressor gene of Drosophila encodes a guanylate kinase homolog localized at septate junctions. Cell 66(3):451–464

Woods DF, Hough C, Peel D, Callaini G, Bryant PJ (1996) Dlg protein is required for junction structure, cell polarity, and proliferation control in Drosophila epithelia. J Cell Biol 134 (6):1469–1482

Woods DF, Wu JW, Bryant PJ (1997) Localization of proteins to the apico-lateral junctions of Drosophila epithelia. Dev Genet 20(2):111–118. doi:10.1002/(SICI)1520-6408(1997)20:2<111::AID-DVG4>3.0.CO;2-A

Wu VM, Schulte J, Hirschi A, Tepass U, Beitel GJ (2004) Sinuous is a Drosophila claudin required for septate junction organization and epithelial tube size control. J Cell Biol 164(2):313–323. doi:10.1083/jcb.200309134

Wu VM, Yu MH, Paik R, Banerjee S, Liang Z, Paul SM, Bhat MA, Beitel GJ (2007) Drosophila Varicose, a member of a new subgroup of basolateral MAGUKs, is required for septate junctions and tracheal morphogenesis. Development 134(5):999–1009. doi:10.1242/dev.02785

Yan J, Lu Q, Fang X, Adler PN (2009) Rho1 has multiple functions in Drosophila wing planar polarity. Dev Biol 333(1):186–199. doi:10.1016/j.ydbio.2009.06.027

Yanagihashi Y, Usui T, Izumi Y, Yonemura S, Sumida M, Tsukita S, Uemura T, Furuse M (2012) Snakeskin, a membrane protein associated with smooth septate junctions, is required for intestinal barrier function in Drosophila. J Cell Sci 125(Pt 8):1980–1990. doi:10.1242/jcs.096800

Zallen JA, Wieschaus E (2004) Patterned gene expression directs bipolar planar polarity in Drosophila. Dev Cell 6(3):343–355

Zhou B, Wu Y, Lin X (2011) Retromer regulates apical-basal polarity through recycling Crumbs. Dev Biol 360(1):87–95. doi:10.1016/j.ydbio.2011.09.009

Chapter 15
Roles of Rab Family Small G Proteins in Formation of the Apical Junctional Complex in Epithelial Cells

Ayuko Sakane and Takuya Sasaki

Abstract Epithelial tissue formation and function require organization of monolayer sheets, which is mediated through the apical junctional complex (AJC). The AJC comprises a diverse set of key factors: adhesion molecules including E-cadherin, claudins, and occludin; polarity proteins that support organization and function of apical and basolateral membranes; and proteins that support adhesion, vesicle transport, actin cytoskeletal rearrangement, and membrane scaffolds. These proteins form a dynamic cooperative network that is engaged in the highly elaborate regulation of AJC. Several lines of evidence indicate that Rab family small G proteins play important roles in the regulation of epithelial apical junctions and that the assembly and disassembly of these junctions can be driven by Rab proteins localized at either endosomes or apical junctions. In this review, we provide an overview of the influence of Rab proteins on AJC functions, focusing especially on the role of the complex containing Rab13 and JRAB/MICAL-L2 (junctional Rab13-binding protein/molecule interacting with CasL-like 2) in the regulation of epithelial apical junctions.

Keywords Actin cytoskeletal reorganization • Apical junctional complex • Membrane traffic • Rab family small G proteins • Rab13-JRAB/MICAL-L2 complex

15.1 Introduction

The principal property of epithelial cells is their ability to form monolayer sheets that can be molded into different shapes. These sheets function as selective and dynamic barriers between compartments within the body, as well as between the environment and the interior of the body. The ability of epithelial cells to organize into monolayer sheets is based on cell–cell adhesion, which is mediated through the apical junctional complex (AJC). Tight junctions (TJs) and adherens junctions

A. Sakane • T. Sasaki (✉)
Department of Biochemistry, Institute of Health Biosciences, The University of Tokushima Graduate School, 3-18-15 Kuramoto-cho, Tokushima 770-8503, Japan
e-mail: sasaki@basic.med.tokushima-u.ac.jp

K. Ebnet (ed.), *Cell Polarity 1*, DOI 10.1007/978-3-319-14463-4_15

(AJs) constitute the AJCs by which neighboring epithelial cells adhere to each other and determine tissue organization. In most epithelial cells, an AJC segregates the apical membrane from the basolateral membrane. This complex contains adhesive elements that bind cells into epithelial sheets, as well as a junction that seals the intercellular space to regulate diffusion of ions or molecules across these sheets. Together with the barrier function, bidirectional active transport allows epithelia to regulate compartment composition and organ function. The AJC is a highly dynamic structure that undergoes continuous remodeling during physiological morphogenesis, as well as under pathological conditions. The assembly and disassembly of the AJC require the dynamic integration of several cellular machineries: (1) vesicle transport of apical junctional proteins that deliver cargo to specific membrane domains; (2) plasma membrane domain-specific scaffolds of polarity proteins that support organization and function of apical and basolateral membranes; and (3) remodeling of the actin cytoskeleton and the associated machinery that support adhesion, vesicle transport, and membrane scaffolds (Bryant and Mostov 2008; Li and Gundersen 2008; Mellman and Nelson 2008). Significant progress has been made in the identification of components of different modules within the AJC network and in understanding the function of individual cellular machineries. However, the mechanisms by which these machineries cooperate in AJC formation remain incompletely understood.

Several lines of evidence indicate that Rab family small G proteins (Rab proteins) are crucially involved in regulation of the AJC. Rab proteins localized at either endosomes or apical junctions can influence the assembly and disassembly of epithelial apical junctions. In this review, following an overview of Rab proteins, we describe how Rab proteins influence epithelial apical junctions. We then discuss our emerging understanding of the role of the Rab13–JRAB/MICAL-L2 complex in regulating cooperation between cytoskeletal organization and vesicle transport during AJC formation.

15.2 Rab Proteins

Rab proteins were first identified in the 1980s as evolutionarily conserved, essential regulators of membrane traffic (Stenmark 2009; Hutagalung and Novick 2011; Mizuno-Yamasaki et al. 2012). They are members of the larger Ras superfamily of small G proteins (Takai et al. 2001); to date, omitting "putative" and Rab-like proteins, more than 60 Rab proteins have been identified in human (Colicelli 2004). Cross-species mammalian sequence analyses have revealed five Rab family (RabF) regions within the sequence that distinguish Rab proteins from other Ras family members and four Rab subfamily-specific (RabSF) regions that are highly conserved within each Rab subfamily (Pereira-Leal and Seabra 2000). Phylogenetic analyses based on RabF and RabSF regions allow classification of the Rab proteins into eleven subfamilies and eight functional groups (Pereira-Leal and Seabra 2000, 2001).

Rab proteins cycle between the GDP-bound inactive and GTP-bound active states. This cycling is regulated by three types of regulators: GDP dissociation inhibitors (GDIs), guanine nucleotide exchange proteins (GEPs), and GTPase-activating proteins (GAPs) (Stenmark 2009; Hutagalung and Novick 2011; Mizuno-Yamasaki et al. 2012). Rab proteins also undergo a cycle of membrane association and dissociation that is coupled to the GTP/GDP cycle. Most Rab proteins are posttranslationally modified with geranylgeranyl moieties that enable their association with membranes. GDIs bind to geranylgeranylated Rab proteins in their GDP-bound form, masking their geranylgeranyl anchor and thereby extracting the Rab proteins from the membranes to maintain them in the cytosol. Therefore, membrane attachment of Rab proteins requires GDI displacement factors (GDFs), which dissociate Rab proteins from Rab–GDI complexes. Once dissociated from GDIs, Rab proteins are converted to their GTP-bound state by specific GEPs. Active, membrane-bound Rab proteins are then able to fulfill a wide range of functions in membrane traffic by binding to their specific effectors. After inactivation by specific GAPs, the GDP-bound Rab proteins can be extracted from the membrane by GDIs and recycled back into the cytosol.

Each Rab effector protein responds to a specific Rab protein and mediates at least one element of its downstream effects (Stenmark 2009; Hutagalung and Novick 2011; Mizuno-Yamasaki et al. 2012). The rapidly growing list of Rab effector proteins has revealed that each Rab protein appears to signal through multiple effector proteins, which act together to translate the signal from the Rab protein to multiple cellular processes. Rab proteins contribute to specificity in membrane traffic and the establishment of specific membrane domains called "Rab domains" (Pfeffer 2003). They also play important regulatory roles in membrane–cytoskeleton interactions by associating with molecular motors and other cytoskeleton-binding proteins (Hammer and Wu 2002; Seabra and Coudrier 2004). In addition, they participate in the regulation of numerous signal transduction pathways (Stenmark 2009; Jean and Kiger 2012).

15.3 Rab Proteins in Membrane Traffic

In the context of membrane traffic, Rab proteins can control cargo collection during transport vesicle formation, enable motor proteins to interact with membranes to drive vesicle motility, and mediate the complicated series of events involved in accurate tethering and fusion of transport vesicles with their target membranes (Stenmark 2009; Hutagalung and Novick 2011). Rab9 is an example of a Rab protein involved in cargo collection during transport vesicle formation. TIP47, a Rab9 effector protein, binds to GTP-bound Rab9 and increases its affinity for mannose 6-phosphate receptor (M6PR), facilitating the capture of M6PR in Rab9-positive transport carrier vesicles (Carroll et al. 2001). In the context of attachment of motors to vesicles, several Rab proteins contribute to interactions along actin filaments (F-actin) and microtubules; for example, GTP-bound Rab6 binds to the microtubule motor Rabkinesin-6 and promotes the delivery of vesicles from the Golgi to endoplasmic reticulum (Echard et al. 1998). Several factors that tether vesicles to target membranes prior to fusion are Rab effectors. p115, a long coiled-coil protein that tethers endoplasmic reticulum-derived vesicles to the Golgi, has been identified as a Rab1 effector protein (Allan et al. 2000). Such Rab effector-mediated vesicle tethering has also been implicated in the subsequent membrane fusion step. GTP-bound Rab5 recruits another long coiled-coil tethering factor, EEA1, onto early endosomes, followed by the interaction between EEA1 and SNARE (soluble *N*-ethylmaleimide-sensitive fusion protein attachment protein receptor) protein, syntaxin 13, which is required for homotypic early endosome fusion (McBride et al. 1999).

A vital aspect of Rab function is the specific localization of each Rab protein to a particular subcellular membrane compartment and its involvement in a specific transport step. Each compartment has a unique set of Rab proteins, which are frequently used as markers of the particular organelle (Seabra and Wasmeier 2004). For example, Rab1 is on the endoplasmic reticulum, Rab6 is on the Golgi, Rab3 is on secretory granules and synaptic vesicles, Rab5 is associated with early endosomes, Rab11 is a marker for recycling endosomes, and Rab7 and Rab9 are primarily associated with the late endosomal compartment. The correct targeting of Rab proteins to their specific membranes was originally thought to depend on the unique, hypervariable C-terminal domains of Rab proteins, which serve as signal sequences. However, several lines of evidence suggest that multiple factors including GDIs, GDFs, GEPs, GAPs, and effector proteins contribute to the specificity of Rab localization and the close coordination of membrane targeting with Rab activation. Rab proteins also contribute to the establishment of specific membrane domains called "Rab domains," which have been well characterized in the endocytic pathway (Zerial and McBride 2001; Pfeffer 2003; Pfeffer 2013). Early endosomes harbor only Rab5 or a combination of Rab4 and Rab5, whereas recycling endosomes carry distinct Rab4 and Rab11 domains, and late endosomes contain distinct Rab7 and Rab9 domains. The Rab5 GEP Rabex-5 activates Rab5 on early endosomes, and activated Rab5 interacts with Rab5 effector Rabaptin-5, which in turn binds to Rabex-5 and increases the exchange activity of Rabex-5 with respect to Rab5 (Horiuchi et al. 1997). This Rabex-5–Rabaptin-5 complex serves as a positive-feedback loop that establishes Rab5 domains on early endosomes. Importantly, Rab5-containing early endosomes can be converted into Rab7-containing late endosomes. During this process, the Mon1/Ccz-1 complex, a Rab7 GEP, plays a key role (Kinchen and Ravichandran 2010). The Mon1 component of this complex is recruited to early endosomes by interaction with phosphatidylinositol 3-phosphate, which is generated by Rab5 effector, phosphatidylinositol 3-OH kinase (Poteryaev et al. 2010). Mon1 also displaces Rabex-5 from early endosomes. These mechanisms contribute to Rab5-to-Rab7 conversion to discriminate early and late endosomes. In addition, there are several mechanisms for achieving specificity in membrane domain and membrane traffic by coupling a downstream Rab GEP or Rab GAP with an upstream Rab effector, e.g., Rab9–

Hps1/Hps4 (Rab32 GEP)–Rab32 and Rab9-RUTBC1 (Rab32 GAP)–Rab32 in late endosome-lysosome-related organelle (Nottingham et al. 2011).

15.4 Apical Junctional Complex

In epithelial cells, cell–cell adhesion is mediated through AJC. The organization of the AJC is defined by TJs, which seal the intercellular space and delineate the boundaries between the apical and basolateral membranes, and AJs, which initiate and maintain cell–cell contacts (Farquhar and Palade 1963). Both TJs and AJs are constructed according to the same architectural principles as other adhesion complexes. In these structures, cell–cell adhesion is mediated by sets of different transmembrane proteins, which are linked to cytosolic plaque proteins that anchor the junction to the cytoskeleton. At TJs, claudins function as the principal transmembrane proteins that form the paracellular diffusion barrier (Tsukita et al. 2001; Furuse 2010). Other transmembrane proteins identified at TJs include occludin, tricellulin, junction adhesion molecules (JAMs), coxsackievirus and adenovirus receptor (CAR), and Crumb3 (CRB3). Occludin is the first identified transmembrane protein at TJs, whose physiological function remains to be established (Furuse et al. 1993). Tricellulin was recently identified as another TJ component specifically localized to tricellular junctions (Ikenouchi et al. 2005). JAMs are IgG-superfamily cell-adhesion molecules that consist of at least five JAM family members. The first JAM to be identified, JAM-A, is involved in the accumulation of a cell polarity protein complex, the Par3/Par6/atypical protein kinase C (aPKC) complex, at TJs (Ebnet et al. 2004; Monteiro and Parkos 2012). CAR associates with JAM-C and mediates attachment and infection by group B coxsackieviruses (CVB) and adenoviruses (Coyne and Bergelson 2005). CRB3 is a member of another cell-polarity protein complex, the CRB3/PALS1/PATJ complex (Shin et al. 2006; Assemat et al. 2008). At AJs, the transmembrane protein E-cadherin is a key cell adhesion molecule. E-cadherin is a member of the cadherin superfamily, which comprises over 100 members, each of which is expressed in non-epithelial cells as well as in epithelial cells (Takeichi 2007; Hulpiau and van Roy 2009; Oda and Takeichi 2011). Nectins, another set of IgG-superfamily cell adhesion molecules, are transmembrane proteins involved in the organization of AJs, either in cooperation with or independently of E-cadherin (Takai et al. 2008a, b). These transmembrane proteins are clustered by AJC plaques, which are in turn linked to the actin cytoskeleton. The AJC plaque is an organizing platform for a variety of scaffolding, signaling, and membrane traffic proteins, including zonula occludens (ZO) proteins (ZO-1, ZO-2, and ZO-3), membrane-associated guanylate kinase inverted (MAGI) proteins (MAGI-1, MAGI-2, and MAGI-3), catenins, Par3/Par6/aPKC, CRB3/PALS1/PATJ complexes, Rab3B, Rab8, and Rab13 (Harris and Tepass 2010; Shin et al. 2006).

15.5 Endocytic Recycling Pathways in Epithelial Cells

Endocytosis regulates the entry of small and large extracellular molecules into cells. It is a complex and multistep process, involving the invagination/budding of the plasma membrane and the formation of membrane vesicles, followed by their delivery and fusion with specific intracellular compartments (Conner and Schmid 2003). Endocytic vesicles initially fuse with a juxtamembrane cytosolic compartment, the early endosome which consists of two spatially separated populations of apical and basolateral early endosome (AEEs and BEEs) in polarized epithelial cells (Fig. 15.1). Some internalized proteins in AEEs or BEEs may return directly to the apical or basolateral membranes or be delivered to multivesicular bodies (MVBs), where they are targeted for degradation in lysosomes. Other proteins in AEEs and BEEs are eventually delivered to common recycling endosomes (CREs), which serve as a sorting station that determines the fate of internalized proteins

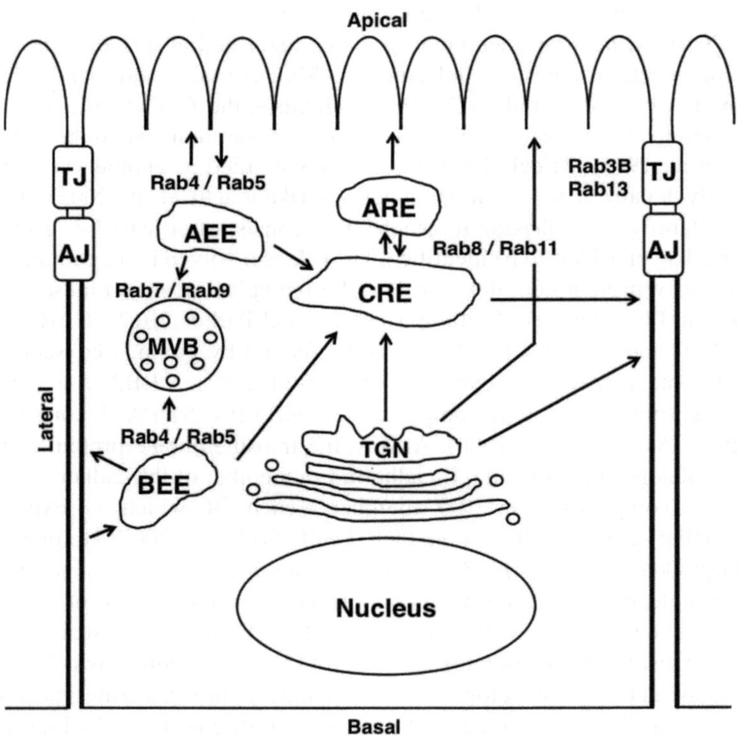

Fig. 15.1 Endocytic recycling pathways in epithelial cells. Multiple Rab proteins are involved in regulation of endocytic recycling in epithelial cells. Some biosynthetic proteins are transported from the trans-Golgi network through common recycling endosomes to apical or basolateral domains. *AEE* apical early endosome, *BEE* basolateral early endosome, *CRE* common recycling endosome, *ARE* apical recycling endosome, *TGN* trans-Golgi network, *MVB* multivesicular body, *TJ* tight junctions, *AJ* adherens junctions

(Hoekstra et al. 2004; Apodaca et al. 2012). Subsequently, proteins may enter apical recycling endosomes (AREs) to return to the apical membrane or instead be delivered directly to the basolateral membrane. It remains controversial whether CREs and AREs are separate entities or merely subdomains of the same structure.

Endocytic recycling pathways are strictly regulated processes requiring elaborate vesicle targeting and fusion machinery. Two critical components of this machinery are the Rab and SNARE proteins. The former are responsible for recognition and initial docking of vesicles with their target compartments, whereas the latter mediate intermembrane fusion (Jahn and Scheller 2006). In humans, there are more than 30 SNARE proteins; these proteins are characterized by the SNARE motif, an evolutionarily conserved stretch of 60–70 amino acids arranged in heptad repeats. Functionally, SNAREs can be classified as v-SNAREs (vesicle SNAREs) and t-SNAREs (target SNAREs). Each endosomal vesicle carries a specific v-SNARE that interacts with a cognate t-SNARE on its target membrane; specific interaction of v-SNARE with the t-SNARE forms a SNARE complex that drives membrane fusion. In polarized epithelial cells, two major t-SNARE proteins, syntaxin 3 and syntaxin 4, appear to be spatially segregated into different plasma membrane domains, with syntaxin 3 confined to the apical surface and syntaxin 4 confined to the basolateral plasma membrane (Carmosino et al. 2010). Apical targeting requires the tetanus neurotoxin (TeNT)-resistant v-SNARE TI-VAMP (VAMP7), whereas basolateral targeting involves the TeNT-sensitive v-SNARE cellubrevin (VAMP3).

Before the SNARE-dependent fusion reaction, endosomal recycling vesicles need to be tethered to the plasma membrane. Rab proteins and a large octameric complex called the exocyst play central roles in the tethering of vesicles to the plasma membrane (Yu and Hughson 2010). The exocyst is composed of Sec3, Sec5, Sec6, Sec8, Sec10, Sec15, Exo70, and Exo84, and it is associated with apical junctions and recycling endosomes in polarized epithelial cells (Grindstaff et al. 1998; Oztan et al. 2007). It interacts with several known regulators of epithelial apical junctions, such as Ral, ARF6, and Rab11, and functions in endocytic recycling as well as basolateral membrane transport (Oztan et al. 2007; He and Guo 2009; Hazelett et al. 2011).

ARF family small G proteins are also crucial regulators of endocytic recycling (D'Souza-Schorey and Chavrier 2006; Donaldson and Jackson 2011). In mammalian cells, the ARF protein family has six members, each of which localizes to distinct subsets of intracellular membranes. ARF6 is a key regulator of the endocytic recycling of cell adhesion molecules (Schweitzer et al. 2011). ARF6 activation results in the internalization of E-cadherin from AJC into early endosomes, whereas interference with ARF6 activation inhibits this internalization and prevents cell–cell dissociation (Palacios et al. 2001; D'Souza-Schorey 2005). Furthermore, ARF6 can interact with the same effector proteins as Rab11 (FIP3 and FIP4; see Sect. 7.5, below), which also regulate an endocytic recycling pathway (Shin et al. 2001; Hickson et al. 2003).

15.6 Remodeling of Apical Junctions

Epithelial cells are very plastic, and they remodel intercellular junctions even within apparently stable, confluent cultured monolayers. To assemble or disassemble epithelial apical junctions, epithelial cells need to regulate the functions of apical junctional proteins at the cell surface. Remodeling of the actin cytoskeleton and endocytic recycling of apical junctional proteins serve important physiological functions. Failure of this regulation is manifested in a variety of diseases, such as tissue fibrosis and tumor invasion/metastasis (Delva and Kowalczyk 2009).

The significance of the actin cytoskeleton in the assembly and maintenance of epithelial apical junctions is demonstrated by the fact that actin-disrupting pharmacological agents such as cytochalasin D and latrunculin A rapidly and efficiently disrupt epithelial apical junctions (Bershadsky 2004; Mege et al. 2006). During the assembly of epithelial apical junctions, the formation of E-cadherin-mediated contacts triggers remodeling of the actin cytoskeleton, and maturation of junctions is accompanied by assembly of a circumferential actin belt and TJs. An established model of E-cadherin-mediated AJ formation predicts that the stable linkage between the E-cadherin/β-catenin complex and the actin cytoskeleton is mediated by α-catenin. However, more recent data shows that α-catenin does not stably couple E-cadherin to the circumferential actin belt, but instead directly regulates F-actin organization by suppressing Arp2/3-mediated actin polymerization (Yamada et al. 2005).

Under basal conditions, apical junctional proteins exist predominantly at the cell surface, but they can be detected in endosomal pools in a variety of cellular contexts (Bryant and Stow 2004; D'Souza-Schorey 2005; Ivanov et al. 2005; Delva and Kowalczyk 2009; Harris and Tepass 2010). For endocytosis of apical junction proteins, four distinct pathways have been revealed. These include macropinocytosis (Bruewer et al. 2005; Coyne et al. 2007), clathrin-dependent endocytosis (Le et al. 1999; Ivanov et al. 2004) (Izumi et al. 2004), caveolin-dependent endocytosis (Akhtar and Hotchin 2001) (Hopkins et al. 2003; Shen and Turner 2005), and clathrin-independent endocytosis (Paterson et al. 2003). Internalized apical junction proteins have also been detected in multiple sites, including Rab5-positive early endosomes (Le et al. 1999), Rab11-positive recycling endosomes (Bruewer et al. 2005; Hopkins et al. 2003; Balzac et al. 2005), Rab7-positive late endosomes (Matsuda et al. 2004), Rab13-positive vesicles (Morimoto et al. 2005), syntaxin 4-positive compartments (Ivanov et al. 2004), and syntaxin 3-positive vacuolar apical compartments (Utech et al. 2005). Although the endocytosed proteins in Rab7-positive late endosomes are usually targeted to lysosomes, others can be recycled from individual compartment back to the plasma membrane.

Indeed, multiple lines of evidence indicate that endocytosis and recycling of E-cadherin are important for maintenance and remodeling of apical junctions (Bryant and Stow 2004; D'Souza-Schorey 2005; Delva and Kowalczyk 2009; Harris and Tepass 2010). It seems reasonable to hypothesize that endocytic recycling should be regulated coordinately with the regulation of actin cytoskeletal organization; to date, however, there has little direct evidence of such a relationship, and no important molecules have yet been shown to be involved. Rab proteins have emerged as leading candidate factors that link these cellular functions: each Rab protein can transport actin-binding proteins, signal transduction molecules, polarity proteins, and scaffold proteins in addition to cell–cell adhesion molecules, leading to the spatiotemporal regulation of actin cytoskeletal reorganization accompanied by remodeling of apical junctions.

15.7 Rab Proteins Involved in Regulation of Apical Junctions

15.7.1 Rab3B

Although Rab3 subfamily proteins (Rab3A, Rab3B, Rab3C, and Rab3D) are enriched in neuronal/secretory cells and control regulated exocytosis through their interactions with the Rab3 effector proteins Rabphilin3, Rim1/2, and Noc2, Rab3B expression has also been detected in other cells (Takai et al. 2001). In epithelial cells, Rab3B is recruited to TJs upon formation of cell–cell contacts, and it is involved in the transport of polymeric immunoglobulin receptor (Weber et al. 1994; van Ijzendoorn et al. 2002). In neuroendocrine PC12 cells, Rab3B also regulates the reorganization of actin cytoskeleton and the targeting of ZO-1 to the plasma membrane through a process involving phosphatidylinositol 3-kinase (PI3K) (Sunshine et al. 2000). Recently, Rab3B has been implicated in the formation of apical membrane domains in epithelial cells (Galvez-Santisteban et al. 2012). In addition, Rab3 GAP1, a GAP for the Rab3 subfamily, regulates exocytosis of claudin-1, resulting in acquisition of the barrier function during epidermal development (Youssef et al. 2013).

15.7.2 Rab4

Rab4 is localized predominantly to the early endosome and, to a lesser extent, to recycling endosome; it is mainly involved in recycling from the early endosome to plasma membrane. In fibroblasts, the short-loop recycling regulated by Rab4 is well-studied. Rab4 regulates interactions between the cell and the extracellular matrix by controlling the PDGF-dependent recycling of $\alpha v\beta 3$ integrin through its interaction with protein kinase D-phosphorylated Rabaptin-5 (Roberts et al. 2001; Christoforides et al. 2012). In Sertoli cells, Rab4 associates with α- and β-catenins as well as with the actin cytoskeleton. Furthermore, it is involved in "ectoplasmic specialization," the disassembly of a testis-specific F-actin-based junctional structure that shares features of TJs, AJs, and focal adhesion (Mruk et al. 2007). Recently, N-myc downregulated gene 1 (NDRG1) was identified as a novel Rab4 effector in several epithelial cancers; this protein is involved in recycling of E-cadherin (Kachhap et al. 2007).

15.7.3 Rab5

Rab5 is a key regulator of transport from the plasma membrane to early endosomes and has also been implicated in macropinocytosis. In CVB-exposed Caco2 cells, Rab5 and its effector Rabankyrin-5 regulate the endocytosis of occludin (Coyne et al. 2007). In MDCK cells, Rab5 activation is involved in the hepatocyte growth factor (HGF)/scatter factor (SF)- or 12-*O*-tetradecanoylphorbol-13-acetate (TPA)-induced disruption of cell–cell adhesion and subsequent cell migration by mediating co-endocytosis of E-cadherin and c-Met (Imamura et al. 1998; Kamei et al. 1999). In HGF/SF-stimulated MDCK cells, Rab5 activation is mediated by the sequential action of c-Met, Ras, and Rab5 GEP RIN2 (Kimura et al. 2006). In v-Src-induced epithelial-to-mesenchymal transitions (EMT), Rab5 activation also mediates the lysosomal targeting of E-cadherin (Palacios et al. 2005). During zebrafish gastrulation, Rab5 controls the Wnt11-dependent endocytosis of E-cadherin and the cohesion of mesendodermal cells (Ulrich et al. 2005). In *Drosophila* epithelial cells, loss of Rab5 results in the cellular accumulation of the cell polarity protein CRB (Lu and Bilder 2005). In *Drosophila* embryonic salivary gland, Rab5-dependent E-cadherin endocytosis contributes to Pak1-regulated lumenal size control (Pirraglia et al. 2010).

15.7.4 Rab8

Rab8 is localized to the trans-Golgi network (TGN), recycling endosome, and cytosolic vesicular structures (Huber et al. 1993a, b; Ang et al. 2003). Rab8 was originally implicated in basolateral sorting events, and it may control the biosynthetic traffic from the TGN through recycling endosomes to the basolateral surface, rather than traffic of endocytosed proteins through recycling endosomes (Henry and Sheff 2008). Recent work has shown that Rab8 is involved in the primary cilia formation (Nachury et al. 2007; Yoshimura et al. 2007), suggesting that this protein plays a role in the sorting of apical proteins. This idea is consistent with the observation that Rab8-knockout mice exhibit missorting of apical peptidase and transporters in the small intestine (Sato et al. 2007). Moreover, Rab8 is required for de novo formation of the apical surface and lumen in MDCK cells (Bryant et al. 2010; Galvez-Santisteban et al. 2012). Rab8 associates with myosin Vb (Roland et al. 2007), optineurin–myosin VI (Au et al. 2007), Rab8 GEP Rabin8

(Hattula et al. 2002), JRAB/MICAL-L2 (Yamamura et al. 2008), optineurin-huntingtin (Hattula and Peranen 2000), and cenexin 1/Odf2 (Yoshimura et al. 2007; Chang et al. 2013); these interactions connect Rab8 with the actin and microtubule cytoskeletons.

Rab8 is also involved in cell–cell adhesion during *Dictyostelium discoideum* development and associates with E-cadherin as well as the actin cytoskeleton in Sertoli cells (Lau and Mruk 2003; Powell and Temesvari 2004). In epithelial cells, Rab8-mediated basolateral transport, dependent on the epithelial-specific adaptor protein complex AP-1B, is observed (Hattula and Peranen 2000; Sahlender et al. 2005; Au et al. 2007). Although E-cadherin was initially recognized as an AP-1B-independent basolateral cargo (Miranda et al. 2001; Folsch 2005), recent studies have demonstrated that E-cadherin is related to AP-1B through its interaction with type Iγ phosphatidylinositol-4-phosphate 5-kinase (Ling et al. 2007). Consistent with this, the Rab8–JRAB/MICAL-L2 complex is involved in E-cadherin transport (Yamamura et al. 2008).

15.7.5 Rab11

The Rab11 subfamily (Rab11A, Rab11B, and Rab11C/Rab25) is distributed across a variety of post-Golgi membranes, but it serves as the most prominent marker of recycling endosomes. Rab11 interacts with Sec15, a component of the exocyst, and is implicated in regulating post-Golgi traffic (Zhang et al. 2004). In *Drosophila* epithelial cells, E-cadherin accumulates in Rab11-positive recycling endosomes upon inactivation of components of the exocyst Sec5, Sec6, and Sec15 (Langevin et al. 2005). In MDCK cells, Rab11A regulates apical traffic and lumen formation through Rabin8, a GEP for Rab8 and Rab11A, which acts through the exocyst to target Par3 to the apical surface (Bryant et al. 2010). Rab11 also interacts with five effectors, the Rab11 family-interacting proteins (FIPs; FIP1–5). FIP3/arfophilin-1 and FIP4/arfophilin-2 are also the effector proteins of ARF6, a key regulator for the endocytic recycling of E-cadherin. In epithelial cells, these proteins control the transport of E-cadherin from the TGN to basolateral membranes via an intermediate compartment, the Rab11-positive recycling endosome (Hickson et al. 2003; Lock and Stow 2005). Furthermore, FIP5 phosphorylation by EGF receptor-activated ERK controls Rab11A endosome distribution and polymeric immunoglobulin A (pIgA)–pIgA receptor transcytosis in MDCK cells (Su et al. 2010).

15.7.6 Rab13

In fibroblasts, Rab13 associates with vesicles throughout the cytosol. In polarized epithelial cells, by contrast, it accumulates at TJs and is recruited to cell–cell contacts from a cytosolic pool at an early stage of junctional complex assembly

(Zahraoui et al. 1994; Sheth et al. 2000). Consistent with this, Rab13 mediates the endocytic recycling of occludin and claudins and is therefore implicated in the assembly of functional TJs in epithelial cells (Marzesco et al. 2002; Morimoto et al. 2005; Yamamura et al. 2008). Rab13 has recently been shown to have more versatile functions in various cell types: in MDCK cells, it regulates the biosynthetic transport of cell-surface proteins from the TGN through the recycling endosome to the plasma membrane (Nokes et al. 2008) as well as scattering in response to TPA (Kanda et al. 2008); in neurons, it regulates neurite outgrowth and neuronal regeneration (Di Giovanni et al. 2005, 2006; Sakane et al. 2010); and in L6 muscle cells, it mediates the translocation of glucose transporter-4 to the cell surface (Sun et al. 2010). cGMP phosphodiesterase δ subunit (δ-PDE), protein kinase A (PKA), JRAB/MICAL-L2, and endospanins have been identified as Rab13-binding proteins. δ-PDE, which possesses two putative carboxyl PDZ binding motifs, regulates the membrane association and disassociation of Rab13 (Marzesco et al. 1998). GTP-bound Rab13 interacts directly with PKA and inhibits the PKA-dependent phosphorylation and TJ recruitment of vasodilator-stimulated phosphoprotein (VASP) (Kohler et al. 2004). Endospanins (endospanin-1 and endospanin-2) are implicated in the cell-surface expression of leptin receptor and growth hormone receptor, but the Rab13-endospanin interaction seems to function in a process other than cell-surface transport (Hirvonen et al. 2013).

15.8 JRAB/MICAL-L2

15.8.1 JRAB/MICAL-L2, a Member of the MICAL Family

JRAB/MICAL-L2, which is one of the effector proteins of Rab13, mediates the endocytic recycling of occludin resulting in the formation of functional TJs (Terai et al. 2006). JRAB/MICAL-L2 belongs to the MICAL family, which consists of MICAL-1, MICAL-2, MICAL-3, and MICAL-L1, in addition to JRAB/MICAL-L2 (Suzuki et al. 2002; Terman et al. 2002). MICAL family proteins are large, multidomain, cytosolic proteins expressed in specific neuronal and nonneuronal cells both during development and in adulthood. All MICAL proteins contain calponin homology (CH), LIM, and coiled-coil (CC) domains, and MICAL-1, MICAL-2, and MICAL-3 also possess a flavin-adenine dinucleotide (FAD)-binding monooxygenase domain. Although MICAL was identified originally in mammals, its function has been studied primarily in *Drosophila* (Terman et al. 2002). *Drosophila* MICAL (D-MICAL) binds to the semaphorin receptor Plexin and functions downstream of this receptor in axon guidance (Terman et al. 2002). In mammals, MICAL-1 was isolated as a novel binding protein of CasL/HEF1/NEDD9, which localizes to focal adhesions and regulates the scattering of epithelial cells and the progression and metastasis of cancer cells (Suzuki et al. 2002; Kim et al. 2006). MICALs also associate with Rab1, vimentin, and microtubules (Suzuki et al. 2002; Fischer et al. 2005). Expression of MICAL-2 isoforms (PVa and PVb) is elevated in prostate cancer cells, and the degree of their expression is correlated with the prostate cancer progression (Ashida et al. 2006). JRAB/MICAL-L2 is involved in the scattering of MDCK cells in response to TPA (Kanda et al. 2008).

15.8.2 Rab8/Rab13–JRAB/MICAL-L2 Complex

JRAB/MICAL-L2 is localized on recycling endosomes, cytoplasmic vesicular structures, and the plasma membrane (Terai et al. 2006). JRAB/MICAL-L2 interacts not only with Rab13 but also with Rab8, via its carboxyl-terminal region (Yamamura et al. 2008). Considering the difference of localization between the Rab13 and Rab8, JRAB/MICAL-L2 shows a closer relationship with Rab13 at the plasma membrane and with Rab8 at the recycling endosome, respectively, indicating the existence of the two distinct JRAB/MICAL-L2 complexes within the cell. JRAB/MICAL-L2 regulates the transport of E-cadherin in addition to claudins and occludin; however, as mentioned above, Rab13 specifically mediates the transport of claudins and occludin, but not E-cadherin, to the plasma membrane (Morimoto et al. 2005; Yamamura et al. 2008). The Rab13-independent transport of E-cadherin is regulated by Rab8 (Yamamura et al. 2008). JRAB/MICAL-L2 organizes both Rab8-dependent E-cadherin transport at perinuclear recycling/storage compartments and Rab13-dependent claudins and occludin transport at the plasma membrane.

Several Rab effector proteins interact with multiple closely related Rab proteins (Hutagalung and Novick 2011; Mizuno-Yamasaki et al. 2012), and this network of interactions contributes to the complexity of membrane traffic. Some of these proteins are divalent Rab effectors that can bind simultaneously to two Rab proteins associated with compartments in dynamic continuity. For example, Rabaptin-5, Rabenosyn-5, and Rabip4′ are able to interact simultaneously with Rab4 and Rab5 and are thus likely involved in the coordination of the endocytic recycling pathway as well as the organization of Rab4 and Rab5 domains on endosomal membranes (de Renzis et al. 2002; Fouraux et al. 2004). JRAB/MICAL-L2 belongs to another type of Rab effector proteins that associate with multiple Rab proteins in mutually exclusive complexes. Specifically, JRAB/MICAL-L2 forms distinct complexes with Rab8 and Rab13 and thereby coordinates the assembly of epithelial apical junctions via at least two different mechanisms (Yamamura et al. 2008).

15.8.3 JRAB/MICAL-L2 and Actin Dynamics During AJC Formation

At the beginning of AJC formation, the assembly of spotlike primordial AJs occurs in parallel with the formation of thick F-actin bundles at the free border. Such bundles produce the force required to pull cells toward their neighbors, allowing them to adhere closely after cell–cell contacts are formed (Yonemura et al. 1995; Vasioukhin et al. 2000; Zhang et al. 2005; Gov 2011). During the maturation of the primordia into belt-like AJs, actin fibers at cell–cell contacts become static, rather than dynamic, in order to maintain stable adhesion. Thus, the regulation of actin cytoskeletal organization must be spatiotemporally coordinated during AJC formation; many players, including actin-binding proteins and signal transduction molecules, are involved in this integration.

Recent studies have demonstrated the importance of the MICAL family in regulating the actin cytoskeleton. It was first reported that D-MICAL directly binds to and disassembles F-actin, resulting in the cellular response to semaphorin/Plexin extracellular repulsive cues (Hung et al. 2010). The redox activity of the FAD domain contributes to this regulation by oxidizing F-actin (Hung et al. 2011). In mammals, MICAL-1 and MICAL-2 are required for normal actin cytoskeletal reorganization in nonneuronal cells (Giridharan et al. 2012). JRAB/MICAL-L2 plays a significant role not only in transport of adhesion molecules but also in the regulation of actin cytoskeletal reorganization (Sakane et al. 2012). Detailed biochemical analyses have revealed that JRAB/MICAL-L2 serves multiple functions in the reorganization of actin cytoskeleton: for instance, JRAB/MICAL-L2 binds directly to F-actin and can form F-actin bundles or stabilize F-actin (i.e., inhibit its depolymerization) via different domains (Sakane et al. 2012). Moreover, JRAB/MICAL-L2 associates with several actin-binding proteins, which are discussed below. Based on the accumulated evidence, it is likely that JRAB/MICAL-L2 regulates actin cytoskeletal reorganization, either directly or indirectly.

15.8.4 Conformational Change of JRAB/MICAL-L2

As discussed above, JRAB/MICAL-L2 consists of an N-terminal domain, containing the CH and LIM subdomains, and the CC domain in the C-terminal portion of the molecule (Terai et al. 2006). The CH+LIM domain engages in an intramolecular interaction with the CC domain (Fig. 15.2) (Sakane et al. 2010). When activated Rab13 binds to JRAB via the C-terminal (CT) region, which follows the CC domain, this interaction is likely to evoke a conformational change in JRAB/MICAL-L2, causing the protein to convert from the closed to the open form (Sakane et al. 2010). This intramolecular interaction of JRAB/MICAL-L2 affects its F-actin-stabilizing activity (Sakane et al. 2012); in particular, the

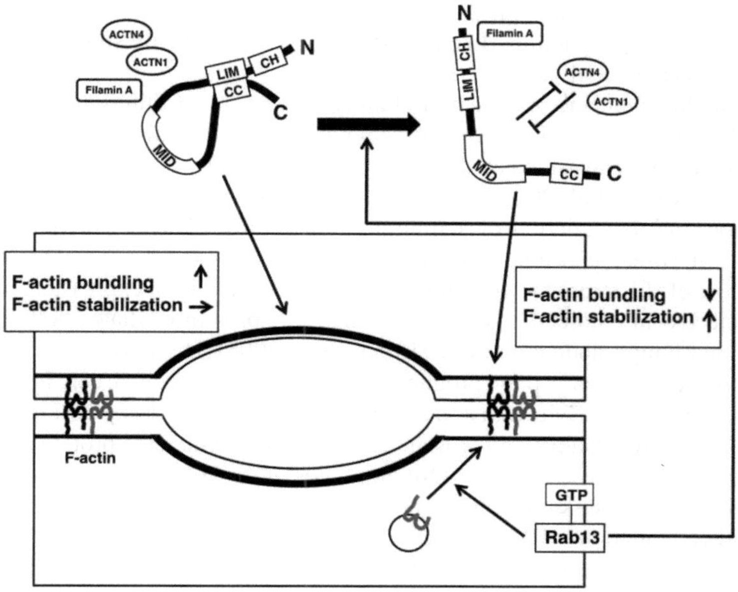

Fig. 15.2 Multiple modes of JRAB/MICAL-L2 function in actin cytoskeletal reorganization during AJC formation regulated by Rab13. JRAB/MICAL-L2 plays a pivotal role in linking vesicle transport of adhesion molecules, mediated via Rab13, to the reorganization of the actin cytoskeleton during AJC formation. At the initial stage, the closed form of JRAB/MICAL-L2 accelerates the formation of F-actin bundles at the free border. Subsequently, the open form of JRAB/MICAL-L2 decreases F-actin bundling via an interaction with actinin-1/4. At the late stage, JRAB/MICAL-L2 stabilizes F-actin at cell–cell adhesion sites via free CH+LIM

stabilizing activity of CH+LIM is inhibited by the intramolecular interaction through the CC domain. These results suggest that the conformational change of JRAB is important for the protein's role in regulating actin cytoskeletal reorganization.

15.8.5 JRAB/MICAL-L2 and Actin-Binding Proteins, Actinin-1 and Actinin-4

JRAB/MICAL-L2, especially in the open form, interacts with two actin-binding proteins, actinin-1 and actinin-4 (Nakatsuji et al. 2008; Sakane et al. 2012). The actinin proteins constitute a family of four closely related gene products (actinin-1, actinin-2, actinin-3, and actinin-4) (Otey and Carpen 2004). Actinin-1 and actinin-4 are expressed in many cell types and are localized at stress fibers, cellular protrusions/leading edges, and cell-adhesion sites (Oikonomou et al. 2011). Actinin-1 and actinin-4 are involved in the cross-linking of F-actin, leading to the anchoring of the actin cytoskeleton to the plasma membrane in non-muscle cells; furthermore, a complex containing actinin-4 and JRAB/MICAL-L2 mutually suppresses F-actin bundling (Sakane et al. 2012).

Among the members of the actinin family, actinin-4 constitutes the cytoplasmic face of cell–cell adhesion sites. In epithelial cells, actinin-4 associates with β-catenin at AJs (Hayashida et al. 2005) and with MAGI-1 and ZO-1 at TJs (Patrie et al. 2002; Chen et al. 2006). In support of the potential role of actinin-4 in AJC formation, the deregulation of actinin-4 expression has been reported in several cancers, in which invasion and metastasis were intimately associated with the disruption of TJ structure and function. Elevated actinin-4 expression has also been detected in colorectal cancers, in which this protein mediates both invasive growth and lymph-node metastasis (Honda et al. 2005). High actinin-4 expression was also identified as a poor prognostic predictor of non-small cell lung cancer (Honda et al. 2004). By contrast, downregulation of actinin-4 expression has been reported in neuroblastoma and prostate cancer cells, in which the protein appears to exert a tumor-suppressive effect (Nikolopoulos et al. 2000; Hara et al. 2007). A recent study, using an excellent in vitro assay to study actin assembly at cadherin-enriched cell junctions of MDCK cells, showed that actinin-4 specifically localizes to sites of actin incorporation and plays an important role in coupling Arp2/3-dependent actin nucleation to assembly at cadherin-based cell–cell adhesive contacts (Tang and Brieher 2012). It is still premature to discuss further the relationship between this actinin-4–JRAB/MICAL-L2–Rab13 complex and AJC formation. However, it seems clear that via a sophisticated mechanism, epithelial cells coordinate many signals to dynamically reorganize AJCs under physiological and pathological conditions.

15.8.6 JRAB/MICAL-L2 and Filamins

Open form of JRAB/MICAL-L2 also interacts with filamins (Sakane et al. 2013), a widely studied family of actin-binding proteins, that can crosslink F-actin to form orthogonal networks (Feng and Walsh 2004; Popowicz et al. 2006; Nakamura et al. 2011). In vertebrates, the filamin family has three members, filamin A, filamin B, and filamin C, which each possess a conserved N-terminal actin-binding domain and 24 Ig-like domains. JRAB/MICAL-L2 binds to all members of the filamin family; in NIH3T3 fibroblasts, this interaction regulates cell spreading (Sakane et al. 2013). Knockout mice lacking filamin A and filamin B exhibit poor development of blood vessels, possibly due to impairments in cell–cell adhesion and migration of endothelial cells (Feng et al. 2006; Zhou et al. 2007). The consequences of filamin B knockdown in MDCK cells suggest that this protein is required for the accumulation of adhesion molecules in the junctional complexes and subsequent formation of an epithelium (Wakamatsu et al. 2011). Thus, JRAB/MICAL-L2 may regulate cell–cell adhesion through interaction of filamins in epithelial cells, although the precise mechanism remains unknown.

The binding site of JRAB/MICAL-L2 to filamins involves its CH domain, which is distinct from the regions involved in binding to actinin-1/actinin-4 and F-actin, indicating that JRAB/MICAL-L2 regulates the actin cytoskeleton by at least three mechanisms: direct, filamin dependent, and actinin dependent. Coordination of these mechanisms may play a crucial role in dynamic actin cytoskeletal reorganization during AJC formation.

15.8.7 Multiple Modes of Function of JRAB/MICAL-L2 in Actin Cytoskeletal Reorganization During AJC Formation

The development of the JRABΔCC mutants, which lacks the CC domain (open form), and the JRABΔCT mutant, which lacks the Rab13-binding CT domain (closed form), could facilitate understanding of the role played by the conformational changes of JRAB/MICAL-L2 (Sakane et al. 2010). JRABΔCT accumulates in the thick F-actin bundles along the free border (Sakane et al. 2012). By contrast, JRABΔCC preferentially localizes to cell–cell adhesion sites, but is only faintly detected at the free edges. Similar results were obtained using the Duolink in situ proximity ligation assay (PLA) system, which consists of a pair of oligonucleotide-labeled secondary antibodies. In summary, JRAB/MICAL-L2 plays numerous roles in actin cytoskeletal regulation at each stage of AJC formation (Fig. 15.2): at the initiation of adhesion, "closed" JRAB/MICAL-L2 accelerates the formation of thick actin bundles at the free border, in cooperation with actinins and filamins. Subsequently, JRAB/MICAL-L2 changes its conformation from closed to open, in a manner that depends on an interaction with GTP-Rab13 carrying adhesion molecules. The open form of JRAB/MICAL-L2 then interacts with actinins and filamins at mature cell–cell adhesion sites. The interaction between open JRAB/MICAL-L2 and actinins suppresses their actin-bundling activities (Sakane et al. 2012); meanwhile, open JRAB/MICAL-L2 activates the stabilization of F-actin via release of CH+LIM, leading to establishment and maintenance of stationary cell–cell adhesions. However, questions remain regarding the functions of JRAB/MICAL-L2–filamin at cell–cell contacts. In one possible model, JRAB/MICAL-L2 has several modes of function in the series of processes involved in AJC formation and thus may perform different roles at different times. Moreover, it is possible that JRAB/MICAL-L2 plays a pivotal role in linking transporting of adhesion molecules to the reorganization of the actin cytoskeleton.

15.9 Concluding Remarks

The AJC of epithelial cells orchestrates cell–cell adhesion, tissue barrier function, and cell polarity. In recent years, major progress has been made in identifying a variety of AJC components and clarifying the crucial importance of the interplay between actin cytoskeletal reorganization and vesicle transport in determining the functions of AJC components. However, many important outstanding questions remain regarding the integration of these phenomena. Our analyses suggest that JRAB/MICAL-L2 serves critical functions both in the regulation of actin dynamics, which is required for processes ranging from initiation to establishment of the AJC, and in the regulation of endocytic recycling of cell-adhesion molecules mediated by Rab13 or Rab8. Moreover, JRAB/MICAL-L2 undergoes a conformational change between the open and closed forms; elucidation of the mechanisms underlying this change could be key to understanding the mechanisms of cross talk between actin cytoskeletal rearrangement and vesicle transport in AJC formation. An important challenge for future investigations will be to uncover the adaptations in molecular pathways that allow the Rab13–JRAB/MICAL-L2 and Rab8–JRAB/MICAL-L2 complexes to contribute to different aspects of AJC formation.

References

Akhtar N, Hotchin NA (2001) RAC1 regulates adherens junctions through endocytosis of E-cadherin. Mol Biol Cell 12(4):847–862

Allan BB, Moyer BD, Balch WE (2000) Rab1 recruitment of p115 into a cis-SNARE complex: programming budding COPII vesicles for fusion. Science 289(5478):444–448

Ang AL, Folsch H, Koivisto UM, Pypaert M, Mellman I (2003) The Rab8 GTPase selectively regulates AP-1B-dependent basolateral transport in polarized Madin-Darby canine kidney cells. J Cell Biol 163(2):339–350. doi:10.1083/jcb.200307046

Apodaca G, Gallo LI, Bryant DM (2012) Role of membrane traffic in the generation of epithelial cell asymmetry. Nat Cell Biol 14(12):1235–1243. doi:10.1038/ncb2635

Ashida S, Furihata M, Katagiri T, Tamura K, Anazawa Y, Yoshioka H, Miki T, Fujioka T, Shuin T, Nakamura Y, Nakagawa H (2006) Expression of novel molecules, MICAL2-PV (MICAL2 prostate cancer variants), increases with high Gleason score and prostate cancer progression. Clin Cancer Res 12(9):2767–2773. doi:10.1158/1078-0432.CCR-05-1995

Assemat E, Bazellieres E, Pallesi-Pocachard E, Le Bivic A, Massey-Harroche D (2008) Polarity complex proteins. Biochim Biophys Acta 1778(3):614–630. doi:10.1016/j.bbamem.2007.08.029

Au JS, Puri C, Ihrke G, Kendrick-Jones J, Buss F (2007) Myosin VI is required for sorting of AP-1B-dependent cargo to the basolateral domain in polarized MDCK cells. J Cell Biol 177(1):103–114. doi:10.1083/jcb.200608126

Balzac F, Avolio M, Degani S, Kaverina I, Torti M, Silengo L, Small JV, Retta SF (2005) E-cadherin endocytosis regulates the activity of Rap1: a traffic light GTPase at the crossroads between cadherin and integrin function. J Cell Sci 118(Pt 20):4765–4783. doi:10.1242/jcs.02584

Bershadsky A (2004) Magic touch: how does cell-cell adhesion trigger actin assembly? Trends Cell Biol 14(11):589–593. doi:10.1016/j.tcb.2004.09.009

Bruewer M, Utech M, Ivanov AI, Hopkins AM, Parkos CA, Nusrat A (2005) Interferon-gamma induces internalization of epithelial tight junction proteins via a macropinocytosis-like process. FASEB J 19(8):923–933. doi:10.1096/fj.04-3260com

Bryant DM, Mostov KE (2008) From cells to organs: building polarized tissue. Nat Rev Mol Cell Biol 9(11):887–901. doi:10.1038/nrm2523

Bryant DM, Stow JL (2004) The ins and outs of E-cadherin trafficking. Trends Cell Biol 14(8): 427–434. doi:10.1016/j.tcb.2004.07.007

Bryant DM, Datta A, Rodriguez-Fraticelli AE, Peranen J, Martin-Belmonte F, Mostov KE (2010) A molecular network for de novo generation of the apical surface and lumen. Nat Cell Biol 12(11):1035–1045. doi:10.1038/ncb2106

Carmosino M, Valenti G, Caplan M, Svelto M (2010) Polarized traffic towards the cell surface: how to find the route. Biol Cell 102(2):75–91. doi:10.1042/BC20090134

Carroll KS, Hanna J, Simon I, Krise J, Barbero P, Pfeffer SR (2001) Role of Rab9 GTPase in facilitating receptor recruitment by TIP47. Science 292(5520):1373–1376. doi:10.1126/science.1056791

Chang J, Seo SG, Lee KH, Nagashima K, Bang JK, Kim BY, Erikson RL, Lee KW, Lee HJ, Park JE, Lee KS (2013) Essential role of Cenexin1, but not Odf2, in ciliogenesis. Cell Cycle 12(4):655–662. doi:10.4161/cc.23585

Chen VC, Li X, Perreault H, Nagy JI (2006) Interaction of zonula occludens-1 (ZO-1) with alpha-actinin-4: application of functional proteomics for identification of PDZ domain-associated proteins. J Proteome Res 5(9):2123–2134. doi:10.1021/pr060216l

Christoforides C, Rainero E, Brown KK, Norman JC, Toker A (2012) PKD controls alphavbeta3 integrin recycling and tumor cell invasive migration through its substrate Rabaptin-5. Dev Cell 23(3):560–572. doi:10.1016/j.devcel.2012.08.008

Colicelli J (2004) Human RAS superfamily proteins and related GTPases. Sci STKE 2004(250): RE13. doi:10.1126/stke.2502004re13

Conner SD, Schmid SL (2003) Regulated portals of entry into the cell. Nature 422(6927):37–44. doi:10.1038/nature01451

Coyne CB, Bergelson JM (2005) CAR: a virus receptor within the tight junction. Adv Drug Deliv Rev 57(6):869–882. doi:10.1016/j.addr.2005.01.007

Coyne CB, Shen L, Turner JR, Bergelson JM (2007) Coxsackievirus entry across epithelial tight junctions requires occludin and the small GTPases Rab34 and Rab5. Cell Host Microbe 2(3): 181–192. doi:10.1016/j.chom.2007.07.003

D'Souza-Schorey C (2005) Disassembling adherens junctions: breaking up is hard to do. Trends Cell Biol 15(1):19–26. doi:10.1016/j.tcb.2004.11.002

D'Souza-Schorey C, Chavrier P (2006) ARF proteins: roles in membrane traffic and beyond. Nat Rev Mol Cell Biol 7(5):347–358. doi:10.1038/nrm1910

de Renzis S, Sonnichsen B, Zerial M (2002) Divalent Rab effectors regulate the sub-compartmental organization and sorting of early endosomes. Nat Cell Biol 4(2): 124–133. doi:10.1038/ncb744

Delva E, Kowalczyk AP (2009) Regulation of cadherin trafficking. Traffic 10(3):259–267. doi:10.1111/j.1600-0854.2008.00862.x

Di Giovanni S, De Biase A, Yakovlev A, Finn T, Beers J, Hoffman EP, Faden AI (2005) In vivo and in vitro characterization of novel neuronal plasticity factors identified following spinal cord injury. J Biol Chem 280(3):2084–2091. doi:10.1074/jbc.M411975200

Di Giovanni S, Knights CD, Rao M, Yakovlev A, Beers J, Catania J, Avantaggiati ML, Faden AI (2006) The tumor suppressor protein p53 is required for neurite outgrowth and axon regeneration. EMBO J 25(17):4084–4096. doi:10.1038/sj.emboj.7601292

Donaldson JG, Jackson CL (2011) ARF family G proteins and their regulators: roles in membrane transport, development and disease. Nat Rev Mol Cell Biol 12(6):362–375. doi:10.1038/nrm3117

Ebnet K, Suzuki A, Ohno S, Vestweber D (2004) Junctional adhesion molecules (JAMs): more molecules with dual functions? J Cell Sci 117(Pt 1):19–29. doi:10.1242/jcs.00930

Echard A, Jollivet F, Martinez O, Lacapere JJ, Rousselet A, Janoueix-Lerosey I, Goud B (1998) Interaction of a Golgi-associated kinesin-like protein with Rab6. Science 279(5350):580–585

Farquhar MG, Palade GE (1963) Junctional complexes in various epithelia. J Cell Biol 17: 375–412

Feng Y, Walsh CA (2004) The many faces of filamin: a versatile molecular scaffold for cell motility and signalling. Nat Cell Biol 6(11):1034–1038. doi:10.1038/ncb1104-1034

Feng Y, Chen MH, Moskowitz IP, Mendonza AM, Vidali L, Nakamura F, Kwiatkowski DJ, Walsh CA (2006) Filamin A (FLNA) is required for cell-cell contact in vascular development and cardiac morphogenesis. Proc Natl Acad Sci U S A 103(52):19836–19841. doi:10.1073/pnas.0609628104

Fischer J, Weide T, Barnekow A (2005) The MICAL proteins and rab1: a possible link to the cytoskeleton? Biochem Biophys Res Commun 328(2):415–423. doi:10.1016/j.bbrc.2004.12.182

Folsch H (2005) The building blocks for basolateral vesicles in polarized epithelial cells. Trends Cell Biol 15(4):222–228. doi:10.1016/j.tcb.2005.02.006

Fouraux MA, Deneka M, Ivan V, van der Heijden A, Raymackers J, van Suylekom D, van Venrooij WJ, van der Sluijs P, Pruijn GJ (2004) Rabip4′ is an effector of rab5 and rab4 and regulates transport through early endosomes. Mol Biol Cell 15(2):611–624. doi:10.1091/mbc.E03-05-0343

Furuse M (2010) Molecular basis of the core structure of tight junctions. Cold Spring Harb Perspect Biol 2(1):a002907. doi:10.1101/cshperspect.a002907

Furuse M, Hirase T, Itoh M, Nagafuchi A, Yonemura S, Tsukita S, Tsukita S (1993) Occludin: a novel integral membrane protein localizing at tight junctions. J Cell Biol 123(6 Pt 2): 1777–1788

Galvez-Santisteban M, Rodriguez-Fraticelli AE, Bryant DM, Vergarajauregui S, Yasuda T, Banon-Rodriguez I, Bernascone I, Datta A, Spivak N, Young K, Slim CL, Brakeman PR, Fukuda M, Mostov KE, Martin-Belmonte F (2012) Synaptotagmin-like proteins control the formation of a single apical membrane domain in epithelial cells. Nat Cell Biol 14(8):838–849. doi:10.1038/ncb2541

Giridharan SS, Rohn JL, Naslavsky N, Caplan S (2012) Differential regulation of actin microfilaments by human MICAL proteins. J Cell Sci 125(Pt 3):614–624. doi:10.1242/jcs.089367

Gov N (2011) Cell mechanics: moving under peer pressure. Nat Mater 10(6):412–414. doi:10.1038/nmat3036

Grindstaff KK, Yeaman C, Anandasabapathy N, Hsu SC, Rodriguez-Boulan E, Scheller RH, Nelson WJ (1998) Sec6/8 complex is recruited to cell-cell contacts and specifies transport vesicle delivery to the basal-lateral membrane in epithelial cells. Cell 93(5):731–740

Hammer JA 3rd, Wu XS (2002) Rabs grab motors: defining the connections between Rab GTPases and motor proteins. Curr Opin Cell Biol 14(1):69–75

Hara T, Honda K, Shitashige M, Ono M, Matsuyama H, Naito K, Hirohashi S, Yamada T (2007) Mass spectrometry analysis of the native protein complex containing actinin-4 in prostate cancer cells. Mol Cell Proteomics 6(3):479–491. doi:10.1074/mcp.M600129-MCP200

Harris TJ, Tepass U (2010) Adherens junctions: from molecules to morphogenesis. Nat Rev Mol Cell Biol 11(7):502–514. doi:10.1038/nrm2927

Hattula K, Peranen J (2000) FIP-2, a coiled-coil protein, links Huntingtin to Rab8 and modulates cellular morphogenesis. Curr Biol 10(24):1603–1606

Hattula K, Furuhjelm J, Arffman A, Peranen J (2002) A Rab8-specific GDP/GTP exchange factor is involved in actin remodeling and polarized membrane transport. Mol Biol Cell 13(9): 3268–3280. doi:10.1091/mbc.E02-03-0143

Hayashida Y, Honda K, Idogawa M, Ino Y, Ono M, Tsuchida A, Aoki T, Hirohashi S, Yamada T (2005) E-cadherin regulates the association between beta-catenin and actinin-4. Cancer Res 65(19):8836–8845. doi:10.1158/0008-5472.CAN-05-0718

Hazelett CC, Sheff D, Yeaman C (2011) RalA and RalB differentially regulate development of epithelial tight junctions. Mol Biol Cell 22(24):4787–4800. doi:10.1091/mbc.E11-07-0657

He B, Guo W (2009) The exocyst complex in polarized exocytosis. Curr Opin Cell Biol 21(4): 537–542. doi:10.1016/j.ceb.2009.04.007

Henry L, Sheff DR (2008) Rab8 regulates basolateral secretory, but not recycling, traffic at the recycling endosome. Mol Biol Cell 19(5):2059–2068. doi:10.1091/mbc.E07-09-0902

Hickson GR, Matheson J, Riggs B, Maier VH, Fielding AB, Prekeris R, Sullivan W, Barr FA, Gould GW (2003) Arfophilins are dual Arf/Rab 11 binding proteins that regulate recycling endosome distribution and are related to Drosophila nuclear fallout. Mol Biol Cell 14(7): 2908–2920. doi:10.1091/mbc.E03-03-0160

Hirvonen MJ, Buki KG, Sun Y, Mulari MT, Harkonen PL, Vaananen KH (2013) Novel interaction of Rab13 and Rab8 with endospanins. FEBS Open Bio 3:83–88. doi:10.1016/j.fob.2013.01.004

Hoekstra D, Tyteca D, van Ijzendoorn SC (2004) The subapical compartment: a traffic center in membrane polarity development. J Cell Sci 117(Pt 11):2183–2192. doi:10.1242/jcs.01217

Honda K, Yamada T, Seike M, Hayashida Y, Idogawa M, Kondo T, Ino Y, Hirohashi S (2004) Alternative splice variant of actinin-4 in small cell lung cancer. Oncogene 23(30):5257–5262. doi:10.1038/sj.onc.1207652

Honda K, Yamada T, Hayashida Y, Idogawa M, Sato S, Hasegawa F, Ino Y, Ono M, Hirohashi S (2005) Actinin-4 increases cell motility and promotes lymph node metastasis of colorectal cancer. Gastroenterology 128(1):51–62

Hopkins AM, Walsh SV, Verkade P, Boquet P, Nusrat A (2003) Constitutive activation of Rho proteins by CNF-1 influences tight junction structure and epithelial barrier function. J Cell Sci 116(Pt 4):725–742

Horiuchi H, Lippe R, McBride HM, Rubino M, Woodman P, Stenmark H, Rybin V, Wilm M, Ashman K, Mann M, Zerial M (1997) A novel Rab5 GDP/GTP exchange factor complexed to Rabaptin-5 links nucleotide exchange to effector recruitment and function. Cell 90(6): 1149–1159

Huber LA, de Hoop MJ, Dupree P, Zerial M, Simons K, Dotti C (1993a) Protein transport to the dendritic plasma membrane of cultured neurons is regulated by rab8p. J Cell Biol 123(1):47–55

Huber LA, Pimplikar S, Parton RG, Virta H, Zerial M, Simons K (1993b) Rab8, a small GTPase involved in vesicular traffic between the TGN and the basolateral plasma membrane. J Cell Biol 123(1):35–45

Hulpiau P, van Roy F (2009) Molecular evolution of the cadherin superfamily. Int J Biochem Cell Biol 41(2):349–369. doi:10.1016/j.biocel.2008.09.027

Hung RJ, Yazdani U, Yoon J, Wu H, Yang T, Gupta N, Huang Z, van Berkel WJ, Terman JR (2010) Mical links semaphorins to F-actin disassembly. Nature 463(7282):823–827. doi:10.1038/nature08724

Hung RJ, Pak CW, Terman JR (2011) Direct redox regulation of F-actin assembly and disassembly by Mical. Science 334(6063):1710–1713. doi:10.1126/science.1211956

Hutagalung AH, Novick PJ (2011) Role of Rab GTPases in membrane traffic and cell physiology. Physiol Rev 91(1):119–149. doi:10.1152/physrev.00059.2009

Ikenouchi J, Furuse M, Furuse K, Sasaki H, Tsukita S, Tsukita S (2005) Tricellulin constitutes a novel barrier at tricellular contacts of epithelial cells. J Cell Biol 171(6):939–945. doi:10.1083/jcb.200510043

Imamura H, Takaishi K, Nakano K, Kodama A, Oishi H, Shiozaki H, Monden M, Sasaki T, Takai Y (1998) Rho and Rab small G proteins coordinately reorganize stress fibers and focal adhesions in MDCK cells. Mol Biol Cell 9(9):2561–2575

Ivanov AI, Nusrat A, Parkos CA (2004) Endocytosis of epithelial apical junctional proteins by a clathrin-mediated pathway into a unique storage compartment. Mol Biol Cell 15(1):176–188. doi:10.1091/mbc.E03-05-0319

Ivanov AI, Nusrat A, Parkos CA (2005) Endocytosis of the apical junctional complex: mechanisms and possible roles in regulation of epithelial barriers. Bioessays 27(4):356–365. doi:10.1002/bies.20203

Izumi G, Sakisaka T, Baba T, Tanaka S, Morimoto K, Takai Y (2004) Endocytosis of E-cadherin regulated by Rac and Cdc42 small G proteins through IQGAP1 and actin filaments. J Cell Biol 166(2):237–248. doi:10.1083/jcb.200401078

Jahn R, Scheller RH (2006) SNAREs—engines for membrane fusion. Nat Rev Mol Cell Biol 7(9): 631–643. doi:10.1038/nrm2002

Jean S, Kiger AA (2012) Coordination between RAB GTPase and phosphoinositide regulation and functions. Nat Rev Mol Cell Biol 13(7):463–470. doi:10.1038/nrm3379

Kachhap SK, Faith D, Qian DZ, Shabbeer S, Galloway NL, Pili R, Denmeade SR, DeMarzo AM, Carducci MA (2007) The N-Myc down regulated Gene1 (NDRG1) is a Rab4a effector involved in vesicular recycling of E-cadherin. PLoS One 2(9):e844. doi:10.1371/journal.pone.0000844

Kamei T, Matozaki T, Sakisaka T, Kodama A, Yokoyama S, Peng YF, Nakano K, Takaishi K, Takai Y (1999) Coendocytosis of cadherin and c-Met coupled to disruption of cell-cell adhesion in MDCK cells—regulation by Rho, Rac and Rab small G proteins. Oncogene 18(48):6776–6784. doi:10.1038/sj.onc.1203114

Kanda I, Nishimura N, Nakatsuji H, Yamamura R, Nakanishi H, Sasaki T (2008) Involvement of Rab13 and JRAB/MICAL-L2 in epithelial cell scattering. Oncogene 27(12):1687–1695. doi:10.1038/sj.onc.1210812

Kim M, Gans JD, Nogueira C, Wang A, Paik JH, Feng B, Brennan C, Hahn WC, Cordon-Cardo C, Wagner SN, Flotte TJ, Duncan LM, Granter SR, Chin L (2006) Comparative oncogenomics identifies NEDD9 as a melanoma metastasis gene. Cell 125(7):1269–1281. doi:10.1016/j.cell.2006.06.008

Kimura T, Sakisaka T, Baba T, Yamada T, Takai Y (2006) Involvement of the Ras-Ras-activated Rab5 guanine nucleotide exchange factor RIN2-Rab5 pathway in the hepatocyte growth factor-induced endocytosis of E-cadherin. J Biol Chem 281(15):10598–10609. doi:10.1074/jbc.M510531200

Kinchen JM, Ravichandran KS (2010) Identification of two evolutionarily conserved genes regulating processing of engulfed apoptotic cells. Nature 464(7289):778–782. doi:10.1038/nature08853

Kohler K, Louvard D, Zahraoui A (2004) Rab13 regulates PKA signaling during tight junction assembly. J Cell Biol 165(2):175–180. doi:10.1083/jcb.200312118

Langevin J, Morgan MJ, Sibarita JB, Aresta S, Murthy M, Schwarz T, Camonis J, Bellaiche Y (2005) Drosophila exocyst components Sec5, Sec6, and Sec15 regulate DE-Cadherin trafficking from recycling endosomes to the plasma membrane. Dev Cell 9(3):365–376

Lau AS, Mruk DD (2003) Rab8B GTPase and junction dynamics in the testis. Endocrinology 144(4):1549–1563. doi:10.1210/en.2002-220893

Le TL, Yap AS, Stow JL (1999) Recycling of E-cadherin: a potential mechanism for regulating cadherin dynamics. J Cell Biol 146(1):219–232

Li R, Gundersen GG (2008) Beyond polymer polarity: how the cytoskeleton builds a polarized cell. Nat Rev Mol Cell Biol 9(11):860–873. doi:10.1038/nrm2522

Ling K, Bairstow SF, Carbonara C, Turbin DA, Huntsman DG, Anderson RA (2007) Type I gamma phosphatidylinositol phosphate kinase modulates adherens junction and E-cadherin trafficking via a direct interaction with mu 1B adaptin. J Cell Biol 176(3):343–353. doi:10.1083/jcb.200606023

Lock JG, Stow JL (2005) Rab11 in recycling endosomes regulates the sorting and basolateral transport of E-cadherin. Mol Biol Cell 16(4):1744–1755. doi:10.1091/mbc.E04-10-0867

Lu H, Bilder D (2005) Endocytic control of epithelial polarity and proliferation in Drosophila. Nat Cell Biol 7(12):1232–1239. doi:10.1038/ncb1324

Marzesco AM, Galli T, Louvard D, Zahraoui A (1998) The rod cGMP phosphodiesterase delta subunit dissociates the small GTPase Rab13 from membranes. J Biol Chem 273(35): 22340–22345

Marzesco AM, Dunia I, Pandjaitan R, Recouvreur M, Dauzonne D, Benedetti EL, Louvard D, Zahraoui A (2002) The small GTPase Rab13 regulates assembly of functional tight junctions in epithelial cells. Mol Biol Cell 13(6):1819–1831, doi:02-02-0029

Matsuda M, Kubo A, Furuse M, Tsukita S (2004) A peculiar internalization of claudins, tight junction-specific adhesion molecules, during the intercellular movement of epithelial cells. J Cell Sci 117(Pt 7):1247–1257. doi:10.1242/jcs.00972

McBride HM, Rybin V, Murphy C, Giner A, Teasdale R, Zerial M (1999) Oligomeric complexes link Rab5 effectors with NSF and drive membrane fusion via interactions between EEA1 and syntaxin 13. Cell 98(3):377–386

Mege RM, Gavard J, Lambert M (2006) Regulation of cell-cell junctions by the cytoskeleton. Curr Opin Cell Biol 18(5):541–548. doi:10.1016/j.ceb.2006.08.004

Mellman I, Nelson WJ (2008) Coordinated protein sorting, targeting and distribution in polarized cells. Nat Rev Mol Cell Biol 9(11):833–845. doi:10.1038/nrm2525

Miranda KC, Khromykh T, Christy P, Le TL, Gottardi CJ, Yap AS, Stow JL, Teasdale RD (2001) A dileucine motif targets E-cadherin to the basolateral cell surface in Madin-Darby canine kidney and LLC-PK1 epithelial cells. J Biol Chem 276(25):22565–22572. doi:10.1074/jbc.M101907200

Mizuno-Yamasaki E, Rivera-Molina F, Novick P (2012) GTPase networks in membrane traffic. Annu Rev Biochem 81:637–659. doi:10.1146/annurev-biochem-052810-093700

Monteiro AC, Parkos CA (2012) Intracellular mediators of JAM-A-dependent epithelial barrier function. Ann N Y Acad Sci 1257:115–124. doi:10.1111/j.1749-6632.2012.06521.x

Morimoto S, Nishimura N, Terai T, Manabe S, Yamamoto Y, Shinahara W, Miyake H, Tashiro S, Shimada M, Sasaki T (2005) Rab13 mediates the continuous endocytic recycling of occludin to the cell surface. J Biol Chem 280(3):2220–2228

Mruk DD, Lau AS, Sarkar O, Xia W (2007) Rab4A GTPase catenin interactions are involved in cell junction dynamics in the testis. J Androl 28(5):742–754. doi:10.2164/jandrol.106.002204

Nachury MV, Loktev AV, Zhang Q, Westlake CJ, Peranen J, Merdes A, Slusarski DC, Scheller RH, Bazan JF, Sheffield VC, Jackson PK (2007) A core complex of BBS proteins cooperates with the GTPase Rab8 to promote ciliary membrane biogenesis. Cell 129(6):1201–1213. doi:10.1016/j.cell.2007.03.053

Nakamura F, Stossel TP, Hartwig JH (2011) The filamins: organizers of cell structure and function. Cell Adh Migr 5(2):160–169

Nakatsuji H, Nishimura N, Yamamura R, Kanayama HO, Sasaki T (2008) Involvement of actinin-4 in the recruitment of JRAB/MICAL-L2 to cell-cell junctions and the formation of functional tight junctions. Mol Cell Biol 28(10):3324–3335

Nikolopoulos SN, Spengler BA, Kisselbach K, Evans AE, Biedler JL, Ross RA (2000) The human non-muscle alpha-actinin protein encoded by the ACTN4 gene suppresses tumorigenicity of human neuroblastoma cells. Oncogene 19(3):380–386. doi:10.1038/sj.onc.1203310

Nokes RL, Fields IC, Collins RN, Folsch H (2008) Rab13 regulates membrane trafficking between TGN and recycling endosomes in polarized epithelial cells. J Cell Biol 182(5):845–853. doi:10.1083/jcb.200802176

Nottingham RM, Ganley IG, Barr FA, Lambright DG, Pfeffer SR (2011) RUTBC1 protein, a Rab9A effector that activates GTP hydrolysis by Rab32 and Rab33B proteins. J Biol Chem 286(38):33213–33222. doi:10.1074/jbc.M111.261115

Oda H, Takeichi M (2011) Evolution: structural and functional diversity of cadherin at the adherens junction. J Cell Biol 193(7):1137–1146. doi:10.1083/jcb.201008173

Oikonomou KG, Zachou K, Dalekos GN (2011) Alpha-actinin: a multidisciplinary protein with important role in B-cell driven autoimmunity. Autoimmun Rev 10(7):389–396. doi:10.1016/j.autrev.2010.12.009

Otey CA, Carpen O (2004) Alpha-actinin revisited: a fresh look at an old player. Cell Motil Cytoskeleton 58(2):104–111. doi:10.1002/cm.20007

Oztan A, Silvis M, Weisz OA, Bradbury NA, Hsu SC, Goldenring JR, Yeaman C, Apodaca G (2007) Exocyst requirement for endocytic traffic directed toward the apical and basolateral

poles of polarized MDCK cells. Mol Biol Cell 18(10):3978–3992. doi:10.1091/mbc.E07-02-0097
Palacios F, Price L, Schweitzer J, Collard JG, D'Souza-Schorey C (2001) An essential role for ARF6-regulated membrane traffic in adherens junction turnover and epithelial cell migration. EMBO J 20(17):4973–4986. doi:10.1093/emboj/20.17.4973
Palacios F, Tushir JS, Fujita Y, D'Souza-Schorey C (2005) Lysosomal targeting of E-cadherin: a unique mechanism for the down-regulation of cell-cell adhesion during epithelial to mesenchymal transitions. Mol Cell Biol 25(1):389–402. doi:10.1128/MCB.25.1.389-402.2005
Paterson AD, Parton RG, Ferguson C, Stow JL, Yap AS (2003) Characterization of E-cadherin endocytosis in isolated MCF-7 and chinese hamster ovary cells: the initial fate of unbound E-cadherin. J Biol Chem 278(23):21050–21057. doi:10.1074/jbc.M300082200
Patrie KM, Drescher AJ, Welihinda A, Mundel P, Margolis B (2002) Interaction of two actin-binding proteins, synaptopodin and alpha-actinin-4, with the tight junction protein MAGI-1. J Biol Chem 277(33):30183–30190. doi:10.1074/jbc.M203072200
Pereira-Leal JB, Seabra MC (2000) The mammalian Rab family of small GTPases: definition of family and subfamily sequence motifs suggests a mechanism for functional specificity in the Ras superfamily. J Mol Biol 301(4):1077–1087. doi:10.1006/jmbi.2000.4010
Pereira-Leal JB, Seabra MC (2001) Evolution of the Rab family of small GTP-binding proteins. J Mol Biol 313(4):889–901. doi:10.1006/jmbi.2001.5072
Pfeffer S (2003) Membrane domains in the secretory and endocytic pathways. Cell 112(4): 507–517
Pfeffer SR (2013) Rab GTPase regulation of membrane identity. Curr Opin Cell Biol 25(4): 414–419. doi:10.1016/j.ceb.2013.04.002
Pirraglia C, Walters J, Myat MM (2010) Pak1 control of E-cadherin endocytosis regulates salivary gland lumen size and shape. Development 137(24):4177–4189. doi:10.1242/dev.048827
Popowicz GM, Schleicher M, Noegel AA, Holak TA (2006) Filamins: promiscuous organizers of the cytoskeleton. Trends Biochem Sci 31(7):411–419. doi:10.1016/j.tibs.2006.05.006
Poteryaev D, Datta S, Ackema K, Zerial M, Spang A (2010) Identification of the switch in early-to-late endosome transition. Cell 141(3):497–508. doi:10.1016/j.cell.2010.03.011
Powell RR, Temesvari LA (2004) Involvement of a Rab8-like protein of Dictyostelium discoideum, Sas1, in the formation of membrane extensions, secretion and adhesion during development. Microbiology 150(Pt 8):2513–2525. doi:10.1099/mic.0.27073-0
Roberts M, Barry S, Woods A, van der Sluijs P, Norman J (2001) PDGF-regulated rab4-dependent recycling of alphavbeta3 integrin from early endosomes is necessary for cell adhesion and spreading. Curr Biol 11(18):1392–1402
Roland JT, Kenworthy AK, Peranen J, Caplan S, Goldenring JR (2007) Myosin Vb interacts with Rab8a on a tubular network containing EHD1 and EHD3. Mol Biol Cell 18(8):2828–2837. doi:10.1091/mbc.E07-02-0169
Sahlender DA, Roberts RC, Arden SD, Spudich G, Taylor MJ, Luzio JP, Kendrick-Jones J, Buss F (2005) Optineurin links myosin VI to the Golgi complex and is involved in Golgi organization and exocytosis. J Cell Biol 169(2):285–295. doi:10.1083/jcb.200501162
Sakane A, Honda K, Sasaki T (2010) Rab13 regulates neurite outgrowth in PC12 cells through its effector protein, JRAB/MICAL-L2. Mol Cell Biol 30(4):1077–1087, doi:MCB.01067-09 [pii] 10.1128/MCB.01067-09
Sakane A, Abdallah AAM, Nakano K, Honda K, Ikeda W, Nishikawa Y, Matsumoto M, Matsushita N, Kitamura T, Sasaki T (2012) Rab13 small G protein and junctional Rab13-binding protein (JRAB) orchestrate actin cytoskeletal organization during epithelial junctional development. J Biol Chem 287(51):42455–42468. doi:10.1074/jbc.M112.383653
Sakane A, Abdallah AAM, Nakano K, Honda K, Kitamura T, Imoto I, Matsushita N, Sasaki T (2013) Junctional Rab13-binding protein (JRAB) regulates cell spreading via filamins. Genes Cells 18(9):810–822. doi:10.1111/gtc.12078

Sato T, Mushiake S, Kato Y, Sato K, Sato M, Takeda N, Ozono K, Miki K, Kubo Y, Tsuji A, Harada R, Harada A (2007) The Rab8 GTPase regulates apical protein localization in intestinal cells. Nature 448(7151):366–369. doi:10.1038/nature05929

Schweitzer JK, Sedgwick AE, D'Souza-Schorey C (2011) ARF6-mediated endocytic recycling impacts cell movement, cell division and lipid homeostasis. Semin Cell Dev Biol 22(1):39–47. doi:10.1016/j.semcdb.2010.09.002

Seabra MC, Coudrier E (2004) Rab GTPases and myosin motors in organelle motility. Traffic 5(6): 393–399. doi:10.1111/j.1398-9219.2004.00190.x

Seabra MC, Wasmeier C (2004) Controlling the location and activation of Rab GTPases. Curr Opin Cell Biol 16(4):451–457. doi:10.1016/j.ceb.2004.06.014

Shen L, Turner JR (2005) Actin depolymerization disrupts tight junctions via caveolae-mediated endocytosis. Mol Biol Cell 16(9):3919–3936. doi:10.1091/mbc.E04-12-1089

Sheth B, Fontaine JJ, Ponza E, McCallum A, Page A, Citi S, Louvard D, Zahraoui A, Fleming TP (2000) Differentiation of the epithelial apical junctional complex during mouse preimplantation development: a role for rab13 in the early maturation of the tight junction. Mech Dev 97(1–2):93–104

Shin OH, Couvillon AD, Exton JH (2001) Arfophilin is a common target of both class II and class III ADP-ribosylation factors. Biochemistry 40(36):10846–10852

Shin K, Fogg VC, Margolis B (2006) Tight junctions and cell polarity. Annu Rev Cell Dev Biol 22:207–235. doi:10.1146/annurev.cellbio.22.010305.104219

Stenmark H (2009) Rab GTPases as coordinators of vesicle traffic. Nat Rev Mol Cell Biol 10(8): 513–525. doi:10.1038/nrm2728

Su T, Bryant DM, Luton F, Verges M, Ulrich SM, Hansen KC, Datta A, Eastburn DJ, Burlingame AL, Shokat KM, Mostov KE (2010) A kinase cascade leading to Rab11-FIP5 controls transcytosis of the polymeric immunoglobulin receptor. Nat Cell Biol 12(12):1143–1153. doi:10.1038/ncb2118

Sun Y, Bilan PJ, Liu Z, Klip A (2010) Rab8A and Rab13 are activated by insulin and regulate GLUT4 translocation in muscle cells. Proc Natl Acad Sci U S A 107(46):19909–19914. doi:10.1073/pnas.1009523107

Sunshine C, Francis S, Kirk KL (2000) Rab3B regulates ZO-1 targeting and actin organization in PC12 neuroendocrine cells. Exp Cell Res 257(1):1–10. doi:10.1006/excr.2000.4855

Suzuki T, Nakamoto T, Ogawa S, Seo S, Matsumura T, Tachibana K, Morimoto C, Hirai H (2002) MICAL, a novel CasL interacting molecule, associates with vimentin. J Biol Chem 277(17): 14933–14941. doi:10.1074/jbc.M111842200

Takai Y, Sasaki T, Matozaki T (2001) Small GTP-binding proteins. Physiol Rev 81(1):153–208

Takai Y, Ikeda W, Ogita H, Rikitake Y (2008a) The immunoglobulin-like cell adhesion molecule nectin and its associated protein afadin. Annu Rev Cell Dev Biol 24:309–342. doi:10.1146/annurev.cellbio.24.110707.175339

Takai Y, Miyoshi J, Ikeda W, Ogita H (2008b) Nectins and nectin-like molecules: roles in contact inhibition of cell movement and proliferation. Nat Rev Mol Cell Biol 9(8):603–615. doi:10.1038/nrm2457

Takeichi M (2007) The cadherin superfamily in neuronal connections and interactions. Nat Rev Neurosci 8(1):11–20. doi:10.1038/nrn2043

Tang VW, Brieher WM (2012) alpha-Actinin-4/FSGS1 is required for Arp2/3-dependent actin assembly at the adherens junction. J Cell Biol 196(1):115–130. doi:10.1083/jcb.201103116

Terai T, Nishimura N, Kanda I, Yasui N, Sasaki T (2006) JRAB/MICAL-L2 is a junctional Rab13-binding protein mediating the endocytic recycling of occludin. Mol Biol Cell 17(5):2465–2475

Terman JR, Mao T, Pasterkamp RJ, Yu HH, Kolodkin AL (2002) MICALs, a family of conserved flavoprotein oxidoreductases, function in plexin-mediated axonal repulsion. Cell 109(7): 887–900

Tsukita S, Furuse M, Itoh M (2001) Multifunctional strands in tight junctions. Nat Rev Mol Cell Biol 2(4):285–293. doi:10.1038/35067088

Ulrich F, Krieg M, Schotz EM, Link V, Castanon I, Schnabel V, Taubenberger A, Mueller D, Puech PH, Heisenberg CP (2005) Wnt11 functions in gastrulation by controlling cell cohesion through Rab5c and E-cadherin. Dev Cell 9(4):555–564. doi:10.1016/j.devcel.2005.08.011

Utech M, Ivanov AI, Samarin SN, Bruewer M, Turner JR, Mrsny RJ, Parkos CA, Nusrat A (2005) Mechanism of IFN-gamma-induced endocytosis of tight junction proteins: myosin II-dependent vacuolarization of the apical plasma membrane. Mol Biol Cell 16(10):5040–5052. doi:10.1091/mbc.E05-03-0193

van Ijzendoorn SC, Tuvim MJ, Weimbs T, Dickey BF, Mostov KE (2002) Direct interaction between Rab3b and the polymeric immunoglobulin receptor controls ligand-stimulated transcytosis in epithelial cells. Dev Cell 2(2):219–228

Vasioukhin V, Bauer C, Yin M, Fuchs E (2000) Directed actin polymerization is the driving force for epithelial cell-cell adhesion. Cell 100(2):209–219, doi:S0092-8674(00)81559-7 [pii]

Wakamatsu Y, Sakai D, Suzuki T, Osumi N (2011) FilaminB is required for the directed localization of cell-cell adhesion molecules in embryonic epithelial development. Dev Dyn 240(1):149–161. doi:10.1002/dvdy.22518

Weber E, Berta G, Tousson A, St John P, Green MW, Gopalokrishnan U, Jilling T, Sorscher EJ, Elton TS, Abrahamson DR et al (1994) Expression and polarized targeting of a rab3 isoform in epithelial cells. J Cell Biol 125(3):583–594

Yamada S, Pokutta S, Drees F, Weis WI, Nelson WJ (2005) Deconstructing the cadherin-catenin-actin complex. Cell 123(5):889–901. doi:10.1016/j.cell.2005.09.020

Yamamura R, Nishimura N, Nakatsuji H, Arase S, Sasaki T (2008) The interaction of JRAB/MICAL-L2 with Rab8 and Rab13 coordinates the assembly of tight junctions and adherens junctions. Mol Biol Cell 19(3):971–983

Yonemura S, Itoh M, Nagafuchi A, Tsukita S (1995) Cell-to-cell adherens junction formation and actin filament organization: similarities and differences between non-polarized fibroblasts and polarized epithelial cells. J Cell Sci 108:127–142

Yoshimura S, Egerer J, Fuchs E, Haas AK, Barr FA (2007) Functional dissection of Rab GTPases involved in primary cilium formation. J Cell Biol 178(3):363–369. doi:10.1083/jcb.200703047

Youssef G, Gerner L, Naeem AS, Ralph O, Ono M, O'Neill CA, O'Shaughnessy RF (2013) Rab3Gap1 mediates exocytosis of Claudin-1 and tight junction formation during epidermal barrier acquisition. Dev Biol 380(2):274–285. doi:10.1016/j.ydbio.2013.04.034

Yu IM, Hughson FM (2010) Tethering factors as organizers of intracellular vesicular traffic. Annu Rev Cell Dev Biol 26:137–156. doi:10.1146/annurev.cellbio.042308.113327

Zahraoui A, Joberty G, Arpin M, Fontaine JJ, Hellio R, Tavitian A, Louvard D (1994) A small rab GTPase is distributed in cytoplasmic vesicles in non polarized cells but colocalizes with the tight junction marker ZO-1 in polarized epithelial cells. J Cell Biol 124(1–2):101–115

Zerial M, McBride H (2001) Rab proteins as membrane organizers. Nat Rev Mol Cell Biol 2(2): 107–117. doi:10.1038/35052055

Zhang XM, Ellis S, Sriratana A, Mitchell CA, Rowe T (2004) Sec15 is an effector for the Rab11 GTPase in mammalian cells. J Biol Chem 279(41):43027–43034. doi:10.1074/jbc.M402264200

Zhang J, Betson M, Erasmus J, Zeikos K, Bailly M, Cramer LP, Braga VM (2005) Actin at cell-cell junctions is composed of two dynamic and functional populations. J Cell Sci 118(Pt 23): 5549–5562, doi:jcs.02639 [pii]10.1242/jcs.02639

Zhou X, Tian F, Sandzen J, Cao R, Flaberg E, Szekely L, Cao Y, Ohlsson C, Bergo MO, Boren J, Akyurek LM (2007) Filamin B deficiency in mice results in skeletal malformations and impaired microvascular development. Proc Natl Acad Sci U S A 104(10):3919–3924. doi:10.1073/pnas.0608360104

Chapter 16
Protein Trafficking in Polarized Epithelial Cells

Sabrina Zink and Ralf Jacob

Abstract The architecture of epithelial cells comprises the separation of the plasma membrane into two separately composed compartments, the apical and the basolateral membrane. This concept is maintained by a sorting and transport machinery which verifies targeting of lipid and protein components to the correct membrane domain. Different intracellular routes, a heterogeneous population of sorting signals, and cellular factors associated with polarized trafficking are discussed.

Keywords Protein sorting • Post Golgi trafficking • Vesicle transport • Epithelial cells

Epithelial cells discriminate between different compartments of the human body. A polarized architecture of the plasma membrane as well as of the underlying cytoskeletal network is maintained by distinct biosynthetic pathways leading to the apical or basolateral membrane compartment and a combination of specific motor proteins that catalyze directed locomotion through the cell body. The intracellular routing of proteins through compartments of the secretory pathway is also denoted "polarized trafficking" (Caplan 1997) and is summarized in Fig. 16.1.

16.1 Trafficking Pathways in Epithelial Cells

Newly synthesized proteins enter this pathway in the endoplasmic reticulum (ER) from where they are delivered to the Golgi apparatus. Passage through this organelle to the *trans*-Golgi network (TGN) has several functions which include the maturation of glycoproteins and, as suggested by several observations, initial

S. Zink • R. Jacob (✉)
Department of Cell Biology and Cell Pathology, Philipps-University Marburg, Robert-Koch-Str. 6, Marburg 35037, Germany
e-mail: zink@staff.uni-marburg.de; jacob@staff.uni-marburg.de

K. Ebnet (ed.), *Cell Polarity 1*, DOI 10.1007/978-3-319-14463-4_16

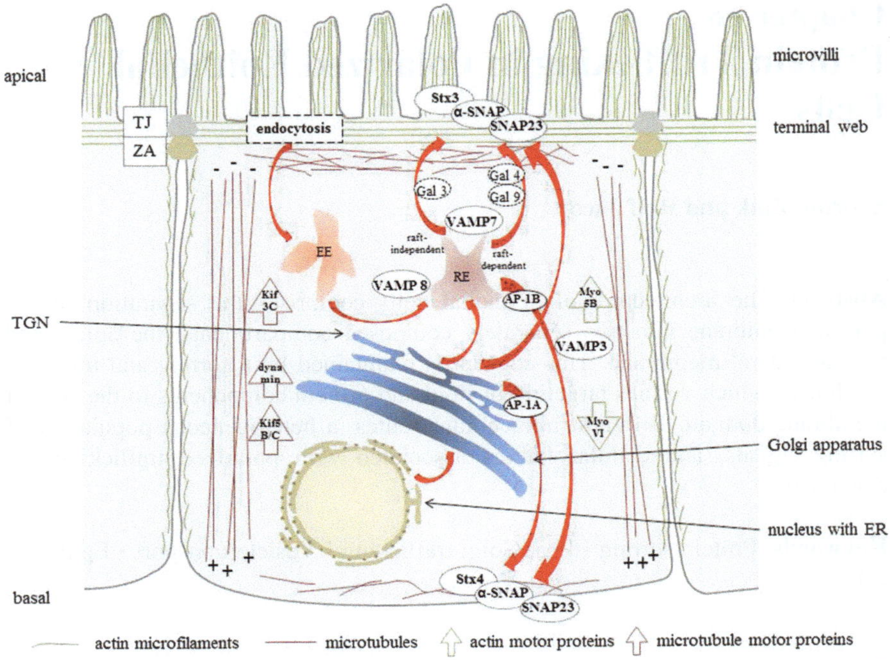

Fig. 16.1 Schematic representation of the main transport routes in epithelial cells. Cytoskeletal elements, motor proteins, SNARE proteins, and tethering complexes are indicated. *AP-1A/B* adaptor protein-1A/B, *EE* early endosome, *ER* endoplasmic reticulum, *Gal* galectin, *Kif* kinesin, *Myo* myosin, *RE* recycling endosome, *SNAP* soluble *N*-ethyl maleimide-sensitive factor attachment protein, *Stx* syntaxin, *TGN trans*-Golgi network, *TJ* tight junctions, *VAMP* vesicle-associated membrane protein, *ZA* zonula adherens

sorting steps (Alfalah et al. 2005; Clermont et al. 1992; Yano et al. 2005; Emr et al. 2009). Interestingly, some apical proteins bypass the Golgi (Tveit et al. 2009). At the TGN newly synthesized cargo is sorted into distinct transport routes. Secretory granules are formed here by epithelial cells, which possess a regulated secretory pathway like bladder umbrella or pancreatic acinar cells. These granules are different from carriers of the constitutive pathway (Fukuda 2008). Lysosomal proteins are diverged into lysosomes at the TGN. Moreover, segregation of apical and basolateral cargo molecules has been described for this last Golgi compartment (Griffiths and Simons 1986; Keller et al. 2001). In the classical view, these molecules are directly transported to their target membrane. More recent data suggest that apical and basolateral cargo molecules traverse endosomal organelles before they reach the cell surface (Ang et al. 2004; Cramm-Behrens et al. 2008; Cresawn et al. 2007).

In these studies, Rab GTPases as membrane organizers (Zerial and McBride 2001) have been used to determine distinct endosomal compartments. Early after TGN-release Rab4-positive common endosomes are the endosomal reception center for biosynthetic cargo (Cramm-Behrens et al. 2008). Apical as well as basolateral polypeptides are then passed through Rab8- and Rab11-positive endosomes (Ang et al. 2004; Cramm-Behrens et al. 2008; Cresawn et al. 2007). Accordingly, knockout of Rab8 in mouse intestine results in intracellular accumulation of apical polypeptides in vacuolar structures, which are very similar to inclusion bodies described for microvillus inclusion disease (Sato et al. 2007). Observations showing that in pancreatic acinar cells Rab8 is required for the formation of zymogen granules destined for the apical membrane (Faust et al. 2008) and that AP-1B-dependent basolateral transport is regulated by Rab8 in MDCK cells (Ang et al. 2003) further support the idea that Rab8-positive endosomes play a pivotal role in polarized biosynthetic trafficking of epithelial cells. The GTPase Rab11 on the other hand is distributed across a variety of intracellular *post*-Golgi membranes, including the TGN and recycling endosomes (Goldenring et al. 1996; Leung et al. 2000). These Rab11-positive recycling endosomes are traversed after passage through Rab8-enriched endosomes and prior to apical membrane delivery, which indicates that endosomal *post*-Golgi trafficking encompasses a subset of endosomes each defined by individual characteristics (Cramm-Behrens et al. 2008). The question is why newly synthesized proteins and lipids are transported through cascades of cellular compartments, which involve the ER, the Golgi, the TGN, and distinct endosomal organelles. Obviously, for each exit of one compartment and entry into the next, a sorting machinery for delivery into the appropriate pathway is ensured. This sequence of sorting events in each pathway could be the basic principle that ends up with a clearly defined distribution of proteins and lipids within the two membrane domains of polarized epithelial cells.

16.2 Sorting Machinery of Epithelial Cells

Current knowledge about components involved in polarized sorting of epithelial cells and the underlying sorting signals is mainly based on studies with epithelial cell lines in culture. These cells provide a useful tool to study basic principles of polarized protein transport since they reproduce characteristic features of epithelial cells in vivo.

16.3 Basolateral Sorting

The basic principle of protein sorting is defined by the interaction of intrinsic motifs and components of the sorting machinery to direct cargo through a specific transport route to the correct membrane domain. Sorting determinants from basolateral proteins, which were initially identified, demonstrated that short amino acid sequences, containing Tyr, Leu–Leu, or Leu–Val motifs, are localized in the cytoplasmic region near the transmembrane domain and direct proteins to the basolateral membrane (Casanova et al. 1991; Matter et al. 1992; Mostov et al. 1986; Rodriguez-Boulan and Powell 1992). These structures form a common β-turn signal for basolateral tethering (Aroeti et al. 1993). More recently, experiments performed with MDCK and FRT cells proposed that basolateral sorting signals comprise a multitude of secondary structures like α-helices or β-sheets and can also be localized in distal parts of the cytoplasmic domain (Beau et al. 2004). Basolateral vesicle formation involves the assembly of clathrin coats by adaptor protein binding (Deborde et al. 2008). Consequently, in some cases basolateral sorting determinants overlap with endocytosis signals (Lin et al. 1997). Proteins containing Tyr motifs bind to the γ1 subunit of the AP-1 adaptor complex (Odorizzi and Trowbridge 1997; Ohno et al. 1995; Bonifacino and Dell'Angelica 1999), and Leu–Leu motifs interact specifically with the β subunit of AP-1 (Rapoport et al. 1998). AP-1 seems to be a major regulator of epithelial sorting (Gravotta et al. 2012). AP-1A sorts cargo proteins from the TGN to the basolateral plasma membrane. Further downstream, within recycling endosomes, a second basolateral sorting step is verified by AP-1B, wherein μ1b adaptin is substituted for μ1a (Folsch et al. 1999). AP-1B-dependent basolateral sorting is not a general mechanism since this adaptor is not expressed in some epithelia, e.g., the liver (Ohno et al. 1999), retinal pigment epithelium (Diaz et al. 2009), and kidney proximal tubule (Schreiner et al. 2010). Another exception is the indirect interaction of AP-1B in basolateral targeting of the LDL receptor (Kang and Folsch 2011). In LLC-PK1 cells, the autosomal recessive hypercholesterolemia protein ARH mediates basolateral exocytosis of the receptor from recycling endosomes. Thus, the composition of the basolateral sorting machinery is cell type and organ specific and determines the fidelity of basolateral targeting. Another adaptor involved in basolateral sorting is AP-4 (Simmen et al. 2002). However, this adaptor does not recruit clathrin (Bonifacino and Traub 2003), and clathrin is required for basolateral delivery of several proteins (Deborde et al. 2008). It seems as if clathrin and AP-1B would not be required for basolateral sorting of TGF-α either. Here, Naked2 acts as cargo recognition and targeting protein to ensure proper targeting, tethering, and fusion of pro-TGFα-containing exocytic vesicles but considerably less is known about its mechanics (Li et al. 2007).

16.4 Apical Sorting

The apical sorting machinery, in contrast to basolateral trafficking, is more complex and involves extracytosolic, intramembrane, and cytosolic interactions. Many of them are still unclear and remain to be identified. Among these interactions is the lipid anchoring by glycosylphosphatidylinositol anchorage (GPI), a *post*translational modification in the endoplasmic reticulum (ER), which is the first apical sorting signal that was described (Brown et al. 1989; Lisanti et al. 1988). GPI-anchored proteins (GPI-APs) are preferentially localized at the apical plasma membrane of epithelial cells and have a high tendency to associate with lipid rafts (Brown and Rose 1992; Garcia et al. 1993; Lisanti et al. 1988; Wilson et al. 1990; Powell et al. 1991). These membrane microdomains are organized by the myelin and lymphocyte protein (MAL) as a key component and may serve as transport and sorting platform (Magal et al. 2009). However, this is not a general mechanism for protein targeting to the apical membrane. Soluble forms of placental alkaline phosphatase (PLAP) or the glycoprotein gp80/clusterin do not associate with lipid rafts but are still apically sorted (Lipardi et al. 2000; Graichen et al. 1996). Thus, alternative apical sorting mechanisms and pathways exist that are raft independent (Jacob and Naim 2001). Moreover, addition of a GPI-anchoring motif to rat growth hormone (rGH), a soluble protein that is secreted in a non-polarized manner, is not sufficient for apical targeting (Benting et al. 1999). Another group of determinants for apical protein delivery relies on the presence of *N*- or *O*-glycans (Alfalah et al. 1999; Benting et al. 1999; Jacob et al. 2000; Yeaman et al. 1997).

Initial ideas on the involvement of *N*-glycosyl chains in apical protein targeting came from studies using *N*-glycosylation inhibitors. Treatment with the GlcNAc-analogue tunicamycin, which inhibits early steps of *N*-glycosylation, missorts apical gp80/clusterin to both membrane domains of MDCK cells (Urban et al. 1987). Experiments with glycosylation-deficient cell lines came to the same conclusion. MDCK cells with a defect in galactose incorporation into glycoproteins mistarget the apical glycoprotein gp114 to the basolateral membrane (Le Bivic et al. 1993). Moreover, recombinant addition of *N*-glycosyl chains to the rat growth hormone (rGH) directs the soluble glycoprotein into the apical medium (Scheiffele et al. 1995). Conversely, removal of one *N*-glycosyl chain from erythropoietin results in impaired apical secretion in MDCK cells (Kitagawa et al. 1994). Further evidence for a role of *N*-glycans in apical protein sorting comes from experiments with endolyn (Ihrke et al. 2001), the glycine transporter GLYT2 (Martinez-Maza et al. 2001), or dipeptidase (Pang et al. 2004). Intriguingly, some *N*-glycosylated basolateral proteins are not delivered to the apical membrane. One reason for this behavior could be the simultaneous presence of cytosolic basolateral targeting determinants that override apical sorting determinants in the extracytosolic part of the protein (Simons and Ikonen 1997). Another explanation would be that the glycan chains for apical protein sorting have to be presented within a specific sterical context to allow the formation of large protein clusters (Delacour et al. 2007).

In addition to *N*-glycans, *O*-linked glycosyl chains may also act as apical targeting signal. This has been shown for the neurotrophin receptor $p75^{NTR}$ and the hydrolase sucrase isomaltase (SI), both of which possess membrane proximal *O*-glycosylated stalk domains. Their deletion results in a reorientation of these polypeptides to both surface domains of epithelial cells (Jacob et al. 2000; Yeaman et al. 1997). These data also indicate the importance of how and where the sugar chains are presented since the position of *O*-glycans in the proximal part of the two stalk domains is crucial for apical sorting (Breuza et al. 2002; Jacob et al. 2000). This supports the idea of particular sterical requirements for the assembly of glycan-dependent apical sorting machinery.

In the meantime, several lectins of the galectin family have been identified that play a role in apical protein sorting. Three galectins, galectin-3, galectin-4, and galectin-9, have been described as coordinators for correct delivery of apical cargo molecules in MDCK cells (Delacour et al. 2005, 2006; Mishra et al. 2010; Mo et al. 2012; Yamamoto et al. 2013). For galectin-3 it has been shown that the lectin forms high-molecular-weight clusters with apical glycoproteins that are not associated with lipid rafts in a *post*-Golgi endosomal compartment (Delacour et al. 2007). Galectin-3 directs these cargo molecules to the apical cell surface from where the lectin is recycled back into endosomes (Schneider et al. 2010; Straube et al. 2013). In addition, knockdown of galectin-3 results in transport defects of intestinal hydrolases in mice (Delacour et al. 2008). Recent data on pH-dependent alterations in the binding and oligomerization characteristics provide additional clues for the underlying mechanistic principles of galectin-3-mediated apical sorting (manuscript submitted).

Finally, also the transmembrane domain of proteins can mediate incorporation into lipid rafts and apical sorting. This has been shown for the transmembrane domains of influenza virus neuraminidase (NA) and hemagglutinin (HA) (Kundu et al. 1996; Lin et al. 1998). In addition to the primary amino acid sequence of the transmembrane domain, its length also seems to be important for raft association and lipid-based sorting (Schuck and Simons 2004). Finally, the cytosolic tails of rhodopsin (Chuang and Sung 1998), megalin (Marzolo et al. 2003; Takeda et al. 2003), and receptor guanylyl cyclases (Hodson et al. 2006) contain apical sorting information. This can be directly linked to motor function as exemplified by the interaction between the dynein light chain Tctex-1 and the C-terminus of rhodopsin (Tai et al. 1999).

16.5 Cytoskeletal Tracks in Apical and Basolateral Trafficking

What are the tracks for apical and basolateral trafficking? The unique organization of epithelial cells is manifested by complex cytoskeletal networks composed of microfilaments, intermediate filaments, and microtubules. Both actin microfilaments and microtubules are essential for apical and basolateral transport routes (reviewed in Cao et al. (2012); Delacour and Jacob (2006); Weisz and Rodriguez-Boulan (2009)).

16.5.1 Actin Microfilaments

Microfilaments, which are composed of G-actin subunits, are associated with manifold components within a cell. They are required to assemble the subcellular architecture of polarized cells like tight and adherens junctions or focal adhesion sites (Kartenbeck et al. 1991; Burridge et al. 1988). In addition, actin filaments build the core of apical microvilli, and these actin bundles insert into the terminal web underneath the apical membrane, which is composed of actin, spectrin, and other cytoskeletal proteins (Mooseker 1985). This distribution is reflected by many cellular actin functions in vesicle-mediated exocytosis, endocytosis, and transcytosis. Here, actin contributes predominantly in the formation, scission, and fusion of these vesicles (Harris and Tepass 2010; Lanzetti 2007; Smythe and Ayscough 2006; Weisz and Rodriguez-Boulan 2009). Experimental data using the actin-inhibitory effects of cytochalasin D indicate the necessity of actin filaments especially for apical transport processes (Jacob et al. 2003; Maples et al. 1997; Ojakian and Schwimmer 1988; Valentijn et al. 1999). However, not all apical trafficking events require an intact actin network, which was demonstrated for apical delivery of endogenous gp80 or exogenously expressed LPH (Jacob et al. 2003; Parczyk et al. 1989). Disruption of the microvillar architecture caused by atrophy and inclusions as is the case in the familial enteropathy microvillus inclusion disease (MID) reveals the importance of the actin terminal web. Here, the apical F-actin layer exhibits a reduced thickness leading to an inhibition of apical protein delivery (Ameen and Salas 2000).

The role of actin in endocytosis was intensively investigated during the last decades and in several polarized tissue cell types: endothelial cells, hepatocytes, MDCK cells, intestinal Caco-2 cells, etc. (Alexander et al. 1998; Apodaca 2001; Durrbach et al. 1996; Gottlieb et al. 1993; Jackman et al. 1994; Kaufman et al. 1990; Kyle et al. 1988; Shurety et al. 1996, 1998). Actin seems to be important for internalization particularly at the apical cell pole, but several experiments indicate an association of actin with basolateral endocytosis as well. Cytochalasin-treated MDCK cells reveal a block in apical endocytosis of membrane markers like VSVG or receptor-bound ferritin and in addition a significant increase in coated pits stuck at the apical surface (Gottlieb et al. 1993). The actin-stabilizing drug jasplakinolide stimulates basolateral endocytosis of the fluid-phase markers FITC-dextran and horseradish peroxidase in MDCK (Jacob et al. 2003; Parczyk et al. 1989; Shurety et al. 1998). Furthermore, postendocytic apical trafficking requires an intact actin network reflected by an inhibition of basolateral to apical transcytosis and apical recycling by cytochalasin treatment (Maples et al. 1997).

Actin microfilaments are associated with myosin motor proteins which have been implicated in apical and basolateral trafficking (Altschuler et al. 2003; Au et al. 2007). The binding mechanism of myosin 5B to Rab proteins may be direct or indirect through Rab effectors providing a complex involvement to the cellular trafficking system. Dependent on interactions with Rab11 and Rab8, myosin 5B was required for apical trafficking, polarization, and lumen formation (Roland et al. 2011; Sato et al. 2007). It is also known that myosin VI is involved in basolateral trafficking of tyrosine motif containing basolateral membrane proteins in polarized MDCK cells, potentially in the AP-1B-dependent pathway (Au et al. 2007).

16.5.2 Microtubules

As we have seen for actin, microtubules are involved in a variety of cellular functions: beside the maintenance of cell shape and mechanical stability, long-distance transport of organelles and vesicles requires a functional microtubule network. These highly dynamic tubules undergo a constant cycle of polymerization and depolymerization to meet all the requirements. The polar structure of microtubules composed of 13 protofilaments consisting of alpha- and beta-tubulin dimers and several posttranslational modifications of the subunits furthermore permits this variety of different functions (Wade 2009).

The microtubule cytoskeleton is subject to extensive changes during the polarization process in epithelial cells. Nonpolar cells comprise microtubule filaments emanating from a centrosome, respectively, a microtubule-organizing center (MTOC) located near the nucleus (Vorobjev and Nadezhdina 1987). Fundamental reorganizations occurring in the course of polarization not only affect microtubules but also the whole intracellular organization. After the release of microtubule minus ends from the centrosome, minus-end binding factors anchor them on cell–cell contacts or at the apical cell pole (Keating et al. 1997; Ligon and Holzbaur 2007; Stehbens et al. 2006). In fully polarized cells, microtubules are present as meshwork underneath the apical membrane and the basal cell pole. In addition, they are organized as bundles along the apical–basal axis (Van Furden et al. 2004; Bartolini and Gundersen 2006; Musch 2004). The apical network consists of horizontally arranged microtubules without a consistent arrangement. Within the vertical microtubule bundles, minus ends can be found at the apical domain, whereas the plus ends are oriented toward the basal membrane (Bacallao et al. 1989; Bre et al. 1990; de Forges et al. 2012; Meads and Schroer 1995; Van Furden et al. 2004). Although microtubules are known to be highly dynamic, it is assumed that these tubules are more stable in polarized cells compared to centrosomal microtubules of non-polarized cells. Stable microtubules are often characterized by different post-translational modifications including acetylation, detyrosination, polyglutamylation, and polyglycylation (reviewed in Hammond et al. (2008); Janke and Kneussel (2010); Wloga and Gaertig (2010); Janke and Bulinski (2011)). Both tubulin subunits can be subject to these modifications, and they underlie significant changes during the polarization process (Quinones et al. 2011; Zink et al. 2012).

The role of microtubules in polarized trafficking seems to be quite complex since, on one hand, microtubule disruption inhibits apical and basolateral *post*-Golgi trafficking and, on the other hand, some proteins are transported to both membrane domains independently of microtubules. In one of the first reports, colchicine and nocodazole as microtubule-depolymerizing agents negatively affected apical transport of influenza virus HA and not basolateral transport of VSVG (Rindler et al. 1987). This observation was confirmed by a number of subsequent studies (Achler et al. 1989; Eilers et al. 1989; Gilbert et al. 1991; Grindstaff et al. 1998a; Lafont et al. 1994; Matter et al. 1990; Ojakian and Schwimmer 1992; Rindler et al. 1987; Salas et al. 1986; van Zeijl and Matlin

1990). Contradictory results, however, were published as well (Grindstaff et al. 1998a; Lafont et al. 1994; Salas et al. 1986). Here, microtubule disassembly did not affect the asymmetric distribution of influenza or vesicular stomatitis virus to the apical or basolateral surface in MDCK cells (Salas et al. 1986). Lafont and coworkers showed that a depletion of the microtubule motor protein kinesin inhibited transport processes to the basolateral membrane (Lafont et al. 1994), and another report revealed a 25–50 % decrease in the delivery of newly synthesized gp135/170 to the apical and of E-cadherin to the basolateral surface (Grindstaff et al. 1998a).

For vesicular protein trafficking along microtubules, motor proteins of the kinesin and dynein family are involved. The movement along microtubules occurs over long distances and is relatively fast with 1–2 μm/s in contrast to movement along actin microfilaments with approximately 0.1 μm/s. Previously, it was supposed that kinesins move their cargoes exclusively plus-end directed, whereas dynein is responsible for the minus-end-directed transport (reviewed, e.g., in Wade (2009)). However, the experimental work during the last decade revealed frequent bidirectional intracellular transport resulting in two models which would allow bidirectional transport: tug-of-war and regulation. The first one postulates that kinesins and dyneins can bind simultaneously to the filament. The following struggle would lead to a stochastic detaching of one polarity marker caused by the force of the other one. The regulation model in contrast assumes cargo binding of both motor proteins, one of which is active to move the cargo along the microtubule into the relevant direction. The activity should be regulated by cofactors (Gross et al. 2002; Gross 2004; Welte 2004; Schuster et al. 2011). Evidences for both models were published recently (Bouzat and Falo 2011; Muller et al. 2008; Assmann and Lenz 2013; Leidel et al. 2012).

The regulation of the motor–cargo–microtubule interaction is thought to occur at several events of the transport process: motor–cargo binding, motor activation, microtubule track selection, cargo release at the destination, and motor recycling (Verhey and Hammond 2009). After successful delivery, cargoes need to be unloaded from the kinesin, and the detached motor protein is then inactivated by an autoinhibition mechanism (Dietrich et al. 2008; Wong et al. 2009; Hammond et al. 2009).

Several evidences suggest a specific participation of kinesin and dynein in the transport and sorting in particular of apical proteins. As indicated above, cytoplasmic dynein interacts with apically transported rhodopsin via its light chain Tctex-1 in MDCK cells (Chuang and Sung 1998; Tai et al. 1999). The minus-end-directed kinesin KifC3 is important for trafficking of influenza virus HA and annexin 13b to the apical membrane (Noda et al. 2001). Later, the plus-end kinesins Kif5B, Kif5C, and dynamin were identified in apical transport of $p75^{NTR}$ in polarized MDCK cells (Astanina and Jacob 2010; Jaulin et al. 2007; Kreitzer et al. 2000). Interestingly, the same transmembrane receptor is transported by members of the kinesin-3 family in non-polarized MDCK cells (Xue et al. 2010), thus suggesting that the subcellular cargo transport machinery changes during cell polarization.

16.6 Vesicle Tethering and the Role of SNAREs

At a final stage, transport carriers need to dock at the target membrane to release their cargo. Here, specific molecules are necessary including tethering complexes, SNARE, and Rab proteins. Tethering complexes seem to be binding factors between two membranes to prepare them for membrane fusion by binding to Rab GTPases and to SNAREs (Sudhof and Rothman 2009; Zerial and McBride 2001).

Beside its role in tethering vesicles at the plasma membrane, the exocyst complex is required for the formation and targeting of vesicles to the basolateral membrane (He and Guo 2009; Munson and Novick 2006; Yeaman et al. 2001). In epithelial cells the exocyst is localized at the Golgi complex, the TGN, recycling endosomes, and the junctional complex and plays a role in endocytosis of basolateral recycling vesicles (Folsch et al. 2003; Prigent et al. 2003; Yeaman et al. 2004). This was demonstrated using overexpressed subunits of the exocyst complex that caused a stimulation of the delivery of basolateral but not apical membrane proteins (Lipschutz et al. 2000). In contrast mutations of the exocyst lead to inhibition of recycling endosome trafficking to the basolateral surface (Langevin et al. 2005), and inhibition with an antibody directed against the exocyst component Sec8 blocks the TGN-to-basolateral membrane transport (Grindstaff et al. 1998b). Further experiments could show that several cofactors are as well involved in these exocyst functions. Thus, the GTPase RalA interacts with different subunits and is required for basolateral transport events (Shipitsin and Feig 2004), and the basolateral cargo–adaptor protein, AP-1B, is able to recruit the exocyst to recycling endosomes (Ang et al. 2004; Folsch et al. 2003). Less is known for the apical transport route; however, several data indicate a role for this complex in apical trafficking. The exocyst is localized to the primary cilium (Rogers et al. 2004), regulates the exocytosis of apical secretory proteins (Lipschutz et al. 2000), and is in different ways connected to apically directed vesicles (Beronja et al. 2005; Barile et al. 2005).

The final vesicle fusion step in the protein-trafficking pathway is mediated by soluble *N*-ethylmaleimide-sensitive factor attachment protein receptors (SNARE proteins). These small molecules are localized in opposing membranes to trigger the fusion of vesicles with their target membrane. To do so they use the free energy released during the formation of a four-helix bundle enabled by the characteristic SNARE motif, a conserved stretch of 60–70 amino acids arranged in heptad repeats. SNARE proteins can be classified into v-SNAREs and t-SNAREs, associated with the vesicle or with the target membrane (reviewed, i.e., in Jahn and Scheller (2006)). For hepatocytes the existence of different t-SNAREs was reported: syntaxin 2 and 3 at the canalicular and syntaxin 4 at the basolateral plasma membrane (Fujita et al. 1998). A similar distribution with syntaxin 3 at the apical and syntaxin 4 at the basolateral membrane was found in polarized MDCK cells (Low et al. 1996). Later Low and coworkers confirmed their results and concluded an involvement of syntaxin 3, synaptosomal-associated protein 23 (SNAP 23), and soluble *N*-ethyl maleimide-sensitive factor attachment protein (α-SNAP) in apical membrane fusion. They could show that the overexpression of syntaxin 3 negatively affects the TGN to apical transport and apical recycling, which causes an accumulation of vesicles underneath the apical membrane. In addition, inhibition of SNAP 23 and α-SNAP by botulinum neurotoxin E or specific antibodies inhibits apical as well as basolateral trafficking. Thus, only components of the SNARE machinery seem to be required for the trafficking specificity in polarized cells (Low et al. 1998). In Caco-2 cells, the SNARE complex at the apical membrane is formed by syntaxin 3, SNAP 23, and the vesicle-associated membrane protein 7 (VAMP 7) (Galli et al. 1998). For the latter a mediation of raft- and non-raft-associated apical transport was shown (Pocard et al. 2007). Furthermore, VAMP 7 is involved in the surface delivery of apical reporter proteins in MDCK cells as well (Lafont et al. 1999). Another example is VAMP 8 which is supposed to operate in the transcytotic apical delivery. In MDCK cells direct apical trafficking does not depend on VAMP 8, however, apical recycling and endocytosis does (Steegmaier et al. 2000). Though different interactions of VAMP 8 were identified, it forms a complex with the apical SNARE syntaxin 3 and SNAP 23 (Pombo et al. 2003), and together with the basolateral SNARE syntaxin 4 and SNAP 23, it seems to regulate exocytosis in the exocrine system (Wang et al. 2007). The requirement of VAMP 8 in endocytic uptake and recycling has also been shown for Caco-2 and Fischer rat thyroid (FRT) cells, whereas direct apical and basolateral trafficking is not affected by VAMP 8 knockdown (Pocard et al. 2007). For AP-1B-dependent basolateral transport, the v-SNARE VAMP 3 (cellubrevin) has been reported, since cleavage of VAMP3 leads to missorting of AP-1B-dependent cargoes (transferrin receptor, LDL receptor) (Fields et al. 2007).

In conclusion, many components involved in vesicle tethering, transport, and fusion show a high degree of specificity for various apical and basolateral transport pathways of epithelial cells.

References

Achler C, Filmer D, Merte C, Drenckhahn D (1989) Role of microtubules in polarized delivery of apical membrane proteins to the brush border of the intestinal epithelium. J Cell Biol 109:179–189

Alexander JS, Jackson SA, Chaney E, Kevil CG, Haselton FR (1998) The role of cadherin endocytosis in endothelial barrier regulation: involvement of protein kinase C and actin-cadherin interactions. Inflammation 22:419–433

Alfalah M, Jacob R, Preuss U, Zimmer KP, Naim H, Naim HY (1999) O-linked glycans mediate apical sorting of human intestinal sucrase-isomaltase through association with lipid rafts. Curr Biol 9:593–596

Alfalah M, Wetzel G, Fischer I, Busche R, Sterchi EE, Zimmer KP, Sallmann HP, Naim HY (2005) A novel type of detergent-resistant membranes may contribute to an early protein sorting event in epithelial cells. J Biol Chem 280(52):42636–42643

Altschuler Y, Hodson C, Milgram SL (2003) The apical compartment: trafficking pathways, regulators and scaffolding proteins. Curr Opin Cell Biol 15:423–429

Ameen NA, Salas PJ (2000) Microvillus inclusion disease: a genetic defect affecting apical membrane protein traffic in intestinal epithelium. Traffic 1:76–83

Ang AL, Folsch H, Koivisto UM, Pypaert M, Mellman I (2003) The Rab8 GTPase selectively regulates AP-1B-dependent basolateral transport in polarized Madin-Darby canine kidney cells. J Cell Biol 163:339–350

Ang AL, Taguchi T, Francis S, Folsch H, Murrells LJ, Pypaert M, Warren G, Mellman I (2004) Recycling endosomes can serve as intermediates during transport from the Golgi to the plasma membrane of MDCK cells. J Cell Biol 167:531–543

Apodaca G (2001) Endocytic traffic in polarized epithelial cells: role of the actin and microtubule cytoskeleton. Traffic 2:149–159

Aroeti B, Kosen PA, Kuntz ID, Cohen FE, Mostov KE (1993) Mutational and secondary structural analysis of the basolateral sorting signal of the polymeric immunoglobulin receptor. J Cell Biol 123:1149–1160

Assmann MA, Lenz P (2013) Characterization of bidirectional molecular motor-assisted transport models. Phys Biol 10:016003

Astanina K, Jacob R (2010) KIF5C, a kinesin motor involved in apical trafficking of MDCK cells. Cell Mol Life Sci 67:1331–1342

Au JS, Puri C, Ihrke G, Kendrick-Jones J, Buss F (2007) Myosin VI is required for sorting of AP-1B-dependent cargo to the basolateral domain in polarized MDCK cells. J Cell Biol 177:103–114

Bacallao R, Antony C, Dotti C, Karsenti E, Stelzer EH, Simons K (1989) The subcellular organization of Madin-Darby canine kidney cells during the formation of a polarized epithelium. J Cell Biol 109:2817–2832

Barile M, Pisitkun T, Yu MJ, Chou CL, Verbalis MJ, Shen RF, Knepper MA (2005) Large scale protein identification in intracellular aquaporin-2 vesicles from renal inner medullary collecting duct. Mol Cell Proteomics 4:1095–1106

Bartolini F, Gundersen GG (2006) Generation of noncentrosomal microtubule arrays. J Cell Sci 119:4155–4163

Beau I, Groyer-Picard MT, Desroches A, Condamine E, Leprince J, Tome JP, Dessen P, Vaudry H, Misrahi M (2004) The basolateral sorting signals of the thyrotropin and luteinizing hormone receptors: an unusual family of signals sharing an unusual distal intracellular localization, but unrelated in their structures. Mol Endocrinol 18:733–746

Benting JH, Rietveld AG, Simons K (1999) N-Glycans mediate the apical sorting of a GPI-anchored, raft-associated protein in Madin-Darby canine kidney cells. J Cell Biol 146:313–320

Beronja S, Laprise P, Papoulas O, Pellikka M, Sisson J, Tepass U (2005) Essential function of Drosophila Sec6 in apical exocytosis of epithelial photoreceptor cells. J Cell Biol 169:635–646

Bonifacino JS, Dell'Angelica EC (1999) Molecular bases for the recognition of tyrosine-based sorting signals. J Cell Biol 145:923–926

Bonifacino JS, Traub LM (2003) Signals for sorting of transmembrane proteins to endosomes and lysosomes. Annu Rev Biochem 72:395–447

Bouzat S, Falo F (2011) Tug of war of molecular motors: the effects of uneven load sharing. Phys Biol 8:066010

Bre MH, Pepperkok R, Hill AM, Levilliers N, Ansorge W, Stelzer EH, Karsenti E (1990) Regulation of microtubule dynamics and nucleation during polarization in MDCK II cells. J Cell Biol 111:3013–3021

Breuza L, Garcia M, Delgrossi MH, Le Bivic A (2002) Role of the membrane-proximal O-glycosylation site in sorting of the human receptor for neurotrophins to the apical membrane of MDCK cells. Exp Cell Res 273:178–186

Brown DA, Rose JK (1992) Sorting of GPI-anchored proteins to glycolipid-enriched membrane subdomains during transport to the apical cell surface. Cell 68:533–544

Brown DA, Crise B, Rose JK (1989) Mechanism of membrane anchoring affects polarized expression of two proteins in MDCK cells. Science 245:1499–1501

Burridge K, Fath K, Kelly T, Nuckolls G, Turner C (1988) Focal adhesions: transmembrane junctions between the extracellular matrix and the cytoskeleton. Annu Rev Cell Biol 4:487–525

Cao X, Surma MA, Simons K (2012) Polarized sorting and trafficking in epithelial cells. Cell Res 22:793–805

Caplan MJ (1997) Membrane polarity in epithelial cells: protein sorting and establishment of polarized domains. Am J Physiol 272:F425–F429

Casanova JE, Apodaca G, Mostov KE (1991) An autonomous signal for basolateral sorting in the cytoplasmic domain of the polymeric immunoglobulin receptor. Cell 66:65–75

Chuang JZ, Sung CH (1998) The cytoplasmic tail of rhodopsin acts as a novel apical sorting signal in polarized MDCK cells. J Cell Biol 142:1245–1256

Clermont Y, Rambourg A, Hermo L (1992) Segregation of secretory material in all elements of the Golgi apparatus in principal epithelial cells of the rat seminal vesicle. Anat Rec 232:349–358

Cramm-Behrens CI, Dienst M, Jacob R (2008) Apical cargo traverses endosomal compartments on the passage to the cell surface. Traffic 9:2206–2220

Cresawn KO, Potter BA, Oztan A, Guerriero CJ, Ihrke G, Goldenring JR, Apodaca G, Weisz OA (2007) Differential involvement of endocytic compartments in the biosynthetic traffic of apical proteins. EMBO J 26:3737–3748

de Forges H, Bouissou A, Perez F (2012) Interplay between microtubule dynamics and intracellular organization. Int J Biochem Cell Biol 44:266–274

Deborde S, Perret E, Gravotta D, Deora A, Salvarezza S, Schreiner R, Rodriguez-Boulan E (2008) Clathrin is a key regulator of basolateral polarity. Nature 452:719–723

Delacour D, Jacob R (2006) Apical protein transport. Cell Mol Life Sci 63:2491–2505

Delacour D, Gouyer V, Zanetta JP, Drobecq H, Leteurtre E, Grard G, Moreau-Hannedouche O, Maes E, Pons A, Andre S, Le Bivic A, Gabius HJ, Manninen A, Simons K, Huet G (2005) Galectin-4 and sulfatides in apical membrane trafficking in enterocyte-like cells. J Cell Biol 169:491–501

Delacour D, Cramm-Behrens CI, Drobecq H, Le Bivic A, Naim HY, Jacob R (2006) Requirement for galectin-3 in apical protein sorting. Curr Biol 16:408–414

Delacour D, Greb C, Koch A, Salomonsson E, Leffler H, Le Bivic A, Jacob R (2007) Apical sorting by galectin-3-dependent glycoprotein clustering. Traffic 8:379–388

Delacour D, Koch A, Ackermann W, Eude-Le Parco I, Elsasser HP, Poirier F, Jacob R (2008) Loss of galectin-3 impairs membrane polarisation of mouse enterocytes *in vivo*. J Cell Sci 121:458–465

Diaz F, Gravotta D, Deora A, Schreiner R, Schoggins J, Falck-Pedersen E, Rodriguez-Boulan E (2009) Clathrin adaptor AP1B controls adenovirus infectivity of epithelial cells. Proc Natl Acad Sci U S A 106:11143–11148

Dietrich KA, Sindelar CV, Brewer PD, Downing KH, Cremo CR, Rice SE (2008) The kinesin-1 motor protein is regulated by a direct interaction of its head and tail. Proc Natl Acad Sci U S A 105:8938–8943

Durrbach A, Louvard D, Coudrier E (1996) Actin filaments facilitate two steps of endocytosis. J Cell Sci 109(Pt 2):457–465

Eilers U, Klumperman J, Hauri HP (1989) Nocodazole, a microtubule-active drug, interferes with apical protein delivery in cultured intestinal epithelial cells (Caco-2). J Cell Biol 108:13–22

Emr S, Glick BS, Linstedt AD, Lippincott-Schwartz J, Luini A, Malhotra V, Marsh BJ, Nakano A, Pfeffer SR, Rabouille C, Rothman JE, Warren G, Wieland FT (2009) Journeys through the Golgi—taking stock in a new era. J Cell Biol 187:449–453

Faust F, Gomez-Lazaro M, Borta H, Agricola B, Schrader M (2008) Rab8 is involved in zymogen granule formation in pancreatic acinar AR42J cells. Traffic 9:964–979

Fields IC, Shteyn E, Pypaert M, Proux-Gillardeaux V, Kang RS, Galli T, Folsch H (2007) v-SNARE cellubrevin is required for basolateral sorting of AP-1B-dependent cargo in polarized epithelial cells. J Cell Biol 177:477–488

Folsch H, Ohno H, Bonifacino JS, Mellman I (1999) A novel clathrin adaptor complex mediates basolateral targeting in polarized epithelial cells. Cell 99:189–198

Folsch H, Pypaert M, Maday S, Pelletier L, Mellman I (2003) The AP-1A and AP-1B clathrin adaptor complexes define biochemically and functionally distinct membrane domains. J Cell Biol 163:351–362

Fujita H, Tuma PL, Finnegan CM, Locco L, Hubbard AL (1998) Endogenous syntaxins 2, 3 and 4 exhibit distinct but overlapping patterns of expression at the hepatocyte plasma membrane. Biochem J 329(Pt 3):527–538

Fukuda M (2008) Regulation of secretory vesicle traffic by Rab small GTPases. Cell Mol Life Sci 65:2801–2813

Galli T, Zahraoui A, Vaidyanathan VV, Raposo G, Tian JM, Karin M, Niemann H, Louvard D (1998) A novel tetanus neurotoxin-insensitive vesicle-associated membrane protein in SNARE complexes of the apical plasma membrane of epithelial cells. Mol Biol Cell 9:1437–1448

Garcia M, Mirre C, Quaroni A, Reggio H, Le Bivic A (1993) GPI-anchored proteins associate to form microdomains during their intracellular transport in Caco-2 cells. J Cell Sci 104:1281–1290

Gilbert T, Le BA, Quaroni A, Rodriguez-Boulan E (1991) Microtubular organization and its involvement in the biogenetic pathways of plasma membrane proteins in Caco-2 intestinal epithelial cells. J Cell Biol 113:275–288

Goldenring JR, Smith J, Vaughan HD, Cameron P, Hawkins W, Navarre J (1996) Rab11 is an apically located small GTP-binding protein in epithelial tissues. Am J Physiol 270:G515–G525

Gottlieb TA, Ivanov IE, Adesnik M, Sabatini DD (1993) Actin microfilaments play a critical role in endocytosis at the apical but not the basolateral surface of polarized epithelial cells. J Cell Biol 120:695–710

Graichen R, Losch A, Appel D, Koch-Brandt C (1996) Glycolipid-independent sorting of a secretory glycoprotein to the apical surface of polarized epithelial cells. J Biol Chem 271:15854–15857

Gravotta D, Carvajal-Gonzalez JM, Mattera R, Deborde S, Banfelder JR, Bonifacino JS, Rodriguez-Boulan E (2012) The clathrin adaptor AP-1A mediates basolateral polarity. Dev Cell 22:811–823

Griffiths G, Simons K (1986) The trans Golgi network: sorting at the exit site of the Golgi complex. Science 234:438–443

Grindstaff KK, Bacallao RL, Nelson WJ (1998a) Apiconuclear organization of microtubules does not specify protein delivery from the trans-Golgi network to different membrane domains in polarized epithelial cells. Mol Biol Cell 9:685–699

Grindstaff KK, Yeaman C, Anandasabapathy N, Hsu SC, Rodriguez-Boulan E, Scheller RH, Nelson WJ (1998b) Sec6/8 complex is recruited to cell-cell contacts and specifies transport vesicle delivery to the basal-lateral membrane in epithelial cells. Cell 93:731–740

Gross SP (2004) Hither and yon: a review of bi-directional microtubule-based transport. Phys Biol 1:R1–R11

Gross SP, Welte MA, Block SM, Wieschaus EF (2002) Coordination of opposite-polarity microtubule motors. J Cell Biol 156:715–724

Hammond JW, Cai D, Verhey KJ (2008) Tubulin modifications and their cellular functions. Curr Opin Cell Biol 20:71–76

Hammond JW, Cai D, Blasius TL, Li Z, Jiang Y, Jih GT, Meyhofer E, Verhey KJ (2009) Mammalian Kinesin-3 motors are dimeric in vivo and move by processive motility upon release of autoinhibition. PLoS Biol 7:e72

Harris KP, Tepass U (2010) Cdc42 and vesicle trafficking in polarized cells. Traffic 11:1272–1279

He B, Guo W (2009) The exocyst complex in polarized exocytosis. Curr Opin Cell Biol 21:537–542

Hodson CA, Ambrogi IG, Scott RO, Mohler PJ, Milgram SL (2006) Polarized apical sorting of guanylyl cyclase C is specified by a cytosolic signal. Traffic 7:456–464

Ihrke G, Bruns JR, Luzio JP, Weisz OA (2001) Competing sorting signals guide endolyn along a novel route to lysosomes in MDCK cells. EMBO J 20:6256–6264

Jackman MR, Shurety W, Ellis JA, Luzio JP (1994) Inhibition of apical but not basolateral endocytosis of ricin and folate in Caco-2 cells by cytochalasin D. J Cell Sci 107 (Pt 9):2547–2556

Jacob R, Naim HY (2001) Apical membrane proteins are transported in distinct vesicular carriers. Curr Biol 11:1444–1450

Jacob R, Alfalah M, Grunberg J, Obendorf M, Naim HY (2000) Structural determinants required for apical sorting of an intestinal brush-border membrane protein. J Biol Chem 275:6566–6572

Jacob R, Heine M, Alfalah M, Naim HY (2003) Distinct cytoskeletal tracks direct individual vesicle populations to the apical membrane of epithelial cells. Curr Biol 13:607–612

Jahn R, Scheller RH (2006) SNAREs—engines for membrane fusion. Nat Rev Mol Cell Biol 7:631–643

Janke C, Bulinski JC (2011) Post-translational regulation of the microtubule cytoskeleton: mechanisms and functions. Nat Rev Mol Cell Biol 12:773–786

Janke C, Kneussel M (2010) Tubulin post-translational modifications: encoding functions on the neuronal microtubule cytoskeleton. Trends Neurosci 33:362–372

Jaulin F, Xue X, Rodriguez-Boulan E, Kreitzer G (2007) Polarization-dependent selective transport to the apical membrane by KIF5B in MDCK cells. Dev Cell 13:511–522

Kang RS, Folsch H (2011) ARH cooperates with AP-1B in the exocytosis of LDLR in polarized epithelial cells. J Cell Biol 193:51–60

Kartenbeck J, Schmelz M, Franke WW, Geiger B (1991) Endocytosis of junctional cadherins in bovine kidney epithelial (MDBK) cells cultured in low Ca^{2+} ion medium. J Cell Biol 113:881–892

Kaufman SS, Blain PL, Park JH, Tuma DJ (1990) Role of microfilaments in asialoglycoprotein processing in adult and developing liver. Am J Physiol 259:G639–G645

Keating TJ, Peloquin JG, Rodionov VI, Momcilovic D, Borisy GG (1997) Microtubule release from the centrosome. Proc Natl Acad Sci U S A 94:5078–5083

Keller P, Toomre D, Diaz E, White J, Simons K (2001) Multicolour imaging of post-Golgi sorting and trafficking in live cells. Nat Cell Biol 3:140–149

Kitagawa Y, Sano Y, Ueda M, Higashio K, Narita H, Okano M, Matsumoto S, Sasaki R (1994) N-glycosylation of erythropoietin is critical for apical secretion by Madin-Darby canine kidney cells. Exp Cell Res 213:449–457

Kreitzer G, Marmorstein A, Okamoto P, Vallee R, Rodriguez-Boulan E (2000) Kinesin and dynamin are required for post-Golgi transport of a plasma-membrane protein. Nat Cell Biol 2(2):125–127

Kundu A, Avalos RT, Sanderson CM, Nayak DP (1996) Transmembrane domain of influenza virus neuraminidase, a type II protein, possesses an apical sorting signal in polarized MDCK cells. J Virol 70:6508–6515

Kyle ME, Nakae D, Sakaida I, Miccadei S, Farber JL (1988) Endocytosis of superoxide dismutase is required in order for the enzyme to protect hepatocytes from the cytotoxicity of hydrogen peroxide. J Biol Chem 263:3784–3789

Lafont F, Burkhardt JK, Simons K (1994) Involvement of microtubule motors in basolateral and apical transport in kidney cells. Nature 372:801–803

Lafont F, Verkade P, Galli T, Wimmer C, Louvard D, Simons K (1999) Raft association of SNAP receptors acting in apical trafficking in Madin-Darby canine kidney cells. Proc Natl Acad Sci U S A 96:3734–3738

Langevin J, Morgan MJ, Sibarita JB, Aresta S, Murthy M, Schwarz T, Camonis J, Bellaiche Y (2005) Drosophila exocyst components Sec5, Sec6, and Sec15 regulate DE-Cadherin trafficking from recycling endosomes to the plasma membrane. Dev Cell 9:365–376

Lanzetti L (2007) Actin in membrane trafficking. Curr Opin Cell Biol 19:453–458

Le Bivic A, Garcia M, Rodriguez-Boulan E (1993) Ricin-resistant Madin-Darby canine kidney cells missort a major endogenous apical sialoglycoprotein. J Biol Chem 268:6909–6916

Leidel C, Longoria RA, Gutierrez FM, Shubeita GT (2012) Measuring molecular motor forces in vivo: implications for tug-of-war models of bidirectional transport. Biophys J 103:492–500

Leung SM, Ruiz WG, Apodaca G (2000) Sorting of membrane and fluid at the apical pole of polarized Madin-Darby canine kidney cells. Mol Biol Cell 11:2131–2150

Li C, Hao M, Cao Z, Ding W, Graves-Deal R, Hu J, Piston DW, Coffey RJ (2007) Naked2 acts as a cargo recognition and targeting protein to ensure proper delivery and fusion of TGF-alpha containing exocytic vesicles at the lower lateral membrane of polarized MDCK cells. Mol Biol Cell 18:3081–3093

Ligon LA, Holzbaur EL (2007) Microtubules tethered at epithelial cell junctions by dynein facilitate efficient junction assembly. Traffic 8:808–819

Lin S, Naim HY, Roth MG (1997) Tyrosine-dependent basolateral sorting signals are distinct from tyrosine-dependent internalization signals. J Biol Chem 272:26300–26305

Lin S, Naim HY, Rodriguez AC, Roth MG (1998) Mutations in the middle of the transmembrane domain reverse the polarity of transport of the influenza virus hemagglutinin in MDCK epithelial cells. J Cell Biol 142:51–57

Lipardi C, Nitsch L, Zurzolo C (2000) Detergent-insoluble GPI-anchored proteins are apically sorted in Fischer rat thyroid cells, but interference with cholesterol or sphingolipids differentially affects detergent insolubility and apical sorting. Mol Biol Cell 11:531–542

Lipschutz JH, Guo W, O'Brien LE, Nguyen YH, Novick P, Mostov KE (2000) Exocyst is involved in cystogenesis and tubulogenesis and acts by modulating synthesis and delivery of basolateral plasma membrane and secretory proteins. Mol Biol Cell 11:4259–4275

Lisanti MP, Sargiacomo M, Graeve L, Saltiel AR, Rodriguez-Boulan E (1988) Polarized apical distribution of glycosyl-phosphatidylinositol-anchored proteins in a renal epithelial cell line. Proc Natl Acad Sci U S A 85:9557–9561

Low SH, Chapin SJ, Weimbs T, Komuves LG, Bennett MK, Mostov KE (1996) Differential localization of syntaxin isoforms in polarized Madin-Darby canine kidney cells. Mol Biol Cell 7:2007–2018

Low SH, Chapin SJ, Wimmer C, Whiteheart SW, Komuves LG, Mostov KE, Weimbs T (1998) The SNARE machinery is involved in apical plasma membrane trafficking in MDCK cells. J Cell Biol 141:1503–1513

Magal LG, Yaffe Y, Shepshelovich J, Aranda JF, de Marco MC, Gaus K, Alonso MA, Hirschberg K (2009) Clustering and lateral concentration of raft lipids by the MAL protein. Mol Biol Cell 20:3751–3762

Maples CJ, Ruiz WG, Apodaca G (1997) Both microtubules and actin filaments are required for efficient postendocytotic traffic of the polymeric immunoglobulin receptor in polarized Madin-Darby canine kidney cells. J Biol Chem 272:6741–6751

Martinez-Maza R, Poyatos I, Lopez-Corcuera B, Nunez E, Gimenez C, Zafra F, Aragon C (2001) The role of N-glycosylation in transport to the plasma membrane and sorting of the neuronal glycine transporter GLYT2. J Biol Chem 276:2168–2173

Marzolo MP, Yuseff MI, Retamal C, Donoso M, Ezquer F, Farfan P, Li Y, Bu G (2003) Differential distribution of low-density lipoprotein-receptor-related protein (LRP) and megalin in polarized epithelial cells is determined by their cytoplasmic domains. Traffic 4:273–288

Matter K, Bucher K, Hauri HP (1990) Microtubule perturbation retards both the direct and the indirect apical pathway but does not affect sorting of plasma membrane proteins in intestinal epithelial cells (Caco-2). EMBO J 9:3163–3170

Matter K, Hunziker W, Mellman I (1992) Basolateral sorting of LDL receptor in MDCK cells: the cytoplasmic domain contains two tyrosine-dependent targeting determinants. Cell 71:741–753

Meads T, Schroer TA (1995) Polarity and nucleation of microtubules in polarized epithelial cells. Cell Motil Cytoskeleton 32:273–288

Mishra R, Grzybek M, Niki T, Hirashima M, Simons K (2010) Galectin-9 trafficking regulates apical-basal polarity in Madin-Darby canine kidney epithelial cells. Proc Natl Acad Sci U S A 107:17633–17638

Mo D, Costa SA, Ihrke G, Youker RT, Pastor-Soler N, Hughey RP, Weisz OA (2012) Sialylation of N-linked glycans mediates apical delivery of endolyn in MDCK cells via a galectin-9-dependent mechanism. Mol Biol Cell 23:3636–3646
Mooseker MS (1985) Organization, chemistry, and assembly of the cytoskeletal apparatus of the intestinal brush border. Annu Rev Cell Biol 1:209–241
Mostov KE, de Bruyn Kops A, Deitcher DL (1986) Deletion of the cytoplasmic domain of the polymeric immunoglobulin receptor prevents basolateral localization and endocytosis. Cell 47:359–364
Muller MJ, Klumpp S, Lipowsky R (2008) Tug-of-war as a cooperative mechanism for bidirectional cargo transport by molecular motors. Proc Natl Acad Sci U S A 105:4609–4614
Munson M, Novick P (2006) The exocyst defrocked, a framework of rods revealed. Nat Struct Mol Biol 13:577–581
Musch A (2004) Microtubule organization and function in epithelial cells. Traffic 5:1–9
Noda Y, Okada Y, Saito N, Setou M, Xu Y, Zhang Z, Hirokawa N (2001) KIFC3, a microtubule minus end-directed motor for the apical transport of annexin XIIIb-associated Triton-insoluble membranes. J Cell Biol 155:77–88
Odorizzi G, Trowbridge IS (1997) Structural requirements for basolateral sorting of the human transferrin receptor in the biosynthetic and endocytic pathways of Madin-Darby canine kidney cells. J Cell Biol 137:1255–1264
Ohno H, Stewart J, Fournier MC, Bosshart H, Rhee I, Miyatake S, Saito T, Gallusser A, Kirchhausen T, Bonifacino JS (1995) Interaction of tyrosine-based sorting signals with clathrin-associated proteins. Science 269:1872–1875
Ohno H, Tomemori T, Nakatsu F, Okazaki Y, Aguilar RC, Foelsch H, Mellman I, Saito T, Shirasawa T, Bonifacino JS (1999) Mu1B, a novel adaptor medium chain expressed in polarized epithelial cells. FEBS Lett 449:215–220
Ojakian GK, Schwimmer R (1988) The polarized distribution of an apical cell surface glycoprotein is maintained by interactions with the cytoskeleton of Madin-Darby canine kidney cells. J Cell Biol 107:2377–2387
Ojakian GK, Schwimmer R (1992) Antimicrotubule drugs inhibit the polarized insertion of an intracellular glycoprotein pool into the apical membrane of Madin-Darby canine kidney (MDCK) cells. J Cell Sci 103(Pt 3):677–687
Pang S, Urquhart P, Hooper NM (2004) N-glycans, not the GPI anchor, mediate the apical targeting of a naturally glycosylated. GPI-anchored protein in polarised epithelial cells. J Cell Sci 117:5079–5086
Parczyk K, Haase W, Kondor-Koch C (1989) Microtubules are involved in the secretion of proteins at the apical cell surface of the polarized epithelial cell, Madin-Darby canine kidney. J Biol Chem 264:16837–16846
Pocard T, Le BA, Galli T, Zurzolo C (2007) Distinct v-SNAREs regulate direct and indirect apical delivery in polarized epithelial cells. J Cell Sci 120:3309–3320
Pombo I, Rivera J, Blank U (2003) Munc18-2/syntaxin3 complexes are spatially separated from syntaxin3-containing SNARE complexes. FEBS Lett 550:144–148
Powell SK, Cunningham BA, Edelman GM, Rodriguez-Boulan E (1991) Targeting of transmembrane and GPI-anchored forms of N-CAM to opposite domains of a polarized epithelial cell. Nature 353:76–77
Prigent M, Dubois T, Raposo G, Derrien V, Tenza D, Rosse C, Camonis J, Chavrier P (2003) ARF6 controls post-endocytic recycling through its downstream exocyst complex effector. J Cell Biol 163:1111–1121
Quinones GB, Danowski BA, Devaraj A, Singh V, Ligon LA (2011) The posttranslational modification of tubulin undergoes a switch from detyrosination to acetylation as epithelial cells become polarized. Mol Biol Cell 22:1045–1057
Rapoport I, Chen YC, Cupers P, Shoelson SE, Kirchhausen T (1998) Dileucine-based sorting signals bind to the beta chain of AP-1 at a site distinct and regulated differently from the tyrosine-based motif-binding site. EMBO J 17:2148–2155

Rindler MJ, Ivanov IE, Sabatini DD (1987) Microtubule-acting drugs lead to the nonpolarized delivery of the influenza hemagglutinin to the cell surface of polarized Madin-Darby canine kidney cells. J Cell Biol 104:231–241

Rodriguez-Boulan E, Powell SK (1992) Polarity of epithelial and neuronal cells. Annu Rev Cell Biol 8:395–427

Rogers KK, Wilson PD, Snyder RW, Zhang X, Guo W, Burrow CR, Lipschutz JH (2004) The exocyst localizes to the primary cilium in MDCK cells. Biochem Biophys Res Commun 319:138–143

Roland JT, Bryant DM, Datta A, Itzen A, Mostov KE, Goldenring JR (2011) Rab GTPase-Myo5B complexes control membrane recycling and epithelial polarization. Proc Natl Acad Sci U S A 108:2789–2794

Salas PJ, Misek DE, Vega-Salas DE, Gundersen D, Cereijido M, Rodriguez-Boulan E (1986) Microtubules and actin filaments are not critically involved in the biogenesis of epithelial cell surface polarity. J Cell Biol 102:1853–1867

Sato T, Mushiake S, Kato Y, Sato K, Sato M, Takeda N, Ozono K, Miki K, Kubo Y, Tsuji A, Harada R, Harada A (2007) The Rab8 GTPase regulates apical protein localization in intestinal cells. Nature 448:366–369

Scheiffele P, Peranen J, Simons K (1995) N-glycans as apical sorting signals in epithelial cells. Nature 378:96–98

Schneider D, Greb C, Koch A, Straube T, Elli A, Delacour D, Jacob R (2010) Trafficking of galectin-3 through endosomal organelles of polarized and non-polarized cells. Eur J Cell Biol 89:788–798

Schreiner R, Frindt G, Diaz F, Carvajal-Gonzalez JM, Perez Bay AE, Palmer LG, Marshansky V, Brown D, Philp NJ, Rodriguez-Boulan E (2010) The absence of a clathrin adapter confers unique polarity essential to proximal tubule function. Kidney Int 78:382–388

Schuck S, Simons K (2004) Polarized sorting in epithelial cells: raft clustering and the biogenesis of the apical membrane. J Cell Sci 117:5955–5964

Schuster M, Lipowsky R, Assmann MA, Lenz P, Steinberg G (2011) Transient binding of dynein controls bidirectional long-range motility of early endosomes. Proc Natl Acad Sci U S A 108:3618–3623

Shipitsin M, Feig LA (2004) RalA but not RalB enhances polarized delivery of membrane proteins to the basolateral surface of epithelial cells. Mol Cell Biol 24:5746–5756

Shurety W, Bright NA, Luzio JP (1996) The effects of cytochalasin D and phorbol myristate acetate on the apical endocytosis of ricin in polarised Caco-2 cells. J Cell Sci 109 (Pt 12):2927–2935

Shurety W, Stewart NL, Stow JL (1998) Fluid-phase markers in the basolateral endocytic pathway accumulate in response to the actin assembly-promoting drug Jasplakinolide. Mol Biol Cell 9:957–975

Simmen T, Honing S, Icking A, Tikkanen R, Hunziker W (2002) AP-4 binds basolateral signals and participates in basolateral sorting in epithelial MDCK cells. Nat Cell Biol 4:154–159

Simons K, Ikonen E (1997) Functional rafts in cell membranes. Nature 387:569–572

Smythe E, Ayscough KR (2006) Actin regulation in endocytosis. J Cell Sci 119:4589–4598

Steegmaier M, Lee KC, Prekeris R, Scheller RH (2000) SNARE protein trafficking in polarized MDCK cells. Traffic 1:553–560

Stehbens SJ, Paterson AD, Crampton MS, Shewan AM, Ferguson C, Akhmanova A, Parton RG, Yap AS (2006) Dynamic microtubules regulate the local concentration of E-cadherin at cell-cell contacts. J Cell Sci 119:1801–1811

Straube T, von Mach T, Honig E, Greb C, Schneider D, Jacob R (2013) pH-dependent recycling of galectin-3 at the apical membrane of epithelial cells. Traffic 14:1014–1027

Sudhof TC, Rothman JE (2009) Membrane fusion: grappling with SNARE and SM proteins. Science 323:474–477

Tai AW, Chuang JZ, Bode C, Wolfrum U, Sung CH (1999) Rhodopsin's carboxy-terminal cytoplasmic tail acts as a membrane receptor for cytoplasmic dynein by binding to the dynein light chain Tctex-1. Cell 97:877–887

Takeda T, Yamazaki H, Farquhar MG (2003) Identification of an apical sorting determinant in the cytoplasmic tail of megalin. Am J Physiol Cell Physiol 284:C1105–C1113

Tveit H, Akslen LK, Fagereng GL, Tranulis MA, Prydz K (2009) A secretory Golgi bypass route to the apical surface domain of epithelial MDCK cells. Traffic 10:1685–1695

Urban J, Parczyk K, Leutz A, Kayne M, Kondor-Koch C (1987) Constitutive apical secretion of an 80-kD sulfated glycoprotein complex in the polarized epithelial Madin-Darby canine kidney cell line. J Cell Biol 105:2735–2743

Valentijn KM, Gumkowski FD, Jamieson JD (1999) The subapical actin cytoskeleton regulates secretion and membrane retrieval in pancreatic acinar cells. J Cell Sci 112(Pt 1):81–96

Van Furden FD, Johnson K, Segbert C, Bossinger O (2004) The C. elegans ezrin-radixin-moesin protein ERM-1 is necessary for apical junction remodelling and tubulogenesis in the intestine. Dev Biol 272:262–276

van Zeijl MJ, Matlin KS (1990) Microtubule perturbation inhibits intracellular transport of an apical membrane glycoprotein in a substrate-dependent manner in polarized Madin-Darby canine kidney epithelial cells. Cell Regul 1:921–936

Verhey KJ, Hammond JW (2009) Traffic control: regulation of kinesin motors. Nat Rev Mol Cell Biol 10:765–777

Vorobjev IA, Nadezhdina ES (1987) The centrosome and its role in the organization of microtubules. Int Rev Cytol 106:227–293

Wade RH (2009) On and around microtubules: an overview. Mol Biotechnol 43:177–191

Wang CC, Shi H, Guo K, Ng CP, Li J, Gan BQ, Chien LH, Leinonen J, Rajaniemi H, Zhou ZH, Zeng Q, Hong W (2007) VAMP8/endobrevin as a general vesicular SNARE for regulated exocytosis of the exocrine system. Mol Biol Cell 18:1056–1063

Weisz OA, Rodriguez-Boulan E (2009) Apical trafficking in epithelial cells: signals, clusters and motors. J Cell Sci 122:4253–4266

Welte MA (2004) Bidirectional transport along microtubules. Curr Biol 14:R525–R537

Wilson JM, Fasel N, Kraehenbuhl JP (1990) Polarity of endogenous and exogenous glycosyl-phosphatidylinositol-anchored membrane proteins in Madin-Darby canine kidney cells. J Cell Sci 96(Pt 1):143–149

Wloga D, Gaertig J (2010) Post-translational modifications of microtubules. J Cell Sci 123:3447–3455

Wong YL, Dietrich KA, Naber N, Cooke R, Rice SE (2009) The Kinesin-1 tail conformationally restricts the nucleotide pocket. Biophys J 96:2799–2807

Xue X, Jaulin F, Espenel C, Kreitzer G (2010) pH-domain-dependent selective transport of p75 by kinesin-3 family motors in non-polarized MDCK cells. J Cell Sci 123:1732–1741

Yamamoto H, Awada C, Hanaki H, Sakane H, Tsujimoto I, Takahashi Y, Takao T, Kikuchi A (2013) The apical and basolateral secretion of Wnt11 and Wnt3a in polarized epithelial cells is regulated by different mechanisms. J Cell Sci 126:2931–2943

Yano H, Yamamoto-Hino M, Abe M, Kuwahara R, Haraguchi S, Kusaka I, Awano W, Kinoshita-Toyoda A, Toyoda H, Goto S (2005) Distinct functional units of the Golgi complex in Drosophila cells. Proc Natl Acad Sci U S A 102:13467–13472

Yeaman C, Le Gall AH, Baldwin AN, Monlauzeur L, Le Bivic A, Rodriguez-Boulan E (1997) The O-glycosylated stalk domain is required for apical sorting of neurotrophin receptors in polarized MDCK cells. J Cell Biol 139:929–940

Yeaman C, Grindstaff KK, Wright JR, Nelson WJ (2001) Sec6/8 complexes on trans-Golgi network and plasma membrane regulate late stages of exocytosis in mammalian cells. J Cell Biol 155:593–604

Yeaman C, Grindstaff KK, Nelson WJ (2004) Mechanism of recruiting Sec6/8 (exocyst) complex to the apical junctional complex during polarization of epithelial cells. J Cell Sci 117:559–570

Zerial M, McBride H (2001) Rab proteins as membrane organizers. Nat Rev Mol Cell Biol 2:107–117

Zink S, Grosse L, Freikamp A, Banfer S, Muksch F, Jacob R (2012) Tubulin detyrosination promotes monolayer formation and apical trafficking in epithelial cells. J Cell Sci 125:5998–6008

Index

K. Ebnet (ed.), *Cell Polarity 1*, DOI 10.1007/978-3-319-14463-4

MIX
Papier aus verantwortungsvollen Quellen
Paper from responsible sources
FSC® C105338

If you have any concerns about our products,
you can contact us on
ProductSafety@springernature.com

In case Publisher is established outside the EU,
the EU authorized representative is:
Springer Nature Customer Service Center GmbH
Europaplatz 3, 69115 Heidelberg, Germany

Printed by Libri Plureos GmbH
in Hamburg, Germany